Pittsburgh-Konstanz Series in the Philosophy and History of Science

Philosophical Problems of the Internal and External Worlds

Essays on the Philosophy of Adolf Grünbaum

EDITED BY John Earman, Allen I. Janis, Gerald J. Massey, and Nicholas Rescher

University of Pittsburgh Press / Universitätsverlag Konstanz

Published in the U.S.A. by the University of Pittsburgh
Press, Pittsburgh, Pa. 15260

Published in Germany by Universitätsverlag Konstanz
GMBH

Manufactured in the United States of America

Library of Congress Cataloging-in-Publication Data

Philosophical problems of the internal and external
worlds : essays on the philosophy of Adolf
Grünbaum / edited by John Earman . . . [et al.].
p. cm. — (Pittsburgh-Konstanz series in the
philosophy of science)
ISBN 0-8229-3738-7
1. Science—Philosophy. 2. Science—History. 3.
Grünbaum, Adolf. I. Earman, John. II. Series.
Q175.3.P444 1993
501—dc20 92-27354
CIP

Die Deutsche Bibliothek – CIP-Einheitsaufnahme

Philosophical problems of the internal and external
worlds : essays on the philosophy of Adolf
Grünbaum / ed. by John Earman . . .–Konstanz :
Univ.–Verl. Konstanz ; Pittsburgh, Pa. : Univ. of
Pittsburgh Press, 1993
(Pittsburgh–Konstanz series in the philosophy
and history of science; I).
ISBN 3-87940-401-1
NE: Earman, John (Hrsg.); GT

A CIP record for this book is available from the
British Library.

Contents

Preface

In the fall of 1960, Adolf Grünbaum left Lehigh University to join the faculty of the University of Pittsburgh as Andrew Mellon Professor of Philosophy and as founding director of the Center for Philosophy of Science. Ten professorships at the University of Pittsburgh had been endowed by the A. W. Mellon Foundation during the 1950s, and for an initial period these chairs were filled on a visiting basis. When the time came to begin to fill these chairs on a permanent basis, the then provost, Charles Peake, in what was to prove a brilliant administrative move, took the bold step of offering the Andrew Mellon chair in philosophy to an unusually promising young scholar, someone so young that the age threshold of forty years for the Mellon Professorships had to be waived in order to secure Grünbaum for the chair. Perhaps no appointment at any university has returned greater dividends than this one.

The administration also gave some assurances for the future, promising a major renovation of the Philosophy Department and the inauguration, with Grünbaum as director, of a Center for Philosophy of Science, under whose aegis an annual lecture series would be offered to provide a "showcase" of enhanced visibility for the university's revitalized commitment to philosophy. With Grünbaum's appointment as director in 1960 (followed in the next year by Nicholas Rescher's appointment as associate director), the Center for Philosophy of Science began a distinguished career whose thirtieth anniversary celebration in 1990 coincided, of course, with the thirtieth anniversary celebrations of Grünbaum's appointments as director of the center and as Andrew Mellon Professor of Philosophy. This triple celebration culminated in a Grünbaumfestivity colloquium, many of whose contributions are among the essays in the present volume.

Like Gaul, Grünbaum's contributions to the philosophy of science are divided into three parts: The first is centered on philosophical problems of space, time, and cosmology; the second concerns the nature of scientific methodology, especially the rationality of scientific inference; and the third concerns the foundations of psychoanalysis and psychiatry. The organization of this volume reflects this division. In addition, a section is included that reflects two further themes that are also to be found in Grünbaum's work and that are natural offshoots of his principal preoccupations: the nature of "free action" in deterministic and indeterministic worlds, and the (non-)implications of big bang cosmology for theism. It is a testimonial to Grünbaum's seminal influence on contemporary philosophy of science that leading lights in each of the areas in which Grünbaum has worked were eager to contribute essays in his honor. But the highest compliment lies in the quality of the essays. Both by his teaching and by his own example Grünbaum has done much to raise the standards of his discipline, and none of the contributors would have dared to turn in anything less than his or her best effort.

As already indicated, many of the essays in this volume grew out of the formal presentations to the three-day Colloquium in Honor of Adolf Grünbaum that took place October 5, 6, and 7, 1990 on the campus of the University of Pittsburgh. The principal sponsors of the colloquium were the Center for Philosophy of Science, the Faculty of Arts and Sciences, and the Health Sciences. Other sponsors were the Department of History and Philosophy of Science, the Department of Philosophy, the Department of Psychiatry, and the provost of the university. (All sponsoring units are divisions of the University of Pittsburgh.) The organizing committee for the colloquium was chaired by Professor Gerald Massey (director of the Center for Philosophy of Science) and included professors Allen Janis (associate director of the center) and Nicholas Rescher (vice-chair of the center). The organizers acknowledge the generosity of the sponsors as well as generous support from the Richard King Mellon Foundation through the Center for Philosophy of Science. Invaluable service in making arrangements for the colloquium was provided by Ms. Linda Butera and Ms. Mary Connor. Ms. Connor was also of great help in the preparation of this volume.

In addition to honoring our esteemed colleague Adolf Grünbaum, this volume also serves to inaugurate a new series. Henceforth, the Pittsburgh Series in the Philosophy of Science becomes the Pittsburgh-Konstanz Series in the Philosophy and History of Science. The new series will be edited by a committee drawn from the Pittsburgh Center for the Philosophy of Science and the Philosophy Faculty from

Space, Time, Cosmology

1

Physical Force or Geometrical Curvature?

Einstein, Grünbaum, and the Measurability of Physical Geometry

Martin Carrier
Philosophy Department, University of Konstanz

Science would be an easy matter if the fundamental states of nature expressed themselves candidly and frankly in experience. In that case we could simply collect the truths lying ready before our eyes. In fact, however, nature is more reserved and shy, and its fundamental states often appear in masquerade. Put less metaphorically, there is no straightforward one-to-one correspondence between a theoretical and an empirical state. One of the reasons for the lack of such a tight connection is that distortions may enter into the relation between theory and evidence, and these distortions may alter the empirical manifestation of a theoretical state. As a result, it is in general a nontrivial task to excavate the underlying state from distorted evidence.

The problem of distortions is by no means restricted to the realm of the physical sciences. On the contrary, the difficulty of guaranteeing the reliability of the data pervades all science, including psychology. Consider the example of Freud's method of free association. This method is supposed to bring to light cognitive material originating in the dark depths of the unconscious. The trustworthiness of the conclusions based on this material is crucially dependent on whether it survives essentially intact the analyst's attempts to lift it to the waking mind. But as Adolf Grünbaum has made convincingly clear, the analyst's "maeutic" interventions in fact contribute to the distortion of the data they purportedly unearth. By emphasizing particular aspects of the reported thoughts and feelings at the expense of others and by posing leading questions, the analyst exercises a suggestive influence on the docile patient on her couch. The result is that the interpreted data are seriously contaminated by the theoretical expectations and thus tend to comply with them (see Grünbaum 1984, 235, 277). So, we have a situation in which the influence of a distorting factor hinders the reliable evaluation of states posited within a theory.

This essay will discuss the problem of extracting theoretical states in the presence of perturbations. I will not be concerned with the difficulty of ascertaining hidden mental states; rather, I address the analogous difficulty as it shows up regarding the measurability of geometric relations. In spite of this difference in content, Grünbaum's work is of substantial help. More specifically, I will consider Grünbaum's discussion of a problem raised by Einstein. Einstein argued that physical geometry cannot be established empirically without invoking the entire system of physical laws. This claim is backed by an analysis of measuring procedures for geometric quantities that is modeled on Reichenbach's argument for the conventionality of geometry. Einstein's point here is the existence of measurement distortions induced by substance-specific effects such as temperature variations. These distortions vitiate the separate determination of physical geometry. Grünbaum, by contrast, holds that there is a way out of the perturbation problem posed by Einstein. He argues that geometric influences and substance-specific distortions can be disentangled empirically. I will present Grünbaum's views on that matter and try to evaluate them in light of some more recent physical and philosophical developments. For that purpose I will first sketch Einstein's argument and relate it to its conceptual context, namely, Reichenbach's conventionality thesis of physical geometry.

Reichenbach: Universal Forces and the Conventionality of Geometry

The problem of the conventionality of geometry originated with Poincaré's consideration of possible distortions of measuring rods. He analyzed the way in which such distortions might influence the results obtained and their physical interpretation, and he argued that certain generally occurring deformations of the rods could easily be explained away by introducing a geometric curvature. Both interpretations would be empirically indistinguishable.

Reichenbach elaborated Poincaré's analysis and gave the conventionality thesis its familiar form. In the first place he coined the distinction between *differential* and *universal* forces. Differential forces are characterized by the fact that they act differently on different substances. Accordingly, the deformation of measuring rods produced by them depends upon the material employed and is thus easily detectable and correctable. Matters are different, however, with respect to universal forces. The latter are supposed to act on all materials alike, and this implies that their presence or intensity is impossible to detect directly by empirical means. However, indirect indications may exist for them. If the universal force field is inhomogeneous (i.e., if its strength is position-dependent), it can induce relative deformations in suitably arranged rods. Hence a gradient of the force field is associated with empirically accessible features; universal forces are not always coincidence-preserving.

Reichenbach's proposal for handling universal forces is to remove them by way of a methodological decision. Though we are not forced by the facts to do so, it is advisable (for reasons that will soon become clear) to set these forces equal to zero. This leads to the definition of the rigid body. A rigid body is a solid body that satisfies the following two conditions: (1) All possibly present differential deformations are corrected, and (2) all universal forces are set equal to zero (see Reichenbach [1928] 1965, 32–39).

Reichenbach's analysis (as far as hitherto presented) can be summarized by the following three assertions: First, the measurement of metric relations is dependent upon a stipulation about the presence or absence of universal forces. Accordingly, a conventional element is involved in empirically establishing the metric. Second, after such a stipulation has been accepted, the metric (together with all geometric relations derived from it) is uniquely fixed. Third, it is recommended that we make this stipulation such that universal forces vanish.

What does it mean to set equal to zero a force that may lead to empirically detectable deformations? The point of Reichenbach's recommendation becomes clear if one considers a serious candidate for a universal force, namely, gravitation. Gravitational forces indeed act on all materials alike. And setting universal forces equal to zero is intended to capture the gist of Einstein's geometrization of gravitation.

This geometrization amounts to interpreting gravitation (at least locally) not as a physical force but as a manifestation of spacetime structure; geometric curvature, not a physical force, underlies the well-known empirical effects of gravitation.

Setting universal forces equal to zero accordingly means that the effects of gravitation are not to be attributed to a deforming force but to the metric structure itself. It means that gravitational effects on measuring rods are not corrected but viewed as veridical indications of the prevailing (nonflat) geometry. After all, it hardly makes sense to correct influences of spacetime on spacetime measurements. It is precisely this geometrization that distinguishes Einstein's approach from Newton's. So Reichenbach's (ibid., 38–39, 293–94) rule is supposed to furnish a justification of the former, although this justification is itself dependent upon a conventional decision.[1]

Now that we know what Reichenbach is getting at, let us turn to the logical structure of his argument. The measurement of the prevailing geometric relations may be distorted by a universal force whose action can only be established or excluded by comparing the empirically obtained (and possibly distorted) relations with the actual ones. This comparison would enable one to correct these distorting influences. Obviously, on the other hand, such a comparison requires that the actual geometry already be known, but one can only come to know it by carrying out measurements. So there is a mutual dependence between two quantities, the universal for F_U and the metric tensor g_{ab}, each of which can only be determined by recourse to the other. This amounts to a correction circularity that is dissolved (if one adopts Reichenbach's rule) by the decision not to correct at all.

In order to make things more perspicuous I want to introduce the concept of a "self-referential" distortion. Effecting the correction of a self-referential distortion requires recourse to the very same quantity whose measurement is to be corrected. Apparently, based on the foregoing discussion, a universal force is a self-referential distortion of the metric. The influence of a universal force on the metric can only be corrected by resorting in turn to the metric. More generally, it is sufficient for a self-referential distortion that there exists a reciprocal dependence between two quantities such that the reliable (i.e., corrected) measurement of either necessarily presupposes a reliable (i.e., corrected) measurement of the other. So Reichenbach's argument amounts to the claim that a self-referential distortion is involved in

the University of Konstanz. In addition to conference proceedings, the series will include single author monographs and books. The University of Pittsburgh Press and the Universitätsverlag Konstanz will serve as joint publishers.

The editors would like to join in the many tributes expressed in this volume, and add their own very best wishes to their long-time colleague and friend, Adolf Grünbaum.

John Earman
Allen I. Janis
Gerald J. Massey
Nicholas Rescher

empirically determining physical geometry, which leads to a circularity. And for that reason, geometry contains a conventional element.

Einstein: Differential Forces and the Conventionality of Geometry

In a discussion of Reichenbach's interpretation of geometry, Einstein objects to Reichenbach's views claiming that even after universal forces have been excluded, physical geometry is still underdetermined. This means Einstein opposes the second of Reichenbach's assertions, namely, the claim that, after a decision about universal forces has been made, the actual geometry is unambiguously given. Einstein's point is that differential forces are not as easily detectable and correctable as Reichenbach assumed. Einstein's argument can be reconstructed to the effect that differential corrections, too, are in fact self-referential. His view is wrapped up in a fictitious dialogue between Poincaré (acting as Einstein's straw man) and Reichenbach. Einstein's Poincaré puts forward the following retort to Reichenbach:

> In gaining the real definition [of the rigid body] improved by yourself [by means of differential corrections] you have made use of physical laws, the formulation of which presupposes (in this case) Euclidean geometry. The verification, of which you have spoken, refers, therefore, not merely to geometry but to the entire system of physical laws which constitute its foundation. An examination of geometry by itself is consequently not thinkable.[2] (Einstein [1949] 1970, 677)

Grünbaum has clarified the meaning of Einstein's argument. First we have to remove reference to Euclidean geometry since it merely reflects the predilection of the historical person Poincaré. With this generalization the argument reads: The concept of length already enters into the differential correction laws, namely, into the laws for correcting deformations of the measuring rods that are due to thermic expansion or mechanical stress, and so on. For this reason, the application of the correction laws presupposes that the metric be already known. Conversely, the metric can only come to be known by resorting to these correction laws. After all, differential effects of this sort are construed as deformations and not as veridical indications of spacetime structure. This reciprocal dependence between the metric and differential distortions induces a circularity that vitiates a separate empirical determination of physical geometry. Whatever the evidence may be, it is always possible to cling to one's pet geometry (see Grünbaum 1960, 80; [1963] 1973, 131–35).

As can be gathered from the passage quoted, Einstein takes his argument as supporting the thesis that geometry is not testable in isolation but only within the context of additional physical laws. Only the combined system of geometry and physics can be confronted with experience. Einstein is, however, not in the least worried about such test restrictions, "Why do the individual concepts which occur in a theory require any specific justification anyway, if they are only indispensable within the framework of the logical structure of the theory, and the theory only in its entirety validates itself?" (Einstein [1949] 1970, 678). A separate test of geometry is as impossible as it is unnecessary.

I should emphasize (*pace* Grünbaum) that Einstein's argument does not simply come down to stressing Duhem's conventionality thesis. The latter concerns the methodological intertwining of hypotheses that are logically independent of each other. This means, in order to test a hypothesis H_1, it is necessary to take recourse to some ancillary hypotheses H_A; but testing these H_A's does not necessarily presuppose H_1. Einstein's argument, by contrast, envisages a situation in which testing H_1 requires reference to H_2, and testing H_2 necessitates recourse to H_1. So we have all the marks of a full-blown circularity that is exactly parallel to the one elaborated by Reichenbach. What Einstein points to here is a differential analogue to Reichenbach-conventionality. And this means geometry remains conventional even after Reichenbach's rule has been applied.

Einstein's Circularity Argument Transferred to Particle Trajectories and Light Rays

Einstein develops his argument with exclusive reference to rigid rods and the problems associated with ascertaining rigidity. This procedure does not, however, constitute the only means for measuring metric relations. On the contrary, in the course of the last two decades it has become far more common to regard the use of particle trajectories and light rays as the theoretically privileged method for that purpose.

The most ingenious implementation of this approach was developed by Ehlers et al. (1972). They proceed roughly as follows: First, coordinate systems are introduced so as to allow light rays and particle trajectories to be tracked. For that purpose, so-called radar coordinates are employed. That is, a spacetime event, such as the spatiotemporal location of a particle, is characterized by sending off a light signal from each of two neighboring world-lines and by recording its emission time and the arrival time of the signal reflected at the event under consideration. "Time" is to be understood here as a nonmetric or order concept, that is, it is only required that the clock run continually and smoothly. Second, the propagation of light is used to attach a light cone, or null cone, to each of the spacetime points under consideration. The null-cone structure thus obtained allows for distinguishing between spacelike and timelike 4-directions, and it can be shown that in this structure the metric is determined up to a positive function of the 4-positions. In the third step the motions of freely falling particles are taken into consideration. These motions determine the straightest lines possible, that is, the timelike geodesics of the respective spacetime. Both results are linked by a compatibility postulate to the effect that these geodesics always remain timelike from the perspective of the null-cone structure. In this way the metric can be ascertained unambiguously.[3]

In this approach the metric is constructed from the paths of free particles and light rays. So one may ask whether Einstein's problem can be solved by doing away with rods and resorting to paths instead. After all, neither the problem of thermic corrections nor of mechanical stresses plays any role in this approach. In fact, however, the trouble reappears at a different level. In order to recognize the analogy to Einstein's problem, we must introduce differential distortions into the Ehlers et al. scenario; that is, we must suspend the condition that our test particles move freely. A relevant differential distortion is electric charge. If the particles used for probing spacetime structure are charged, the influence of possibly present electromagnetic fields on their trajectories has to be evaluated and (as the case may be) corrected. Such corrections can be carried out by applying Lorentz's force law.

The problem with that procedure is that in the generally covariant formulation of that law the metric appears:

$$f^a = qg^{ab}F_{bc}(dx^c/d\tau)$$

(with f^a denoting the correcting Lorentz force, and F_{bc} the electromagnetic field strength tensor, see Weinberg 1972, 125). As can be gathered from this equation, a reciprocal dependence exists between the metric g^{ab} to be determined and the correcting electromagnetic force f^a. Calculating this force from Lorentz's law requires knowledge of the metric, and conversely, the metric can in general only be determined by applying this force law and thereby coping with differential trajectory deformations. This situation thus apparently leads into a circle that parallels the one in Einstein's rod scenario: In order to reliably measure the metric, we have to carry out corrections that, in turn, make use of the metric. This means that, from the perspective of the trajectory method, distortions induced by lack of neutrality are in fact self-referential. Correspondingly, Einstein's differential analogue to Reichenbach-conventionality has a general bearing on space-time measurements.

Grünbaum: Dissolving Einstein's Circularity by Way of Successive Approximations

In the remainder of this essay, I discuss the import of Einstein's circularity argument on the measurability of physical geometry. Mutual dependence of the sort described by Einstein inevitably creates the impression that a vicious circularity is present. In that event, one of the relevant quantities (the metric tensor, for example) would have to be chosen arbitrarily and the other one (electromagnetic force, for instance) would have to be correspondingly adjusted. After all, this is to be done in Reichenbach's universal-force scenario. In fact, however, reciprocal dependence does not necessarily entail circularity. We have two possible ways out of the difficulty, namely, the method of *successive approximations* and the *selection of undistorted instances*. The first procedure attacks the problem head-on and the second one circumvents it. Each method can be applied to the original rod-variant of the argument, for one, and to its updated trajectory-version, for another. This leaves us with a total of four options for coping with Einstein's argument. I will now address each of these options in turn.

We owe to Grünbaum an elaboration of the application of the method of successive approximations to the rod-variant. He considers a situation in which only thermic distortions are present. In that case one starts with an arbitrary geometry in the correction law for thermic expansion and determines the geometry with the help of rods corrected by this law. The geometry obtained does not in general coincide with the one employed in the correction. So one uses this measured geometry for effecting the correction a second time and applies again the rods corrected in this improved fashion to determine the prevailing metric relations. This procedure is to be repeated until an agreement is reached between the geometry that enters into the correction law and the geometry that is obtained by measurements with the corrected rods (see Grünbaum [1963] 1973, 140–46).

If this procedure indeed converges to a result that is independent of the geometry used at the start, it is sensible to indentify this resulting geometry with *the* geometry of spacetime. The problem is, however, that there are lots of circumstances under which this procedure is not convergent. A necessary precondition for convergence is, first, that curvature is spatiotemporally constant. That is, in order for the method to work successfully, the curvature of the region under scrutiny must be neither time- nor position-dependent (Grünbaum is well aware of this limitation, see Grünbaum [1963] 1973, 146). As a matter of fact, however, this is not generally true in our universe. The general theory of relativity associates spatiotemporal curvature variations with a large number of gravitational effects.

Second, the reliability of this method presupposes that the distorting temperature variations remain unchanged during its application. Clearly, if the strength of the perturbing influence changes in the course of the correction process, we cannot expect to obtain a convergent series of results. Hence, this approach relies on the condition that there is a stable spatial pattern of distortions. Only on that condition can this pattern be extracted from successive measurements.

Third, the rods used in performing the measurements must be guaranteed to be perfectly elastic. It is requisite that the possibly changing intensity of the deforming influences be faithfully reflected in corresponding changes in the length of the rods. This requirement is violated if a "hysteresis" occurs due to inelastic deformations. In that case the rod does not regain its original length after the distorting influence has ceased but rather retains a somewhat altered length. Consequently, the rod would indicate the presence of a distortion where in fact none exists. The problem here is that judging whether the condition of perfect elasticity is met demands recourse to reliable length measurements which in turn were supposed to be established by the whole procedure.

These three restrictions make Grünbaum's proposal unsuitable as a general means for disentangling differential distortions and geometric effects. His ingenious application of the method of successive approximations to rigid rods, useful as it is in some cases, is not generally applicable.

I should point out, however, that the limitation to the exclusive presence of temperature variations is inessential. Grünbaum's method is successful even when several differential distortions occur simultaneously. Suppose that in addition to temperature variations a (constant) mechanical deformation is present. Such deformations are governed by Hooke's law, and from that it appears that the evaluation of the distorting force necessarily makes use of length measurements. So we encounter again a circle analogous to that involved in thermic corrections. In this case, however, we can provisionally assume a geometry, carry out *all* relevant corrections unambiguously, and measure the resulting geometric relations. Because all possibly occurring deformations can be evaluated on the basis of a posited geometry, we can effectively carry through the successive approximation procedure. The simultaneous presence of several differential distortions leads to only technical rather than fundamental complications.

Furthermore, the viability of Grünbaum's proposal remains unaffected by an epistemological objection of the following kind: The method of successive approximations works satisfactorily only if we know in advance which distortions are present and, correspondingly, which correction laws are to be applied. In particular, there may occur hitherto unknown perturbations that consequently escape our corrective procedure. But the nature of the distorting influences cannot be established by means of the experimental set-up under consideration. The reliability of this approach thus crucially depends on conditions whose validity cannot be verified within its framework.

This objection does indeed hold, but is irrelevant in the present context. It would only apply to a situation in which we wished to use Grünbaum's method to determine the prevailing geometric relations from scratch. That is, it refers to a situation in which we are only given some possibly distorted length measurements and are asked to disentangle geometric and differential effects. In that situation Grünbaum's method clearly fails. But this problem is distinct from the one created by Einstein's argument. The latter proceeds on the assumption that the relevant distorting influences are known and denies that even under these circumstances the metric relations can be unambiguously determined. Einstein's argument has nothing to do with the difficulty of ascertaining what kinds of distortions are present but rather with the problem of quantitatively evaluating their impact on length measurements. Even if we know what kinds of perturbations spoil our measurements, the ensuing correction circularity prevents us from unambiguously sifting out the disturbing effects. So neither Einstein's problem nor Grünbaum's proposal to its solution is affected by the epistemological difficulty sketched.

The conclusion is that Grünbaum's implementation of the first option is helpful in some special cases of Einstein's problem, but it does not provide a means for its general solution.

Successive Approximation Applied to the Trajectory Case

Virtually the same conclusion emerges from the application of the method of successive approximation to the trajectory case. At first, to be sure, that method appears to break down right at the start. In order to see this let us once again consider Lorentz's force law. In addition to the previously mentioned reciprocal dependence between f^a and g^{ab}, there apears another reciprocal dependence in that law. The charge q enters into the calculation of the correcting force, and this charge can in general only be determined by recourse to electromagnetic fields, that is, by recourse to Lorentz's force law. Charge relates the field strength to the force and this implies that we need the value of the charge to calculate the value of the force. Conversely, the value of the charge can in general only come to be known by the force exerted on that charge in an electromagnetic field. So there is a reciprocal dependence between q and f^a in addition. Because of this double reciprocity, the application of the method of successive approximation is blocked. There is too much free play for the quantities involved.

But there exists a remedy for this difficulty. For the moment let us restrict our attention to just one particle. In that case we may assume that—barring collision processes—it retains its charge value so that the charge dependence of the force need not be taken into account. Then we start an adaptation procedure between geometry and force. That is, we posit a geometric structure and explore on that basis the motion of a particle of unknown (but constant) charge. This is done by calculating the particle acceleration, that is, the deviation of the observed particle trajectory from the posited geodesics, and by trying to account for that acceleration by introducing appropriate F_{bc}-functions. Next we take another particle of likewise unknown but constant charge and examine whether its motion can be accommodated as well by the formerly assumed geometric and field structure. This trial-and-error procedure is repeated until the acceleration values can be explained by ascribing a parameter to each particle that does not in fact change spatiotemporally. This set of parameters may then legitimately be identified with the respective particles' charge values.

This approach tries to avoid the previously mentioned double reciprocity problem by considering a series of experiments in which the critical charge parameter can be considered fixed and by applying the condition that the theoretical entities "geometry" and "field" be adjusted such that the ensuing charge ratios of any two particles indeed come out as spatiotemporally invariant. Systematically speaking, we take advantage of a constraint imposed on charge by electromagnetic theory: Charge is a property inherent to a particle which consequently remains unaltered unless the nature of the particle is changed. By virtue of this constraint, the circularity threatening the applicability of the method of successive approximation does not occur. It is clear, on the other hand, that this constraint must be presupposed for that method to work and thus cannot be tested empirically. If unaccounted charge fluctuations occur, we are heading right into a mess. Either we obain no consistent set of charge parameters or an inaccurate one. But since our problem is not the determination of the prevailing geometry from scratch, this untestability objection does not carry much weight. We are allowed to avail ourselves of the pertinent theoretical apparatus.

This procedure suffers from a more serious shortcoming, however. Another constancy assumption is involved: Neither the geometry nor the field is allowed to change in time. If there are any fluctuations of either quantity while the procedure is carried out, the results obtained are clearly mistaken. And again, the validity of this constancy assumption has to be presupposed and cannot be tested. What makes this assumption appear much more problematical than its charge analogue is that, in contrast to the latter case, it is by no means guaranteed theoretically that the geometry and the field retain their values. This constancy is a contingent feature of the particular situation under consideration and not a (supposedly) general trait of nature.

Hence the conclusion is that we are not better off now than we were before. The method of successive approximation offers a basis for tackling special cases of the differential circularity problem, but it fails to bring us closer to its general solution. Accordingly, Einstein's problem cannot be overcome in a general fashion by relying on this method.

Grünbaum Again: The Selection of Undistorted Rods

If distortions cannot be adequately treated, they can perhaps be discounted. So let us turn to the second type of options, that is, to the selection of ideal cases. The rod variant of this option has again been studied in detail by Grünbaum. In order for that method to work noncircularly, it is necessary to identify undistorted cases without having a metric at our disposal. For the rod scenario this amounts to empirically securing the absence of differential deformations without availing ourselves of metric relations. Under such circumstances we cannot uniquely compare lengths numerically on the same world-line (i.e., we cannot give length ratios), and we can in no way compare lengths on different world-lines. What we can do, however, is to judge whether or not a rod has the same length as another one placed next to it in the same direction. Local congruence can be established nonmetrically. And this capacity suffices for ascertaining the absence of perturbations. If any two rods of different chemical composition that coincide locally always remain locally congruent when being jointly transported to different locations, this reliably indicates that the region under consideration is perturbation-free (see Grünbaum [1963] 1973, 136).

In that event it is safe to conclude that the geometric relations obtained truthfully indicate the spacetime structure. But cheers are premature. A serious limitation of this method is due to the possible presence of tidal forces. A curved spacetime induces relative accelerations (and consequently deformations) in a particle ensemble. In the case of an extended rod this amounts to a substance-unspecific force that tends to deform the rod and to alter its length. But since the actual deformation of the rod is the joint effect of the tidal force and the substance-specific force of cohesion, this resulting deformation depends on the material employed. As a result of the different elasticity of different materials, changing tidal forces may induce responses of different strength in different materials. This implies that chemically distinct rods may not remain locally congruent under transport because curvature variations occur, despite the fact that the region under consideration is perturbation-free.

Consequently, Grünbaum's perturbation test may be falsely positive. The test is not only sensitive to differential distortions but also to changing curvature (or changing tidal forces). It may indicate the presence of differential distortions where there are in fact none. Grünbaum's test is indeed sufficient to ascertain that a certain region is perturbation-free (namely, if local congruence is retained). The reverse, however, does not hold. One cannot reliably detect the presence of perturbations, and one cannot make sure, consequently, that all perturbation-free regions are correctly identified.[4]

A second limitation is involved in Grünbaum's approach. Suppose that the perturbation test is rightly positive and let us ask what can be derived from that result. The answer is: precisely nothing. More specifically, we cannot extract the prevailing geometry from distorted cases. That is, we cannot derive the geometry present in distorted situations from the geometry realized in the undistorted ones. Such a derivation would require a comparison between the distorted and an undistorted region, and in that case we can never rule out that the geometry is actually different in both regions. So the transition from a perturbation-free state of affairs to a distorted one is blocked here. Clearly, however, differential distortions are almost ubiquitous in our universe. Accordingly, the selection of pure instances, when applied to the rod scenario, is of only limited help.

Selection of Pure Instances in the Trajectory Case

Our last resort is to apply the selection method to the trajectory scenario. Our task is, then, to identify free-particle trajectories without relying on metric procedures. If we follow Reichenbach's recommendation and set universal forces equal to zero, a particle is free if its path is at most influenced by gravitational effects. In that case its trajectory is a geodesic. So our task comes down to nonmetrically identifying geodesics. If this can be accomplished, deviations from the geodesic path are to be attributed to the presence of differential distortions. And by this means the latter could be evaluated and consequently corrected.

A generally applicable test exists for singling out geodesics nonmetrically. So now the cheers are in order. A geodesic can be characterized by three conditions. It is, first, timelike and, second, unique in the sense that the entire trajectory is uniquely determined by specifying the 4-direction of the trajectory at one point on it. This condition is obvious in the special case of a straight line in flat spacetime, and the just-given form is a natural generalization of that special case to situations with nonvanishing curvature. These two conditions, however, are not only satisfied by geodesic motions but also by particle motions under the influence of electromagnetic fields. So the third condition must carry the bulk of the weight. This condition invokes a local microsymmetry of geodesics that can be characterized roughly and intuitively as follows. Consider a collection of path elements or curves that are all passing through a given event. Disregard, moreover, the special parametrizations of these path elements or, put the other way around, construe every path element as an equivalence class of differently parametrized curves. Then introduce a "dilation operation" that radially stretches or contracts the path elements away from or toward the event under consideration. If this operation leaves the paths invariant, that is, if it amounts to a map of each path onto itself, these paths are geodesics. After all, it is intuitively clear that the only paths that are left unchanged by such a dilation procedure are radial straight lines, and it can be shown that the same feature is present in their analogues in nonflat geometries, namely, geodesics. A microsymmetry of that sort is sufficient to single out geodesics in every spacetime manifold (see Ehlers 1988, 155–56; Ehlers and Köhler 1977, 2014, 2017). Trajectories of small gravitational monopoles indeed pass this geodesics test.

Now we are almost done. The last step only consists in realizing that what we have here is an effective criterion for the absence of differential distortions. This implies that we are now in a position to reliably set apart the motions of neutral and charged particles. Only the former, and not the latter, pass the geodesics test. It is now easily possible to evaluate electromagnetic perturbations. For that purpose we only have to compare the paths of neutral particles to those of charged ones in the same spacetime region. The latter's deviations indicate the intensity of the distortions present which can consequently be quantified and corrected. Unlike the rod scenario, the comparison between the undistorted and the distorted case can here be effectively carried through.

It deserves emphasis that this procedure is not restricted to electromagnetic distortions. It is not confined to situations in which the nature of all possibly occurring differential distortions is known beforehand. The reason is that the pure state is not identified by a stepwise exclusion of distortions but rather by means of a direct operational criterion. After an empirically applicable criterion for geodesics is at hand, we can simply count every particle whose motion deviates—whatever the cause may be—from the thus distinguished paths as distorted. In order to carry out this procedure we need not know in which way its motion is distorted.

Conclusion

This leads to the overall conclusion that there is indeed a separate access to physical geometry. The latter can be obtained, even in distorted cases, without making use of additional physical correction laws. That is, pure instances can be selected or self-referential distortions can be avoided without loss of generality. This works because we can arrange things such that the perturbing influences have no effects. Neutral (small, spinless, and so on) gravitational monopoles follow unflustered their geodesic paths, regardless of all turmoil on the differential level. This peculiarity has no analogue in the rod case and this is why the latter method is of no general avail.

This conclusion implies that Reichenbach and Grünbaum are right and that Einstein is wrong. Physical geometry can be established and tested in isolation from the differential correction laws so that the geometry is uniquely determined after a decision about universal forces has been made. On the other hand, Reichenbach and Grünbaum are only right for contingent reasons. That is, a separate access to geometry is in general only possible because undistorted motions do in fact exist. If, for instance, there were only charged particles, no general solution to Einstein's differential circularity problem could be given. Fortunately enough, nature does not hide its pure states entirely.

This discussion neither constitutes an argument for the nonconventionality of geometry nor for an antiholistic or non-Duhemian interpretation of geometry. As to the first item, the present argument clearly does not extend to universal forces so that Reichenbach's conventionality thesis remains untouched. The result is only that the decision to regard gravitationally influenced particle motions as free can be unambiguously put into experimental practice. Second, this is at the same time one of the reasons why Duhem's thesis is unscathed by the present argumentation. We are always free to introduce universal forces and thus to create a conceptually distinct but empirically equivalent description. The line of demarcation between geometry and physical force can be shifted at will. Moreover, the foregoing discussion does not imply any restrictions as to the structure of the differential correction laws themselves. How differential distortions are to be accommodated theoretically, that is, what kinds of distortions are to be assumed and by what laws they are supposed to be governed, is left open. So there is still enough room for theoretical holism. The upshot merely is that we cannot tamper with differential distortions in order to hold fast to our favorite geometry. So one special option to uphold one special element of a conceptual structure is blocked. And this is certainly not enough antiholism to bother even the staunchest Duhemian.

It appears, then, that the overall thrust of the argumentation as well as the results arrived at match with Grünbaum's approach. Although the details of the argument and the interpretation of the results are not always coincident, there is perfect agreement with the general conclusions reached by Grünbaum almost three decades ago. This is all the more remarkable since a lot of new developments have meanwhile occurred in spacetime philosophy. We may thus conclude that Adolf Grünbaum's work has stood the test of time.

NOTES

I would like to thank my colleague Claus Lämmerzahl for advice in matters of physics.

1. Reichenbach's recommendation refers to 4-forces, not to 3-forces such as the Newtonian gravitational force. For the relation of both quantities see Earman and Friedman (1973, 355). For brevity's sake I will omit such physical complexities.

2. A preliminary stage of the argument is found in Einstein [1921] 1984, 122–23.

3. In fact, the metric is still not completely fixed. An ambiguity remains that is due to the possible presence of second clock effects. (I will not dwell on that problem here. For a more detailed discussion of this approach, see Carrier 1990).

4. Millman argues that Grünbaum's test is not even sufficient for establishing freedom from perturbation, but his argument is less than persuasive. He imagines a situation in which rods of different chemical composition respond in an equal fashion to a differential force. In that case the perturbation test would be falsely negative (see Millman 1990, 29–30). The trouble with this argument is that it is either irrelevant or incoherent. If it is supposed to refer to a situation in which all actually employed (but by no means all existing) materials show an equal response, the failure to detect the differential distortion is simply due to the use of a biased sample and thus eventually due to inductive uncertainty. But Grünbaum does not purport to resolve the latter. If, on the other hand, Millman's argument is to refer to a situation in which all existing materials respond likewise, we no longer deal with a differential force. The same reply holds with respect to a conspiracy version of that argument: The superposition of several differential forces brings about a substance-independent deformation (see ibid., 31). But either there exist situations in which these conspiring forces can be empirically disentangled or we are not dealing with differential forces at all.

REFERENCES

Carrier, M. 1990. "Constructing or Completing Physical Geometry? On the Relation Between Theory and Evidence in Accounts of Space-Time Structure." *Philosophy of Science* 57: 369–94.

Earman, J., and M. Friedman. 1973. "The Meaning and Status of Newton's Law of Inertia and the Nature of Gravitational Forces." *Philosophy of Science 40:* 329–59.

Ehlers, J. 1988. "Einführung in die Raum-Zeit-Struktur mittels Lichtstrahlen und Teilchen." In J. Audretsch and K. Mainzer, eds., *Philosophie und Physik der Raum-Zeit.* Mannheim: Bibliographisches Institut, pp. 145–62.

Ehlers, J., and E. Köhler. 1977. "Path Structures on Manifolds." *Journal of Mathematical Physics* 18: 2014–18.

Ehlers, J.; F.A.E. Pirani; and A. Schild. 1972. "The Geometry of Free Fall and Light Propagation." In L. O'Raifeartaigh, ed., *General Relativity: Papers in Honour of J.L. Synge.* Oxford: Clarendon, pp. 63–84.

Einstein, A. [1949] 1970. "Remarks Concerning the Essays Brought Together in this Cooperative Volume." In P. A. Schilpp, ed., *Albert Einstein: Philosopher—Scientist.* 3d ed. La Salle, IL: Open Court, pp. 665–88.

———[1921] 1984. "Geometrie und Erfahrung." In C. Seelig, ed., *Albert Einstein. Mein Weltbild.* Frankfurt: Ullstein, pp. 119–27.

Grünbaum, A. 1960. "The Duhemian Argument." *Philosophy of Science* 27: 75–87.

———[1963] 1973. *Philosophical Problems of Space and Time.* 2d ed. Dordrecht: Reidel.

———1984. *The Foundations of Psychoanalysis: A Philosophical Critique.* Berkeley and Los Angeles: University of California Press.

Millman, A. B. 1990. "Falsification and Grünbaum's Duhemian Thesis." *Synthese* 82: 23–52.

Reichenbach, H. [1928] 1965. *The Philosophy of Space and Time.* New York: Dover.

Weinberg, S. 1972. *Gravitation and Cosmology: Principles and Applications of the General Theory of Relativity.* New York: Wiley.

2

Substantivalism and the Hole Argument

Carl Hoefer
Department of Philosophy, University of California at Riverside

Nancy Cartwright
Department of Philosophy, London School of Economics

Adolf Grünbaum's *Philosophical Problems of Space and Time* (1973) set the agenda for studies of these topics for mid-twentieth century analytic philosophy. It was an agenda with a pronounced point of view: a firm empiricism combined with a rigorous understanding of contemporary spacetime physics. The recent "hole argument" of John Earman and John Norton against spacetime substantivalism follows closely in this tradition. Earman and Norton argue roughly as follows: If points of the spacetime manifold were real, that would imply that any generally covariant theory, such as the general theory of relativity, would be indeterministic; but the issue of determinism versus indeterminism is an empirical question, and should not be settled by metaphysical assumptions.

We will argue that Earman and Norton deploy empiricism in an inappropriate way. Their argument is based on the observation that facts about the identities and locations (relative to material events and objects) of spacetime points have no empirical consequences in general relativity, nor for that matter in any other current theory of physics. The proper empiricist attitude in a situation like this is agnosticism: Without an empirical test we must not admit entities into our ontology. Nor can we rule them out, declare them nonexistent, as Earman and Norton do. Empiricists have no special glasses to see what is metaphysical and what is not.

More ambitiously, we want to suggest that the Earman and Norton argument rests on a bad adaptation of a common empiricist characterization of determinism. In order to avoid untestable claims about necessitation, empiricists these days tend to define determinism of a theory in terms of its models. The idea is that any models which are identical up to a given time should continue to be identical thereafter if the theory is deterministic. The definition is a descendent of the original characterization of Laplace, and it borrows much of its plausibility from its past uses. In the normal case this kind of characterization has been employed to treat theories that presuppose an ontology of continuants—individuals that exist through periods of time and for which notions of change, interaction, and evolution are appropriate. If we lay the conventional definition of determinism, which Earman and Norton presuppose, on top of an ontology of continuants, it works very well to characterize an interesting concept that matches our intuitive idea of determinism. We maintain, however, that without this underlying ontology the definition is too thin to pick out a significant notion, and certainly not a notion that can do the work that Earman and Norton require.

They insist that spacetime manifold points cannot exist because that would automatically settle the question of determinism. Even if we agreed that determinism could play such a crucial role, we should not reject spacetime points on account of their argument. For the concept of determinism that the hole argument deploys is not the rich, intuitively important concept that matters. Empiricists can admit a concept of continuance (that is what work on genidentity has shown), and we can start from that basis to provide the kind of explicit characterization of a fuller notion of determinism that empiricists demand. The hole argument proceeds from a definition that does only half the job.

The Hole Argument of Earman and Norton

Earman and Norton (1987) presented their hole argument which is very similar to an argument of Einstein's in 1913–1914 of the same name (in German, *Lochbetrachtung*).[1] Later we will look at the structure of their argument in detail, particularly the crucial final steps. First we will review the technical apparatus used in the hole argument; readers unfamiliar with the hole arguments are urged to look at Butterfield (1989), Earman and Norton (1987), Maudlin (1990), and Norton (1987). The argument relies on facts about *general covariance* (GC), which is a feature possessed by all well-known spacetime theories, including, Newtonian mechanics and the special theory of relativity, according to Earman and Norton (1987).[2] General covariance is best known as a feature of the general theory of relativity (GTR), however, and in the discussion to follow we will restrict our attention to GTR and its models. The active version of GC is used in the hole argument:

(Active) **General Covariance:** Let *M* be a 4-D differentiable manifold, g be a metric tensor defined on *M*, and T be the stress-energy tensor defined on *M*, giving the material contents of spacetime. Then given a model ⟨*M*,g,T⟩ of GTR, the spacetime model generated by applying an arbitrary (smooth, "one-to-one") transformation (diffeomorphism) to the points of *M* is also a model of GTR.

Notice that on the standard picture, to provide a model for GTR one specifies a manifold and then fixes a metric and a stress-energy tensor for a compatible with the field equations. This means that, so far as GTR is concerned, all the "empirical facts"—all the facts about the world that the theory describes or predicts—are expressed as facts about the values of these tensors on various manifold points (see Earman 1989, esp. chap. 9, sec. 12). This will be important for our later concerns.

Now, turning to the hole argument, let *h* be a nontrivial diffeomorphism; since *h* is applied to the points of *M* rather than the contents of *M*, *h* can be thought of as "shifting around" or rearranging the points of *M* (in a continuous fashion), *under* the contents g and T (see figure 2.1 which is largely borrowed from Butterfield's 1989 version). We can also think of this action in an "inverse" way, as holding the manifold fixed, and shifting, that is rearranging the contents g and T "on top of" the manifold. So the action of *h* on a model is equivalent to the action of the "drag-along" transformation h^* applied to the contents:

$$\langle M, \mathrm{g}, \mathrm{T} \rangle \rightarrow \text{New Model: } \langle M, h^*\mathrm{g}, h^*\mathrm{T} \rangle.$$

The hole construction proceeds very simply. Let *h* be a transformation that is the identity function ($h(p) = p$) everywhere on the manifold except for a small region, which is called the hole (for historical reasons that do not matter for our purposes; the hole can be hole-shaped, but it need not be), But let *h* *not* be the identity inside the hole; *h* rearranges the points inside the hole underneath the contents **g** and **T** (see figure 2.1).

This construction can be used to attack a particular kind of substantivalism that Earman and Norton say is the only kind viable in GTR:

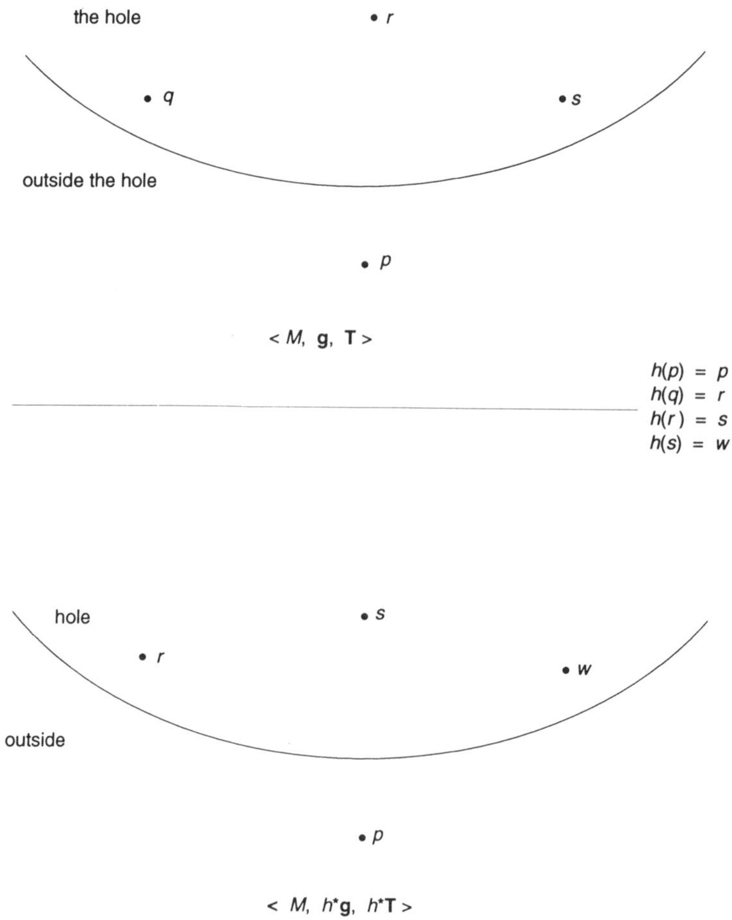

Figure 2.1 Diffeomorphism "shifts" points inside the hole.

Manifold Substantivalism is the view that the four-dimensional point manifold *M* represents a substantival spacetime whose points and regions exist and are distinct individual, substantial entities.

For a manifold substantivalist, two models related by a hole diffeomorphism (call them model 1 and model 2) differ physically inside the hole; they have different metric and stress-energy properties and relations at point *p* and its neighborhood, for example. Suppose that, given some typical coordinatization of *M*, prior to some time t_0 there is no physical difference between model 1 and model 2. If the hole is a small region of *M*, this will be the case. This shows that two models of GTR can be completely identical up to some time t_0, and disagree about physical facts after that time. This is a clear violation of determinism on one straightforward way of defining determinism for a theory in terms of its models. For concreteness, here is such a definition:

Determinism: A theory is deterministic if and only if any two of its models (representing worlds) that agree on all physical facts prior to time t_0 also agree on all physical facts thereafter (i.e., on the whole model).

Earman and Norton's hole argument starts by showing that for manifold substantivalists, and given a straightforward definition of determinism, GTR is radically indeterministic.[3] The crucial part of their argument is the claim that this indeterminism is unacceptable:

Our argument does not stem from a conviction that determinism is or ought to be true. There are many ways in which determinism can and may in fact fail. . . . [Here they cite several difficulties for determinism in classical physics, GTR and in quantum mechanics.] Rather, our point is this. If a metaphysics, which forces all our theories to be deterministic, is unacceptable, then equally a metaphysics, which automatically decides in favor of indeterminism, is also unacceptable. Determinism may fail, but if it fails it should fail for reasons of physics, not because of commitment to substantival properties which can be eradicated without affecting the empirical consequences of the theory. (1987, 524)

The last sentence alludes to the fact that models related by a transformation *h* are empirically indistinguishable: If we decided to give up manifold substantivalism and stop viewing them as distinct physical situations, GTR would lose none of the empirical predictive power that it possesses. We will return to this point later.

To further clarify Earman and Norton's ideas, we will cite other summaries of this part of the argument, beginning with Earman (1989):

> The argument in the proceeding section [the hole argument] does not rest on the assumption that determinism is true, much less on the assumption that it is true *a priori,* but only on the presumption that it be given a fighting chance. To put the matter slightly differently, the demand is that if determinism fails, it should fail for a reason of physics. We have already seen a couple of such reasons. For example, the laws of physics might allow arbitrarily fast causal signals, as in Newtonian mechanics, opening the way for space invaders, or the laws might impose a finite upper bound on causal propagation but, as in some general-relativistic cosmological models, still permit analogues of space invaders. . . . But the failure of determinism contemplated in section 3 is of an entirely different order. Rule out space invaders; make the structure of space-time as congenial to determinism as you like by requiring, say, the existence of a Cauchy surface; ban stochastic elements; pile law upon law upon law, making the sum total as strong as you like, as long as logical consistency and general covariance are maintained. And still determinism fails if the proffered form of space-time substantivalism holds. Here my sentiments turn against substantivalism. (Pp. 180–81)

Butterfield (1989) agrees with Earman and Norton about the salience of the hole argument; after briefly describing the problem for determinism that it exposes, he writes: "[W]hat is wrong with substantivalism is not simply that it rules out determinism, but that it does so at a stroke for a wide class of theories—and furthermore without affecting the predictions of the theories" (p. 3). In a paper that discusses the problem of inertial structure as well as the hole argument, Teller (1991) affirms the Earman and Norton line as well, "As Earman and Norton say, if determinism fails it should fail for reasons which take empirical investigation to establish, and not for this sort of trivial, *a priori* reason" (p. 29).

These passages display the essential point of the part of the hole argument directed against substantivalism: Since manifold substantivalism saddles us with a commitment to indeterminism for metaphysical or a priori reasons rather than empirical reasons ("reasons of physics"), it should be rejected. In a later section, we will criticize this argument and establish that the hole argument gives no new reasons to reject substantivalism. What the hole argument does do is force us to examine and clarify the commitments of substantivalism and its relation to determinism.

Before moving to the criticisms of the hole argument, we will first sketch how a similar argument can be generated against substantivalism. This argument, which we will call "the new hole argument," also falls prey to the criticisms later presented, so it is not a more compelling argument against substantivalism. It is worth looking at, however, because it illustrates key aspects of the substantivalist ontology.

New Hole Argument

The possibility of this kind of argument was first pointed out by Christopher Ray. (For a discussion of topological holes, see Ray 1987, chap. 4; see also Geroch and Horowitz 1979.) In the new hole argument, a pair of models of GTR are again generated that are identical before a certain time, but not thereafter. Before, it was the fact of general covariance that let us stipulate models 1 and 2, related by a hole-diffeomorphism; here a different fact is going to be used (Earman 1989, chap. 8):

Hole Fact: If $\langle M, g, T\rangle$ is a model of GTR, then the mathematical object obtained by deleting part of the manifold M (in an appropriate way) together with its contents, $\langle M^{-}, g^{-}, T^{-}\rangle$, is also a model of GTR. (See figure 2.2)

The constraint a model has to satisfy in order to be a model of GTR is that Einstein's equations—which are *local* equations—hold at every point on the manifold. But if these equations held in the first model, they will also hold on the new, reduced model: We have discarded some of the points of M, but on every point that remains (in M^{-}) everything is exactly the same, and so Einstein's equations must still hold at every point that exists in the new model.[4]

With pairs of models like 1′ and 2′, indeterminism of a curious nature follows. Models 1′ and 2′ agree on all physical facts before time t_0 (tomorrow), and disagree on physical facts thereafter, in the following sense. The Stanford quad, say, fails to exist starting tomorrow, in model 2′. This form of indeterminism may strike us as more radical than that of Earman and Norton's argument, because the differences between the models are more physically significant in a certain sense. The models disagree not just about what material stuff will be at what points tomorrow; they disagree about whether large scale things (like the Quad) will even exist tomorrow! Here we have cut a relatively small hole in spacetime. Obviously, there is no limit to

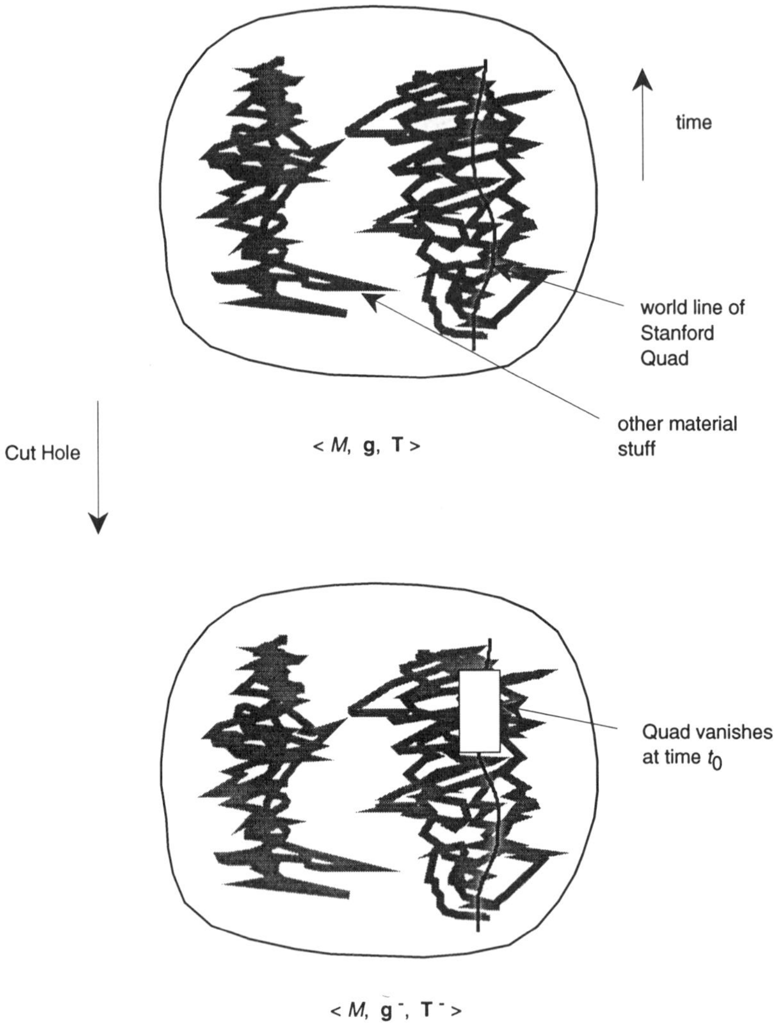

Figure 2.2 Topological hole.

the size of the part of the manifold that can be deleted. So, for example, the entire future half of the manifold could be cut away; the world could stop existing tomorrow. More interestingly, the past half could be cut away; we can model the old epistomological puzzle case in which the universe is supposed to have come into being yesterday (or one minute ago).[5]

Substantivalism is crucial in generating this indeterminism. It is the view that *M* and its points are substantial entities that makes it possible to suppose that some part of *M* might not have existed: The points and regions of *M*, being individual entities, have contingent existence. Nothing in the definition of manifold substantivalism above entails that points are taken to exist only contingently, but this is the view most consonant with the claim that spacetime points are individual physical objects. Paul Teller (1987, 426) assumes the opposite, namely, that substantivalists must take spacetime points to be necessary existents. Perhaps he had in mind the problems of holes that can otherwise arise, but Teller notes that this necessity then sits uneasily with the substantivalist's view that spacetime points and regions are physical objects. Earman simply claims that manifold substantivalists ascribe only contingent existence to their points, and he points to new-hole type constructions used by physicists as evidence for his claim (1989, 14).

We can now present the same form of argument that Earman and Norton make. The new hole argument shows a radical breakdown of determinism that comes only from a certain ontological commitment, not from "reasons of physics," and for this reason, we should give up that ontological commitment, i.e., give up manifold substantivalism. Given that Earman (1989) discusses topological holes at length, one may wonder why he does not raise the possibility of this analogous hole argument himself. The reasons probably have to do with the relative avoidability of the new hole problems for substantivalists. In working with spacetime models, one can stipulate that only inextendible models should be used to represent physically possible worlds; general relativists would perhaps find this a plausible restriction. This has the effect of ruling out models generated from prior models by the hole-cutting operation. Still another way to avoid models with holes might be the way Teller suggests, namely, claiming necessary existence for spacetime points. Earman views his and Norton's hole argument as forceful partly because no such escapes as these are open to the manifold substantivalist confronted with their hole argument. But this distinction should not be of much comfort to manifold substantivalists. Unless they can supply independent cogent reasons for either ruling out extendible models or ascribing necessary existence to points—reasons compatible with the spirit of their substantivalism—the new hole argument has no less force against manifold substantivalism than the hole argument.

The new hole argument shows that general covariance is not necessary for substantivalism to be subject to indeterminism; the contingent existence of spacetime points leads to indeterminism also. In the hole argument, unlike the new hole argument, it is crucial that spacetime points have individual identities, and that GTR does not regulate these identities. This feature of spacetime points can be exploited in another way to give rise to indeterminism. Instead of using active general covariance to shift around the points of the manifold, we can construct models of the theory that have manifolds made from *distinct* point sets. That is, we can make an analog of the hole argument using a diffeomorphism from M onto M^*, where M^* is (in part) a different manifold.

Manifolds are constructed out of "base point sets," so we can "build" two diferent manifolds, M and M^*, from point sets that *overlap* in much, but not all, of their membership.[6] Then we construct a diffeomorphism between them that takes the points that exist in both sets (i.e., the overlap points) into themselves. Suppose these overlap points comprise all of the manifold except for "the holes." Then the second model differs from the first only in the hole—but the difference now is not that the manifold points have been shifted around; instead, they have been replaced by altogether distinct points. The indeterminism this shows is essentially no different from that of the hole argument, though as in the new hole argument some assumptions about contingency must be made. This argument assumes that in addition to the spacetime points that actually exist, other individual spacetime points exist in other possible worlds.

Manifold substantivalism can, in conjunction with other assumptions, yield forms of indeterminism in GTR. One way is with the assumption of active general covariance; and in this section we have seen two other ways. Now we turn to the question of whether indeterminism can be used to give us grounds for rejecting manifold substantivalism.

Criticism of Earman and Norton's Argument

The hole argument may seem persuasive. It seems to show an unexpected and counterintuitive way in which determinism breaks

down if we accept manifold substantivalism. We now argue that Earman and Norton give us no clear reason to use indeterminism as a reason to reject manifold substantivalism. In order to make our responses as clear as possible, we will first give a step-by-step reconstruction of the crucial part of Earman and Norton's argument:

1. The issue of determinism should be decided by "reasons of physics," that is, the nonmetaphysical, contingent, a posteriori truths of physics.
2. Therefore, any metaphysical commitment that forces a verdict on the issue of determinism is unacceptable.
3. Manifold substantivalism is a metaphysical commitment, as we can see from the fact that its rejection does not affect the empirical, predictive power of GTR.
4. In any generally covariant theory, manifold substantivalism forces us to accept a radical indeterminism.
5. Therefore, manifold substantivalism is unacceptable.

We will give two direct challenges to Earman and Norton's argument. The first centers on the part of the argument we have represented through premise (3).

Remember that manifold substantivalism is an ontological claim: Manifold substantivalists believe that spacetime is a physically existent thing, composed of points and/or regions that are themselves physical, existing things. Ontological claims are not "metaphysical" in general—consider the claim that electrons exist, or perhaps quarks. Mot philosophers would count these as empirical claims rather than metaphysical doctrines. What makes manifold points different? According to Earman and Norton, "[Facts about substantival points] can be eradicated without affecting the empirical consequences of the theory" (1987, 524). They seem to take this as sufficient proof that manifold substantivalism is a metaphysical doctrine rather than a "reason of physics."

Manifold substantivalists, however, see matters another way. Manifold points exist and are distinct from one another. That is a fact about the world as it is; hence, an empirical fact. If GTR does not, or cannot, adequately represent this fact, then GTR is incomplete. That should be no surprise; after all, GTR does not purport to tell us much about the world around us. It is a theory entirely about the shape of spacetime and its relation to the motions and properties of the material substance that is in it. It is not, nor does it purport to be, a theory about which spacetime points exist, or how they got to be where they are.

Should there be such a theory? Indeed, is there any phenomenon about which to have such a theory? That of course is the question at issue. Earman and Norton clearly suppose not. Perhaps they have in mind the fact that GTR physicists routinely treat diffeomorphic models as the same, that is, as representing the same physical world or situation (the assumption Earman and Norton call "Leibniz Equivalence."). This practice of physicists is understandable and unproblematic, but physicists have different goals and issues in mind than philosophers. Philosophers who are manifold substantivalists often adopt that view, in part at least, because the point manifold seems to be an intrinsic part of the description of the world through GTR—*not* something that seems superfluous or eliminable. While point facts can be ignored in the use of GTR, this view is far from showing that the point manifold is an eliminable part of the description of the physical world in GTR. Earman (1989) stresses the apparent difficulty, if not impossibility, of eliminating the point manifold from the mathematics of GTR. If the use of the point manifold is not eliminable in GTR, then many scientific realists will believe in the existence of substantial points regardless of the fact that their individual identities make no difference to the relations that GTR studies.

The real issue in premise (3) is "What is to count as empirical?" The history and identity of bits of the manifold are not regulated by GTR, but that is no argument that these are not matters of empirical fact. Nor is the fact that no current physics studies them. That might be a failing of current physics. That would not mean that manifold point facts are not empirical.

We have *one* good, rigorous criterion of the empirical, which Grünbaum deploys so trenchantly in his studies of psychoanalysis. That is the criterion of testability. It is a stringent requirement and it may be no service to the cause of GTR to invoke it. It is in no way clear how much of GTR—or indeed most of the highly abstract theories of contemporary mathematical physics—would stand up to the kind of detailed and penetrating inquiry about exactly what tests are available, and exactly which claims they test, to which Grünbaum has subjected psychoanalysis. Moreover, though we may subscribe to it, testability is not the only criterion in the running. Many philosophers of science reject the demand that each claim we take to be true must be tested individually. Rather, theories as a whole are to be judged as confirmed or disconfirmed. If we were to adopt this less stringent view, we would be faced with the sticky question, already raised, of what the exact role of the manifold is in GTR "as a whole." Let us restrict our attention though to considerations that bear on some kind of testability requirement, since that will suffice to make clear the point we want to stress.

Imagine we agree to admit as empirical only what is testable. We do not know what is testable. To a large extent what is testable will depend on what there is, and not the other way around. We know fairly well what we can test now, so we know some things we can reasonably count as empirical now. But we do not know what cannot be tested. We may agree that a stringent testability criterion (as opposed to the more "holistic" point of view of many scientific realists previously alluded to) provides no grounds for admitting manifold points as part of empirical reality. Earman and Norton's argument requires the much stronger assumption that they are *not* part of empirical reality. Issues of determinism, they argue, should be settled not by the dictates of metaphysics, but rather by the nature of the empirical world. But it is exactly a question of the nature of the empirical world that is at stake. GTR does not show us how to observe manifold points, nor does it tell us why they are located where they are. Perhaps we can take GTR and its successes as a tentative reason for counting some abstract quantities, like stress-energy, as empirically real. But we must not let its gigantic silences dictate what is *not*.

Our second line of criticism is directed at the combination of premises (1) and (2). Together they seem to constitute a kind of metaphilosophical premise to the effect that truths of physics—contingent, a posteriori truths—should be the adjudicators on issues of determinism, as opposed to metaphysics. This is not spelled out explicitly in Earman and Norton's paper, but it seems to be their intent; and recall Teller's formulation, which emphasizes that determinism should be decided by "reasons which take empirical investigation to establish." How could such a metaphilosophical premise be supported? And, given the history of the subject of determinism, is it even plausible?

Determinism can be understood as a property of a theory, or as a property of the world. The technical part of the hole argument shows that GTR as a theory is indeterministic in a certain sense—full stop. Hole-diffeomorphic models are models of the theory that are identical up to a certain time, and not identical thereafter. Earman and Norton presumably object to extending automatically this indeterminism verdict to the *world* through the adoption of manifold substantivalism. Now, considering the determinism or indeterminism of the world, why should we believe that questions of metaphysics cannot play a decisive role?

It is not hard to find issues of metaphysics that should have much to do with the question of determinism. The question of free will, and the mind-body problem, are two that come to mind. An even better example is the issue of the status of *laws of nature*—whether there are such things, and if so, how they should be understood. Whatever position one adopts on the question of laws is likely to influence the question of the possibility of determinism, even to the point of deciding the issue.[7] One might want the question of determinism to be isolated from questions of metaphysics, but as long as it is so intimately linked to the issue of the status of laws of nature, such isolation is not possible.[8]

Indeed, we think that the concerns about metaphysics need to be turned around. Whether a theory is deterministic depends not on the actual history, not on the events that occur in the material world, the events that could enter our experience. Determinism versus indeterminism depends on possible histories, on what models—besides the one most like the actual world—are allowed by the theory. Earman and Norton do not want to let an important "empirical" question like determinism-indeterminism be settled by a matter of metaphysics. We, by contrast, do not want to let an important scientific question, what there is in the world as it actually occurs, be affected by modal issues, which, even under typical empiricist renderings, are questions about what our best theories say may happen in some other merely possible worlds.

One can of course pursue more stages in this argument. On some empiricist accounts, the laws to be accepted in our best theories can in principle be "read off" the actual history of the world. For example, they might be the "best" summary of the facts, where "best" involves a number of possibly competing criteria such as elegance, efficiency, and breadth of scope. At this point we recall our earlier caution that the hole argument must not beg the question against

manifold substantivalism. Manifold substantivalists assume that manifold point facts are (or, for the stricter empiricists among them, may eventually be observed to be) among the facts of the world. If they are among the facts, then the "best" summary of those facts had better not rule them out. We also want to make a different, more general point about the kinds of laws we get with this procedure. We are concerned that, when laws are rendered as "best summaries"—that is, when that is taken to be the basis for counting something "necessary"—the concept of possibility loses much of its usual sense, and correlatively, its usual interest. Therefore, so does the issue of determinism. If this issue is settled by complicated facts about the actual world without reference to possibilities in any stronger sense, then the question no longer amounts to what we thought it did, and it is unclear why we shold pay it special attention.

This brings us to our last concern. What in any case is the significance of the question of *determinism* versus *indeterminism* when it is the existence of spacetime points that is at issue? These are terms rich in association, most of them bearing on worries about human spontaneity, freedom, and responsibility. Grünbaum has vigorously resisted these associations, and that is the path appropriate to empiricism. But once these associations are stripped away from the issue of determinism, what interest is left to it?

We think that there are some interesting questions about how individual entities evolve over time. In a given domain of characteristics, what kind, if any, of simple descriptions can be found for how individuals with those characteristics change when left on their own, and for how they change in interaction with other individuals? Those questions, though, have no bearing on spacetime points, points without pasts or futures, without histories, that experience no change. Our underlying hypothesis is that, once issues of human freedom and necessitation have been laid aside, as the empiricist program urges, the remaining interest in determinism versus various kinds of indeterminism presupposes an ontology of what W. E. Johnson (1921–1924) called "continuants"—individuals with changing properties but identity over time interacting with other individuals in their environment.

None of this, of course, is built into the family of definitions of determinism employed in discussions of the hole argument. These basically look at models which are identical on some timelike surface, or over some thicker region of time, and ask if they are identical everywhere. These definitions provide a neat way to adapt traditional questions of determinism both to meet empiricist strictures against talking of necessity, and to enable them to apply to Minkowski-type world pictures where not space points, but only spacetime points, may be found. But just on account of their success at this latter task, the definitions are too thin. If we apply them to an ontology of continuants—or, more to the point of Minkowski-type theories, an ontology where some surrogate for continuants is presupposed, such as genidentity—they work very well to pick out the relations we expect of determinism. But *by themselves* these definitions do not characterize features of particular philosophical interest.

For this reason, we return to questions about continuants and how they change over time as the locus for concerns about determinism and different kinds of indeterminism. Here we have a story about how the world is made up and about what happens in it that has a traditional interest and is compatible with empiricism. Claims that we can map out a sense of individual identity across time, correlatively with simple rules of how characteristics in individuals evolve and change, are the meat of empirical research. We can do it for massive objects—we have learned that empirically. We cannot do it for photons—that is what we have learned from photon bunching and antibunching experiments.[9] It is not ruled out *tout court* to look for determinism and indeterminism in Minkowski-type spacetime theories. A good deal of empiricist literature is provided on the notion of genidentity of events across separate spacetime points. A strip of genidentical events is a good stand-in for a continuant in a Minkowski-type theory, and hence a good starting point for interest in determinism. What we want to urge is that, where genidentity—or some stronger notion of a continuant—does not apply, these questions of determinism are without much point.

This is the case, we believe, with the question of the existence or location of individual spacetime points on which the hole argument depends.[10] If manifold points do exist, then where they exist is "indeterministic" in GTR; the definition of indeterminism in terms of models of the theory is satisfied. But what work does the definition do in this case? The spacetime points are not continuants, nor does GTR give us a way to make them genidentical. They are not the kind of things that can evolve over time, either uniformly or probabilistically, or in some more chaotic way. The "thicker" concepts of determinism and indeterminism that we usually have in mind when we use these terms do not apply to them one way or another.

For example, imagine that we are given a box of white stickers cut into simple shapes, such as a star, circle, moon slice, pentagram, and diamond. For concreteness, let there be 18 such stickers, all different. First we assign coordinates to the stickers, keeping in mind that we want to end up with a simple timelike structure, say 6 time slices. Otherwise, the assignment of coordinates is arbitrary. Now each sticker has a label: $\langle i,j \rangle$, $1 \leq i \leq 6, 1 \leq j \leq 3$. This is our analogue of the manifold with coordinates. We need also a characteristic analogous to the metric or affine connection properties, a propety defined at every sticker point. Let it be color distribution, taken from the color wheel in figure 2.3.

On the wheel the three *primary* colors are separated by the three *secondary* colors. We have a simple theory for this universe: Only primary colors are allowed, or only secondary colors are allowed; and at any "time" each sticker must be associated wtih a different color. This theory is *deterministic* with respect to primary-versus-secondary-coloredness: Given the fact that a time slice of a model is primary- (or secondary-) colored, every admissible model sharing this time slice will be the same everwhere else with respect to this property as well. The theory, however, is indeterministic with respect to which shape has which of these allowed colors. For instance, a red square, blue triangle, and yellow parallelogram can be followed by a red moon slice, a blue diamond and a yellow circle; or by a blue circle, a red diamond, and a yellow moon; and so on (see figure 2.4).

The point is that this is in no way disturbing or remarkable. We see the primary or secondary colors appear again and again. In the theory, *some* shape must occur at a given colored vane, but which shape is entirely arbitrary, and of no consequence. The role of the shaped stickers in this theory mirrors the role of manifold points in GTR.

Consider a contrasting case. Instead of a box of geometric shapes, we are given a "Gavagai" box which contains six rabbit slices, six hedgehog slices, and six fox slices. We assign coordinates so that each time surface has one rabbit slice, one fox slice, and one hedgehog slice. The theory is as before, but now the question of determinism for the colors of the individual stickers is one of real interest—not

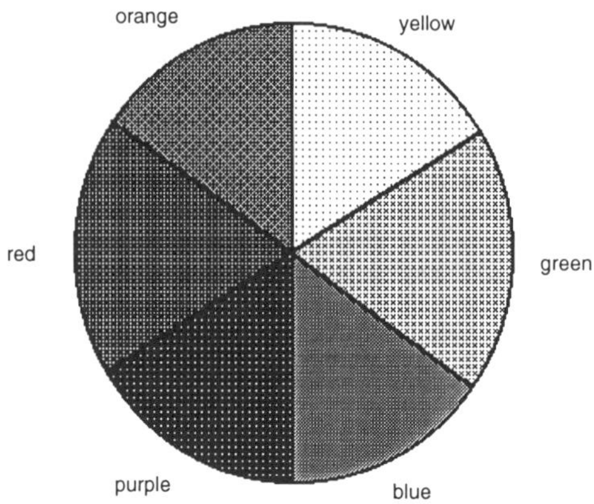

Figure 2.3

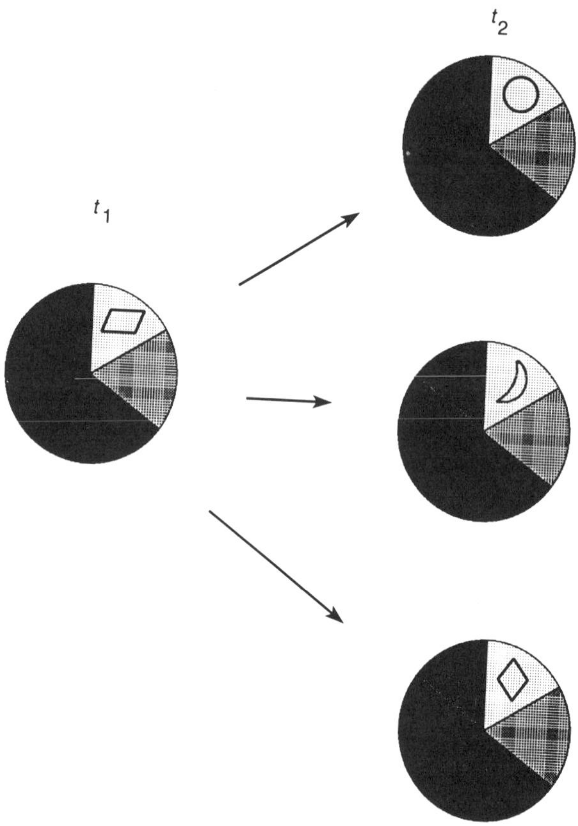

Figure 2.4

just the crude distinction, determinism versus indeterminism, but the more detailed question of what kind of indeterminism. Is there some pattern to how the animals change (or do not chage) their colors? Can we find a rule under which they evolve naturally? If no simple rule emerges, can we imagine them to be interacting, and find a system of more complicated rules? The geometric shapes of the first example have nothing to do with each other, and each occurs only once. There is no natural sense in which one shape is the descendent of another. Once we do have this sense of inheritance (or genidentity: the event of redness at t_2 is genidentical with the event of blueness at t_1 since it is the rabbit that is both blue at t_1 and red at t_2) it makes sense to ask about the patterns of change. This analogy is unfavorable to the hole argument. Of course the earlier universe of geometric shapes meets the definition for indeterminism, but in a trivial way. There is nothing in this universe which "continues" across time, nothing about which one can ask "Is the way it changes orderly or disorderly" or "If disorderly, how disorderly?"

We think that the hole argument give philosophers, empiricists or otherwise, no compelling reason to abandon the ontological position of manifold substantivalism. The hole argument, however, raises interesting problems about our understanding of determinism. When the concept of determinism is applied in the context of an ontology of evanescent entities, rather than an ontology of continuants, we believe that the concept loses most of its usual interest. For this reason as well as others advanced earlier, it is inappropriate to try to settle the ontological issue of the status of spacetime points in GTR based on its relation to determinism. If spacetime points do exist as substantial entities in the way manifold substantivalism claims, then a kind of indeterminism does obtain in the GTR description of the world. Philosophers, though, must look for other grounds, including empiricist grounds, with which to try to settle this ontological question.

NOTES

1. Einstein used "indeterminism," or rather a failure of uniqueness of the metric field given the stress-energy field, to argue against general covariance. See Norton (1987) for the fascinating history of this argument in Einstein's development of GTR.

2. On this point, Earman seems to have recently changed his mind. Earman (1989) presents the hole argument as applying only to GTR. Formally, he says, earlier theories such as Newtonian mechanics could be presented as generally covariant, but to do so "papers over some important distinctions" (Earman and Norton 1987, 516), namely, the intrinsic structure of absolute space that should limit what diffeomorphic transformations we are allowed to use to generate new models.

3. Obviously, one way to try to escape the hole argument is to deny the version of determinism needed to support the argument, and to propose an alternative. This is part of Jeremy Butterfield's (1989) strategy.

4. We are ignoring some technical complications here, such as the requirement that the hole include its own boundary points so that the remaining manifold is still everywhere differentiable. These complications do not affect the point of the argument.

5. These remarks sound as if they are presupposing a global difinition of absolute simultaneity, but there is no such presupposition. At most, existence of a global coordinate system that is locally inertial (around the points in question, e.g., the Quad), and whose *t* coordinate is everywhere timelike, is here assumed. This constraint is satisfied in many physically interesting GTR models.

6. It might seem natural to object at this point, that any "two" manifolds with the same topology are really the same, and that this talk of "different" manifolds composed out of overlapping point sets is nonsense. Mathematically this view may be correct, but it cannot be presupposed in the present context where manifolds as *physical* entities are at issue. The substantival individuality of spacetime points that we are appealing to here is just the individuality needed to make the hole argument work in the first place.

7. For example, van Fraassen's (1989) rejection of the metaphysical concept of laws of nature entails that determinism can *at most* be a feature of theories, not a feature of the world. This decides the issue of determinism of the world, in a sense, by ruling that it is a pseudoquestion.

8. The two lines of criticism we have raised complement each other: The first argues that manifold substantivalism is a doctrine that perhaps should not be classified as "metaphysical" in the sense of "metaphysics" that Earman and Norton want to keep isolated from determinism; the second argues that, even if we regard substantivalism as metaphysics, there is no clear reason to suppose that we can or should keep metaphysics isolated from determinism. These two lines of criticism can be brought together, in a sense, in a turnabout maneuver: Why should we not regard the technical part of the hole argument as a *counterexample* to the claim that metaphysical considerations cannot decide the issue of determinism? The technical part of the hole argument shows that if manifold substantivalism is true, and GTR as well, then determinism is false. Only if we adopted an explicit commitment to the *truth* of determinism could the hole argument be used to rule out manifold substantivalism.

9. If we wanted to treat photons as continuants with simple laws of evolution and interaction, we would have to allow that one photon could interfere with another even though the latter had already been observed.

10. When we speak of the "location" of a spacetime point, we mean its location relative to material things in spacetime, or relative to other points.

REFERENCES

Butterfield, J. 1989. "The Hole Truth." *The British Journal for the Philosophy of Science 40:* 1–28.

Earman, J. 1989. *World Enough and Space-time.* Cambridge, Mass.: MIT Press.

Earman, J., and J. Norton. 1987. "What Price Spacetime Substantivalism? The Hole Story." *The British Journal for the Philosophy of Science 38:* 515–25.

Geroch, R., and G. T. Horowitz. 1979. "Global Structure of Spacetimes." In S. W. Hawking, and W. Israel, eds. *General Relativity: An Einstein Centenary Survey.* Cambridge, England: Cambridge University Press.

Grünbaum, A. 1973. *Philosophical Problems of Space and Time.* 2d ed. Dordrecht: Reidel.

Johnson, W. E. 1921–1924. *Logic.* Cambridge, England: Cambridge University Press.

Maudlin, T. 1990. "Substances and Space-time: What Aristotle Would Have Said to Einstein." *Studies in History and Philosophy of Science 21:* 531–61.

Norton, J. 1987. "Einstein, the Hole Argument, and the Objectivity of Space." In J. Forge, ed., *Measurement, Realism, and Objectivity.* Dordrecht: Reidel, pp. 153–88.

Ray, C. 1987. *The Evolution of Relativity.* Philadelphia: Hilger.

Teller, P. 1987. "Space-time as a Physical Quantity." In R. Kargon, and P. Achinstein, eds., *Kelvin's Baltimore Lectures and Modern Theoretical Physics.* Cambridge, Mass.: MIT Press, pp. 425–48.

———. 1991. "Substance, Relations and Arguments About the Nature of Spacetime." *Philosophical Review 100:* 353–97.

van Fraassen, B. C. *Laws and Symmetry.* 1989. New York: Oxford University Press.

3

The Cosmic Censorship Hypothesis

John Earman
Department of History and Philosophy of Science, University of Pittsburgh

> It is one of the little ironies of our times that while the layman was being indoctrinated with the stereotype image of black holes as the ultimate cookie monsters, the professionals have been swinging around to the almost directly opposing view that black holes, like growing old, are really not so bad when you consider the alternative. —**Werner Israel (1986)**

The idea of cosmic censorship was introduced over twenty years ago by Roger Penrose (1969). About a decade later Penrose noted that it was not then known "whether some form of cosmic censorship principle is actually a consequence of general relativity" (1978, 230), to which he added, "In fact, we may regard this as possibly the most important unsolved problem of classical general relativity theory" (ibid.). This sentiment has been echoed by Stephen Hawking (1979, 1047), Werner Israel (1984, 1049), Robert Wald (1984, 303), Frank Tipler (1985, 499), Douglas Eardley (1987, 229), and many others. Thus, if an "important problem" in physics is one which is deemed to be important by leading research workers in the field, then the problem of cosmic censorship is undoubtedly near the top of the list for nonquantum general relativity. One of my goals here is to show why it is important in a more substantive sense. I also want to indicate why it is that despite the intense effort that has been devoted to this problem, it remains unsolved. Indeed, the very statement of the problem remains open to debate.

A study of this topic can lead to payoffs in several areas of philosophy of science, two of which I will mention and one of which I will actually pursue. In Earman (1986) I attempted to deflate the popular image of determinism as unproblematically at work outside of the nonquantum domain. My message fell largely on deaf ears. The failure of cosmic censorship could well herald a breakdown in classical predictability and determinism of such proportions that it could not be ignored.[1] Second, the growing band of philosophers of science who are turning toward an increasingly sociological stance will find the history of the cosmic censorship hypothesis a fascinating case study in the dynamics of a research program. Particularly interesting is how, in a subject with very few hard results, the intuitions and pronouncements of a small number of people have shaped and directed the research. I leave the investigation of such matters to more capable hands.

Cozying Up to Singularities

Prior to the 1960s, spacetime singularities were regarded as a minor embarrassment for the general theory of relativity (GTR). They constituted an embarrassment because it was thought by many that a true singularity, a singularity in the very fabric of spacetime itself, was an absurdity.[2] The embarrassment, though, seemed to be a minor one that could be swept under the rug; for the models of GTR known to contain singularities all embodied very special and physically unrealistic features.[3] Two developments forced a major shift in attitude. First, the observation of the cosmic low temperature blackbody radiation lent credence to the notion that our universe originated in a big bang singularity. Second, and even more importantly, a series of theorems due to Stephen Hawking and Roger Penrose indicated that singularities cannot be relegated to the distant past because under very general conditions they can, according to GTR, be expected to occur both in cosmology and in the gravitational collapse of stars (see Tipler et al. 1980 and Wald 1984, chap. 9). Thus, singularities cannot be swept under the rug; they are, so to speak, woven into the pattern of the rug. Of course, these theorems might have been taken as turning what was initially only a minor embarrassment into a major scandal. Instead what occurred was a 180-degree reorientation in point of view: Singularities were no longer to be thought of as lepers to be relegated to obscurity; rather, they must be recognized as a central feature of the GTR which calls attention to a new aspect of reality that was neglected in all previous physical theories, Newtonian and special relativistic alike. Thus we can hope to get definitive confirmation of GTR by observing these new objects.

Before we get carried away in our newfound enthusiasm for singularities, we should pause to contemplate a potential disaster. If the singularities that occur in Nature were naked, then chaos would seem to threaten. Since the spacetime structure breaks down at singularities and since (*pace* Kant) physical laws presuppose space and time, it would seem that these naked singularities are sources of lawlessness. The worry is illustrated (albeit somewhat tongue in cheek) in figure 3.1 where all sorts of nasty things—such as TV sets showing Nixon's "Checkers speech," green slime, and Japanese horror movie monsters—emerge helter-skelter from the singularity.

The point can be put more formally in terms of the breakdown in predictability and determinism. If S is a spacelike surface of a general relativistic spacetime M, g_{ab},[4] the *future* (respectively, *past*) *domain of dependence* $D^+(S)$ ($D^-(S)$) of S is defined to be the collection of all points $p \in M$ such that every causal curve which passes through p and which has no past (respectively, future) endpoint meets S. If $p \notin D^+(S)$ (respectively, $D^-(S)$), then no amount of initial data on S will suffice for a sure prediction (retrodiction) of events at p since there are possible causal influences which can affect events at p but which do not register on S.

To illustrate how naked singularities can lead to a breakdown in predictability, let us start with two-dimensional Minkowski spacetime $\mathbb{R}^2$, η_{ab} and consider the spacelike hypersurface S corresponding to the level surface $t = 0$ of some inertial time coordinate t. $D^+(S)$ encompasses the entire future of S. Now remove from $\mathbb{R}^2$ a closed set of point C on the future side of S. The resulting spacetime has a naked singularity, the presence of which excludes the shaded region of figure 3.2 from $D^+(S)$. The future boundary of $D^+(S)$, called the *future Cauchy horizon* of S, is labeled as $H^+(S)$. This naked singularity is rather trivial since it can be removed by extending the surgically mutilated spacetime back to full Minkowski spacetime. To make the example less trivial, one can choose a positive valued scalar field Ω that goes rapidly to zero as the missing region C is approached. The new conformally related spacetime M, g_{ab}, where $M = \mathbb{R}^2 - C$ and $g_{ab} = \Omega\eta_{ab}$, is inextendible and so its naked singularity is irremovable.

Cosmic censorship is an attempt to have one's proverbial cake and eat it too. The idea is that we can cozy up to singularities without fear

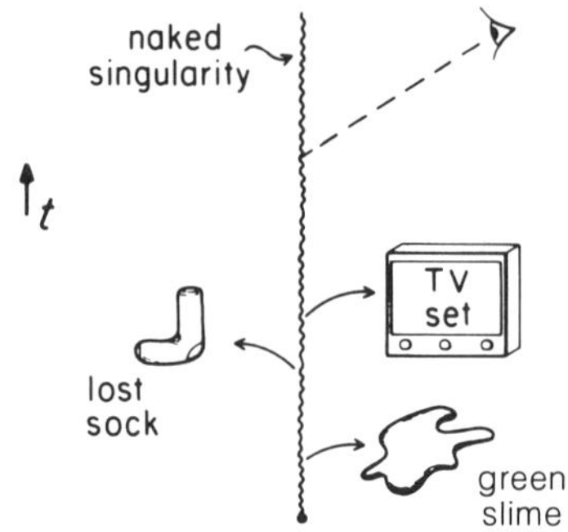

Figure 3.1

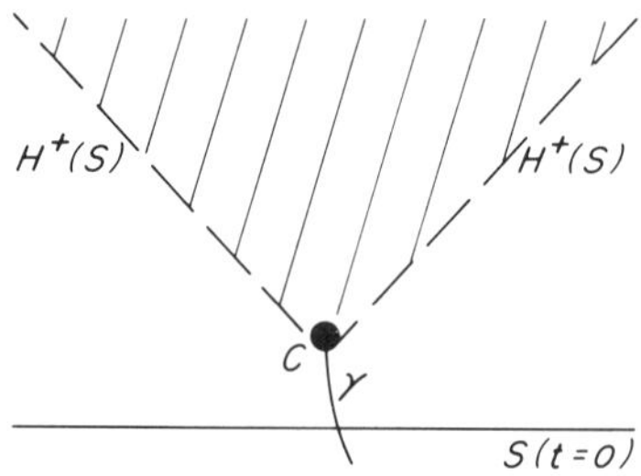

Figure 3.2

of being infected by the ghastly pathologies of naked singularities since GTR implies that, under reasonable conditions, Nature exercises modesty and presents us only with singularities that have been clothed in some appropriate sense. The task of the following section is to try to understand in more precise terms what this means. Before turning to that task, however, I need to attend to the prior one of getting a better grip on the notion of spacetime singularity.

I will begin on the latter task by returning to the Hawking-Penrose singularity theorems. Though rightly regarded as one of the major achievements in the theoretical analysis of classical GTR, the theorems nevertheless had two perceived shortcomings. Since a hypothesis of the main theorem (Hawking and Penrose 1970) was that the spacetime does not violate causality with closed timelike curves, it was left open that the formation of singularities could be prevented by the presence of acausal features. To some extent this gap has been filled by the results of Tipler (1976, 1977) and Kriele (1990). For present purposes the gap may be viewed as nonexistent since on at least some versions of cosmic censorship, the development of acausal features would be counted as a violation of censorship.

The second shortcoming is that the theorems proved the existence of singularities in the form of geodesic incompleteness, and no information was provided about the connection to more intuitive notions of singularities, such as the blowup of curvature scalars, unbounded tidal forces, and the like. This is a highly complex matter which I will not attempt to address except insofar as it affects the formulation of cosmic censorship (see Tipler et al. 1980 and Wald 1984, chap. 9). In particular, we need to clarify what a spacetime singularity is and how geodesic incompleteness signals that the spacetime is singular.

In trying to explicate the notion of a spacetime singularity in GTR we have to confront an apparent conundrum. By definition, a general relativistic spacetime consists of a manifold M equipped with a Lorentz signature metric g_{ab} defined at every point of M and typically assumed to be C^∞ (or at least to satisfy some nice differentiability conditions). But then the intuitive notion of a spacetime singularity as a place where the metric becomes undefined or nondifferentiable is excluded ab initio. The way out of this conundrum is not to try to think of singular points as elements of M but as missing points which have been cut out of the manifold in order to leave a metric which is well behaved at all of the remaining points.[5] The existence of, say, a future incomplete timelike or null geodesic (that is, a geodesic which is inextendible in the future direction and which has finite affine length as measured from any point on the geodesic and moving into the future) can be taken as a sign of such missing points. The timelike geodesic γ pictured in figure 3.2 is an example of such a future incomplete geodesic that terminates on a naked singularity.

Unfortunately geodesic incompleteness is not a sure sign in either the sense of a necessary or a sufficient condition. It is not necessary because there exist spacetimes that are timelike and null geodesically complete but which may nevertheless be regarded as singular because they contain inextendible timelike curves which have bounded acceleration and finite proper length (Geroch 1968). This point needs to be kept in mind when in the following section various formulations of cosmic censorship are considered. In the other direction, geodesic incompleteness is not a sufficient sign for singularities in the intended sense, and this for two reasons. First, there are compact spacetimes which may be regarded as ipso facto nonsingular (because, for example, every infinite sequence of points has a limit point) but which are geodesically incomplete (Misner 1963). However, such spacetimes could be dismissed on the grounds of causal pathology since they all contain closed timelike curves. Second, and more seriously, geodesic incompleteness may signal only the existence of missing regular points, as would be the case in figure 3.2 if we deleted the region C and restricted the Minkowski metric to the remainder. The obvious remedy is to restrict attention to spacetimes that are maximal in that they cannot be isometrically embedded as a proper subset of another spacetime. This is a legitimate move since every spacetime can be shown to extend to a maximal spacetime. Unfortunately even inextendible spacetimes can have missing regular points, as is illustrated by again deleting the region C from (two-dimensional) Minkowski spacetime and taking the universal covering spacetime. The missing regular points in such cases have to be detected by a criterion of local rather than global inextendibility. Several such criteria are available,[6] but for present purposes the most satisfactory seems to be to assume what Geroch (1977) terms *hole freeness*. The spacetime M, g_{ab} is said to have a hole just in case there is a (not necessarily global) spacelike surface $S \subset M$ and an isometric embedding ψ: $D(S) \rightarrow M'$ of $D(S)$ into another spacetime M', g'_{ab} in such a way that $\psi(D(S))$ is a proper subset of $D(\psi(S))$. From here on I will assume that all the spacetimes to be discussed are inextendible and hole free, even though this assumption begs to some extent the problem of cosmic censorship (see the next section).

What Is a Naked Singularity? What Is Cosmic Censorship?

When Penrose first raised the problem of cosmic censorship, it was not clear what to include under the notion of a naked singularity. For example, Penrose said that "[i]n a sense, a 'cosmic censor' can be shown *not* to exist. For . . . the 'big bang' singularity is, in principle, observable" (1969, 274). Today standard big bang cosmologies would not be regarded as naked singular and, thus, not in conflict with cosmic censorship. In what follows I will describe three attempts to pinpoint the correlative notions of cosmic censorship and naked singularities.

Approach 1

Since the key concern with the development of naked singularities is the breakdown of predictability and determinism, cosmic censorship may be formulated by imposing conditions that assure that no such unwanted behavior occurs.

This approach leads to a strong and to a weak version of censorship. For future reference it is helpful to begin with the notion of a *partial Cauchy surface* S for a spacetime M, g_{ab}: an acausal spacelike hypersurface without edges. Such an S is the relativistic analogue of a Newtonian "constant time" slice and the appropriate basis for specifying instantaneous initial data that, hopefully, will allow the future to be predicted and the past retrodicted. Spacetimes we are dealing with are assumed to be temporally orientable, which implies that S is two-sided.[7] Thus, we may say that S is a *future* (respectively, *past*) Cauchy surface for M, g_{ab} just in case $D^+(S)$ ($D^-(S)$) includes all of that part of M that lies on the future (past) side of S. (Thus, if S is future—respectively, past—Cauchy, then the future—past—Cauchy horizon of S is empty.) The spacelike surface S is a *Cauchy surface* simpliciter just in case it is both past and future Cauchy, that is, $D(S) \equiv D^-(S) \cup D^+(S) = M$. We may now take strong cosmic censorship (SCC) to hold for M, g_{ab} just in case M, g_{ab} possesses a Cauchy surface. The standard big bang cosmological models satisfy this statement of SCC, and thus the big bang singularity is not counted as naked.

The existence of a Cauchy surface is a very strong condition, and one can wonder whether it is too heavy-handed a way to capture the intuitive idea that there are no naked singularities. A relevant example is provided by the universal covering of the anti-de Sitter spacetime which is represented schematically in figure 3.3a and whose Penrose diagram is given in figure 3.3b.[8] This spacetime violates the proposed formulation of SCC but it is arguably singularity free; for example, it is geodesically complete. The proponent of the present approach could concede the point but maintain that since the key worry raised by naked singularities is the breakdown in predictability, no harm is done by formulating a version of cosmic censorship that is strong enough to assuage the worry whatever the source, naked singularities or no. Even if we share this sentiment, a need for a separate

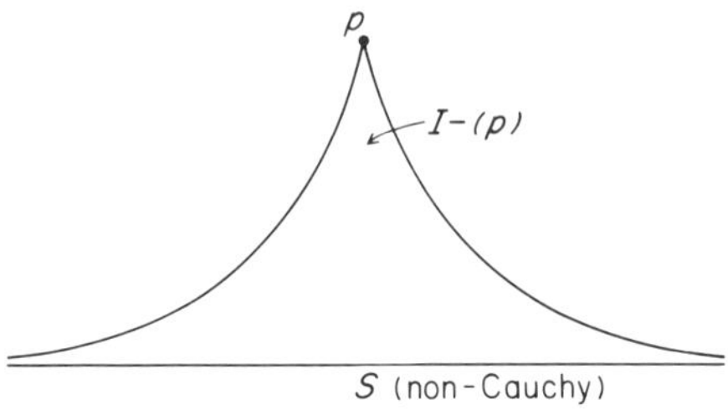

Figure 3.3a

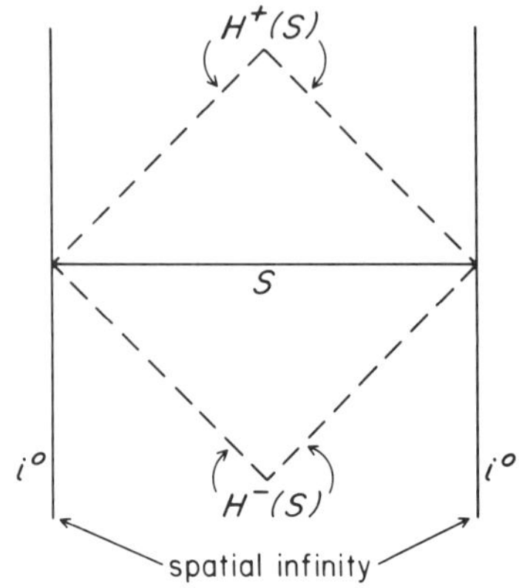

Figure 3.3b

definition of naked singularity is still evident. The second and third approaches described take up this challenge.

We also want to be able to distinguish breakdowns in predictability that are not too drastic. In particular, Nature may practice a form of modesty by hiding singularities in "black holes," the exterior regions of which may be predictable because they admit future Cauchy surfaces. If so, weak cosmic censorship (WCC) is said to hold. In asymptotically flat spacetimes this idea can be made precise by appealing to the notion of *future null infinity* $\mathscr{I}^+$, the terminus of outgoing null geodesics that escape to spatial infinity (see Hawking and Ellis 1973 and Wald 1984 for precise definitions). The interior of a *black hole* is then defined to be the complement of $J^-(\mathscr{I}^+)$, the events that can be seen from future null infinity.[9] The boundary E of $J^-(\mathscr{I}^+)$ is the *absolute event horizon;* this is the modesty curtain that hides the goings-on in the interior of the back hole from external observers. These concepts are illustrated in the familiar (maximally extended) Schwarzschild spacetime (figure 3.4a) and its Penrose diagram (figure 3.4b). Frank Tipler (1979) has proposed a more general definition of black hole that does not assume asymptotic flatness and is supposed to apply to any stably causal spacetime. Until this or some substitute definition is shown to be satisfactory, we cannot claim to have a general formulation of WCC in the cosmological setting.

The Schwarzschild case is not only an example of a black hole and WCC but it also displays SCC since it possesses a Cauchy surface.

The difference between the strong and weak versions of cosmic censorship is illustrated schematically in figures 3.5a–3.5b. In figure 3.5a the singularity that develops in gravitational collapse is hidden from external observers but is visible to observers within the black hole. In figure 3.5b the black hole is even blacker because even those unfortunate observers who fall into the hole cannot "see" the singularity, though they may well feel and, indeed may be torn apart by the tidal forces as they are inevitably sucked into the singularity. Nevertheless, they may take some cold comfort in the fact that SCC holds.

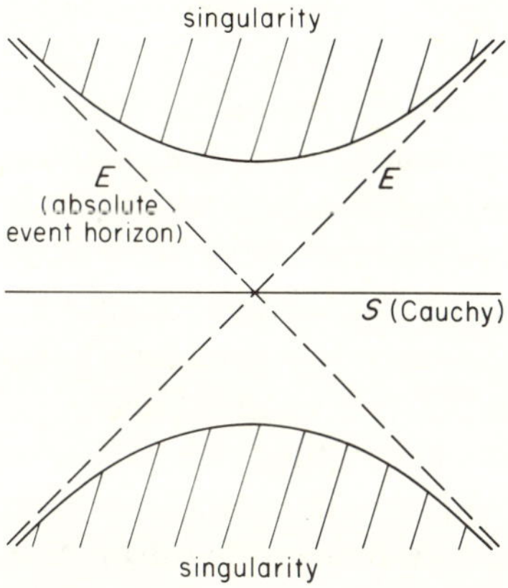

Figure 3.4a

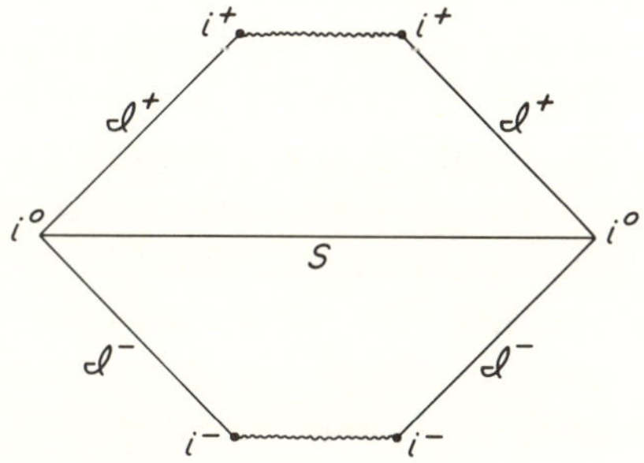

Figure 3.4b

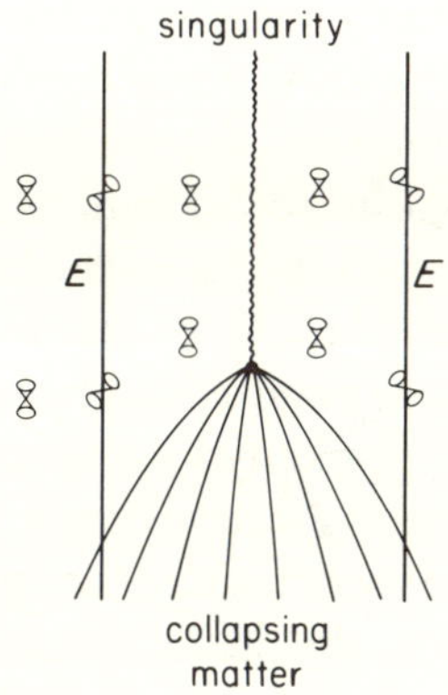

Figure 3.5a

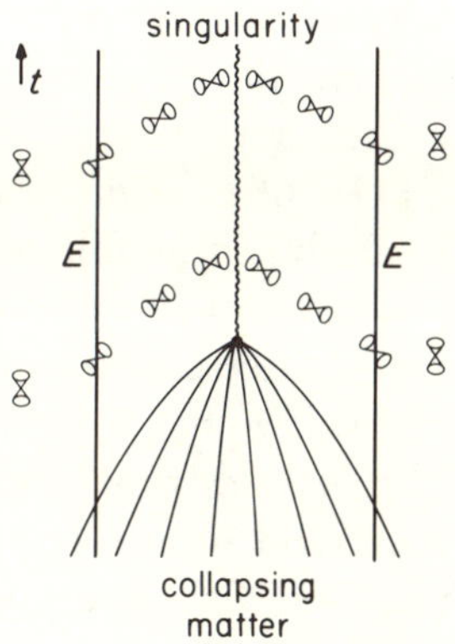

Figure 3.5b

If SCC holds there is a sense in which, paradoxically, the singularity never occurs. It follows from a result of Geroch (1970) that if M, g_{ab} admits a Cauchy surface, then it also admits a global time function, a map $t: M \to \mathbb{R}$ such that t increases along every future directed timelike curve. Furthermore, t can be chosen so that each t = constant surface is Cauchy. Since no Cauchy surface can intersect the singularity, we can conclude that there is no time t at which the singularity exists. (To test their understanding, readers should draw the foliation of Cauchy surfaces for the spacetime pictured in figure 3.5b.) Of course, the statement that no singularity exists will be hotly disputed by the ghosts of the observers who have been sucked into the black hole and have been snuffed after only a finite existence (as measured by their respective proper times).

It is instructive to take the time reverses of the processes pictured in figures 3.5a and 3.5b to produce "white holes" where the singularities explode into expanding ordinary matter (figures 3.6a and 3.6b). If figure 3.5b deserves to be called a black-black hole, then figure 3.6b would seem to deserve the label of a white-white hole since it is visible both to local and to distant observers. But since figure 3.6b is the time reverse of figure 3.5b and since 3.5b possesses a Cauchy surface, 3.6b also possesses a Cauchy surface and, thus, obeys our formulation of SCC. In this sense the singularity in figure 3.6b is not naked even though it is highly visible (as is the case with the big bang singularity of the Friedman-Robertson-Walker models).

But does predictability really hold in the situation pictured in figure 3.6b? Penrose has argued that "the future behavior of such a white hole does not, in any *sensible* way, seem to be determined by its past. In particular, the precise moment at which the white hole explodes into ordinary matter seems to be entirely of its own 'choosing' "(1979, 601). Penrose's point seems to be this. The spacetime in figure 3.6b can be foliated by Cauchy surfaces. But the singularity lies to the past of any such surface, which means that any such surface must intersect the ordinary matter. So the explosion cannot properly be said to be predicted from any such surface. There are other spacelike surfaces from which one can properly speak of predicting the explosion since it lies to their future, but since these surfaces do not have the Cauchy property, the prediction cannot be of the deterministic variety. If Penrose's worry is to be taken seriously and if

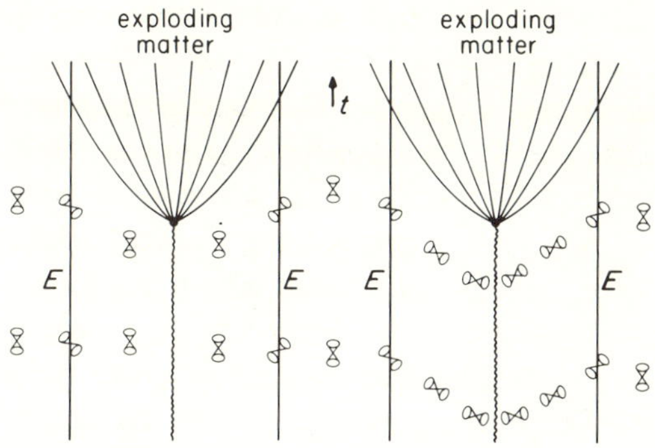

Figure 3.6a

Figure 3.6b

it is to be assuaged by some form of cosmic censorship, then an even stronger version of SCC than the one previously given is needed. Since it seems that this way lies disaster for cosmic censorship, I will not pursue the matter.

Figures 3.5a,b and 3.6a,b also raise another problem, not so much for the statement of cosmic censorship as for the validity and proof of the hypothesis that GTR contains a mechanism for enforcing cosmic censorship. Suppose that in typical cases of gravitational collapse SCC fails while WCC holds, that is, the process in figure 3.5a is what we should expect. Since Einstein's field equations are time reversal invariant, every solution of the type of figure 3.5a is matched by a solution of the type of figure 3.6a. So if black holes that violate SCC but satisfy WCC are a pervasive feature of general relativistic models, then white holes that violate WCC would also seem to be a pervasive feature.

One can take the attitude here that what is needed is a division of labor. The initial effort should be devoted to proving (or refuting) the conjecture that naked singularities do not occur in reasonable models of gravitational collapse. Then attention can be turned to the problem of white holes, which may be regarded as an aspect of the more general problem of time's asymmetries. In other branches of physics—electromagnetism and mechanics, for example—the fundamental laws are also time reversal invariant. But we find that while certain types of solutions are commonly encountered, their time reverse counterparts never or very seldomly occur (e.g., we often encounter electromagnmetic waves expanding from a center to spatial infinity but we never encounter waves converging on a center). So it is hardly unexpected that an analogous situation occurs in gravitational physics. The fact that the origin of time's arrow remains an unsolved problem is nothing to boast about, but it is not a special problem for gravitational physics and it should not prevent work from going ahead on the issue of cosmic censorship.[10]

Although the present approach to cosmic censorship has yielded some valuable insights, it is subject to some serious shortcomings. Most obviously, although it provides a sufficient condition (modulo Penrose's worry) for ruling out naked singularities, it has not told us directly what a naked singularity is, nor has it told us how the violation of cosmic censorship leads to singularities in some intuitive sense.

Joshi and Saraykar (1987) provide some information on the latter issue. We know from the result of Geroch mentioned earlier that SCC in the guise of the existence of a Cauchy surface S implies that the topology of space does not change with time in the sense that the spacetime manifold is topologically $S \times \mathbb{R}$. So we ask: If SCC is violated and there is a change in topology, what can we expect about the existence of singularities? To make this question more precise, define a partial Cauchy surface S of M, g_{ab} to be *maximal* just in case $D(S)$ is maximal on the set of all partial Cauchy surfaces for M, g_{ab}. If S and S' are both maximal, $S' \subset I^+(S)$, and $S' \not\approx S$, then a topology change is said to take place (see Gowdy 1977).[11] Since S cannot be Cauchy surface, $H^+(S)$ cannot be empty. Joshi and Saraykar show that if in addition a weak energy condition (see the following section) is satisfied and if all timelike trajectories encounter some nonzero matter or energy, then a singularity will occur in that a null generator of $H^+(S)$ will be past incomplete.[12]

Nevertheless, we still want to have a direct definition of "naked singularity" and a demonstration that the approach just explored is on the right track because the existence of a Cauchy surface entails the nonexistence of naked singularities so defined.

Approach 2

The second approach seeks to define a naked singularity in terms of its detectability. Cosmic censorship then becomes the statement that such singularities do not occur.

Penrose's (1974, 1976, 1978, 1979) version of this approach emphasizes local detectability:

[I]t seems to me to be quite unreasonable to suppose that the physics of a comparatively local region of spacetime should really 'care' whether a light ray setting out from a singularity should ultimately escape to 'infinity' or not. To put things another way, some observer . . . might intercept the light ray and see the singularity as 'naked', though he be not actually situated at infinity. . . . The unpredictability entailed by the presence of naked singularities which is so abhorrent to many people would be present just as much for this local observer . . . as for an observer at infinity.

It seems to me to be comparatively unimportant whether the observer himself can escape to infinity. Classical general relativity is a scale-independent theory, so if locally naked singularities can occur on a very tiny scale, they should also, in principle, occur on a very large scale. . . .

It would seem, therefore, that if cosmic censorship is a principle of Nature, it should be formulated in such a way as to preclude such *locally* naked singularities. (1979, 618–19)

Penrose's technical explication of a locally naked singularity uses the concepts of TIFs and TIPs (see Penrose 1978). I will present a related explication that follows an idea of Geroch and Horowitz (1979). The form of a definition for the set $\mathfrak{N}$ of points from which the spacetime M, g_{ab} can be detected to be singular can be stated as follows:

DEFINITION 3.1. *$\mathfrak{N} \equiv \{p \in M$: $I^-(p)$ contains a timelike curve γ which has no future endpoint and which ______$\}$.*

The reason, intuitively speaking, that γ has no future endpoint is that it runs into a singularity. Since this fact is directly detectable from p, the spacetime is nakedly singular as viewed from p. Thus, the statement that M, g_{ab} harbors naked singularities becomes: $\mathfrak{N} \neq \varnothing$; and conversely, cosmic censorship becomes the statement: $\mathfrak{N} = \varnothing$.[13]

The strongest version of cosmic censorship is obtained by putting no further restrictions in the blank. Then $\mathfrak{N} = \varnothing$ is equivalent to the existence of a Cauchy surface, a result that dovetails nicely with Approach 1. If we are somewhat more restrictive and fill in the blank with "is a geodesic," then $\mathfrak{N} = \varnothing$ no longer entails the existence of a Cauchy surface. Depending upon one's point of view, it could be counted as a benefit of this latter version of cosmic censorship that anti-de Sitter spacetime is no longer counted as nakedly singular. If one wants the focus of cosmic censorship to be curvature singularities, then the blank should be filled with restrictions that guarantee that γ terminates on what one chooses to regard as a curvature singularity. In this way we obtain several versions of cosmic censorship.

If one does not share Penrose's sentiments about local detectability, the present approach can still be adapted to a weaker statement of cosmic censorship by requiring only that $\mathfrak{N} = \varnothing$ for the region exterior to black holes.

Approach 3

As part of establishing cosmic censorship, one would like to prove that in reasonable models of gravitational collapse naked singularities do not develop from regular initial data. Thus, on this approach one would not regard the negative mass Schwarzschild model[14] as a counterexample to cosmic censorship since although it is nakedly singular, the singularity has existed for all time.

Before we can start on this project we need a definition that isolates naked singularities that can be said to develop from regular initial data, as illustrated schematically in figure 3.7. The ideas of Geroch and Horowitz (1979) and Horowitz (1979) suggest

DEFINITION 3.2. *A spacetime M, g_{ab} is future nakedly singular in the first sense (FNS_1) with respect to the partial Cauchy surface $S \subset M$ just in case $\exists\, p \in H^+(S)$ such that $\overline{I^-(p)} \cap S$ is compact.*

Newman (1984a) works with a somewhat stronger criterion:

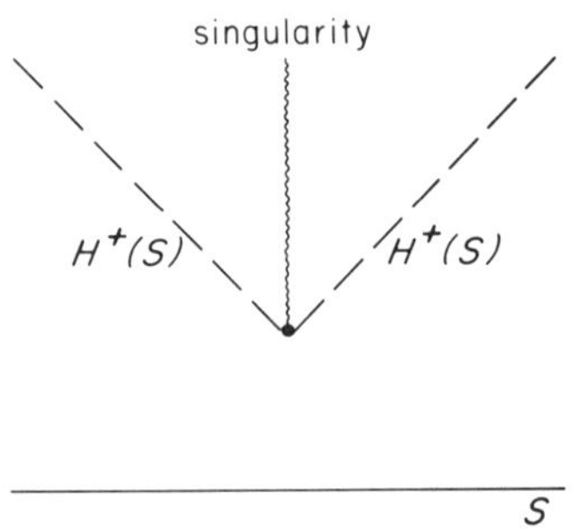

Figure 3.7

DEFINITION 3.3. *A spacetime M, g_{ab} is future nakedly singular in the second sense (FNS_2) with respect to the partial Cauchy surface $S \subset M$ just in case $\exists\, p \in H^+(S)$ such that $\overline{I^-(S)} \cap S$ is compact and the generator of $H^+(S)$ through p is past incomplete.*

The spacetime pictured in figure 3.7 is both FNS_1 and FNS_2 with respect to S. The two-dimensional Minkowski spacetime pictured in figure 3.8 is neither FNS_1 nor FNS_2 with respect to the spacelike but asymptotically null surface S. Likewise anti-de Sitter spacetime is not future nakedly singular in either sense. The spacetime shown in figure 3.9 is FNS_1 but is not FNS_2 if the metric can be chosen so that the closed null geodesic which lies in $H^+(S)$ which winds around and around *ad infinitum* is past complete. (Figure 3.9 superficially resembles Misner's 1963 two-dimensional model which displays some of the causal properties of Taub-NUT spacetime—see Hawking and Ellis 1973. However, Misner's model is FNS_2. Although I know of no specific models that illustrate the difference between FNS_1 and FNS_2, it seems plausible that such examples exist.) Thus, the first definition is more appropriate to capturing the notion of a breakdown in predictability due to whatever cause, whereas the second definition is more relevant to identifying the development of naked singularities of the type that are indicated by geodesic incompleteness. However, as noted in the first section, such singularities need not be due to unbounded curvature. So if curvature singularities are to be ruled out by the statement of cosmic censorship, a third definition is needed which strengthens the second by requiring that, in some appropriate sense, the generator of $H^+(S)$ through p encounters a curvature singularity in the past direction.

Before proceeding further we should make sure that a future nakedly singular spacetime as just defined really is singular in some minimal sense. This would mean showing that if M, g_{ab} is future nakedly singular with respect to S, then the region of M that is on the future side of S and that is causally accessible from S is not globally hyperbolic.[15] Suppose for purposes of contradiction that $J^+(S)$ is globally hyperbolic. Then on either of the above definitions there is a point $p \in H^+(S)$ such that $\overline{I^-(p)} \cap S$ is compact. So $J^-(p) \cap J^+(\overline{I^-(p)} \cap S)$ is a compact set (Hawking and Ellis 1973, Cor. to Prop. 6.6.1). Since S has no edge, the generator α of $H^+(S)$ through p is past

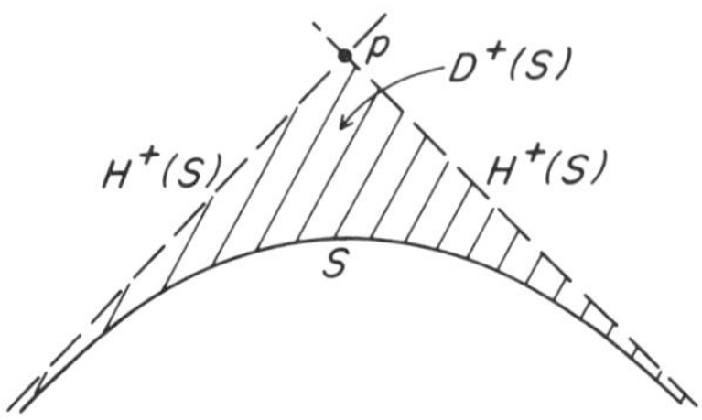

Figure 3.8

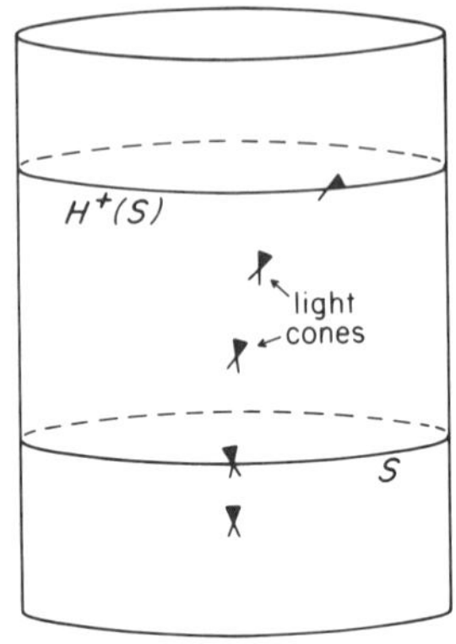

Figure 3.9

endless. And since $\alpha \in (J^-(p) \cap J^+(\overline{I^-(p)} \cap S))$, we have a past endless null curve imprisoned in a compact set. Thus $J^+(S)$ is not strongly causal and, a fortiori, is not globally hyperbolic (Hawking and Ellis 1973, Prop. 6.4.7).[16]

Conversely, can we be sure that if M, g_{ab} is not future nakedly singular with respect to S then $J^+(S)$ is globally hyperbolic? The answer is yes if S is compact and "future nakedly singular" is taken in the first sense. Indeed, S itself must be a Cauchy surface. If S were not a Cauchy surface, $H^+(S)$ would be nonempty. Since S is compact, $\overline{I^-(p)} \cap S$ is compact for any $p \in H^+(S)$, so that the spacetime is FNS_1 with respect to S. On the other hand, this result does not hold if S is not compact, as we already know from the case of anti-de Sitter spacetime. But an even worse counterexample is provided by Reissner-Nordstrom spacetime (a piece of whose Penrose diagram is shown in figure 3.10) since this model possesses a naked curvature singularity to the future of a partial Cauchy surface. One could argue, however, that the very feature which allows this spacetime to escape the previously given definitions makes it physically unreasonable (see Wald 1984, 304). Namely, for any partial Cauchy surface S lying below the singularity and for any $p \in H^+(S)$, $I^-(p)$ contains an infinite (i.e., noncompact portion of) S. As a result a small perturbation on S can produce an "infinite blue-shift" singularity on $H^+(S)$ (Chandrasekhar and Hartle 1982). This line of reasoning, however, has the defect of failing to respect what would seem a natural division of labor: First we need a definition of what it is for

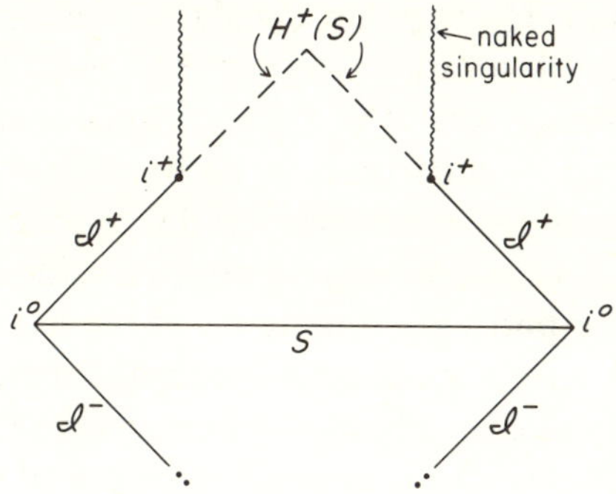

Figure 3.10

a spacetime to be future nakedly singular with respect to S, and then we try to show that such behavior does not occur in physically acceptable models. Perhaps in the end this division cannot be maintained, but it seems wrong to give up at the beginning.

If we do want to try to respect the division of labor, we are faced with a version of the by now familiar tension. On one hand we can try to target various forms of singular behavior that in some sufficiently tight sense are traceable to the development of initial data from S; that is the course taken in the previously given definitions. On the other hand we can try to frame the definition of "future nakedly singular with respect to S" so that being *not* future nakedly singular with respect to S entails that $J^+(S)$ is free of any taint of singularity. But the only way to assure the latter is to require that $J^+(S)$ is globally hyperbolic, and in turn the only sure way to guarantee that the development of initial data from S makes $J^+(S)$ globally hyperbolic is to require that $D^+(S)$ includes $J^+(S)$, which is to say that S is a future Cauchy surface and which is to collapse back to Approach 1.

The Cosmic Censorship Hypothesis

The cosmic censorship hypothesis (CCH) is the claim that the only naked singularities that occur in the models of GTR are harmless (here I am using the terminology of Israel 1984). A *model* of GTR is a triple M, g_{ab}, T^{ab}, where M, g_{ab} is a relativistic spacetime, T^{ab} is a symmetric second-rank tensor called the *stress-energy-momentum tensor,* and g_{ab} and T^{ab} together satisfy Einstein's field equations without cosmological constant: $G_{ab} \equiv R_{ab} - (1/2)Rg_{ab} = 8\pi T_{ab}$ (R_{ab} is the Ricci tensor and R is the Kretchmann curvature scalar). Examples of naked singularities in such models are easily constructed, as we know from the previous discussion. But for one reason or another such examples may be brushed aside as *harmless*. The principal reason for putting a singularity in the harmless category is that the model in which it occurs has features that, apart from the singularity itself, make it "physically unreasonable." The literature on cosmic censorship can be confusing to the casual reader because it mixes together a number of different senses in which a model can be physically unreasonable, among which are the following: the model is literally physically impossible; the model involves unrealistic idealizations, and there is no reason to expect that more realistic counterparts will also have naked singularities; the model is physically possible but involves such rare features as to leave no reason to think that anything like the model will be actually encountered.

Whatever specific content is given to the physically reasonable-physically unreasonable distinction, two boundary conditions should be satisfied. First, "physically unreasonable" should not be used as an elastic label that can be stretched to include any ad hoc way of discrediting putative counterexamples to the CCH. Second, the conception we settle on must permit the CCH to be stated in a precise enough form that it lends itself to proof (or refutation). Some aspects of the physically reasonable-unreasonable distinction can be stated in advance. Others emerge only in the process of assessing potential counterexamples to the CCH. This, of course, raises worry about the first boundary condition. But as we will see, the real worry is about satisfying the second boundary condition while at the same time making the CCH not obviously false and also general enough to cover the situations that can be expected to occur in the actual universe.

Among the constraints on a physically reasonable model of GTR, three, at least in part, can be motivated independently of cosmic censorship.

(i) *Energy conditions* Start with a spacetime M, g_{ab} and compute the Einstein tensor G_{ab} associated with g_{ab}. Then define $T_{ab} \equiv (1/8\pi)G_{ab}$. In this way we create, at least formally, innumerable models of GTR, among which will be many that violate cosmic censorship (here I am following Geroch and Horowitz 1979). To return to an example from the first section, we can start with empty two-dimensional Minkowski spacetime $\mathbb{R}^2$, η_{ab}, remove a compact set $C \subset \mathbb{R}^2$, and choose a conformal factor Ω that goes rapidly to 0 as C is approached. The resulting triple M ($\equiv \mathbb{R}^2 - C$), g_{ab} ($\equiv \Omega\eta_{ab}$), T^{ab} ($\equiv (1/8\pi)G^{ab}(g_{ab})$) is a nakedly singular model. Of course, we may not get a model in the intended sense that the stress-energy-momentum tensor arises from normal sources, such as massive particles, electromagnetic fields, and the like. This intention is difficult to state in terms of precise formal conditions on T^{ab}, but at a minimum we can impose one or another energy condition that we expect normal sources to satisfy.

The *weak energy condition* requires that $T_{ab}V^aV^b \geqslant 0$ for any timelike vector V^a. This means that the energy density as measured by any observer is nonnegative. A stronger requirement is the *dominant energy condition* which says that for any timelike V^a, $T^a_b V^b$ is a future-directed timelike or null vector, which means that the flow of energy-momentum as measured by any observer does not exceed the speed of light. Finally, the so-called *strong energy condition* requires that for any unit timelike V^a, $T_{ab}V^aV^b \geqslant -1/2T$, $T \equiv T^a_a$. To see in more detail what these conditions mean physically, consider a perfect fluid whose stress-energy-momentum tensor has the form $T_{ab} = (\mu + p)\, U_aU_b + pg_{ab}$, where μ and p are respectively the energy density and pressure of the fluid and U^a is the unit tangent to the flow lines of the fluid. The strong energy condition says that $\mu + p \geqslant 0$ and $\mu + 3p \geqslant 0$. The weak energy condition requires that $\mu \geqslant 0$ and $\mu + p \geqslant 0$. The dominant energy condition says that $\mu \geqslant 0$ and $\mu > |p|$.

I conjecture that these energy conditions rule out all of the artificial examples of naked singularities constructed by the method given two paragraphs above, but I know of no formal proof of this.

We can appreciate the importance of the stipulation of a zero cosmological constant. If this stipulation is removed, Einstein's field equations generalize from $G_{ab} = 8\pi T_{ab}$ to $G_{ab} + \Lambda g_{ab} = 8\pi T_{ab}$, where Λ is the cosmological constant. Anti-de Sitter spacetime has constant scalar curvature $R < 0$ and Einstein tensor $G_{ab} = -1/4Rg_{ab}$. With $\Lambda = 0$ we can interpret this as a solution with a perfect fluid source of constant density $(-R/32\pi) > 0$ and constant pressure $(R/32\pi) < 0$. However, it is ruled out by the strong energy condition. If, however, the cosmological constant is allowed to be nonzero, then anti-de Sitter spacetime can be interpreted as an empty-space solution ($T_{ab} = 0$) with $\Lambda = 1/4R$. Thus, without the stipulation of a zero cosmological constant, it will be much more difficult to satisfy various versions of cosmic censorship.

(ii) *Causality conditions* For both physical and philosophical reasons we might require that a reasonable model does not contain closed or almost closed causal curves. (The condition of strong causality introduced rules out almost closed causal curves.) But imposing causality conditions by fiat is a little awkward for present purposes since it smacks of assuming what we want to prove, at least for versions of cosmic censorship that seek to censure breakdowns of predictability that occur because of the development of acausal features *after* the specification of initial data. Thus, in Taub-NUT spacetime we can choose a partial Cauchy surface S such that in some neighborhood of S things are causally as nice as you like; but further to the future of S things turn causally nasty, and as a result a Cauchy horizon for S develops. At the present stage of discussion, however, fiat is needed to secure cosmic censorship since Taub-NUT spacetime is a vacuum solution to Einstein's field equation so that the energy conditions are trivially satisfied.

(iii) *Determinism conditions.* In addition to requiring that T^{ab} satisfy energy conditions, Wald (1984, 303) stipulates that the sources giving rise to T^{ab} should fulfill another condition; namely, the coupled Einstein-source equations should admit a well-posed initial data problem. (Formally, the coupled equations are required to form a second-order, quasi-linear, diagonal, hyperbolic system for which we have local existence and uniqueness theorems.) Two motivations for this addition can be produced. First, in evaluating the CCH we are only interested in breakdowns in determinism due to the development from regular initial data of a pathology in the spacetime structure and not in the kind of indeterminism already present in an ill-posed initial value problem. The weakness of this motivation is that fields with a nondeterministic evolution may in fact be present, and these fields may contribute to the appearance of a naked singularity. The second motivation comes from Wald's idea that cosmic censorship should only be concerned with "fundamental fields." Since those fundamental classical fields, such as electromagnetism and gravity, for which we have an accurate description do admit a well-posed initial value problem separately and jointly, it is natural to think that any other classical field which has pretensions of being fundamental should behave similarly. While I resonate to this motivation, I have to note that it threatens to derail the investigation of cosmic censorship among a wide class of cosmological models that posit a perfect fluid source. Assuming the energy conditions, the coupled Einstein-Euler equations do form a second-order, quasi-linear, diagonal, hyperbolic system, but the equations contain a μ^{-1} term, which means that for a finite fluid body (μ with compact spatial support) existence and uniqueness for the initial value problem cannot be proven in the standard way. We could try to maneuver around this difficulty by idealizing a fluid body by a μ which is strictly positive but which falls off rapidly outside of a compact spatial region. Unfortunately, the time for which the solution can be proven to exist may also decay rapidly.[17] This may only be a technical difficulty that can be overcome by a different proof technique. If not, one may still be well advised to conclude that the macroscopic description of matter as a perfect fluid is revealed to be not fundamental enough for purposes at hand and, therefore, that a more fundamental description that promotes a well-posed initial value problem is called for. Again, while I am sympathetic to this attitude, I note that the price to be paid is that in our present state of knowledge the cosmic censorship question would be largely mooted.

With only conditions (i)–(ii) in place, the CCH fails, as is shown by the "shell-focusing" singularities that may arise in spherically symmetric dust collapse (see Yodzis et al. 1973). The collapse is arranged so that the outer shells of dust fall inward faster than the inner shells. A black hole eventually develops, but not before the crossing of shells produces an infinite density singularity that is visible from both near and far. Hawking (1979) and others have suggested that such naked singularities are harmless because they are relatively mild; in fact, the solution can be continued through the Cauchy horizon as a generalized distributional solution (Papapetrou and Hamouri 1967). However, Seifert (1979) has cautioned that harmlessness in the relevant sense has not been demonstrated if our concern with cosmic censorship is over the potential breakdown in predictability, for uniqueness theorems for such generalized solutions are not in place. In any case this method of trying to render harmless naked singularities does not have a very long reach since, as will be discussed, stronger and irremovable singularities threaten it.

A second strategy for dealing with shell-focusing singularities and similar examples is to impose (iv) *Further conditions on T^{ab} and the equations of state*. These further conditions are supposed to assure that the sources are sufficiently realistic. In particular, it could be demanded that pressure effects (neglected in dust models) be taken into account, and further that if matter is treated as a perfect fluid, the equation of state must specify that $p = p(\mu)$ is an increasing function of density and that the pressure becomes unbounded as the density becomes unbounded. This would rule out some of the early shell-focusing counterexamples. But (i), (ii), and (iv) are not sufficient to prevent violations of cosmic censorship in self-similar gravitational collapse[18] with an equation of state $p = a\mu$, with $0 < a < 1$ a constant (see Ori and Piran 1987, 1988, 1990). Perhaps reasons can be found to deem this "soft" equation of state physically unreasonable. Or perhaps it should be demanded that to be realistic the fluid description should incorporate viscosity and, thus, shear stresses into T^{ab} (see Seifert 1983). Further violations of cosmic censorship satisfying these additional demands would drain much of the interest in this tack for trying to render harmless potential counterexamples.

A rather different tack is taken by Eardley (1987) who demands (v) *Realistic equations of motion for the sources*. In analyzing the naked shell-focusing singularities that emerge in the Tolman-Bondi models of spherical gravitational collapse as matter piles up at the center of symmetry, Eardley finds that the tacit assumption underlying the model is that the dust shells cannot pass through the origin. The objection is not to the idealization of matter as a pressureless, viscosity free dust but rather to the unreasonable assumption that dust shells behave completely inelastically. He conjectures that if the motion of the dust is treated more realistically, for example, by specifying elastic recoil when the shells collide at the origin, then naked singularities will not develop. Settling this and related conjectures seems to be one of the more important items on the agenda for evaluating the prospects for the CCH. However, I worry that the present approach, while possibly effective in dealing with putative counterexamples to cosmic censorship on a case by case basis, may not lead to a neat formal statement of the CCH that can be proven to hold for a wide class of models.

Tipler (1985) and Newman (1986) have suggested a way to avoid the delicate and contentious question of what counts as a physically reasonable source by concentrating instead on (vi) *Strength of singularities*. The idea is that for a source to create a physically realistic singularity, the singularity must be strong enough to crush ordinary matter out of existence by squashing it to zero volume. A formal definition is provided by Tipler (1985). Clarke and Krolak (1985) show that a geodesic with tangent V^a and affine parameter λ terminates in such a strong curvature singularity at $\lambda = 0$ if $\lim_{\lambda \to 0} \lambda^2 R_{ab} V^a(\lambda) V^b(\lambda) = 0$. This condition is satisfied for the Schwarzschild singularity and for the big bang and big crunch singularities of the Friedman-Robertson-Walker models.

On this approach we can give the CCH a clean and precise formulation: The *strong* (respectively, *weak*) *CCH* holds just in case strong (weak) censorship holds in any model which satisfies the energy conditions (i) and the causality conditions (ii) and in which the only singularities that occur are of the strong curvature type. Unfortunately, the virtues of simplicity and precision are not rewarded by truth, for this version of the CCH is in fact false, as is shown by the presence of strong naked singularities in the models of self-similar gravitational collapse (Lake 1988, Ori and Piran 1990).

The final constraint on physically reasonable models I will consider is Penrose's (1974, 1978) idea of (vii) *Stability under perturbations of initial conditions and equations of state*. The idea is often illustrated by an example mentioned in the preceding section, the Reissner-Nordström model where small perturbations on a partial Cauchy surface S in the initial data for the Einstein-Maxwell equations can build until the perturbation becomes an infinite blue-shift singularity on $H^+(S)$. But exactly what does such a demonstration show? Wald takes it to show that "there is good reason to believe that in a physically reasonable case where the shell is not exactly spherical, the Cauchy horizon . . . will become a true physical singularity, thereby producing an 'all encompassing' singularity inside the black hole formed by the collapse . . . " (1984, 318). If this is the moral of the perturbational analysis, then one should be able to drop reference to instability under perturbations and say directly (either in terms of the above ideas or in terms of some altogether different ideas) what a physically reasonable model is and then proceed to prove that physically reasonable models so characterized obey cosmic censorship.

However, what the talk about instability is supposed to point to is the notion that naked singularities that develop from regular initial data are relatively rare within the set of all models of GTR. To make this precise one would either need to put a measure on the set of all solutions to Einstein's field equations and then show that the target set is "measure zero," or else define a topology on the set of solutions and show that the target set is the complement of an open dense set. Given our present limited knowledge of generic features of solutions to Einstein's field equations, the project of establishing this version of the CCH seems rather grandiose. In the meantime we can lower our sights and investigate particular families of models whose members can be parametrized in a natural way and try to show that cosmic censorship holds for almost all parameter values. Or failing this we can challenge potential counterexamples by showing instability under small perturbations and take this as a sign that in some yet-to-be-made-precise sense a measure zero result should be forthcoming. The bad news here is that various counterexamples have been shown to be free of blue-shift instabilities, and the formation of naked singularities in some cases of spherical collapse have been shown to be stable under spherically symmetric perturbations. If the formation of naked singularities can be shown to be stable under generic perturbations, then we will have a strong indication that naked singularities are not rare, and the CCH will have to avail itself of some other escape route. If on the contrary it can be shown that naked singularities are rare, there still remains the nagging question of what we would say if we found that the actual universe we inhabit is one of the rare ones.

The good news from this review for philosophers of science is that the hunting ground for the CCH contains a rich array of examples for studying the related concepts of "physically possible," "physically reasonable," and "physically realistic." The bad news for advocates of cosmic censorship is that the CCH does not yield easily to a formulation that is not obviously false, is reasonably precise, and is such that one could hope to demonstrate its truth. There is the very real danger that the CCH will forever remain a hypothesis, and a rather vague hypothesis at that.

Is the Cosmic Censorship Hypothesis True?

Since the literature does not agree on a precise statement of the CCH, the question is a murky one. Not surprisingly, the murk has provided fertile soil from which have sprung a wide variety of clashing opinions. Rather than review these opinions, it seems to me more productive to try to state the strongest cases on either side of the issue.

On the negative side we can begin by noting that faith in the CCH depends on the notion that the GTR has some built-in mechanism for preserving modesty by clothing singularities. In this respect the original Oppenheimer-Synder-Volkoff model of gravitational collapse (Oppenheimer and Synder 1939, Oppenheimer and Volkoff 1939) was badly misleading in creating a false sense of security. Indeed, subsequent analysis has revealed a number of different mechanisms by means of which cosmic censorship can be violated. Steinmüller, King, and Lasota (1975) show that a momentarily naked singularity (see figure 3.11) can be produced by having a collapsing star radiate away its mass in such a way that the star never forms a black hole since it remains outside of its Schwarzschild radius (see also Lake and Hellaby 1981). Then there are the persistently naked strong curvature singularities that form in self-similar gravitational collapses, either as a result of the shell-focusing of dust (see figure 3.12) as in Tolman-Bondi models (see Gorini et al. 1989) or as a result of the spherical collapse of an adiabatic fluid (see Ori and Piran 1990). Joshi and Dwivedi (1992) show that the non-self-similar collapse of radiation can also form naked strong curvature singularities. Roberts (1989) shows how naked singularities can be produced in imploding spacetimes with a massless scalar field as the source. To be sure, each of these models involves some special and artificial features. Neverthe-

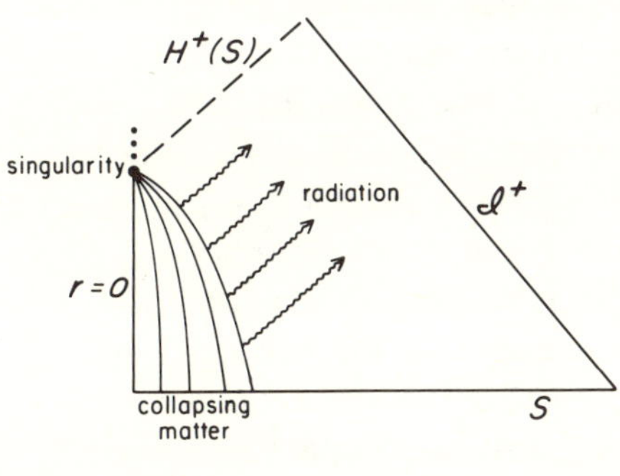

Figure 3.11

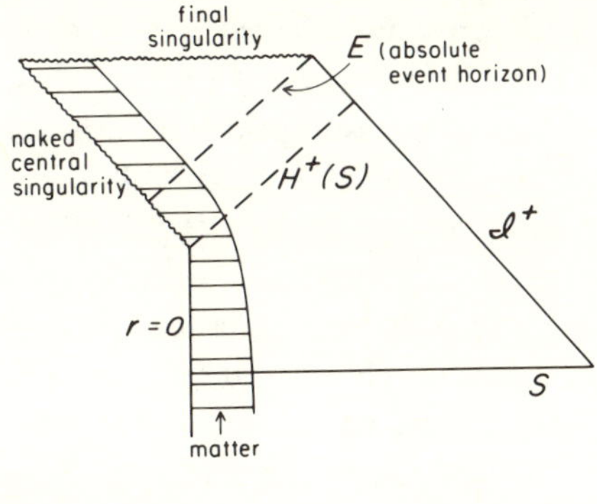

Figure 3.12

less, the fact that naked singularities can arise in such a variety of ways should shake one's faith that GTR does have a modesty mechanism at work. This faith suffers another apparent blow from the computer simulation studies of Shapiro and Teukolsky (1991). They interpret their results to indicate that a naked singularity can emerge from the nonspherical collapse of a collisionless gas. However, this interpretation is currently a matter of dispute. In particular, Shapiro and Teukolsky take the absence of an apparent horizon in their simulations to serve as evidence for the absence of an event horizon clothing the singularity. The example of Wald and Iyer (1991) shows why such an inference is dangerous.[19]

Furthermore, despite the intense efforts devoted to the question of cosmic censorship, very little in the way of positive results in the form of censorship theorems has resulted. Penrose (1973) notes that the CCH can be used to derive inequalities, involving areas of trapped surfaces and available masses, on the behavior of black holes. He tried but failed to find physically plausible ways to violate these inequalities. The failure constitutes only weak evidence for cosmic censorship since, as Penrose himself notes, if cosmic censorship does not hold, a naked singularity may form without a trapped surface also forming.[20] Wald's (1973) investigation starts with the inequality necessary for the Kerr-Newman solutions to represent a black hole: $M^2 \geqslant Q^2 + J^2/M^2$, where M, Q, and J are respectively the mass, electric charge, and the angular momentum. He found that various ways of trying to violate this inequality by injecting charge and angular momentum all fail. Again this gives only weak support to the CCH since although it confirms the stability of stationary black holes once they form, it gives no confidence that a black hole rather than a naked singularity will form in gravitational collapse. Krolak (1986, 1987a, 1987b) has offered various censorship theorems, but they rely on the dubious assumption of the existence of marginally outgoing null geodesics.[21] Newman (1984b) proved that conformally flat, null convergent spacetimes ($R_{ab}k^a k^b \geqslant 0$ for every null vector k^a) are not FNS_1.[22] But this is a purely geometric result which does not rely on Einstein's field equations, and by itself is not indicative of the kind of behavior that can be expected in the models of GTR. Chrusciel, Isenberg, and Moncrief (1990) have proven a censorship theorem for polarized Gowdy spacetimes. Roughly, "most" such spacetimes satisfy cosmic censorship in that the spacetimes which are developed from initial data on a partial Cauchy surface S and which cannot be extended beyond $D(S)$ form an open dense set in a natural parametrization space. But this inextendibility results from the fact that timelike curves terminate in a big crunch singularity, and not because of the formation of a persistent singularity within a black hole. So while the result does engender some faith in cosmic censorship at the cosmological level, it does not support the existence of a modesty mechanism for gravitational collapse that must be the guts of the CCH.

The supporters of cosmic censorship can respond that it is significant that despite the intense effort to refute the CCH, no knockdown counterexamples have been found; indeed, while many potential counterexamples exist, each can be faulted for the sorts of reasons discussed in the previous section. Moreover, there is a respect in which the Oppenheimer-Snyder-Volkoff model is suggestive of a feature which may be generic to gravitational collapse; namely, when a sufficient amount of matter is compacted into a small enough volume, an event horizon forms. Of course, this event horizon conjecture (EHC), even if true, does not by itself suffice to establish either the strong or the weak form of the CCH. Strong cosmic censorship may still fail because within the event horizon a locally naked singularity can develop. And weak cosmic censorship can fail because, as illustrated in figure 3.12, the formation of the event horizon may be preceded by the development of a globally naked singularity. Nevertheless, proving the EHC would be a big step toward proving the CCH. Israel (1986) has established a preliminary version of the EHC by assuming that a trapped surface develops and that its cylindrical extension remains nonsingular.

A crucial test case for the CCH concerns vacuum solutions ($T^{ab} = 0$) to Einstein's field equations. For then worries about whether the coupled Einstein-matter equations admit a well-posed initial value problem vanish, as do worries about what conditions T^{ab} and the equations of state must satisfy in order to qualify as physically realistic. Some care is still needed in formulating the CCH in this setting. For example, the Taub-NUT spacetime falsifies the most naive attempt to formulate the CCH in this setting. *Conjecture O:* Among vacuum solutions to Einstein's field equations, future naked singularities (on any reasonable definition) do not develop from regular initial data.

One of two directions can be taken to maneuver around this counterexample. The first is to exclude by fiat acausal solutions. *Conjecture 1:* Let M, g_{ab} be a vacuum solution to Einstein's field equations. If $S \subset M$ is a partial Cauchy surface, then the spacetime is not future nakedly singular with respect to S unless strong causality is violated at some $p \in H^+(S)$. If "future nakedly singular" is taken as FNS_1, then from the point of view of evolution we can formulate this conjecture as it applies to a compact initial data surface as follows: The maximal future Cauchy development of the appropriate initial data for a vacuum solution to Einstein's field equation prescribed for a compact S (without edges) is not extendible as a solution of the field equations unless strong causality is violated at some $p \in H^+(S)$. (This follows from the same kind of argument used in an earlier section.) More generally, in a globally hyperbolic spacetime any compact partial Cauchy surface is a Cauchy surface (see Budic et al. 1978). So if cosmic censorship in the form of global hyperbolicity is to hold, $D^+(S)$ for a compact partial Cauchy S cannot be extendible.

The alternative is to eschew spatially closed universes, leading to *Conjecture 2:* Let M, g_{ab} be a vacuum solution to Einstein's field equations. If M, g_{ab} does not admit a compact partial Cauchy surface then it is not *FNS*. It would be encouraging to establish a somewhat weakened version of this conjecture that applies to asymptotically flat spacetimes that represent the behavior of isolated gravitating systems.

This retreat from Conjecture 0 to Conjectures 1 and 2 is less than satisfying. Conjecture 1 rules out by fiat one sort of breakdown in predictability due to the emergence of acausal features from a past that may have been causally pure, while Conjecture 2 refuses to consider spatially closed universes. At present, however, no more attractive alternatives are available.

A failure of Conjecture 1 or 2 need not be seen as fatal to cosmic censorship since one could retreat to the version of censorship that asserts only that such failures are sparse among the vacuum field solutions. Genericity considerations can also be used to fill the gap in the above conjectures. Moving from Conjecture 0 to Conjecture 1 amounts to ignoring the violation of cosmic censorship that arises in Taub-NUT spacetime. That such ignorance may not be bliss was suggested by Moncrief's (1981) finding that there is an infinite dimensional family of vacuum solutions with NUT-type extensions.

However, Moncrief (1983) also established that such behavior is nevertheless special. For a partial Cauchy surface S in Taub-NUT spacetime, $H^+(S)$ is a compact null surface ruled by closed null curves. Moncrief's result is that analytic vacuum solutions possessing such a null surface must also possess a Killing field.

My own conclusion from this review is that it is much too soon to pronounce the CCH dead, but the prognosis is not particularly cheerful.

What If Cosmic Censorship Should Fail?

How much of a disaster for physics would it be if the CCH should prove to be wrong? Early in the investigation of the problem of cosmic censorship, Penrose posed this question and sketched a preliminary response:

> It is sometimes said that if naked singularities do occur, then this would be disastrous for physics. I do not share this view. We have already had the example of the big-bang singularity in the remote past, which seems not to be avoidable. The "disaster" to physics occurred right at the beginning. Surely the presence of naked singularities arising occasionally in collapse under more "controlled" circumstances would be the very reverse of a disaster. The effects of such singular occurrences could then be accessible to observation *now*. Theories of singularities would be open to observational test. The initial mystery of creation, therefore, would no longer be able to hide in the obscurity afforded by its supposed uniqueness. (1973, 133)

The reference to the big bang singularity shows that the visibility of the singularity by itself portends no disaster; but this point has already been codified in the various definitions of naked singularities, which exclude the initial singularities in standard big bang cosmologies. Only those singularities that entail a breakdown in prediction-retrodiction are counted as naked. Now for whatever psychological reasons, we are much less disturbed by an inability to retrodict the past than by an inability to predict the future. The effects of past naked singularities would imply a breakdown in retrodiction; but, as Penrose says, the effects of such singularities would be accessible to observation now, and given a knowledge of these effects we can hope to make an accurate prognostication of the future. The hope vanishes if the spacetime is future nakedly singular. This would seem to be a disaster for physics in its current state since current principles do not tell us whether the naked singularity will passively absorb what falls into it or will regurgitate television sets, green slime, or whatever.

Penrose has cautioned that not just anything goes:

> If we envisage an isolated naked singularity as a source of new matter in the universe, then we do not *quite* have unlimited freedom in this! For although in the neighborhood of the singularity we have no equations, we still have normal physics in the space-time *surrounding* the singularity. From the mass-energy flux theorems of Bondi et al. and Sachs, it follows that it is *not* possible for *more* mass to be ejected from a singularity than the original total mass of the system, *unless* we are allowed to be left with a singularity of *negative* total mass. (Such a singularity would *repel* other bodies, but would still be attracted by them!) (1969, 274)

The "unless" clause of the penultimate sentence of the quotation gives the game away. For what is or is not allowed for the behavior of naked singularities is not specified by the current rules of the game. A parade of horribles resulting from negative mass singularities—for example, hook a negative mass to a positive mass by means of a rigid rod and one has a self-accelerating machine—do not provide a *reductio* of such possibilities but only serve to underline how unconstrained physics becomes if naked singularities are allowed.

The point can be made in a more mundane fashion by returning to the normal physics referred to in the quotation. The conclusion that the energy radiated away by an isolated gravitating system in an asymptotically flat spacetime is bounded by the total mass-energy rests on three results: first, that the energy flux of gravitational radiation is positive; second, that the total mass-energy (as measured by the ADM mass), is positive; and third, that the Bondi energy, which measures the amount of mass-energy left in the system after radiation, is also positive.[23] The proofs of the latter two results assume no naked singularities; for instance, the negative mass Schwarzschild solution, which has a globally naked singularity, has negative ADM mass.[24]

Does it follow that if naked singularities occur we must simply throw up our hands and wait nervously for weirdness to unfold? I think not, for two types of considerations can be brought to bear, one of which belongs to familiar GTR, the other of which lies beyond the bounds of the theory but not so far as to extend into fantasy land. We can begin by coming to grips with questions such as the following. If a naked singularity develops from such-and-such initial data on S, how can $D^+(S)$ be extended as a solution to Einstein's field equations? What types of inequivalent (that is, nonisometric) extensions are allowed? How are the different types of extensions related to the behavior of the singularities? For example, if the singularity is quiescent in some appropriate sense, is the extension unique (up to isometry)? Though they may prove difficult to answer, these questions do not involve techniques that go beyond standard GTR.

Next we can heed what was perhaps the main message from the first quotation from Penrose and try to discern what regularities naked singularities display. For example, are the singularities that develop in certain situations quiescent? Do those that develop in other situations all ooze green slime, and if so do they ooze it at a regular rate? The attitude that physics is hopeless if naked singularities occur stems from what may be termed GTR chauvinism—the notion that Einstein and his followers discovered all of the laws relevant to classical gravitation. If we acknowledge that laws of nature are simply codifications of certain deep regularities (see Earman 1986, chap. 5), then we should be prepared to discover through observation that naked singularities obey laws of their own. If we are lucky, these additional laws, when conjoined with the laws of standard GTR, will restore predictability and determinism. Even if we are not so lucky they may still give us some interesting physics. Of course, we must be prepared for the eventuality that naked singularities exhibit no interesting regularities at all, in which case they would indeed be a disaster for physics. But at present only GTR chauvinism would lead us to fixate on this worst-case scenario.

Conclusion

The prospects seem dim for proving a wide-ranging censorship theorem; indeed, it is not even clear how to state the censorship hypothesis in a form that is precise, that has wide scope, and that escapes all of the known counterexamples. For both the proponents and opponents of cosmic censorship this means that there is no easy road to victory. The proponents have to be content with proving limited censorship theorems for special cases, the opponents have to pile up more and more counterexamples, and each side has to hope that eventually it will wear the other down. It is too soon to try to predict the outcome of this battle of attrition, but two things seem certain. First, whatever the final resolution, a good deal of interesting physics and interesting philosophy of physics will be generated by the process. Second, the battle is not apt to fizzle out for lack of interest, for the business of cosmic censorship is too important to ignore. In addition to the problems of predictability and determinism I have emphasized, cosmic censorship is connected with a number of other fundamental issues in classical general relativity. I have mentioned the matter of the positivity of the total mass of a gravitating system. In addition, cosmic censorship is an assumption underlying black hole thermodynamics and, in particular, Hawking's area law.[25] Theoreticians would dearly love to have some suitable form of censorship to combine with the "no hair" theorems for black holes (see Wald 1984, chap. 12) in order to conclude that the final state of gravitational collapse is characterized by just three parameters: mass, charge, and angular momentum.

In closing I should note that my entire discussion has assumed that classical general relativity remains valid right down to the singularities, if any. Ultimately, however, we must take into account the implications of quantum effects which surely come into play in the regimes of extreme density and curvature that precede the formation of singularities according to the classical story. Perhaps the upshot will be the avoidance of all singularities, clothed as well as naked. If not, the threat to cosmic censorship becomes even worse since the evaporation of black holes via Hawking radiation leads to the formation of a naked singularity (see Kodama 1979).[26]

NOTES

I am grateful for help received from Al Janis, Robert Geroch, and Frank Tipler. Of course, there should be no presumption that their opinions agree with those expressed here.

1. In Earman (1986) I argued that for certain purposes it is essential to distinguish between predictability and determinism. For present purposes that message will be ignored.

2. Peter Bergmann (1980, 156) characterized Einstein's attitude toward singularities as follows: "It seems that Einstein always was of the opinion that singularities in classical field theory are intolerable. They are intolerable from the point of view of classical field theory because a singular region represents a breakdown of the postulated laws of nature. I think one can turn this argument around and say that a theory that involves singularities and involves them unavoidably, moreover, carries within itself the seeds of its own destruction."

3. Lifshitz and Khalatnikov (1963) tried to show that singularities do not occur in a fully general solution of Einstein's field equations.

4. The variable M is a differentiable manifold without boundary and g_{ab} is a Lorentz signature metric defined on all of M. In the case of four-dimensional Minkowski spacetime $M = \mathbb{R}^4$ and g_{ab} is the Minkowski metric η_{ab}.

5. Various constructions attempt to represent singularities as boundary points attached to the spacetime manifold M, but each of these constructions has counterintuitive consequences; see Tipler et al. (1980) and Wald (1984, chap. 9).

6. The criterion of local inextendibility offered in Hawking and Ellis (1973, 59) is too strong since, for example, standard Minkowski spacetime is counted as locally extendible.

7. The spacetime M, g_{ab} is *temporally orientable* just in case it admits a globally consistent time direction. Technically this means that there exists a continuous everywhere defined timelike vector field on M. (Intuitively, the arrows point in the "future direction.") Let S be a partial Cauchy surface in a temporally orientable spacetime. Since S is spacelike, the timelike vector field that defines the temporal orientation is everywhere nontangent to S, which means that S is two-sided. Not every relativistic spacetime is temporally orientable. And not every temporally orientable spacetime admits a partial Cauchy surface; Gödel spacetime is a counterexample.

8. Penrose diagrams bring infinity into a finite distance by means of a conformal transformation that preserves causal relations; see Hawking and Ellis (1973) for details.

9. For a spacetime M, g_{ab}, the *causal future* $J^+(X)$ (respectively, *causal past* $J^-(X)$) of $X \subset M$ is defined as the set of all points which can be reached from X by a future-directed (respectively, past-directed) causal curve. The *chronological future* $I^+(X)$ (chronological past $I^-(X)$) of $X \subset M$ is defined as the set of all points which can be reached from X by a nontrivial, future-directed (respectively, past-directed) causal curve.

10. Roger Penrose (1979) has put forward the "Weyl curvature hypothesis" which rules out the type of white holes that violate cosmic censorship. He conjectures that this asymmetry must be underwritten by a fundamental law of physics that is not time reversal invariant.

11. The reader is invited to construct an example to show why the definition of topology change must be restricted to maximal partial Cauchy surfaces.

12. The generators of the null surface $H^+(S)$ are null geodesics. The generators are either past endless or else have a past endpoint in the edge of S.

13. If $\mathfrak{N} = \varnothing$ because there is a $p \in M$ such that $I^-(p)$ contains a timelike curve γ of infinite future proper length, then we have an example of a so-called Pitowsky spacetime. Such a spacetime would apparently allow the observer whose world-line is γ to complete an infinite computational task and to transmit the result to another observer whose world-line passes through p. This matter is discussed in Earman and Norton (1993).

14. In Schwarzschild coordinates the line element has the form $ds^2 = -(1 - 2M/r)dt^2 + (1 - 2M/r)dr^2 + r^2(d\theta^2 + \sin^2\theta d\phi^2)$ where M is the total

mass. If $M < 0$, the metric components are regular down to $r = 0$ where a naked singularity resides.

15. The technical definition of global hyperbolicity can be found in Hawking and Ellis (1973). It is equivalent to the existence of a Cauchy surface. The demonstration that follows comes from Newman (1984b).

16. M, g_{ab} violates *strong causality* at $p \in M$ just in case there is a neighborhood $N(p)$ such that for any subneighborhood $N'(p) \subseteq N(p)$ there is a causal curve that leaves and reenters $N'(p)$.

17. I am grateful to A.D. Rendall for information concerning these matters. Recently Rendall (1992b) has obtained existence results for time symmetric initial data for a finite fluid body obeying the Einstein-Euler equations.

18. A spacetime is *self-similar* just in case there is a timelike vector field V_a such that $\nabla_a V_b - \nabla_b V_a = \Theta g_{ab}$, Θ = constant, where ∇_a is the derivative operator of g_{ab}. For a spherically symmetric spacetime, this means that under a coordinate transformation $t \to \bar{t} = at$, $r \to \bar{r} = ar$, $\theta \to \bar{\theta} = \theta$, the metric components g_{ij} $(i, j = 1, 2, 3, 4)$ transform as $\bar{g}_{ij}(\bar{r}, \bar{t}) = (1/a^2)g_{ij}(r, t)$.

19. An *apparent horizon* is the outer boundary of a trapped region (see Hawking and Ellis 1973, sec. 9.2, for precise definitions). The relevant point here is that the existence of an apparent horizon entails the existence of an event horizon but not conversely. Another reason for doubting the significance of the Shapiro and Teukolsky results is given by Rendall (1992a). Shapiro and Teukolsky use the Vlasov equation to describe the distribution of the dust particles in their model. Rendall argues that it is unreasonable to expect that more general solutions to this equation will exhibit the singular behavior found in the Shapiro-Teukolsky model.

20. A *trapped surface* $\mathcal{T}$ is a closed two-surface such that both ingoing and outgoing null geodesics orthogonal to $\mathcal{T}$ are converging.

21. A null geodesic of M, g_{ab} that forms an achronal set and has an endpoint in $\mathcal{I}^+$ is called *outgoing*. It is *marginally outgoing* if it is a limit curve of all outgoing null geodesics.

22. The spacetime M, g_{ab} is said to be *conformally flat* just in case $g_{ab} = \Omega\eta_{ab}$, where the metric η_{ab} has vanishing Riemann tensor and the conformal factor Ω is a C^∞ map from M to $(0, +\infty)$. Newman's theorem requires that Ω be a proper map in the sense that the inverse image of any compact set of $(0, +\infty)$ is a compact set of M.

23. ADM stands for Arnowitt, Deser, and Misner. For an overview of the positive mass theorems, see Wald (1984, Chap. 11).

24. This is a vacuum solution ($T^{ab} = 0$ everywhere) so that the energy conditions are trivially satisfied. As already noted, this example would not be regarded as a violation of the form of cosmic censorship that excludes only those naked singularities that develop from regular initial data.

25. Roughly, the area of a black hole does not decrease with time.

26. The threat is mitigated by the fact that the evaporation of a black hole with a mass comparable to that of the sun takes many times longer than the age of the universe.

REFERENCES

Bergmann, P. 1980. "Open Discussion, Following Papers by S. W. Hawking and W. G. Unruth." In H. Woolf, ed., *Some Strangeness in the Proportion*. Reading, Mass.: Addison-Wesley, p. 156.

Budic, R.; J. Isenberg; L. Lindblom; and P. B. Yasskin. 1978. "On the Determination of Cauchy Surfaces from Intrinsic Properties." *Communications in Mathematical Physics 61*:87–95.

Chandrasekhar, S., and J. B. Hartle. (1982). "On Crossing the Cauchy Horizon of a Reissner-Nordström Black Hole." *Proceedings of the Royal Society (London) A34:* 301–15.

Chrusciel, P. T.; J. Isenberg; and V. Moncrief. 1990. "Strong Cosmic Censorship in Polarized Gowdy Spacetimes." *Classical and Quantum Gravity 7:* 1671–80.

Clarke, C.J.S., and A. Krolak. 1985. "Conditions for the Occurrence of Strong Curvature Singularities." *Journal of Geometry and Physics 2:* 127–45.

Eardley, D. M. 1987. "Naked Singularities in Spherical Gravitational Collapse." In B. Carter, and J. B. Hartle, eds., *Gravitation and Astrophysics, Cargese 1987*. New York: Plenum Press, pp. 229–35.

Earman, J. 1986. *A Primer on Determinism*. Dordrecht: Reidel.

Earman, J., and J. Norton. 1993. "Forever Is a Day: Supertasks in Pitowsky and Hogarth-Malament Spacetimes." *Philosophy of Science 60:* 22–42.

Geroch, R. P. 1968. "What Is a Singularity in General Relativity?" *Annals of Physics 48:* 526–40.

———. 1970. "Domains of Dependence." *Journal of Mathematical Physics 11:* 437–49.

———. 1977. "Prediction in General Relativity." In J. Earman, C. Glymour, and J. Stachel, eds., *Foundations of Space-Time Theories*. Minneapolis: University of Minnesota Press, pp. 81–92.

Geroch, R. P., and G. T. Horowitz. 1979. "Global Structure of Spacetimes." In S. W. Hawking and W. Israel, eds., *General Relativity: An Einstein Centenary Survey*. Cambridge, England: Cambridge University Press, pp. 212–93.

Gorini, V.; G. Grillo; and M. Pelizza. 1989. "Cosmic Censorship and Tolman-Bondi Spacetimes." *Physics Letters A135:* 154–58.

Gowdy, R. H. 1977. "Instantaneous Cauchy Surfaces, Topology Change, and Exploding Black Holes." *Journal of Mathematical Physics 18:* 1798–1801.

Hawking, S. W. 1979. "Comments on Cosmic Censorship." *General Relativity and Gravitation 10:* 1047–49.

Hawking, S. W., and G.F.R. Ellis. 1973. *The Large Scale Structure of Space-Time*. Cambridge, England: Cambridge University Press.

Hawking, S. W., and R. Penrose. 1970. "The Singularities of Gravitational Collapse and Cosmology." *Proceedings of the Royal Society (London) A314:* 529–48.

Horowitz, G. T. 1979. "Finding a Statement of Cosmic Censorship." *General Relativity and Gravitation 10:* 1057–61.

Israel, W. 1984. "Does a Cosmic Censor Exist?" *Foundations of Physics 14:* 1049–59.

———. 1986. "The Formation of Black Holes in Nonspherical Collapse and Cosmic Censorship." *Canadian Journal of Physics 64:* 120–27.

Joshi, P. S., and I. H. Dwivedi. 1992. "Naked Singularities in Non-Self-Similar Gravitational Collapse of Radiation Shells." *Physical Review D 45:* 2147–50.

Joshi, P. S., and R. V. Saraykar. 1987. "Cosmic Censorship and Topology Change in General Relativity." *Physics Letters A120:* 111–14.

Kodama, H. 1979. "Inevitability of a Naked Singularity Associated with the Black Hole Evaporation." *Progress in Theoretical Physics 62:* 1434–35.

Kriele, M. 1990. "Causality Violations and Singularities." *General Relativity and Gravitation 22:* 619–23.

Krolak, A. 1986. "Towards a Proof of the Cosmic Censorship Hypothesis." *Classical and Quantum Gravity 3:* 267–80.

———. 1987a. "Strong Cosmic Censorship and the Strong Curvature Singularities." *Journal of Mathematical Physics 28:* 2685–87.

———. 1987b. "Towards a Proof of the Cosmic Censorship Hypothesis in Cosmological Space-Times." *Journal of Mathematical Physics 28:* 138–41.

Kuroda, Y. 1984. "Naked Singularities in the Vaidya Spacetime." *Progress in Theoretical Physics 72:* 63–72.

Lake, K. 1988. 'Comment on "Naked Singularities in Self-Similar Spherical Gravitational Collapse.' " *Physical Review Letters 60:* 241.

Lake, K., and C. Hellaby. 1981. "Collapse of Radiating Fluid Spheres." *Physical Review D24:* 3019–22.

Lifshitz, E. M., and I. M. Khalatnikov. 1963. "Investigations in Relativistic Cosmology." *Advances in Physics 12:* 185–249.

Misner, C. W. 1963. "The Flatter Regions of Newman, Unti, and Tamburino's Generalized Schwarzschild Space." *Journal of Mathematical Physics 4:* 924–37.

Moncrief, V. 1981. "Infinite-Dimensional Family of Vacuum Cosmological Models with Taub-NUT (Newman-Unti-Tamburino)-type Extensions." *Physical Review D 23:* 312–15.

———. 1983. "Symmetries of Cosmological Horizons." *Communications in Mathematical Physics 89:* 387–413.

Newman, R. P. A. C. 1984a. "A Theorem of Cosmic Censorship: A Necessary and Sufficient Condition for Future Asymptotic Predictability." *General Relativity and Gravitation 16:* 175–93.

———. 1984b. "Cosmic Censorship and Conformal Transformation." *General Relativity and Gravitation 16:* 943–53.

———. 1986. "Cosmic Censorship and the Strengths of Singularities." In P. G. Bergmann and V. De Sabbata, eds., *Topological Properties and Global Structure of Space-Time.* New York: Plenum Press, pp. 153–68.

Oppenheimer, J. R., and H. Synder. 1939. "On Continued Gravitational Collapse." *Physical Review 56:* 455–59.

Oppenheimer, J. R., and G. M. Volkoff. 1939. "On Massive Neutron Stars." *Physical Review 55:* 374–81.

Ori, A., and T. Piran. 1987. "Naked Singularities in Self-Similar Spherical Gravitational Collapse." *Physical Review Letters 59:* 2137–40.

———. 1988. "Self-Similar Spherical Gravitational Collapse and the Cosmic Censorship Hypothesis." *General Relativity and Gravitation 20:* 7–13.

———. 1990. "Naked Singularities and Other Features of Self-Similar General-Relativistic Gravitational Collapse." *Physical Review 42:* 1068–90.

Papapetrou, A. 1985. "Formation of a Singularity and Causality." In N. Dadhich, J. Krishna Rao, J.V. Narlikar, and C.V. Vishveshwara, eds., *A Random Walk in Relativity and Cosmology.* New York: Wiley, pp. 184–91.

Papapetrou, A., and A. Hamouri. 1967. "Surfaces caustiques degenerees dans la solution de Tolman. La singularite physique en relativite generale." *Annales de Institut Henri Poincaré A6:* 343–64.

Penrose, R. 1969. "Gravitational Collapse: The Role of General Relativity." *Revisita del Nuovo Cimento.* Serie 1, *1* Numero Speciale: 252–76.

———. 1973. "Naked Singularities." *Annals of the New York Academy of Sciences 224:* 125–34.

———. 1974. "Gravitational Collapse." In C. DeWitt, ed., *Gravitational Radiation and Gravitational Collapse.* Dordrecht: Reidel, pp. 82–90.

———. 1976. "Space-Time Singularities." In R. Ruffini, ed., *Proceedings of the 1975 Marcel Grossman Conference.* Amsterdam: North-Holland, pp. 173–81.

———. 1978. "Singularities of Spacetime." In N. R. Lebovitz, W. H. Reid, and P. O. Vandervoort, eds., *Theoretical Principles in Astrophysics and Relativity.* Chicago: University of Chicago Press, pp. 217–43.

———. 1979. "Singularities and Time-Asymmetry." In S. W. Hawking and W. Israel, eds., *General Relativity: An Einstein Centenary Survey.* Cambridge, England: Cambridge University Press, pp. 581–638.

Rendall, A. D. 1992a. "Cosmic Censorship and the Vlasov Equation." *Classical and Quantum Gravity 9:* L99–L104.

———. 1992b. "The Initial Value Problem for a Class of General Relativistic Fluid Bodies." *Journal of Mathematical Physics 33:* 1047–53.

Roberts, M. D. 1989. "Scalar Field Counterexamples to the Cosmic Censorship Hypothesis." *General Relativity and Gravitation 21:* 907–39.

Seifert, H. J. 1979. "Naked Singularities and Cosmic Censorship: Comment on the Current Situation." *General Relativity and Gravitation 10:* 1065–67.

———. 1983. "Black Holes, Singularities, and Topology." In E. S. Schmutzer, ed., *Proceedings of the Ninth International Conference on General Relativity and Gravitation.* Cambridge, England: Cambridge University Press, pp. 581–638.

Shapiro, S. L., and S. A. Teukolsky. 1991. "Formation of Naked Singularities: The Violation of Cosmic Censorship." *Physical Review Letters 66:* 994–97.

Steinmüller, B.; A. R. King; and J. P. Lasota. 1975. "Radiating Bodies and Naked Singularities." *Physics Letters A51:* 191–92.

Tipler, F. 1976. "Causality Violation in Asymptotically Flat Space-Times." *Physical Review Letters 37:* 879–82.

———. 1977. "Singularities and Causality Violation." *Annals of Physics 108:* 1–36.
———. 1979. "What Is a Black Hole?" *General Relativity and Gravitation 10:* 1063–67.
———. 1985. "Note on Cosmic Censorship." *General Relativity and Gravitation 17:* 499–507.
Tipler, F.; C.J.S. Clarke; and G.F.R. Ellis. 1980. "Singularities and Horizons—A Review Article." In A. Held, ed., *General Relativity and Gravitation*, vol. 2. New York: Plenum Press, pp. 97–206.
Wald, R. M. 1973. "On Perturbations of a Kerr Black Hole." *Journal of Mathematical Physics 14:* 1453–61.
———. 1984. *General Relativity.* Chicago: University of Chicago Press.
Wald, R. M., and V. Iyer. 1991. "Trapped Surfaces in the Schwarzschild Geometry and Cosmic Censorship." *Physical Review D44:* 3719–22.
Yodzis, P.; H. J. Seifert; and H. Müller zum Hagen. 1973. "On the Occurrence of Naked Singularities in General Relativity." *Communications in Mathematical Physics 34:* 135–48.

4

From Time to Time

Remarks on the Difference Between the Time of Nature and the Time of Man

Jürgen Mittelstrass
Philosophy Department, University of Konstanz

Time is something familiar and something puzzling. It is the reality of hope, of waiting and of forgetting; it is the motion of the hands of a clock, and it is what Augustine (1955, 628–29) explained he could not explain and what Kant (1956–1964, vol. 1, 753) later called the still unexplained. Even in mythology time is so remarkable that it takes on divine form. Chronos is its name with the Greeks. According to Sophocles (Soph. *El.* 179) Chronos sees and hears all, brings everything to light and conceals it again. He is a helpful God. The Day is the daughter of Chronos and the Night (*Bakchyl.* 7,1), just as are Justice and Truth (Eurip. *frag.* 223 N. Gell. 12,11,7). Very early Chronos is identified with Kronos, the youngest of the six sons and daughters of Uranos and Gaia, who castrates his father, devours his children and finally after his banishment to Tartaros is reconciled with his son Zeus, who had escaped him, and rules over the Islands of the Blessed. Thus, the God of time is associated with the cosmogonic notion of the separation of heaven and earth as well as with the notion of the creative and destructive power of time.

For humans, too, time is a genuine power over us, pitiless and at the same time inscrutable—as in Momo's encounter with the grey gentlemen. Master Hora explains to Momo that these gentlemen live off the dead, "You know they live off the life-time of people. But this time literally dies if it is torn away from its real owner. For everyone has *his* time. And only as long as it is really his does it remain alive"

(Ende 1973, 152). That is why the grey gentlemen are dead, too. They have no time of their own; and without the stolen time, "they would have to go back to the nothingness they came from" (ibid., 153). Momo understands. Time, to which Greek mythology gives divine shape and cosmogonic dimension, is not a *property* of things but a *shape* or *gestalt* of things.[1] Without this gestalt the things are nothing. Life, too, which according to Master Hora's words lives in the heart, has its time. When it loses its time, it becomes cold and grey and passes into nothingness.

Attached to the philosophy of time, we find the poetry of time, but this is not our subject here. I deal with the *scientific* concept of time, the concept of time in physics, and with the concept of time in *everyday life,* explicated in the concept of a *time of action* and—approaching Momo again—with a gestalttime. It will become clear that the concepts of time in physics and everyday life have little to do with one another. However, the time constructions of everyday physics directly affect the lifeworld. Thus the need to coordinate *concrete* times, among them times of nature, of action and of life, and to pass from one temporal gestalt to another, and to comprehend other gestalten of time leads to an "abstract" time which is everywhere the *same* and everywhere *one* time. Time theories which are in the everyday physical sense theories of an abstract time have this *practical* background. The paradigms of the transition from concrete times—and thus from a gestalt model of time—to an abstract time are clocks. These are derived from the gestalt notion of time (astronomical models of the cosmos are also clocks) and lead to the continuum notion or the time's arrow notion of time. In Plato's astronomy (see *Tim.* 37d–38e), for instance, the "circle of the same" (the rotation of the heavens on its axis) represents pure periodicity (like the dial of an analogue clock), and the "circle of the diverse" (the motion of the planets on the ecliptic) represents a celestial calendar which makes it possible to count the days.

The Anisotropic Time of Science

The natural sciences, it seems, are bound to assign a prominent role to time. After all, lots of things change in the world; and change needs time to unfold. The reverse also holds true; as the saying goes, "the times they are a-changing." Thus, there appears to be ample use for time. This apparent evidence of the senses notwithstanding, the view has gained acceptance in some quarters that in reality there is no change. The true world is timeless. Generation and corruption are mere illusions of the deceived mind and have no foundation in nature.

Let me illustrate this point of view by one example. As Meyerson claims, the development of scientific theories brings about the elimination of time. His chief argument is that in science changes are expressed by means of equations and conservation laws. Consider chemistry, for instance. The description of a chemical process with the concomitant changes in the qualitative properties of the substances involved is given by a chemical *equation.* By virtue of being an equation, its left-hand and right-hand side express the same quantity in different form. Describing change in terms of equations thus conveys the impression that in reality nothing has been created and nothing has been lost. This means that the very notion of change evaporates under the grip of scientific explanation, "On the whole, as far as our explanation reaches, *nothing has happened.* And since phenomenon is only change, it is clear that according as we have explained it we have made it disappear. Every explained part of a phenomenon is a part denied; to explain is to explain away" (Meyerson [1930] 1976, 261). In Meyerson's view, science does thus away with change and becoming altogether; it relegates them to the realm of deceptions—comparable to optical illusions—which are exposed and overcome by rational examination. Science imposes a static worldview upon us, according to which looking at the world produces continual déjà-vu experiences. We appear to be trapped in a boring Parmenidean block universe in which nothing new ever occurs.

One need not be a wire service reporter to consider this scenario a nightmare. It stands out among Grünbaum's achievements to have exorcized these Parmenidean nightly specters. I definitely take this to be a real service, thus rejecting the Freudian claim that underlying every overt fear there is a covert wish (see Grünbaum 1983, 330–31). Put less metaphorically, we owe to Grünbaum a perceptive analysis and an important conceptual clarification of the nature of physical time. What is particularly relevant here is his analysis of the physical basis of the "anisotropy" and the lack of physical basis for the "arrow" of time. In this section I discuss Grünbaum's views on the anisotropic character of physical time, while in the following sections I turn to the problem of the arrow of time and psychological time in general.

Clearly, the basis of physical time is to be sought in physical processes (as Aristotle already knew). The processes at issue are *irreversible* in kind. Irreversible processes are temporally directed in that they go off only one way; their temporal inverse does not occur. That is, no counterprocesses exist that would be suited to restore the original type of state. In addition, we should distinguish between two kinds of irreversible processes. Processes are *nomologically* irreversible if the realization of the temporal inverse is excluded by a law of nature. They are de facto irreversible if their temporal inversion requires that particular initial or boundary conditions be realized which, as a matter of fact, do not obtain (see Grünbaum [1963] 1973, 209–10; [1971] 1976, 474).

Grünbaum claims that the occurrence of de facto irreversible processes is sufficient to establish the *anisotropy* of physical time. Anisotropy means that there is a structural distinction between the two opposite time directions. If some states always follow one another in a fixed sequence we are in a position to set apart earlier and later states. Irreversible processes thus provide a basis for a physical implementation of the temporal relations "earlier than" or "later than," and this is precisely what the anisotropy of time comes down to. An isotropic time would require that no such physical implementation exists, which would in turn demand that all processes can actually be inverted. For that reason the mere existence of irreversible processes, of whatever origin, confers anisotropy to time. And this is why the occurrence of de facto irreversibility is sufficient for anisotropy to emerge (see Grünbaum [1963] 1973, 211–12; [1971] 1976, 475).

What are the irreversible processes that might be suited to confer anisotropy to time? Consider processes such as the development of apparently homogeneous mixtures out of heterogeneous components or as the equalization of temperature differences. If cream is poured into a cup of coffee, both liquids intermingle as time passes; the reverse process, that is, the spontaneous separation of cream and coffee, has never been observed. Analogously, the differently heated ends of a metal rod will acquire the same temperature in the course of time; nobody has come across the spontaneous generation of a temperature gradient. Irreversible processes of this kind are described by the so-called *second law of thermodynamics*. According to that law, a quantity called *entropy* exists which in every closed system either increases in time or, in the case of equilibrium systems, remains unaltered. The second law thus appears to express the existence of irreversible processes and to afford, consequently, the physical basis for the anisotropy of time. For every closed, nonequilibrium system, higher entropy states are *later* than lower ones. In what follows I focus on this thermodynamic basis of the anisotropy of time.

The just outlined, seemingly smooth solution is, in fact, afflicted with a serious difficulty. Namely, the entropic behavior of macroscopic thermodynamic systems should evidently be rooted in the behavior of their microscopic constituents. It should be derivable from the motions of atoms or molecules. Boltzmann attempted such a derivation and came up with his famous *H*-theorem. He applied the laws of mechanics to colliding gas molecules and assumed that the molecular motions proceed in a random fashion; that is, he supposed that on the mechanical level there are no preferred states of motion. On that basis he arrived at a microscopically definable quantity which could be related to the macroscopic entropy, and which, correspondingly, exhibited the same temporally asymmetric behavior. Since the randomness premise entering this deduction is, however, of an essentially statistical nature, the resulting version of the second law is likewise statistical in character. In contrast to its thermodynamic model, the mechanically derived second law does not rigorously preclude decreasing entropy values; it merely says that such cases are less probable and occur less frequently than the contrary cases of increasing entropy. Moreover, on the statistical variant, the entropy does not remain constant in equilibrium states but fluctuates irregularly around a fixed value. This implies that in these states the entropy drops and rises with equal frequency.

Grünbaum has shown, elaborating and substantially improving an earlier conception of Reichenbach's, how we can take advantage of this attenuated version of the second law so as to specify a physical basis for temporal asymmetry. It follows from the above considerations that we must resort to nonequilibrium systems for that purpose. Suitable systems of that kind are *branch systems*. Such systems branch off from their environment in a state of comparatively low entropy, remain closed for some time and finally merge again with the surrounding wider system. The low entropy state realized in branch systems is not the result of a statistical entropy fluctuation; rather, it is the product of an interaction with some outside agency of natural or human origin. Take an iced drink as an example. By immersing an ice cube into a lukewarm liquid we create such a low-entropy system. Subsequently, as the ice cube melts, the whole drink assumes a uniform temperature that finally agrees with the temperature of the environment. This process of continual equalization of temperature represents an increase in entropy with the result that the branch subsystem is eventually reintegrated into the larger system from which it was initially detached by the intervention. The temporal evolution of the entropy in branch systems is thus markedly asymmetric.

But we have not yet taken into account the statistical nature of the second law. After all, due to that statistical nature it cannot be ruled out that some (untypical) branch system actually undergoes a decrease in entropy. This problem can be overcome by resorting to a large number of like branch systems. We employ a class of such systems characterized by the same initial macroscopic entropy states, and we *demand* that the microscopic realizations of these initial states be distributed at random. As a consequence of the second law, a typical representative of such a class will display a temporal increase in entropy. This feature constitutes the physical basis for the anisotropy of time (see Grünbaum [1963] 1973, 254–60, for more detail).

The next question to be addressed is as follows: From what precisely does this temporally asymmetric behavior of branch systems arise? The relevant point is the peculiar status of the irreversibility as expressed by the statistical variant of the second law. As already indicated, this law does not follow from the laws of mechanics alone. After all, the latter are perfectly reversible and do not exclude the time reversal of any process. This reversibility of the mechanical laws is reflected in the temporally symmetric entropic behavior of equilibrium systems. What we need in addition in order to arrive at irreversible processes is reference to the prevailing *initial or boundary conditions*. The derivation of the statistical version of the second law requires premises that are not lawlike in character but rather refer to actual circumstances as they *happen to occur*. Accordingly, the second law is not a fundamental law of nature; it arises from more basic, time-symmetric laws only by having recourse to particular states of affairs. This peculiarity is reflected in the branch-systems approach, for this method likewise relies essentially on the realization of particular nonequilibrium states that are supposed to exhibit microscopic randomness. Consequently, the anisotropy arising from this kind of irreversible processes is not nomological but merely de facto.

Accordingly, the anisotropy of time is not built into nature's foundations; it occupies a somewhat peripheral position instead. Grünbaum is, of course, well aware of, and even stresses, this de facto character of temporal anisotropy but he downplays the significance of this feature:

> [A]n asymmetry is no less an asymmetry for depending on de facto, nomologically contingent boundary conditions rather than being assured by a *law* alone. (Grünbaum [1963] 1973, 258)

This claim is further emphasized by the following consideration:

> [H]uman hopes for an eternal biological life are no less surely frustrated if all men are indeed de facto moral, i.e., mortal on the strength of "boundary conditions" which do obtain permanently, than if man's mortality were assured by some law. (Ibid., 272)

This is correct in regard to the fatal consequences, but wrong, I think, as regards the philosophical interpretation.

The mere de facto character of temporal anisotropy certainly *is* significant when the place of time in nature is at issue. This is all the more true if we take into account that every law known in physics is time-symmetric.[2] Apart from the second law there is not even a prima-facie candidate for a temporally anisotropic law. The laws of nature thus do not recognize the difference between earlier and later. With respect to them, future and past are completely alike. But this generates a conspicuous disparity between the central importance of the notion of a directed time for the human condition and the human life, on the one hand, and the peripheral place of a directed time in nature. There is a physical counterpart to directed time, to be sure, but it is of a markedly ephemeral character. These aspects of Grünbaum's work thus uncover one more Ptolemaean illusion: We would have expected beforehand that those traits of nature that seem to be essential from our lifeworld perspective should also play an important role in nature's nomological machinery. The disappointment of this expectation constitutes another refutation of the belief that Man is the measure of all things.

The conclusions drawn in philosophy of science often depend crucially on the particulars of the pertinent scientific theories. One exciting development in recent physics *might* induce a significant alteration as regards the philosophical interpretation of the directedness of time, which is the rise of irreversible thermodynamics that mainly occurred after Grünbaum had laid the spacetime subtleties aside and plunged into the depths of the human mind. The relevant point here is the *status of the second law.* Underlying the more traditional picture outlined so far is the view that the laws of mechanics are more fundamental than the principles of thermodynamics. This is certainly not an implausible premise since the thermal properties are supposed to arise form the statistical behavior of microparticles whose motions are in turn governed by mechanical laws. It is precisely by way of this premise that the only limited derivability of the second law from the laws of mechanics leads to the interpretation that the second law, along with the temporal anisotropy based on it, cannot be ascribed a fundamental status.

In the wake of the development of irreversible thermodynamics this ordering as to the fundamental character of mechanics and thermodynamics is no longer unanimously accepted. The crucial phenomenon here is the so-called *deterministic chaos.* In that case, the laws governing the temporal evolution of a system are strictly deterministic but the interaction between the component parts of the system introduces an extreme sensitivity to differences in the initial conditions. Slightly different initial states of the component parts eventually result in radically different behavior of the system. Consequently, the system's future behavior, though being strictly determined nomologically, is utterly unpredictable from the microscopic initial conditions. An example is the occurrence of multiparticle collisions as they are realized, for instance, by the molecular collisions in a gas. Arbitrarily small changes in the initial position-momentum values of the particles involved subsequently generate vastly different molecular motions.

Such intricate cases can be treated, however, by having recourse to suitable distribution functions as is done in the thermodynamic approach. Chaotic behavior on the microscopic level is perfectly compatible with ordered and regular behavior on the macroscopic level. The point is that in such cases the thermodynamic description is *the only appropriate one.* A statistical description in terms of distribution functions can thus not always be avoided in favor of a mechanical description in terms of trajectories. For that reason, as it is argued, we should abandon the primacy of mechanics. On this view, a statistical, thermodynamic description loses its derivative character. Mechanics and thermodynamics are complementary in that they are both necessary for an exhaustive elucidation of the relevant phenomena. As a result, the second law gains a new role as a fundamental law of nature (see Prigogine 1980, 196–99, 206–08, 212; Rae 1986, 103–06; and Carrier and Mittelstrass 1991, 258–66).

If this assessment turned out to be warranted, the anisotropy of time would acquire a more fundamental character than hitherto attributed to it. It would indeed be rooted in the nomological structure of nature. It seems to be premature, however, to consider this interpretation at present a viable and worked-out alternative to Grünbaum's views. After all, there is a somewhat dubious inference from *epistemic* limitations of mechanics to the *ontological* emancipation of thermodynamics involved here. I only wished to call attention to a novel strand of scientific theorizing that may possibly have a revolutionary impact on the philosophical interpretation of the anisotropy of physical time.

The Unidirectional Time of the Lifeworld

There is, however, more to the philosophical problem of time, as Grünbaum is the first to admit. He emphasizes time and again that *becoming,* in contrast to change, cannot be accommodated by any physical theory. The truth-value of this claim is in no way dependent on the subtleties of contemporary science. As Grünbaum points out, the notion of becoming is associated with the idea of a distinguished point in time, the *"Now."* The Now separates the past from the future and thus marks the point where processes and events "come into being." Moreover, the Now shifts along the time axis in one preferred direction, namely, the future; by means of this shifting, it indicates the "passage" or the "flow" of time. This notion is expressed metaphorically by the "arrow of time," showing the "advance" of this transient Now.

Grünbaum is at pains to set apart this notion of becoming from the conception of anisotropy as explained in the foregoing section. Whereas the latter involves a mere *difference* between two opposite time senses without singling out one of them as *the* direction of time, the former introduces precisely such a preferred time direction. The arrow of time is definitely not directed into the past; it points into the future. Becoming thus embraces the idea of a *unidirectional* time.

The notion of becoming thus goes beyond the idea of change. So, the question suggests itself whether there is any physical basis for becoming. This question is to be put as follows: Is there any physical process that allows for distinguishing between different Nows? Grünbaum emphatically denies this. I refrain from following his detailed survey and examination of the proposed solutions to that problem and just cite his result: There is no physical counterpart to the Now. The concept of "Now" has meaning only in the context of an experiencing ego, and its meaning is that an event happening "now" occurs simultaneously to a conscious organism's becoming aware of it (see Grünbaum [1963] 1973, 217–18, 314–16, 324–26; [1971] 1976, 475–82).

Since the Now denotes the division between past and future, this division has no factual significance as well. This implies in turn that there is no physical basis for the unidirectionality of time. This means that the notion of the passing and flowing time belongs exclusively to psychological time. "Coming into being" is essentially an epistemic, not an ontological, category. In sum, physics can tell us what is earlier and what is later but it cannot tell us what is now. Physics can tell us that there is a difference between past and future but it cannot tell us what is distinctive about the future. There is no physical reason for the fact that we care more for our offspring than for our ancestors. The time of Man can only incompletely and insufficiently be reduced to the time of nature. This is why the foregoing considerations need to be supplemented with a philosophical anthropology of time.

The Time of Man

At the border between science and the lifeworld lurks a philosophical trap, which has been the ruin of many a brave philosophical conception. I am referring to the *psychologism* or *mentalism* offered (not only in regard to time) as a philosophical alternative to *physicalism*.

What is not in nature is *in us,* reads the alternative. Beside physical time enters, in the form of a "stream of consciousness," *experienced* time, which appears as something internalized, exclusively subjective. Between it and physical time lies only the public time of clocks and train schedules that make time a social institution. According to such a notion the space between a physics of time and a psychology of time settles the fate of an anthropology of time.

Augustine was the first to think this way. For him time was not a property of the world (or a property of the worldly) but a property (or an action) of the *soul.* Time, understood as *lasting presence,* is not "measured" by motions in the world or in nature, but is "measured" by the soul (1955, 632ff). *Memoria* as presence of the past, *contuitus* as presence of the present, and *expectatio* as presence of the future are psychological or mental performances (ibid., 640ff) to which no temporal order of the world corresponds, "I measure the time, true; but I do not measure the future time because it is not yet; I do not measure the present because it has no extension; I do not measure the past because it is no longer. What then do I measure? Perhaps the passing time that is not yet past?" (ibid., 656–57). Augustine answers, "It is in you, my mind [my soul] that I measure my times. . . . Do not interrupt me with the tumult of your impressions. In you, I say, I measure the times. The impression which passing things make in you remain in you when these have passed. This I measure as the present, not those which have passed that it be made; this I measure when I measure times. Thus either these are the times or I do not measure the times at all" (ibid., 660–91; on Augustine's theory of time as contrasted with the Aristotelian theory, see Janich 1980, 259–71).

The soul measures time by measuring *impressions* (including impressions of past impressions). A "cosmological" paradigm of time such as Plato and Aristotle propounded is replaced by a "psychological" or "mental" paradigm that is independent of clocks and replaces the concept of physical time by the concept of experienced time. No longer does the world, nature, (in the form, say, of periodic planetary motions) "have" time; only the soul "has" time. Thus, the soul is also the measure of time. That means that time has a *subjective* structure, not some time or other—in the sense of: the (physical) world has its time and Man has his time or the soul has its time—but *the one and only time.* For Augustine the remembrances of humans, their acts of bringing to mind (*memoria, contuitus, expectatio*), constitute time itself. Both the answer to the question "Where is the world?" as well as the answer to the question "Where is time?" are, "In the soul." In questions of philosophy of nature Augustine remains true to his motto, "God and my soul I wish to know. Nothing more? Nothing more!" (1986, 18–19).

The dilemma seems perfect: on the one side, *physical* time, which as far as the essence of the laws of nature is concerned does not actually exist for modern physics or exists only in the boundary conditions of nature; on the other side, *mental* time, whose adherents say that there is no other time. Between these two notions of time—epistemologically speaking: between a philosophical physicalism and a philosophical psychologism—time itself seems to lose its own proper reality. Where we expect time to be, somewhere between the time in nature and the time in us, it is not to be found (if we are to believe modern physics and Augustine and his philosophical friends). Or perhaps it is. Philosophy does not give up that easily.

My thesis at this point is that temporal structures are taken from *action structures*. Neither nature nor the ego (the "soul") hold the key to time, but rather the way humans orient themselves in and through their actions. What I mean is the following: Actions occur not so much *in* time—this aspect is external to actions, for example, when we consider them relative to a time measured by a clock—rather they have their *own* time. That is, they have their own *duration* and their own *order.* Let us take three examples. (1) Children build a snowman. No child decides to make a snowman in ten minutes, and no child decides to begin the snowman with the carrot nose. The action itself, here making a snowman, determines its duration (until the snowman is finished) and its ordering according to earlier and later. (2) We play chess. Here, too, a context of action has its duration and its order (in this case determined by rules). That the game of chess as other games can also be played against the clock is rather an external arbitrary element. Games have their own time. Wherever they are adapted to the clock's time they lose a part of their essence. (3) The talk of lovers. Here the loss of time is almost proverbial. This talk neither serves a particular purpose, as in the examples of the snowman and chess, nor is it subject to any particular rules, as in the example of chess, nor is it even conscious of itself as action.

In all three examples the duration of an action can be expressed in (measured) time. However, this only grasps part of the actions that is *transferable* and thus is not a constitutive element of the action itself. In other words, every action, that is, every context of action, has so to speak two times, its *own* time and an *alien* time. Its own time is primary, the alien time secondary, What is constitutive of actions as such is the existence of an own time. If actions did not have their own times, we would not understand what time means. The "universal," that is, the *public* or, as Heidegger put it, the "world time" (1977, 406ff) has a derivative mode. Among the legitimating structures of this universal time is the need—itself related to action—to compare and coordinate actions *and their times* (for instance in the form of making appointments).

At this point the objection could be raised that there is not only the time of actions but also the time of nature, natural time (for instance, day and night, summer and winter, youth and old age). That is, what we *do* and what we *can* do has its time and what we *do not* do and what we *cannot* do has its time. Natural time determines (for instance, the order of days and nights) all actions. Is it not possible then, that a *natural* time is also the time *in* which our actions (with their times) take place?

This "natural" time—which, let it be noted, is not the time sought after or stipulated by a physics of time—can be expressed in terms of structures of action. To do this, we must realize that we are not first *actionless* in nature (in the world) and then join it with our actions or our actions with it. Nature is in the first instance nothing but the *natural* side of our actions. The experience that we make with our actions and with nature is *one* experience. Thus, in walking, the ground (as a piece of nature or, if changed by labor, a piece of the world) is part of the action, walking (in walking we do not distinguish between our feet and the ground). In speaking, the vocal chords (as a piece of nature) are part of the action, speaking (in speaking we do not distinguish between our vocal chords and what we say). In building a snowman, snow (as a piece of nature) is part of the action, building a snowman (in building a snowman we do not distinguish between the snow and our building actions). In all three cases nature (the world), as "substrate," "material," "resistance," and "space" (in which we, for instance, build), is part, and integrally part of actions. Actions themselves do not distinguish between what is action in them and what is nature in them. This holds for natural time as an integral part of actions as well, for example, the time between waking up and getting up (a difficult time); the time between two glasses of beer among friends (an easy time); the time between thought and action, wanting and giving up (a forgetful time).

Actions not only have their *own* time, we get to know *in them*—both in the form of action and experience—what is nature in us and in the world. Wanting to see the world without recourse to (our own) actions is something which we can perhaps continue with but nothing that we can begin. Therefore nature is also something that we originally *have* (in the form of action) and then lose; it is not something whose primary form of acquaintance is appropriation (as technological cultures attempt to persuade us). This means that what originally constituted a unity (nature as the natural side of actions) can "in retrospect" be analyzed conceptually: for example, into a time of action and a time of nature without identifying the former with the time of the "soul" and the latter with the time of physics. What is decisive is rather that we are dealing with a concept of time that does not suffer the previously mentioned dilemma between physicalism and psychologism and itself has a (temporarily) clock-free gestalt character. Such gestaltlike forms of time are the time to go home, the time to say good bye, the time of love and the time of happiness, and also the monsoon season, the age of Aquarius, and the strawberry season.

What I have tried to clarify about the concept of action can also be explicated on the concept of *life*. Just as it is not time that creates action but action that creates its time, so too it is not time that creates life, but life that creates its time. Moreover, all things have their time, though in different ways, for instance with regard to the past, "Every living thing whether animal or plant *is* its past—retroactively, that is, in a fashion *mediated* by its future. Here lies the difference to inanimate structures. A mineral, a mountain, an entire landscape: each is also its past, but each is immediately constructed from it, consists of it. The living on the other hand is more than just what it was" (Plessner [1928] 1981, 351). In the terminology of Plessner (and Heidegger), the living "is the being that is before it" (ibid.). The present of the living does not consist only in its past; it consists also in the contours of the future recognizable in its forms of life. This applies especially to human life, which in its actions, plans, wishes, and anticipations cannot be described or comprehended through the categories of past and present alone.

Human time is also reflected in the gestalten of life. These are, in Plessner's words, the "fateful forms" of life, because—like youth, maturity and old age—"they are essential to the process of development. Fateful forms are not the forms *of* what is but *for* what is; being submits to them and endures them" (ibid., 211–12). Youth and age, but also farewell and happiness, are not properties of (individual) life but rather *forms,* gestalten, *times* that life submits to. In the "nature" of life lie the determinants of the time of life; life is not a temporal process. Or in another formulation, *the time of life is its times.*

Once again, what I have just said has nothing to do with a physics of time that has lost sight of time, nor with an internalization of time that declares the soul to be the custodian of time. It is closer to Kant's notion that time is a (pure) form or shape that underlies all our intuitions, experiences and actions.[3] We find time, which is our time, human time, neither in a four-dimensional manifold nor in the soul.

A Historical Remark

When philosophy invented time as a philosophical problem, it was dealing with cosmological questions. Time, according to Plato, has a cosmic nature. It arose with the cosmos, namely, with its "uncorruptible" aeonian nature (*Tim.* 37c/d; on the concept of *aeon* see Fauth 1975). Aeon (αἰών), that is, life, fulfilled time reflects eternity, which is the essence of an ideal cosmos (the essence of its idea in Plato's sense); the "moved" time (χρόνος) reflects the aeon. Plato distinguishes between the time forms of an ideal cosmos, called the "eternal animal" (ζῷον ἀίδιον, *Tim.* 37d), a created cosmos, which also has an incorruptible structure, and the "moved" or measured time, which is the reflection of the astronomical nature of the cosmos and like this has a periodic structure (see *Tim.* 37d–38a; on the "astronomy" of time *Tim.* 35a–36d). The "moved" time (χρόνος) here means not the "essence" of the natural world (such as "everything is temporal" in the sense of "everything comes into being and passes away"), but the representation of the *aeonic* essence of the world, in which eternity "temporalizes" itself.

From the beginning the concept of time was connected to the concept of life: *Aeon is life.* Furthermore, the cosmos is the totality of different forms of time, among them the forms of natural time, action time and life time. This is also clear in Aristotle who does not otherwise think very much like Plato in cosmological affairs, "[The cosmos] continues through its entire aeon unalterable and unmodified, living the best and most self-sufficient of lives. As a matter of fact this word aeon possessed a divine significance for the ancients; for the fulfillment (τέλος) which includes the *time of life* of any creature outside of which no natural development can fall has been called its aeon" (*de cael.* A9.279a20–25). That means not only the cosmos has its time, its aeon, but also every living creature. But just this had been expressed before—without the "aeonic" terminology—in the assertion that time does not create life but life has its time (Aristotle: its aeon) (see Böhme 1974, 80–81; Böhme also gives detailed analyses of Plato's and Aristotle's conceptions of time). This time is at the same time a cosmological, "natural" time. Nature and time are still interwoven without our having to view this as particularly mythological.

In Plato's conception this is expressed by saying that it is the "parts" of the moved time that "imitate" the structure of the whole of cosmic aeon. As the "time of the cosmos" has days and nights, months and years, so too human time has, for example, the periods of childhood, youth, adulthood, and old age. The life as well as the time of humans consists of structured wholes, that is, (temporal) gestalten. It has no temporal structure in the sense of dissolution in a temporal continuum or an arrow of time on which we are supposedly riding; it has purely temporal gestalten. The whole is life (of the cosmos and of people in the mythological pictures of Plato's cosmology). Therefore, life consists not of time but of *times,* and these in turn "imitate" the temporal structure of the cosmos. Human time and human life represent cosmic time and cosmic life. Physics and anthropology (still) have much to say to one another.

That this was not an invention of Greek philosophy is clear from earlier Greek sources. Here, Chronos is a God but not a real subject of discussion. It is not Time but *Day* that is the focus of attention. It is the Day that brings happiness or doom. People are called ἐφήμεροι, those exposed to the changing fates of the day. The day is divided in turn into further temporal shapes: early morning, morning (the time of the market), noon, and evening. Temporal modes are mentioned in archaic simplicity: Kalchas "knew what is, what would be or was before" (Homer *Il.* I,70). Time is not a continuous flowing but a "returning" of times. Alongside the day and its times we find the seasons and ages of life (Böhme 1974, 125ff).

The connection between life and aeon, considered as the totality of life, also appears clearly in Homer, "Could I only rob you of the soul [ψμχή] and aeon" (*Od.* IX,523); "but as soon as the soul [ψμχή] and aeon leave him, then hand him over to death and the quieting sleep . . . " (*Il.* XVI,453). Lived and cosmic time join hands in the gestalt character of time. Life has temporal gestalten which follow upon or relieve one another like night and day, morning and noon, fall and winter. Thus, Achilles can remain forever young—his life "has" only the shape of youth. Therefore, every temporal gestalt harbors different experiences (according to Aristotle, for instance, mathematics for youth and metaphysics for old age). It is senseless to mark boundaries for the temporal gestalten of life; the invention of the clock only leads us astray on this head. An alternative would be a continuum model of time or the picture of time's arrow, but both of these belong to the world of physics. It is not *time* that flows but *things* that flow and change, albeit gestaltlike in time. Even Aristotle's theory of time does not allow such notions as are expressed in the phrase "time flies." For Aristotle, time is a form of motion (*Phys.* Δ11.219b23). However, in motion time does not move; only *duration* grows analogously to the space traversed in a motion (see Janich 1980, 255).

Thus the Platonic and the Aristotelian constructions of time correspond to the lived and (in mythical and everday form) represented time of their time. What might seem archaic here is rather the expression of lifeworld experience in dealing with human time, which has not yet made its way either into an abstract physical theory or into the soul. That we seem today to have other experiences lies in the need mentioned earlier to coordinate concrete times. Modern constructions of time are doubtless theoretically superior to the notions of classical antiquity; but in lifeworld terms from the point of view of human time they are, in spite of their practical character, perhaps inferior. In this regard we are only misled by the fact that analogue watches that simulate the motion of the sun have become old fashioned and digital watches which represent nothing but only "count" are on the march. At the same time the insight of the Greeks, that life is not time, processual time, but rather has times, gestaltlike times, is fading. Modernity, even when it on occasion bets on the wrong horse or on the wrong clock, is contagious.

NOTES

The distinction between the time of nature and the time of Man was the subject of an earlier essay "Die menschliche Zeit. Bemerkungen zur Philosophie der Lebensalter" (1989). In the following I draw on this paper and expand it with a discussion of Adolf Grünbaum's fundamental analyses of physical time. My thanks go to Martin Carrier (University of Konstanz) for good advice and expert assistance in the sections on physics and to Peter McLaughlin (University of Konstanz) for help with the English text.

1. I use the term *gestalt* for the shape things take. In doing so I do not intend any of the holistic connotations often associated with the term in psychology.

2. This applies equally to the metrogenic irreversibility that shows up in quantum mechanics. The point is that the relevant process here, that is, the so-called "reduction of the wave function," is not governed by any known law, let alone a temporally asymmetric one.

3. In Kant's terminology time is not a discursive concept but rather a "pure form of sensible intuition" (1956–1964, vol. 2, 78). That is, time is not an element of the physical world like heaven and earth; rather it has to do with our ways of experiencing and representing. Empirical time, that is, that which is the element of our experience, is formed *in* time which is not a part, an essential part, but the *form* of this experience. Heidegger (1977, 323ff, 355ff) later formulated this, using the concept of the temporality of existence.

REFERENCES

Augustinus, A. 1955. *Confessiones*. Edited by J. Bernhart. Munich: Koesel.

———. 1986. *Soliloquia/Selbstgespräche von der Unsterblichkeit der Seele*. Edited by H. Müller. Munich: Artemis.

Böhme, G. 1974. *Zeit und Zahl: Studien zur Zeittheorie bei Platon, Aristoteles, Leibniz und Kant*. Frankfurt: Klostermann.

Carrier, M., and Mittelstrass, J. 1991. *Mind, Brain, Behavior: The Mind-Body Problem and the Philosophy of Psychology*. New York: de Gruyter.

Ende, M. 1973. *Momo oder Die seltsame Geschichte von den Zeit-Dieben und dem Kind, das den Menschen die gestohlene Zeit zurückbrachte*. Stuttgart: Thienemanns.

Fauth, W. 1975. *Aion*. In K. Ziegler and W. Sontheimer, eds., *Der Kleine Pauly. Lexikon der Antike*. Vol. 1. Munich: Artemis, pp. 185–88.

Grünbaum, A. [1963] 1973. *Boston Studies in the Philosophy of Science*. Vol. 12, *Philosophical Problems of Space and Time*. 2d ed. Dordrecht: Reidel.

———. [1971] 1976. "The Exclusion of Becoming from the Physical World." In M. Čapek, ed., *Boston Studies in the Philosophy of Science*. Vol. 22, *The Concepts of Space and Time: Their Structure and Their Development*. Dordrecht: Reidel, pp. 471–500.

———. 1983. "Retrospective Versus Prospective Testing of Aetiological Hypotheses in Freudian Theory." In J. Earman, ed., *Minnesota Studies in the Philosophy of Science.* Vol. 10, *Testing Scientific Theories.* Minneapolis: University of Minnesota Press, pp. 315–47.

Heidegger, M. 1977. *Sein und Zeit.* 14th ed. Tübingen: Niemeyer.

Janich, P. 1980. *Die Protophysik der Zeit: Konstruktive Begründung und Geschichte der Zeitmessung.* Frankfurt: Suhrkamp.

Kant, I. 1956–1964. *Werke in sechs Bänden.* Edited by W. Weischedel. Darmstadt: Wissenschaftliche Buchgesellschaft.

Meyerson, E. [1930] 1976. "The Elimination of Time in Classical Science." In M. Čapek, ed., *Boston Studies in the Philosophy of Science.* Vol. 22, *The Concepts of Space and Time: Their Structure and Their Development.* Dordrecht: Reidel, pp. 255–64.

Mittelstrass, J. 1989. "Die menschliche Zeit. Bemerkungen zur Philosophie der Lebensalter." In *Akademie der Wissenschaften zu Berlin. Jahrbuch 1988.* Berlin: de Gruyter, pp. 306–35.

Plessner, H. [1928] 1981. *Die Stufen des Organischen und der Mensch: Einleitung in die philosophische Anthropologie.* Gesammelte Schriften, vol. 4. Frankfurt: Suhrkamp.

Prigogine, I. 1980. *From Being to Becoming: Time and Complexity in the Physical Sciences.* San Francisco: Freeman.

Rae, A. 1986. *Quantum Physics: Illusion or Reality?* Cambridge, England: Cambridge University Press.

5

The Conventionality of Simultaneity

Michael Redhead
Department of History and Philosophy of Science, University of Cambridge

There is an honorable tradition in the philosophy of time that sees simultaneity as an equivalence relation defined on the set of all actual or indeed possible events,[1] and time itself as the linearly ordered quotient set induced by this equivalence relation. The equivalence classes of simultaneous events provide successive boundaries between a "fixed" past where present action can produce no change and an "open" future, a limbo of unrealized possibilities; the "moving" present breathes momentary reality on events before they are consigned to the other limbo of "dead" facts. Such is the rhetoric of *A*-theorists as opposed to *B*-theorists, who take the first part of this analysis with its two-place predicate of temporal precedence ordering the simultaneity classes but reject the ontological role of the moving present. Classical physics, with its notion of absolute time, provided uncritical support for the first part of the analysis, and while not actually requiring the additional *A*-theory role for the "present" simultaneity class, was certainly not inconsistent with it. The object of this essay is to consider the import of the special theory of relativity (SR) as putatively undermining these classical notions, which it must be admitted reflect the untutored intuition of our legendary friend, the man on the Clapham omnibus!

Three versions of how we should understand the notion of simultaneity in SR are as follows:

1. SR gives up the notion of simultaneity as an equivalence relation, more particularly rejecting transitivity, and replaces the classical conception by what Grünbaum somewhat confusingly calls "topological simultaneity," two events being topologically simultaneous if each lies outside the light cone constructed at the other.[2]
2. SR replaces the notion of simultaneity as a two-place relation between events with a three-place relation signifying simultaneity with respect to an inertial reference frame.
3. The classical notion still survives, since only one of the relativized notions introduced in (2) is "real" simultaneity.

We will now discuss these three approaches in more detail, and address the various arguments that have been produced in favor of each. As we will see, the first option has led to the notion of the conventionality of what Grünbaum refers to as metrical simultaneity.

Option 1: Topological Simultaneity

The classical arguments in the literature for this position are due to Reichenbach (1957) and Grünbaum (1973). (The arguments of course really go back to Einstein 1905 who declares the definitional status of his own choice of synchrony.)

Reichenbach's argument is essentially as follows. How could we tell that two spatially separated events were simultaneous? We have two identically constructed clocks situated at A and B. How can we check that when the A-clock reads a time T_A, then at the selfsame time the B-clock reads T_B which is numerically the same as T_A? We must in some way communicate between A and B so as to establish the required synchrony. Suppose we send a signal from A to B with a velocity v. It leaves A at spacetime point a, and reaches location B at spacetime point b, as shown in figure 5.1. If d is the spatial separation of A and B, then clearly we must set our clock at B so that the time reading t_b at b is numerically equal to the time reading $t_{a'}$ for the A-clock, where $t_{a'}$ is later than t_a, the time at which the signal is sent, by an amount d/v. But, says Reichenbach, this method will only work if we can measure v. That already requires us to measure lapse of time across a spatial interval and hence presupposes the very notion of synchrony we are trying to establish. The situation, however, is not totally hopeless. Let the signal return from b to a''. Then t_b must lie between t_a and $t_{a''}$, otherwise the signal would not be propagating

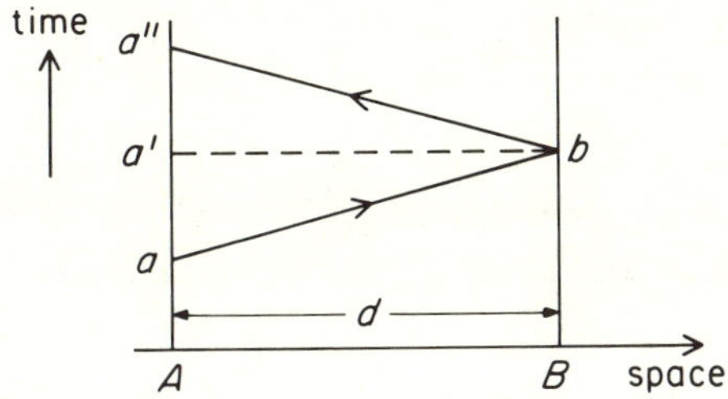

Figure 5.1

forwards in time. The interval $(t_a, t_{a''})$ can be measured just by the A-clock, so identifying this interval does not presuppose any knowledge of synchrony between locations A and B. In particular $2d/t_{a''} - t_a$ measures a unique two-way or average speed $\bar{v}$ of the signal in the round-trip journey. Now suppose we could increase the two-way speed $\bar{v}$ without limit. Then we could squeeze the ambiguity $t_{a''} - t_a$ in assigning a time at A in synchrony with t_b at B without limit, so establishing, in the limit, a unique synchrony between A and B. This is the situation in classical physics. Suppose that even in the classical world there was a de facto "first signal", that is, an upper bound to $\bar{v}$, then even in the classical world, we would be left with an unbridgeable ambiguity $(t_a, t_{a''})$ if this now corresponded to the emission and reception of the "first" or fastest signal. Reichenbach now makes a verificationist move. If we cannot tell which time instant in the interval $(t_a, t_{a''})$ is simultaneous with t_b, then there is no fact-of-the-matter at stake, and we are left with a conventional choice of which instant in the interval $(t_a, t_{a''})$ to identify as simultaneous with t_b. If we write

$$t_b = t_a + \epsilon(t_{a''} - t_a)$$
$$\text{where } 0 < \epsilon < 1 \qquad (5.1)$$

then this conventional ambiguity translates into a choice for ϵ. In turn this dictates a conventional choice for the one-way speeds of the first signal.[3]

$$\overrightarrow{v} = \frac{d}{t_b - t_a} = \frac{\bar{v}}{2\epsilon}$$
$$\overleftarrow{v} = \frac{d}{t_a'' - t_b} = \frac{\bar{v}}{2(1 - \epsilon)}. \qquad (5.2)$$

So $\overrightarrow{v} = \overleftarrow{v}$ iff $\epsilon = ½$.

This makes the choice $\epsilon = ½$ in some sense the simplest, but according to Reichenbach there is nothing "factual" about such a choice.

Suppose, to make contact with SR, the maximum value of $\bar{v}$ is the speed of light in the vacuum, then the ambiguity aa'' sections the exterior of the light cone constructed at b. The event b is topologically simultaneous with the whole interval (a,a'') instead of a purely conventional metrical simultaneity of the event b with some arbitrarily chosen member of the set (a,a'').

We have talked of a notion of topological simultaneity even in a classical world. But in such a world the two-way speed of the first signal depends in general on the inertial frame relative to which it is measured. This of course is the conceptual origin of the Michelson-Morley experiment. If the two-way speed of the first signal was independent of inertial frame, then in such a world the topological latitude between two locations at rest in a given inertial frame would depend only on the distance d between the two locations, as measured in that frame, and in this sense we would have a frame-independent generalization of the classical conception of simultaneity.

The situation in SR, it is claimed, is that there is both a first signal, and an invariant two-way speed, so the conditions are now met, in a principled way, for introducing the generalized notion of topological simultaneity in the manner described.

Grünbaum accepts the conclusion of the Reichenbach argument about the conventionality of simultaneity in SR, but his reasons are based on ontological rather than epistemological considerations. Adopting a relational theory of time, Grünbaum considers that the ordinal properties of the temporal continuum must be based on, and indeed reduced to, relations of causal precedence between events. Grünbaum assumes that a *symmetric* relation of causal connectability between events can be established in terms of an antecedently understood notion of a continuous genidentical set of events which comprise the spacetime locus of a material particle or (quasi-genidentical) of a ray of light. Temporal betweenness for events at different spatial locations is dictated essentially by the following definitions (we follow here the reconstruction of Grünbaum's ideas provided in Redhead 1983). Let e_1 and e_2 be events at location A and

e_3 an event at location B. Then e_3 is temporally between e_1 and e_2 (written as $e_3Be_1e_2$) if and only if the following conditions hold:

1. e_1,e_2,e_3 are members of a possible genidentical set G.
2. If e_3 is deleted from G, no continuous genidentical subset of G exists that contains both e_1 and e_2.

Formally, using G as a variable ranging over all possible complete (and hence continuous) genidentical sets and S as a variable ranging over all possible continuous (but not necessarily complete) genidentical sets, so $\forall S \exists G(S \subseteq G)$, we have

$$e_3Be_1e_2 \text{ iff } \exists G[G(e_1,e_2,e_3) \wedge \forall S(S \subseteq G - \{e_3\} \rightarrow \sim S(e_1,e_2)]$$

where $(e_1 \in G) \wedge (e_2 \in G) \wedge (e_3 \in G)$ is written $G(e_1,e_2,e_3)$ and similarly for $S(e_1,e_2)$. We notice that the relation B may fail to obtain for two reasons. If condition (1) holds but not condition (2) in the above definition then we may say that e_3 is absolutely outside the temporal interval (e_1,e_2). But if condition (1) does not hold then the relation B fails to obtain in a sense that we may want to express by saying that a betweenness relation is not really now applicable because e_3 is in an absolute sense simultaneous with either or both of the events e_1 and e_2. To capture this idea of absolute or topological simultaneity between an event e_3 at location B and any event, call it e_4, at location A, Grünbaum makes effectively the following definition:

$$e_3S_Ge_4 \text{ iff } \forall G(\sim G(e_3,e_4)).$$

We use the notation S_G for this Grünbaum definition of simultaneity. The relations B and S_G now mesh in the following way:

The Meshing Condition:

$$\sim \exists G(G(e_1,e_2,e_3)) \text{ iff } e_1S_Ge_3 \text{ or } e_2S_Ge_3.$$

The Meshing Condition involves two claims:

$$\exists G(G(e_1,e_2,e_3)) \rightarrow \exists G(G(e_1,e_3)) \wedge \exists G(G(e_2,e_3))$$

and

$$\exists\, G(G(e_1,e_3)) \wedge \exists\, G(G(e_2,e_3)) \rightarrow \exists\, G(G(e_1,e_2,e_3)).$$

The first part, being a logical truth, is not controversial. As we will see, the second part of the Meshing Condition breaks down in the presence of so-called tachyons.

Employing our notation, we can formulate the First Signal Principle. Using now e_1,e_2, and e_4 as variables ranging over events at location A, we assert

$$\exists\, e_1 \exists\, e_2[e_3Be_1,e_2 \wedge \forall\, e_4(e_4Be_1e_2 \rightarrow e_3S_Ge_4)].$$

Let $\bar{e}_1,\bar{e}_2$ denote the events at location A corresponding to the sending and return of the "First Signal" which arrives at B as event e_3. Then, according to Grünbaum, e_3 is topologically simultaneous with any event e_4 at location A which lies temporally between $\bar{e}_1$ and $\bar{e}_2$. Any particular choice from this range is, for Grünbaum, a matter of convention: the conventionality of metrical simultaneity.

Option 2: The Relativity of Simultaneity

If we adopt the standard Einstein synchrony convention of $\epsilon = \frac{1}{2}$ in two relatively moving inertial frames, then, clearly, simultaneity becomes relativized to an inertial frame. This has nothing to do with relativity theory, but would arise in a Newtonian world if we were sufficiently perverse to use the Einstein convention in such a world. The exact significance of calling such a use "perverse" will be explained. To keep things general, let us write down some general transformations, which will include the Einsteinian world and the Newtonian world as special cases. Consider an inertial reference frame K' moving with velocity v along the positive X-axis of another inertial frame K. Confining ourselves for the moment to the simple case of two-dimensional spacetime, the most general linear relationship[4] between the coordinates (x',t') of an event relative to K' and the coordinates (x,t) of the same event relative to K would be of the form

$$\begin{aligned} x' &= Ax + Bt \\ t' &= Ct + Dx. \end{aligned} \tag{5.3}$$

For $x' = 0, x = vt$, so $B = -Av$ and defining $m = -D/C$, we obtain

$$x' = A(x - vt)$$
$$t' = C(t - mx). \tag{5.4}$$

Now we introduce some substantive physical assumptions about the behavior of moving rods and clocks; namely, a rod stationary in K' is contracted relative to K by a factor F, while a moving clock is similarly dilated in time interval by a factor G, then it follows immediately that

$$A = \frac{1}{F} \text{ and } C = \frac{1}{G} \cdot (1 - mv)^{-1}.$$

So we now have

$$x' = \frac{1}{F} \cdot (x - vt)$$
$$t' = \frac{1}{G}(1 - mv)^{-1} \cdot (t - mx) \tag{5.5}$$

In a Newtonian world $F = G = 1$ while in an Einsteinian world $F = 1/G = \sqrt{1 - v^2/c^2}$. The Galilean transformations correspond to taking $m = 0$ in a Newtonian world. But suppose we choose m so as to satisfy the Einstein synchrony convention that light, assumed to propagate isotropically with velocity c, continues to propagate isotropically (although not with the same magnitude as c) relative to K'. It is easy to show that this requires $m = v/c^2$, leading to the transformation equations

$$x' = x - vt$$
$$t' = \left(1 - \frac{v^2}{c^2}\right)^{-1}\left(t - \frac{vx}{c^2}\right). \tag{5.6}$$

This particular result, using the Einstein convention in a Newtonian world, was first explicitly presented by Zahar (1977). The parameter m, which we may call the simultaneity parameter, determines the locus in K of events simultaneous with the origin in K' as the line

$t = mx$. Notice that the X'-axis and T'-axis in K' are Minkowski orthogonal even in this Newtonian world. The condition for the Minkowski orthogonality is simply $m_t m_s = 1/c^2$, where m_t is the slope of a timelike and m_s the slope of a spacelike line as viewed from K.[5] The T'-axis has slope $1/v$ in K while the X'-axis has slope v/c^2, so the condition is satisfied. Of course the geometry of Newtonian spacetime is in no sense Minkowskian since the Minkowski metric is not generally invariant under the transformation (5.6). The important point to notice is that Einstein synchrony in a moving frame is specified by Minkowski orthogonality to the world-line of the origin of the moving frame, whatever sort of world we live in, as specified by the "physical" functions F and G. We will return to this point, but first we want to explain why we cannot allow Einstein synchrony as a *principled* definition in a classical world. In such a world there is *in fact* no upper limit on signal speeds relative to K and hence, as we explained, synchrony in K is uniquely specified in the topological sense. Now defenders of factual metrical simultaneity must in general allow that K$'$ cannot declare metrically simultaneous events which are not topologically simultaneous relative to K. Hence, in a classical world K' must choose $m = 0$; the choice $m = v/c^2$ contravenes the above constraint on the relationship between topological simultaneity relative to K and metrical simultaneity relative to K'. In the Einstein world, with a finite upper limit in signal speed provided by the propagation of light in the vacuum, this constraint does not rule out the Einstein convention for any velocity v up to the maximum c. In other words, in SR the topological latitude specified by the boundaries of the future and past light cone with slopes $\pm 1/c$ exactly permit the full range of Einstein synchrony in moving frames defined by slopes v/c^2 with $|v| < c$. To put the point a little differently, nonstandard synchrony in K matches up in a one-to-one relationship with standard synchrony in an appropriately moving frame—we just have to choose the velocity of the frame so that $v/c^2 = m$, where m is the simultaneity parameter lying in the interal $(-1/c, +1/c)$, specifying the nonstandard synchrony in K.

So the proponents of option (2) cannot be ruled out of court in the Einsteinian world in the way we used for the Newtonian world.

So much is agreed between the proponents of option (1) and option (2). But option (2) now deploys a number of arguments to show that the Einstein convention in all inertial frames is not just a *permissible* conventional choice, but is actually the fact-of-the-matter.

I. First, arguments exist that purport to show that nonstandard synchronies are unsuitable candidates for defining simultaneity because they do not specify an equivalence relation on events. If we choose a fixed Reichenbach ϵ-parameter to specify nonstandard synchrony, we can easily show that simultaneity so defined is neither symmetric nor transitive. To see how this comes about, we first note the relation between the parameters ϵ and m for two-dimensional spacetime. We can easily check that, in the frame K, for $x > 0$, $\epsilon = \frac{1}{2}(1 - mc)$ while for $x < 0$, $\epsilon = \frac{1}{2}(1 + mc)$. So if ϵ has a fixed value, we must replace m by $-m$ for the negative part of the line of simultaneity. So, for fixed ϵ, the line of simultaneity of fixed slope m becomes replaced by a pair of lines of slopes $\pm m$. Only for the choice $m = 0$, which defines standard synchrony in K, does a fixed ϵ, namely ½, correspond to a single line. These results are readily generalized to the realistic four-dimensional spacetime. For fixed ϵ, the surface of simultaneity is a hypercone, which only degenerates into a plane when $\epsilon = \frac{1}{2}$. Only when the surface of simultaniety is a hyperplane can we recover symmetry and transitivity and hence an equivalence relation. But the remedy is simple; choose ϵ to be an appropriate function of direction so as to recover a hyperplane. The details are just a generalization of the results already given. (See Torretti 1983, 225–26, or Friedman 1983, 171–72. Note that directional dependence of ϵ was mooted already by Reichenbach 1957, 162.) Then nonstandard synchrony just corresponds to hyperplanes of simultaneity that are not Minkowski arthogonal to the time axis of the inertial frame. At one time the literature covered a great deal of discussion establishing the nontransitivity of nonstandard synchrony (compare, for example, Grünbaum 1969, Quinn 1974, and Roxburgh 1975). But this we submit has no irremediable relevance to the discussion of the viability of nonstandard synchrony.

II. An important consideration is the fact that slow-clock-transport serves to pick out standard synchrony in an Einsteinian world. Of course, in general, in SR if two standard clocks are synchronized at the same location and then moved along different spacetime trajectories until they are again in spatial coincidence, they will no longer be in synchrony, that is the origin of the twin paradox, which we will discuss in more detail. But Bridgman (1962, 64–67) and Ellis and Bowman (1967) have stressed that if a clock was moved "infinitely slowly" from location A to location B, fixed locations relative to a frame K, then the clock, on arrival at B, would agree in reading with standard synchrony defined relative to *K* (see also Eddington 1923, 15). We must first determine what we mean by "infinitely slowly." If the clock did not move at all it would never reach *B*! So we must consider a limiting procedure for a sequence of clocks moving more and more slowly from *A* to *B*. But how do we define the sequence of velocities, the specification of which clearly presupposes a synchrony convention? The most satisfactory procedure would seem to be to specify the velocities using an arbitrary ϵ-parameter and then show that if we take the double limit, first letting the velocities tend to zero, and then letting $\epsilon \rightarrow \frac{1}{2}$ the discrepancy between the time at *B* recorded by the moving clock and the time at *B* specified by ϵ-synchrony itself tends to zero (compare Friedman 1977).

One can make two responses to this admitted mathematical fact about slow-clock-transport in SR. First, one can follow Grünbaum (see Grünbaum 1969), who identifies a lingering convention in the assumption that the infinitely slow clock measures the same durational time interval between leaving *A* and arriving at *B* as does the clock remaining at *A* and in standard synchrony with clocks at *B*. Grünbaum's qualms on this score arise from his general arguments for the metrical amorphousness of a linear continuum. We do not, however, have to resort to Grünbaum's potentially controversial thesis about metrical amorphousness to resist the slow-clock-transport argument for the facticity of standard synchrony. (For discussion of Grünbaum's thesis, see in particular Torretti 1983 and references detailed therin; see also Friedman 1983 and Grünbaum's 1970 own reply to his critics.) We could instead, argue as follows. There is no absolute factual sense in the term "slow." If we estimate "slow" relative to a moving frame K', then slow-clock-transport will pick out standard synchrony in K', but this, as we have already noted, corresponds to nonstandard synchrony in *K*.

In brief, *if you are already committed to option (2),* so that simultaneity is relativized to a frame, then slow-clock-transport can arguably pick out the Einstein convention as the correct way to implement the relativization. The defenders of option (1) will see nothing in this argument to show that simultaneity *should* be factually relativized to an inertial frame, and will conclude that the choice between *K*'s slow-clock-transport and K''s slow-clock-transport is indeed a conventional one. They will claim that the conventional latitude permitted by the totality of slow-clock-transport criteria with respect to all *pos-*

sible frames exactly corresponds to the latitude of topological simultaneity as specified on causal criteria by Reichenbach and Grünbaum. So anybody resisting the causal reductionism of these authors' arguments can *still* defend the conventionality of simultaneity using the very metrical apparatus of the option (2) proponents against themselves!

III. The proponents of the facticity of standard synchrony have made great play with the fact that the standard synchrony, unlike nonstandard synchrony, is causally definable (see Robb 1914, 1921, 1936; Mehlberg 1980; Zeeman 1964; Winnie 1977; Malament 1977; and Janis 1983). The basic idea is very simple. We have already noted that standard synchrony in a frame K' corresponds to Minkowski orthogonality between the line of simultaneity, that is, the X'-axis, and the world-line of the origin of the frame, that is, the T'-axis.

Minkowski orthogonality, however, is easily shown to be definable in causal terms by the following construction (see figure 5.2), due originally to Robb. We first note a simple lemma of plane *Euclidian* coordinate geometry.

LEMMA .1. *If ABCD is a parallelogram whose adjacent sides AD,AB have slopes* $(+m)$ *and* $(-m)$, *then, if the diagonals AC and BD have slopes* m_1 *and* m_2 *respectively, the quantities* m, m_1 *and* m_2 *satisfy the relation* $m_1 m_2 = m^2$.

Now apply this result to two-dimensional spacetime (i.e., replace the Y-axis with a T-axis), and let AB,AD,BC and DC be the spacetime loci of rays of light, so $m = 1/c$, then the construction shows that for the timelike line AC of slope m_1 and spacelike line BD of slope m_2, we have $m_1 m_2 = 1/c^2$, but this is just the necessary and sufficient

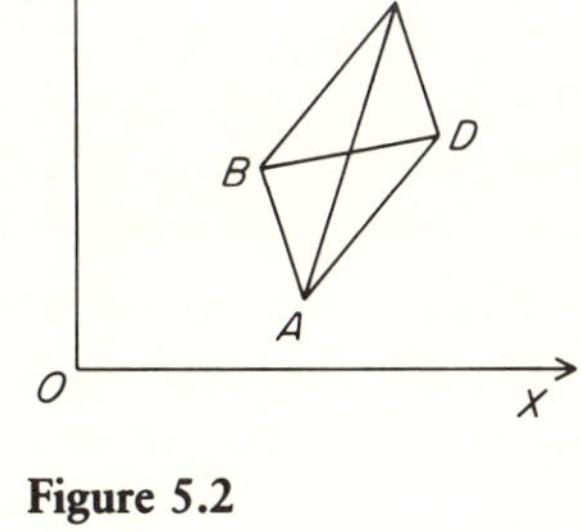

Figure 5.2

condition for Minkowski orthogonality. Although we have demonstrated the construction for two-dimensional spacetime, it generalizes immediately to the four-dimensional case.

Malament (1977) has proven the remarkable result that standard synchrony is the only nontrivial equivalence relation even implicitly definable from the relation of causal connectibility and the world-line of the origin of an inertial frame of reference. We can use two arguments against the significance of Malament's result. First, we can dispute that simultaneity need be an equivalence relation, which leads back in the direction of option (1), and the clearly causally definable notion of topological simultaneity. Second, we can use the same argument as we did for slow-clock-transport to demonstrate that we are faced with a conventional choice between standard synchronies defined à la Malament in all possible inertial frames, and this latitude, as we have seen, leads us back to just the same latitude as countenanced by topological simultaneity. This line of argument has been particularly urged by Janis (1983) in his discussion of the significance of Malament's result.

Option 3: The Neo-Lorentzians

A steady stream of literature over the years has been promoting a neo-Lorentzian understanding of SR (see, for example, Prokhovnik 1967). There is a privileged "aether" frame, with respect to which the propagation of light is "really" isotropic. The apparent isotropic propagation of light, and the invariance of the two-way speed of light in moving frames, is due to the fact that moving rods and clocks are affected according to the Einstein-Lorentz contraction and dilatational factors. So nature conspires to hide from us the empirical detection of the aether frame. But the neo-Lorentzians argue that only a crude positivism would lead us to believe that because we cannot detect the aether frame, that such a frame does not exist. They point to the role of the aether frame hypothesis in actually *explaining* the contraction and dilatation effects, which Einstein, so they claim, just assumes rather than explains. Accepting the neo-Lorentzian approach depends on one's understanding of what constitutes a scientific explanation. The neo-Lorentzians seem to be arguing on the basis of a crude causal approach to explanation, whereas modern phys ics has followed the unification approach that a good explanation is one that subsumes many disparate empirical regularities under the deductive closure of a unified set of theoretical principles (see Redhead 1990). On this account the Einstein approach is clearly superior to the Lorentzian; indeed Einstein's own contribution to scientific methodology largely consists in arguing from formal general principles rather than detailed causal mechanisms.

Another line of argument has been used to justify the existence of a privileged aether frame based, paradoxically enough, on *relativistic* ideas about the possible existence of tachyons, particles that travel faster than light. If tachyons exist then Grünbaum (1973, 827) himself has argued that they could be used to pick out a unique synchrony relative to an effective "aether" frame. This argument has been examined critically in Redhead (1983). Since this whole question is of considerable conceptual importance in understanding the significance of SR, let us summarize the discussion here. The thought that SR actually licences the possibility of "superluminal" particles arises from the view that SR is committed only to an invariance principle (compare Nerlich 1982), namely, the invariance of the four-dimensional Minkowski interval, which implies in particular the invariance of the light sphere in velocity space.[6] Lorentz transformations allow any point inside the light sphere to be transformed to any other interior point, the corresponding motions of the interior points corresponding to Lobachevsky's hyperbolic geometry (first pointed out by Varicák 1910; see also Brouwer 1975, 114). The region exterior to the light sphere in velocity space is also invariant under Lorentz transformations. The significance of these mathematical facts is that the light sphere is an impenetrable barrier; a particle with speed less than that of light can never be Lorentz-boosted across the light sphere. This is the sense in which SR incorporates a First Signal Principle; no material particle, or more generally localized energy packet, can be made to move faster than light. But suppose a particle, a so-called tachyon, moved *always* faster than light; Lorentz transformations could never slow it down so that it moved to the interior of the light sphere, which again functions as an impenetrable barrier, but now from fast to slow rather than slow to fast. Clearly, if tachyons move at *any* speed greater than that of light, then an appropriate Lorentz boost can generate any velocity however great; that is, the speed of a tachyon has no upper limit. Grünbaum argues that if there were tachyons, and hence infinite velocity tachyons (speaking loosely), then we would be in the same situation as in the Newtonian world and these "infinite velocity" tachyons would pick out a unique synchrony, by the usual causal argument, and the conventional latitude of topological simultaneity would be eliminated.

Let us look at the case of "acoustic relativity." Consider a sound wave which propagates with velocity $\pm w$ along the positive and negative X-axis relative to a frame K, where $w > c$. To fix our ideas, suppose the equation for the pressure ϕ of the sound wave relative to K (assuming for simplicity two-dimensional spacetime) is of the form

$$\frac{1}{w^2}\frac{\partial^2\phi}{\partial t^2} - \frac{\partial^2\phi}{\partial x^2} = 0. \tag{5.7}$$

This superluminal sound could be used to identify a latitude in topological simultaneity at a distance x of $\pm x/w$ about the event distinguished as metrically simultaneous at this location by invoking the acoustic synchronization convention of isotropic sound propagation in both directions along the X-axis. Suppose we introduce the Einstein "optical" synchronization in such a universe in a moving frame K'. The latter will now contravene the ontologically specified acoustic latitude when $vx/c^2 > x/w$, that is, $v > (c/w)c$. As $w \to \infty$, the Einstein convention will be ontologically disallowed for any $v > 0$; that is, the moving observer is forced to use the synchronization which is now specified uniquely in K. In such a situation the transformation equations between K and K' are easily shown to reduce to the Tangherlini (1961) form:

$$x' = \frac{1}{\sqrt{1 - v^2/c^2}}(x - vt)$$
$$t' = \sqrt{1 - v^2/c^2} \cdot t. \tag{5.8}$$

Grünbaum (1973), together with Sjödin (1979) and Salmon (1975), apparently assume that this is the analysis that applies in the presence of tachyons. But the situation is not that simple. Tachyons arise from consideration of a Lorentz-invariant equation, typically the Klein-Gordon equation with an imaginary mass-parameter, namely,

$$\frac{1}{c^2}\frac{\partial^2\phi}{\partial t^2} - \frac{\partial^2\phi}{\partial x^2} - \mu^2\phi = 0. \tag{5.9}$$

The solutions of this equation have properties different from the non-invariant equation (5.7). Whan subjected to quantization, equation (5.9) describes techyons moving with any velocity between c and ∞. Equation (5.7) by constrast describes the propagation of sound waves with a unique velocity $w > c$. Equation (5.9) also accommodates tachyons propagating backwards in time, with velocities between c and ∞, but these turn out to have negative energies and a Reinterpretation Principle is standardly invoked in which tachyons formally described as propagating backwards in time are reinterpreted as antitachyons travelling forwards in time with positive energies. The notion of emitting or absorbing a tachyon is not an absolute one, with the emission of a tachyon in one frame being identified with the absorption of an antitachyon in other frames, and vice versa. So the possibility of generating a causal loop paradox via a chain of emission and absorption events becomes very problematic. However, this does not rule out the possibility that tachyonic interaction might sustain a *symmetric* relation of connectibility between events at spacelike separation, although an asymmetric cause-effect relation could not be established. Since Grünbaum grounds the notion of temporal betweenness in terms of causal connectibility, we will discuss how this analysis fares in the hypothesized presence of tachyons. First, however, we want to explain the problem involved in using tachyons to establish "absolute" synchrony.

Suppose an observer stationary in the frame K attempts to establish absolute synchrony using infinite velocity tachyons (usually referred to in the literature as *transcendent* tachyons). We suppose that these transcendent tachyons are identified as transcendent relative to the adoption of the Einstein convention in the K frame. In figure 5.3 we illustrate to-and-fro signaling between A and B using "nearly transcendent" tachyons, where the event e_4 at A, which is metrically simultaneous with e_3 at B is "squeezed" between the emission (e_1) and return (e_2) of the tachyon signal. Now consider the description of

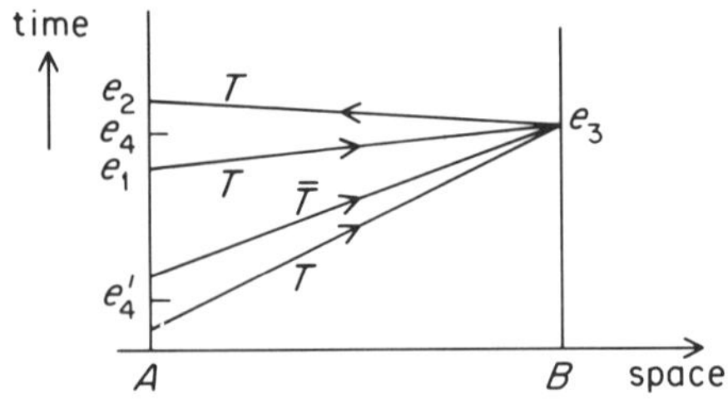

Figure 5.3 Tachyon synchronization as viewed from K (T = tachyon, $\bar{T}$ = antitachyon)

what is going on from the point of view of observers attached to the moving K', who also adopt the Einstein convention (see figure 5.4). They will deny that the physical processes being described relative to K establish simultaneity between e_3 and e_4 at all. What is happening as seen from K' is that a tachyon-antitachyon pair is emitted at e_3, the tachyon arriving at e_2 and the antitachyon at e_1. The reason, from the point of view of K', no simultaneity relation is being established between e_3 and e_4 is that the chain of events $e_1e_3e_2$ is not a genidentical set since two particles, not one, are clearly involved. Of course relative to K', there is another physical process, which observers attached to K' describe as identifying e_3 as simultaneous with an event e_4', squeezed between tachyons travelling from A to B and back again with nearly transcendent velocities as viewed from K'. But observers attached to K will resist the claim that e_3 and e_4' are truly simultaneous by pointing out that what they see is a tachyon-antitachyon pair annihilating as e_3 as illustrated in figure 5.3. Note the symmetry in the situation. Observers in K cannot *disallow* the description of

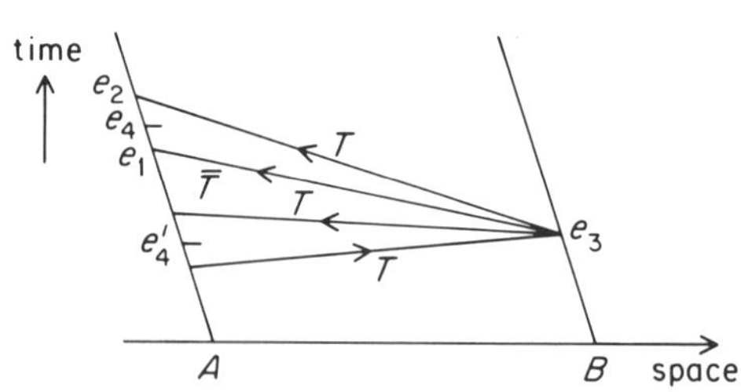

Figure 5.4 Tachyon synchronization as viewed from K'

their simultaneity claim by observers in K' since the former are going to invoke the Reinterpretation Principle to circumvent the K' observers' claim that e_3 and e_4' are simultaneous! In brief both sets of observers must allow that what counts as a genidentical path e_1 to e_3 and back to e_2 is a notion that depends on the observer and has no absolute significance. Only with either subluminal or photon signaling is an unambiguous ascription of genidentity for the to-and-fro signal available.

The difficulty with the Grünbaum notion of simultaneity, the relation we characterized as S_G, is now apparent. The betweenness relation B and the simultaneity relation S_G no longer satisfy the Meshing Condition. More specifically, the second part of the Meshing Condition fails when tachyon interactions are taken into account. Redhead (1983) argues that the satisfaction of the Meshing Condition is a necessary requirement in capturing the notion of simultaneity. A suggested revised definition of simultaneity, S_R, is specified as follows:

Let e_5 be a variable ranging over events at location A.

$$e_3 S_R e_4 \text{ iff } \forall\, e_5\, \forall\, G\, (\sim G(e_3,e_4,e_5)).$$

Clearly $e_3 S_G e_4 \rightarrow e_3 S_R e_4$ but not vice versa. The revised definition S_R will now satisfy the Meshing Condition provided we allow the general validity of

$$\exists\, e_5\, \exists\, G\, (G(e_1,e_3,e_5)) \wedge \exists\, e_5\, \exists\, G\, (G(e_2,e_3,e_5)) \\ \rightarrow \exists\, G(G(e_1,e_2,e_3)).$$

This condition is perfectly acceptable for interactions mediated by tachyons, photons or "subluminal" particles (often referred to in the literature as bradyons).

With the revised definition the range of events at location A, which are simultaneous with event e_3 at location B, is exactly the same as with photon signaling, and this is just the right latitude to justify us in employing the Einstein convention in any (subluminal) frame K' and so underpins, ontologically speaking, the usual Lorentz transformations and the associated Invariance Principle for the Minkowski metric. Note that we are still able to allow two events to be causally connectible via tachyon interaction even at spacelike separation provided we distinguish causal connectibility from the violation of a simultaneity relation in the way we have sketched. In Redhead (1983) these ideas are applied, albeit tentatively, to reconciling nonlocality in quantum mechanics and the constraints of SR.

Following Redhead (1983) let us stress again the key distinction between tachyons and our hypothetical superluminal phonons (acoustic particles). With "infinite velocity" phonons we can establish absolute synchrony. It is true, of course, that relative to some other frame K' (which employs the Einstein optical synchronization convention), these phonons will no longer appear to have infinite velocity, but because the phonon equation is not Lorentz invariant, these nontranscendent phonons cannot *also* occur as possible physical processes relative to the K frame. This latter possibility gave rise to the symmetrical situation as between K and K' in the tachyon case when disagreement arose as to whether e_3 was simultaneous with e_4 or $e_{4'}$. In other words, transcendent phonons pick out uniquely an absolute frame of reference relative to which they all have the same (infinite) velocity in all spatial directions. Tachyons, because they satisfy Lorentz-invariant equations, fail to pick out a preferred frame.

Conclusion: Some Remarks on the Twin Paradox

It is often argued that adopting option (1) and the associated thesis of the conventionality of metrical simultaneity depends on adopting a reductionist causal theory of time, and it is rightly contended that notions such as continuous genidentical sets of events seem to be grounded themselves in notions of temporal order, while furthermore the whole apparatus of genidentity, continuous trajectories, and so on, become suspect at the quantum-mechanical level. However, our discussion has shown that option (1) can be regarded as standing on its own as a generalization of classical simultaneity. Although we give up transitivity, we crucially retain the feature of invariance, that topological simultaneity is a relation that is agreed on by *all* observers, just as in the case of the classical concept. The terminology "relativity" for Einstein's theory is arguably misleading. The ontological ingredients are still picked out as invariants, but the invariants are different from those arising in classical physics. After all, in replacing the Galilean transformations by the Lorentz transformations as our invariance group, it is only to be expected that a different class of invariants is going to result. All the factual, experimentally well confirmed, content of SR can be expressed in terms of invariants, that is, in a coodinate-free formulation.

As an example of the conceptual clarification afforded by option (1) we consider briefly the celebrated twin paradox. Initially the two twins coincide at spacetime location O. Twin α remains at rest relative to a frame K while twin β moves to spacetime location B, with velocity V relative to standard synchrony in frame K, and then returns with velocity $-V$ to meet α again at spacetime location A. If $2T$ is the time lapse shown by a standard clock fixed to α, then everyone is agreed that the time lapse measured by a standard clock attached to β is $2T\sqrt{1 - v^2/c^2}$. This arises very simply from the fact that the clocks measure the invariant Minkowski interval, call it s, and, relative to K, we calculate at once (see figure 5.5):

$$s(OA) = 2T$$
$$s(OB) = s(BA) = T\sqrt{1 - v^2/c^2}.$$
$$\text{So } s(OBA) = 2T\sqrt{1 - v^2/c^2}.$$

Again relative to K, the lines of simultaneity are orthogonal to OA, and if we plot proper time $\tau_\beta(P)$ along β's trajectory *up to a point declared by α* to be simultaneous with P on α's trajectory, against the proper time $\tau_a(P)$ measured up to P along α's trajectory, we obtain the following simple result (see figure 5.6): a straight line OA, where $OM = 2T$ and $AM = 2T\sqrt{1 - v^2/c^2}$. This much also is agreed by everyone. The arguments all start with "explaining" these results from β's point of view. But approaching the problem using option (1)

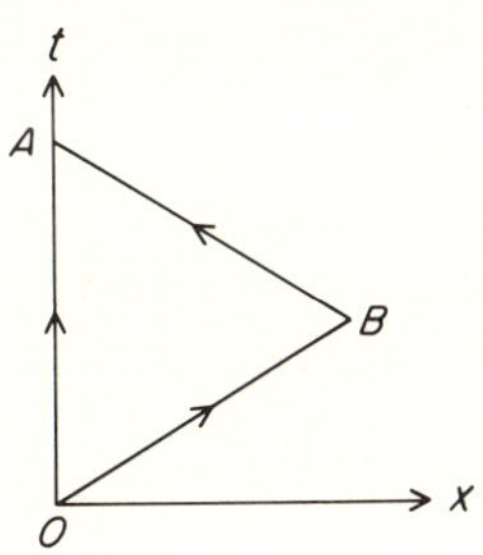

Figure 5.5

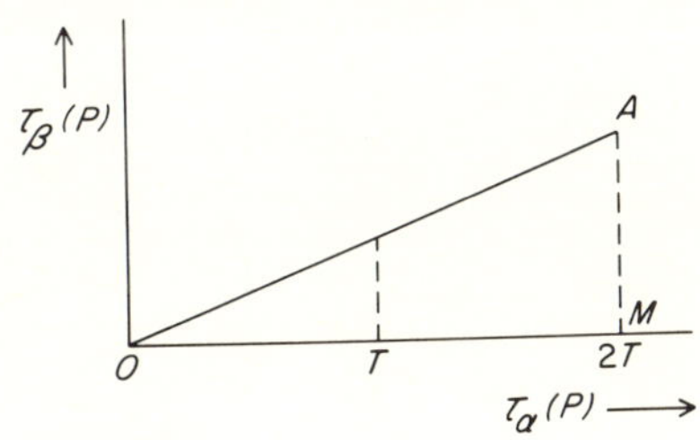

Figure 5.6

it is all very simple. For each value of $\tau_\alpha(P)$ there is a latitude in possible values for $\tau_\beta(P)$ depending on *which* point on β's trajectory is taken to be metrically simultaneous with the point P on α's trajectory. So there will be an upper and lower limit $\tau^u_\beta(P)$ and $\tau^l_\beta(P)$ for each value of $\tau_\alpha(P)$ dictated by the constraints of topological simultaneity. It remins to calculate the functions $\tau^u_\beta(P)$ and $\tau^l_\beta(P)$, considered, that is, as functions of $\tau_\alpha(P)$. Imagine the point P to move from O along the world-line of α, that is, along OA in figure 5.7, and consider the intersection of the light cone at P in various positions, with the world-line of β, namely, OBA. For $O \leq \tau_\alpha(P) \leq T(1 - v/c)$ the intersection of both forward and backward light cones is on the segment OB. For $T(1 - v/c) \leq \tau_\alpha(P) \leq T(1 + v/c)$ the backward light cone intersects OB and the forward light cone BA. For $T(1 + v/c) \leq \tau_\alpha(P) \leq 2T$, both in intersections are with the segment BA (see figure 5.7, cases (i), (ii) and (iii)). Then $\tau^l_\beta(P) = s(OL_1)$ and $\tau^u_\beta(P) = s(OL_2)$, the interval being computed along the trajectory OBA. The calculations are straightforward and produce the diagram shown in figure 5.8. $\tau^u_\beta(P)$ is given by the sides OS, SA of the parallelogram

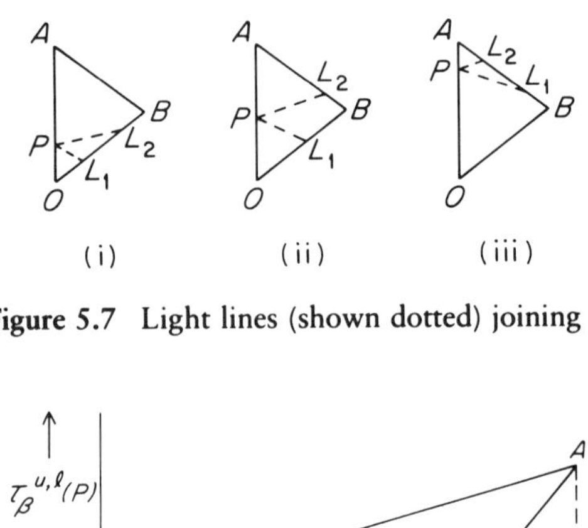

Figure 5.7 Light lines (shown dotted) joining P to OBA

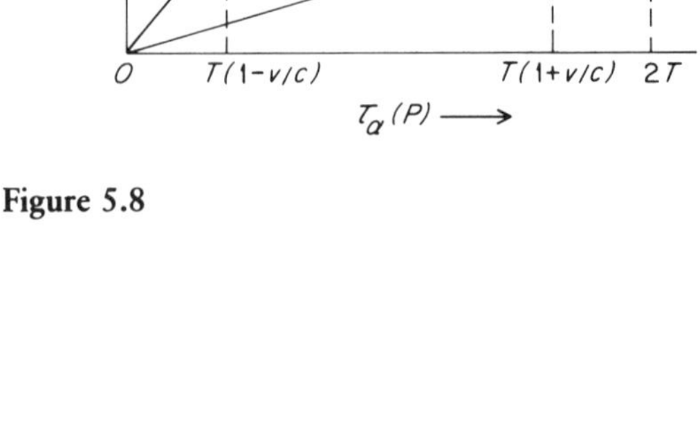

Figure 5.8

OQAS. The slope of *OS* is $\sqrt{(1 + v/c)/1 - v/c}$ and the slope of *SA* is $\sqrt{(1 - v/c)/1 + v/c}$, the abscissa of *S* being $T(1 - v/c)$ and of *Q*, $T(1 + v/c)$. The lower limit $\tau^{l}_{\beta}(P)$ is given by the remaining sides *OQ*, *QA* of the parallelogram. Note that *S* and *Q* correspond to the same value of τ_β, namely, $T\sqrt{1 - v^2/c^2}$. The line *OA* of figure 5.6 is simply the diagonal of the parallelogram *OQAS* in figure 5.8. But, exploiting the conventionality of simultaneity means that any nondecreasing curve joining *O* and *A* and lying within the parallelogram *OQAS* is an acceptable "story" of how the clock readings are related at various readings of α's clock.

Let us take two examples that have been discussed in the literature. (1) Take the average of $\tau^{u}_{\beta}(P)$ and $\tau^{l}_{\beta}(P)$ up to $\tau_\alpha(P) = T(1 - v/c)$ and extrapolate to $\tau_\alpha(P) = T(1 - v^2/c^2)$. Do the same for $\tau_\alpha(P)$ beyond $T(1 + v/c)$ and extrapolate back to $T(1 + v^2/c^2)$. Complete with a *horizontal* section between $\tau_\alpha(P) = T(1 - v^2/c^2)$ and $\tau_\alpha(P) = T(1 + v^2c^2)$. The resulting curve is *OFGA*, where the slopes of *OF* and *GA* are $1/\sqrt{1 - v^2/c^2}$ (see figure 5.9). (For a clear discussion of the virtues of this approach to the twin paradox see, for example, Salmon 1975, 98.) This shows β's clock running ahead of α's along the sections *OF* and *GA*, but mysteriously standing still along *FG*, allowing the differential aging of α to take place. Note that the points *SFGQ* lie on the same horizontal line in the diagram.

(2)Take the average of $\tau^{u}_{\beta}(P)$ and $\tau^{l}_{\beta}(P)$ everywhere.[7] This is illustrated in figure 5.10. Along OF, β is gaining (the slope of *OF* is 1/ 1 − v2/c2, *similarly along* GA. But along *FG*, β is losing (the slope of *FG* is just $\sqrt{(1 - v/c)/(1 + v/c)}$. A noteworthy feature of this solution is that β's estimate of α's distance, namely, $(1/2c)\ (\tau^{u}_{\beta}(P) - \tau^{l}_{\beta}(P))$, re-

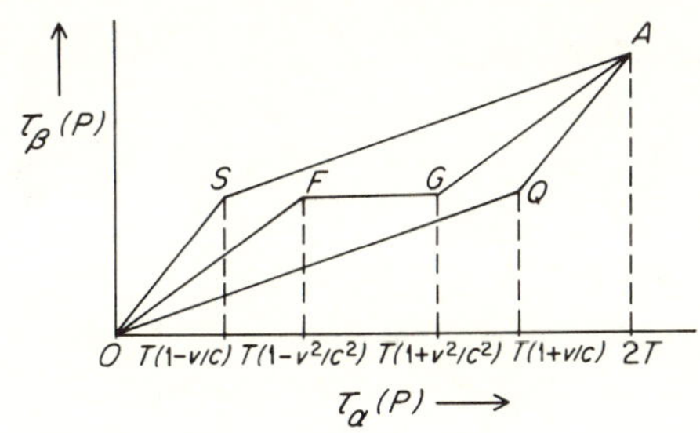

Figure 5.9

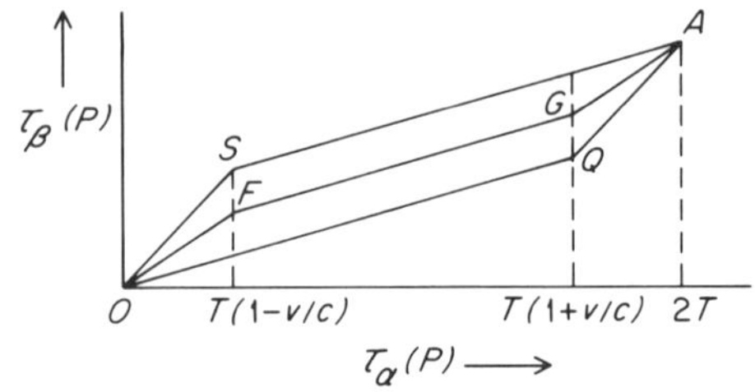

Figure 5.10

mains constant between $\tau_\alpha(P) = T(1 - v/c)$ and $\tau_\alpha(P) = T(1 + v/c)$, a curious result first noticed by Kilmister, but which is trivially obvious from the geometry of figure 5.10.

Discussions of the twin paradox in the context of general relativity typically adopt the analysis of example (1) but with more realistic description of the accelerations involved, in particular the reversal of β's velocity. In the limit of infinite acceleration the graph of figure 5.9 is then recovered, the horizontal section *FG* being then ascribed to the "infinite" slowing down of β's clock by the effective gravitational field generated by the Equivalence Principle (see Møller 1952, 258ff.) But since the spacetime is flat, introducing general relativity is a red herring since the calculation of the relevant proper times for clocks attached to α and β are, of course, independent of the coordinate system used.

The analysis we have given can also be applied directly to illuminate the twin paradox in cylindrical spacetime. The twin α travels along *OA*, while β moves now with uniform velocity along *OBA* to meet α again at *A* (see figure 5.11). To find the latitude in proper time of β that may be judged topologically simultaneous with a given lapse of proper time of α, we now have to take account of light propagation in both a clockwise and anticlockwise sense. Unwrapping the surface of the cylinder we obtain figure 5.12, where the points *A* and *A′* are to be identified. Folding about the vertical line through *B* brings us back to a diagram like figure 5.5. It is easily checked that the constraints of clockwise and anticlockwise light propagation on topological simultaneity exactly correspond to the situation sketched in figure 5.7, cases (i), (ii) and (iii). So we recover again a parallelogram like *OQAS* in figure 5.8. For fixed $\tau_\beta(P)$ this diagram identifies the range of $\tau_\alpha(P)$ that can be judged topologically simultaneous, and this

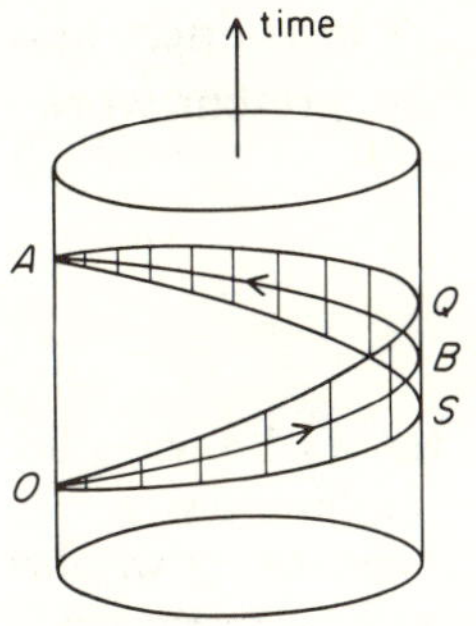

Figure 5.11

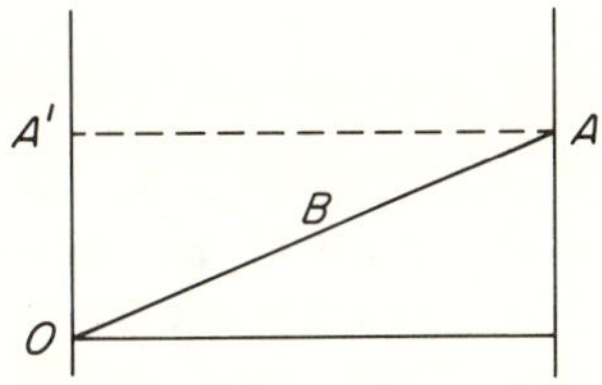

Figure 5.12

variable range can be represented on figure 5.11 by wrapping the (stretched) parallelogram appropriately back around the cylinder as indicated by the region shaded vertically. Any more detailed story of which clock reading on β is to be related to which clock reading on α is again a matter of convention. So option (1) gives a very satisfactory account of these matters.

But what about the *A*-theorists and the role of the moving present? This is definitely ruled out on the basis of option (1) since the non-transitivity of topological simultaneity coupled with the transitivity of co-reality would imply that all events are co-real. This is the real significance of Putnam's (1967) famous discussion of this issue: to show the inconsistency of option (1) with the *A*-theory of time. The only way to rescue an *A*-theory is to go for option (2) and relativize co-reality to a frame, in my view a very unattractive option, or to use option (3) to pick out a privileged present. Since this essay has come down firmly in favor of option (1), our own position is that SR *is* inconsistent with the *A*-theory of time.

Finally, what about general relativity (GR)? Many of the specific arguments that work in SR fail in GR. For example, the conformal structure no longer specifies the metric (modulo dilatations); we need to know also the projective structure. The question of specifying a global time at all depends on the possibility of a foliation of the spacetime manifold by a family of spacelike surfaces, which may not, in general, be possible. Locally, GR has the structure of SR (in the precise sense that the tangent spaces to the manifold have the structure of Minkowski spacetime). So the analysis we have given for SR shows how the *local* analysis of simultaneity works in GR. It is in the global aspects of the concept of simultaneity that GR introduces new conceptual puzzles that lie beyond the horizon of the present work.

NOTES

I am grateful to Jeremy Butterfield, Mark Hogarth, Harvey Brown, Clive Kilmister and Al Janis for helpful discussions on the topic of this essay.

1. This essay will not be concerned with the important question of whether this relation is itself induced by a primary relation that obtains between spacetime points, interpreted in some sense "substantivally." For a recent comprehensive discussion see Earman (1989).

2. Better terminology might be "absolute simultaneity" since the relation is not a general topological invariant of Minkowski spacetime. But we will continue to use the well-entrenched Grünbaum locution.

3. Much literature addresses the possibility of measuring the one-way speed of light. For a representative refutation of a number of such schemes see Salmon (1977).

4. The fact that the relationship is a linear one can be justified on the basis of the homogeneity of space and time. See for example Podlaha and Navrátil (1966). The following treatment of spacetime transformations relies on Redhead (1983). See also Podlaha and Navrátil (1967) and Sjödin (1979).

5. Choosing units so that $c = 1$ gives the condition for Minkowski orthogonality in our two-dimensional world as $m_s m_t = +1$, as constrasted with the usual Cartesian condition $m_s m_t = -1$.

6. The light sphere is not, of course, pointwise invariant, hence the relativistic explanation of stellar aberration. The motion of points in velocity space *on* the light sphere leads naturally to the spin representations of the Lorentz group.

7. See Kilmister (1990) who adopts essentially this prescription. Twin β synchronizes with twin α by employing a "radar rule" everywhere along his trajectory.

REFERENCES

Bridgman, P. W. 1962. A Sophisticate's Primer of Relativity. Middletown, Conn.: Wesleyan University Press.

Brouwer, L. E. J. 1975. *Collected Works.* Vol. 1. Edited by A. Heyting. Amsterdam: North-Holland.

Earman, J. 1989. *World Enough and Space-Time: Absolute versus Relational Theories of Space and Time.* Cambridge, Mass.: MIT Press.

Eddington, A. S. 1923. *The Mathematical Theory of Relativity.* Cambridge, England: Cambridge University Press.

Einstein, A. 1905. "Zur Electrodynamik bewegten Körper." *Annalen der Physik 17:* 891–921.

Ellis, B., and P. Bowman 1967. "Conventionality in Distant Simultaneity." *Philosophy of Science 34:* 116–36.

Friedman, M. 1977. "Simultaneity in Newtonian Mechanics and Special Relativity." In J. Earman, C. Glymour, and J. Stachel, eds. *Minnesota Studies in the Philosophy of Science.* Vol. 8, *Foundations of Space-Time Theories.* Minneapolis: University of Minnesota Press, pp. 403–32.

———. 1983. *Foundations of Space-Time Theories: Relativistic Physics and Philosophy of Science.* Princeton: Princeton University Press.

Grünbaum, A. 1969. "Simultaneity by Slow Clock Transport in the Speical Theory of Relativity. "Philosophy of Science 36: 5–43.

———. 1970. "Space, Time and Falsifiability. Critical Exposition and Reply to 'A Panel Discussion of Grünbaum's Philosophy of Science.' " *Philosophy of Science 37:* 469–588.

———. 1973. *Philosophical Problems of Space and Time.* 2d ed. Dordrecht: Reidel.

Janis, A. I. 1983. "Simultaneity and Conventionality," In R. S. Cohen, and L. Laudan, eds., *Physics, Philosophy and Psychoanalysis.* Dordrecht: Reidel, pp. 101–10.

Kilmister, C. W. 1990. "Discrete Physics and the Lorentz Transformation." Unpublished manuscript.

Malament, D. 1977. "Causal Theories of Time and the Conventionality of Simultaneity." *Noûs 11:* 293–300.

Mehlberg, H. 1980. *Time, Causality and the Quantum Theory.* Dordrecht: Reidel.

Møller, C. 1952. *The Theory of Relativity.* Oxford: Clarendon Press.

Nerlich, G. 1982. "Special Relativity is not Based on Causality." *The British Journal for the Philosophy of Science 33:* 361–88.

Podlaha, M., and E. Navrátil 1966. "On the Linearity of the Lorentz Transformations." *Acta Physica Austriaca 24:* 99–100.

———. 1967. "Formulation of the Axioms for Deriving the Lorentz Transformations, Galileo Transformations and Palicios Transformations." *Revista de la Real Academia de Ciencias Exactas, Fisica y Naturalis de Madrid 61:* 555–61.

Prokhovnic, S. J. 1967. *The Logic of Special Relativity.* Cambridge, England: Cambridge University Press.

Putnam, H. 1967. "Time and Physical Geometry." *Journal of Philosophy 64:* 240–47.

Quinn, P. L. 1974. "The Transitivity of Non-Standard Synchronisms." *The British Journal for the Philosophy of Science 25:* 78–82.

Redhead, M. L. G. 1983. "Nonlocality and Peaceful Coexistence." In R. Swinburne, ed., *Space, Time and Causality.* Dordrecht: Reidel, pp. 151–89.

———. 1990. "Explanation." In D. Knowles, ed., *Explanation and Its Limits.* Cambridge, England: Cambridge University Press, pp. 135–54.

Reichenbach, H. 1957. *The Philosophy of Space and Time*. English translation by M. Reichenbach and J. Freund. New York: Dover.
Robb, A. A. 1914. *A Theory of Time and Space*. Cambridge, England: Cambridge University Press.
———. 1921. *The Absolute Relations of Time and Space*. Cambridge, England: Cambridge University Press.
———. 1936. *The Geometry of Space and Time*. Cambridge, England: Cambridge University Press.
Roxburgh, I. W. 1975. "The Intransitivity of Non-Standard Synchronisms." *The British Journal for the Philosophy of Science 26:* 47–49.
Salmon, W. C. 1975. *Space, Time and Motion*. Encino Calif.: Dickenson.
———. 1977. "The Philosophical Significance of the One-Way Speed of Light." *Noûs 11:* 253–92.
Sjödin, T. 1979. "Sychronization in Special Relativity and Related Theories." *Il. Nuovo Cimento 51B:* 229–46.
Tangherlini, F. R. 1961. "An Introduction to the General Theory of Relativity." *Supplemento Nuovo Cimento 20:* 1–86.
Torretti, R. 1983. *Relativity and Geometry*. Oxford: Pergamon.
Varicák, V. 1910. "Anwendung der Lobatschefskijschen Geometrie in der Relativtheorie." *Physikalische Zeitschrift 11:* 93–96.
Winnie, J. A. 1977. "The Causal Theory of Space-Time." In J. Earman, C. Glymour, and J. Stachel, eds., *Minnesota Studies in the Philosophy of Science*. Vol. 8, *Foundations of Space-Time Theories*. Minneapolis: University of Minnesota Press, pp. 134–205.
Zahar, E. 1977. "Mach, Einstein and the Rise of Modern Science." *The British Journal for the Philosophy of Science 28:* 195–213.
Zeeman, E. C. 1964. "Causality Implies the Lorentz Group." *Journal of Mathematical Physics 5:* 490–93.

6

The Meaning of General Covariance

The Hole Story

John Stachel
Center for Einstein Studies and Department of Physics, Boston University

Let me begin by explaining my title and the aims of this essay. Between 1913 and 1915 Einstein gave several versions of an argument for rejecting what he referred to as "general covariance" as a criterion for the selection of the field equations of the gravitational theory he was then developing. This argument is based on consideration of the gravitational field within a region of spacetime devoid of all other matter and fields—a region that he called a "hole" [*Loch*]. Einstein afterwards referred to the argument itself as the "hole argument" [*Lochbetrachtung*]. Together with Marcel Grossman, he proposed a metric but nongenerally covariant gravitational theory (see Einstein and Grossmann 1913); but toward the end of 1915, a series of difficulties with that theory led him to reconsider generally covariant field equations. He found the way around his hole argument against them, and soon gave a definitive formulation of what we now call the general theory of relativity (for detailed historical accounts, see Stachel 1989 and Norton 1984).

In 1980, I gave an account of the significance of Einstein's hole argument that differed from the prevailing viewpoint among the few physicists and historians of relativity who had considered this episode (published in Stachel 1989). My account suggests an interpretation of the meaning and significance of the concept of general covariance, for which I do not claim great originality[1]—and which is certainly not universally accepted by other physicists—but which was received favorably by a number of philosophers of science (see, for example, Torretti 1983, sec. 5.6, 162–68; and the Introduction to Norton 1984).

John Earman and John Norton (1987) used a version of the hole argument against a certain class of absolute or substantival interpretations of spacetime.[2] In their version, the hole argument is applicable not only to theories such as general relativity, all models of which obey generally covariant field equations, but to *any* physical theory if it is suitable formulated. In particular, they apply it to all pregeneral–relativistic spacetime theories, both Newtonian and special–relativistic.

The title of my essay is meant dialectically as a tribute to their ingenious pun ("The Hole Story") and a rejection of their arguments, for I cannot accept their conclusion, but still maintain that the import of the hole argument is restricted to theories such as general relativity that are generally covariant in the proper sense of that term. The structure of my essay is as follows: After some mathematical preliminaries, I explain my interpretation of the hole argument and the concept of general covariance to which it leads. Then I indicate why this concept—and the hole argument—are not applicable to pregeneral–relativistic theories. Although I include a few historical comments, I will not discuss here whether my account is a historically accurate reconstruction of Einstein's views (for such a discussion, see Stachel 1989). However interesting in itself, this question is not relevant to the philosophical issues addressed here.

Before embarking on the exposition proper, it seems important to say a few words about my approach to such issues in the philosophy of science as the "true nature" of spacetime. Apart from the network of concepts and theories that we weave in our attempts to comprehend and act upon the world, we have no immediate access to "reality." So, in my opinion, one cannot fruitfully discuss questions such as the "true" or "real" nature of space and time apart from consideration of some physical theory (or class of such theories) and its (or their) attendant spatiotemporal concepts. Rather than being answered as such, these questions should be transformed into investigations of the spacetime structures associated with various physical theories.

There remain, of course, the more general problems of the grounds we have for accepting or rejecting such theories at various stages in the development of physics, and of the ontological status that acceptance of these theories implies for the concepts they invoke. The question of the reasons that lead us to adopt, modify, or reject a physical theory is a very important one, as is the question of the ontological status of various theoretical concepts, but these questions go far beyond my limited topic. Here, I will confine myself to the question of the spatiotemporal mathematical structures employed in various theories, and some possible implications of these structures for the concepts of space, time, and spacetime associated with these theories. In particular, I will critically examine the claim of Earman and Norton that the hole argument effectively precludes a substantivalist account of pregeneral-relativistic physical theories.

Manifolds, Diffeomorphisms, and Dragged-Along Tensor Fields

As mentioned earlier, the hole argument is based on consideration of a region of spacetime that is the site of nothing but a gravitational field. In mathematical terms, this means a region of a four-dimensional differentiable manifold with nothing but a metric tensor field given on this region. The argument involves diffeomorphisms of the manifold that reduce to the identity outside the hole, and tensor fields dragged along by such diffeomorphisms.

For those not already familiar with these mathematical concepts, the brief and imprecise descriptions given here should suffice for an understanding of the rest of the essay. A differentiable manifold M is a topological space with some additional structures. Such a space (as I will call it for short) consists of a set of entities, referred to as points, that are related to each other by certain continuity conditions, which (among other things) imply that the space has a certain dimension—finite or infinite.[3] In the case of spacetime theories, the spaces used generally have four dimensions,[4] and I henceforth confine myself to that number. In the case of a differentiable manifold, the points of the space are labeled by a family of charts. A chart associates an ordered set of four numbers, called a coordinate system, with each point of some region of the space; that is, it sets up a one-to-one bicontinuous correspondence between the points of that region of the space and some region of R^4, the space of all ordered quadruples of real numbers. All the charts in the family, often called an atlas, taken together, must cover the entire topological space. Two charts may overlap in the sense that they have some region of the space in common. If so, the transformation connecting the coordinate system of one chart and that of the other in the overlap region must not only be continuous but differentiable to some order (finite or infinite). Indeed, the important thing about a differentiable manifold is that the coordinate systems allow us to take derivatives of functions defined on such a manifold—hence the name *differentiable* manifold—in a way that does not really depend on which chart in the atlas we happen to use. A tensor field on a differentiable manifold may be defined as an entity such that, in each chart, at each point of the manifold, it will be represented by a certain set of numbers called the components of the tensor at that point in that coordinate system. There are various types of tensors, each being characterized by a certain number of so-called covariant and contravariant indices. For each type of tensor, there is a simple rule relating the set of numbers defining the tensor field at a point in one chart to the set of numbers defining the same tensor field at the same point in another chart. As we move from point to point in a region covered by the chart, the components change; by moving from chart to chart if necessary, we can cover the entire space. In sum, the components of a tensor field correspond to a set of functions defined in each chart at each point of the differentiable manifold.

In a topological space, one can carry out arbitrary bicontinuous point transformations, or homeomorphisms, that take each point of the manifold into some other point of the space. In a differentiable manifold, these homeomorphisms are further required to satisfy certain differentiability requirements, and hence are called diffeomorphisms. I must emphasize that I am discussing point transformations that take one point of the manifold into another, and not coordinate transformations, which just relate the coordinates used to describe the same point in different charts. Symbolically, if D is the action of a certain diffeomorphism and x represents the coordinates of a point of the manifold in some chart, then Dx represents the coordinates of the point into which x is dragged along by D in the same chart (assuming both points to lie in a common chart—we will not discuss possible global complications). A coordinate transformation would just produce a change in the coordinates attached to the same point x as a result of a change of *charts*, not points; the new coordinates of the same point in the new chart can be symbolized by adding a prime to the symbol for the point x'.

A coordinate transformation is a relabeling or renaming of the points of a manifold (subject to the differentiability requirements mentioned). Such a passive transformation does not have any physical significance, since it amounts to nothing more than a mathematical redescription of the same model of a physical theory. A point transformation, as an active transformation, is of potential physical significance since, as we will see, under certain circumstances it can be used to turn one physical model into another.

The important conceptual distinction between point and coordinate transformations is often blurred due to the following circumstance: In each region covered by two or more charts (we will neglect possible global complications), there is a one-to-one correspondence between the two types of transformations. More explicitly, for each point transformation taking a point x into a new point Dx, there is a coordinate transformation taking the (unprimed) coordinates of that point in the original chart into the (primed) coordinates of the new point in the new chart with the same numerical values. (For short, the *new* point in the *new* coordinate system has the *same* coordinates as the *old* point in the *old* coordinate system.) Symbolically, this means $(Dx)' = x$. The new coordinates are said to have been dragged (or carried) along with the diffeomorphism. Thus, it is possible to talk about coordinate transformations in a way that parallels the way that one talks about point transformations (a circumstance that has sometimes led to considerable confusion).[5]

Given any tensor field on a differentiable manifold, a diffeomorphism D on the manifold can be used to generate a new tensor field, called the drag-along (or carry-along) of the original tensor field under D. The new field can be defined without reference to coordinate systems, but it is simpler to use the definition of the dragged-along coordinate system to define it: The components of the drag-along field at the *new* point in the *new* (dragged-along) coordinate system have the same numerical values as the components of the *original* tensor field at the *old* point in the *old* coordinate system. Symbolically let $F(M)$ represent the tensor field, $F(x)$ represent the values of its components at the point x in the original coordinate system, and $[F(x)]'$ represent the values of its components at the same point in the new coordinate system. Then the drag-along field, symbolized by $DF(M)$, is defined by the condition that $[DF(Dx)]' = F(x)$.

Spacetime Structures

A dynamical physical theory purporting to describe events that can be characterized by their positions in space at certain moments of time employs certain mathematical structures.[6] Some of these structures, used to characterize the behavior of (ideal) clocks and measuring rods,[7] are called chronometrical and geometrical structures, respectively, or collectively, the chronogeometrical structures of the theory. Other structures, used to characterize the behavior of freely falling (ideal, force-free, spherically-symmetrical, nonrotating) test particles are referred to as affine structures. (In both Newtonian and special-relativistic kinematics, the behavior of such freely falling test particles is described by the law of inertia—Newton's first law; so it would be fitting to describe the corresponding spacetime structures as "inertial." Once gravitation is taken into account, and the principle of equivalence is applied, there is no way either in Newtonian gravitational theory or general relativity to make an absolute distinction between inertial and gravitational effects on the motion of such freely falling test particles; so the corresponding spacetime structures would best be described as "inertio-gravitational." In order to have a name that covers both cases, I have used the name "affine structures" since it describes the mathmatical objects that are used in both cases to represent these physical structures.) Both these structures are defined over some space, which for the moment we assume to be a differentiable manifold. Collectively, we refer to the chronogeometrical and affine structures, together with any relations between the two needed to assure their compatibility, as the spacetime structures of the theory.[8] These spacetime structures may have a symmetry group. That is, there may be a group of diffeomorphisms that leave the spacetime structures invariant. This means that the drag-along field of each spacetime structure under any diffeomorphism of the symmetry group coincides with the original field.[9] We shall see later, for example, that the inhomogeneous Galilei group and the Poincaré (or inhomogeneous Lorentz) group are the symmetry groups of Newtonian spacetime without gravitation and of special-relativistic spacetime, respectively.

A dynamical theory will also contain other mathematical structures, referred to as its dynamical structures, characterizing the behavior of physical fields and/or particles in spacetime. (Whether any of the spacetime structures are also dynamical is a question that is central to my discussion of the hole argument). This behavior is usually specified by requiring that the dynamical variables be subject to a set of differential equations (partial for fields, ordinary for particles), which we may call the dynamical equations of the theory. These equations involve not only the dynamical variables but their derivatives up to some order (generally first or second), so we need the language of fiber bundles and jet extensions to describe the theory itself. I will avoid that complication by discussing only models of the theory. A model consists of a set of dynamical variables that constitute a particular solution to the dynamical equations, together with any nondynamical structures needed in the theory (which, for nongenerally covariant theories, will include some or all of the spacetime structures). A model of a theory is thus a possible world according to that theory.

In all pregeneral-relativistic theories (with the partial exception of Newton's theory of gravitation),[10] it is possible to abstract all dynamical structures from the models of a theory, leaving a set of nondynamical spacetime structures that are common to each model of the theory. When a differentiable manifold is endowed with these nondynamical spacetime structures, its points may be directly identified as points of spacetime itself so that we may speak of a dynamical event as taking place at a given point of spacetime.

Whether or not one can abstract all dynamical structures without affecting the spacetime structures, there is a further sequence of abstractions that can be performed on a model of a theory. One can abstract both dynamical and spacetime structures, while preserving the atlas of charts, leaving a differentiable manifold. Next, one can abstract the atlas of charts while preserving the continuity relations between the points, leaving a topological space. Finally, one can abstract the continuity relations, leaving a point set.

At what stage in this sequence of abstractions does one pass from the realm of physics to that of pure mathematics? I argue that, once one has abstracted the spacetime structures, whether they are dynamical or not, leaving a bare differentiable manifold, one has passed beyond the limits of physics. In other words, I assert that the points of a differentiable manifold have no spatiotemporal properties unless and until at least some spacetime structures have been introduced.

In any physical theory that is either globally special–relativistic, like special relativity, or locally special–relativistic, like general relativity, there is a certain ten–component tensor field encoding all information about the chronogeometrical and affine structures of spacetime. It is called the metric tensor field, or "the metric" for short. In general relativity, the affine structure of spacetime is identical with the inertio-gravitational field. Consequently, the metric tensor plays a dual role in general relativity: It engenders not only all the spacetime structures, but also all the gravitational structures of the theory. It is worthwhile to emphasize this point since it is crucial to the hole story: Spacetime and gravitational field in general relativity are like love and marriage in the old song. In its role as source of all gravitational structures, the metric is a dynamical structure, obeying certain gravitational field equations. In its chronogeometric role, this implies that all the spacetime structures of the general theory are also dynamical structures.

The Hole Argument

We are finally ready to discuss the hole argument in general relativity. By a hole H, I mean an open subspace of a manifold M with no physical structure defined on it except a gravitational field. As indicated in the previous section, the gravitational field is represented mathematically by the metric tensor field. Let x symbolize the coordinates of a point of the manifold in some chart including that point, and $G(M)$ symbolize a gravitational-cum-metric tensor field over the manifold. Then $G(x)$ symbolizes the components of that metric tensor field at the point x with respect to that chart. It is physically reasonable to impose the following uniqueness requirement on any theory of gravitation: There must be *some* set of conditions that we can place on the gravitational field *outside* of H that will determine uniquely the field *inside* of H. This physical uniqueness requirement may seemingly be expressed by the following mathematical uniqueness condition: The field equations obeyed by the metric tensor field, together with specification of all sources of the gravitational field outside of H and some set of suitable boundary (or other) conditions that may be imposed on the field outside of and on the boundary of H, should have a unique solution $G(H)$ within H.

The hole argument shows that, if the field equations obeyed by the metric are generally covariant, it is impossible to fulfill this mathematical uniqueness requirement. But first, we must define generally covariant field equations. Consider a tensor field (or set of tensor fields) on the manifold, symbolized by $F(M)$. If D symbolizes the action of some diffeomorphism of the manifold, then $DF(M)$ is the drag-along of $F(M)$ under the diffeomorphism D. We call a set of field equations for these fields generally covariant if and only if, for every solution $F(M)$, $DF(M)$ is also a solution to these equations for every diffeomorphism D. Applying this definition to the metric tensor field, this means that, if $G(M)$ is a solution to a set of generally covariant field equations, so is $DG(M)$. (Note that *both* solutions are written in the *same* coordinate system; i.e., at a point x we are talking about $G(x)$ and $DG(x)$.)

Now for the hole argument: Consider a diffeomorphism D that reduces to the identity transformation (i.e., leaves everything unchanged) outside of and on the boundary of H, but does *not* reduce to the identity inside H. Applied to $G(M)$, D will produce a new solution $DG(M)$. Since D reduces to the identity outside and on the boundary of H, outside of the hole the new solution will be identical with the old. That is, it will have the same sources, boundary conditions, and so on, as $G(M)$; indeed, $G(M - H) = DG(M - H)$. Yet D *does* change the solution inside the hole: Since D is *not* the identity there, $DG(H)$ is *not* equal to $G(H)$. What is more, $DG(H)$ is different for each one of the infinity of inequivalent diffeomorphisms D that can be carried out inside H, even though nothing has changed outside or on the boundary of H. So, if $G(M)$ obeys generally covariant field equations, the uniqueness requirement cannot be satisfied in the mathematical form stated.

Let me emphasize again that this argument has nothing to do with different coordinate systems: $G(x)$ and $DG(x)$ are the components of *different* tensor fields at the *same* point in the *same* coordinate system. In this connection, I cannot refrain from a historical comment: The notation $G(x)$ is taken from Einstein (1914); he uses $G'(x')$ to denote the components of the *same* metric tensor field in a *different* coordinate system, and explicitly states that $G(x)$ and $G'(x')$ represent the *same* gravitational field. So, whether my version of the hole argument is a faithful reconstruction of Einstein's argument, *his* version *cannot* have been based on the trivial mistaken idea that a mere coordinate transformation changes the gravitational field, as has often been asserted in the literature.

We have reached the apparent dilemma: either generally covariant field equations for the gravitational field at the price of abandonment of the uniqueness requirement for solutions (at least for the case of a pure gravitational field in any hole) or uniqueness of solutions at the price of abandonment of generally covariant field equations. As mentioned earlier, Einstein's initial response to this dilemma was to give up general covariance, a position that he held for over two years.

A modern relativist will presumably see the way out of this dilemma at once: Although they differ *mathematically,* $G(M)$ and $DG(M)$ must represent the *same* gravitational field *physically.* (However, A. Logunov and his coworkers have recently used what is essentially the hole argument to argue against the generally covariant interpretation of general relativity. See Logunov et al. 1988 and Logunov and Mestvirishvili 1989. For a rebuttal, see Zel'dovich and Grishchuk 1988.) In other words, a gravitational field corresponds not to a *single* metric tensor field, but to an equivalence class of metric tensor fields; the members of such a class are equivalent if related to one another by diffeomorphisms. Of course I agree with this resolution of the dilemma, but I suggest that we examine its apparent triviality to see what it implies about the nature of spacetime in a general-relativistic theory, and to consider why what appears trivial today was not so trivial from 1913–1915 to Einstein, a person who was presumably at least as clever then as we are now.

Individuating Fields

To assert that $G(M)$ and $DG(M)$ represent one and the same gravitational field is to imply that the *mathematical* individuation of the points of a differentiable manifold (for example, by their coordinate labels in some chart that includes the point in question) has no *physical* content, and indeed can be given no physical content, apart from and until a metrical tensor field on that manifold is specified. For, if $G(x)$ and $DG(x)$ are to represent the same gravitational field, they must not be physically distinguishable in *any* way. So it must be the case that, when we act on $G(M)$ with the diffeomorphism D to create the drag-along field $DG(M)$, we leave behind *nothing* of physical significance, including nothing that would identify a point of the differentiable manifold as the *same* point of spacetime for both $G(M)$ and $DG(M)$. For example, the point of $DG(M)$ that corresponds *physically* to the point x of $G(x)$ is *not* x, but Dx, the drag-along of the point x. One may say that the points of the manifold *inherit* all of their physical properties, including their spatiotemporal individuation, from the metric tensor field.[11]

Once having decided to adopt generally covariant field equations, does the absence of all physical individuation of points of the underlying manifold follow as an obvious or trivial corollary? I will demonstrate by an example that it does not. Riemann introduced the concept of an "*n*-fold extended magnitude" or "manifold" in his famous *Habilitationsschrift* of 1854 (see Riemann [1929] 1953, 272–87, or Riemann 1919). He gives two examples of three-dimensional manifolds that arise on the basis of everyday experience: "The positions of sense objects [i.e., space] and the colors" ([1929] 1953, 274).[12] Russell ([1897] 1956, 67) points out an important difference between these two examples, "There is no inherent quality in a single point as there is in a single color, by which it can be quantitatively distinguished from another." Actually, we need three quantities to characterize the perceived colors: hue, brightness and saturation, for example (for further discussion of color space, with references to the literature, see Stachel 1986a, sec. 2, 236–49).

As this example of the color manifold shows, it is entirely possible for the points of a manifold to be individuated by some property or properties that characterize(s) each of the points. I call any structure on a differentiable manifold that provides a set of such properties an *individuating field*. Suppose that there are additional structures on a manifold with such an individuating field; for example, suppose we put a metric on the color manifold.[13] Consider a diffeomorphism D on this manifold. We may consider the drag-along of the color metric together with the original (undragged) individuating field that specifies the color of each point. This certainly corresponds to a *new* color metric on the manifold. Equivalently, we may consider the drag-along of the individuating field under the *inverse* diffeomorphism D^{-1} together with the original color metric; both combinations correspond to the *same* new color metric.

Obviously, if one considers the drag-along of *both* the color metric *and* the individuating field—in other words, if there is *no* relative dragging—then there is *no* change in the color metric. For, by assumption, the individuating field is the *only* structure that encodes the distinction *as colors* between the otherwise identical (as regards colors, even if mathematically distinct) points of the manifold. In summary, any *relative* dragging between color metric and individuating field produces a *new* color metric on the manifold.

Suppose the color metric were not given once and for all, but were subject to a set of generally covariant field equations. If we carry out a diffeomorphism of the color manifold and consider the drag-along of a color metric that obeys these equations, then by the definition of generally covariant field equations (see the previous section of this article), the drag-along color metric will also be a solution. The color analogue of the hole argument is now inescapable. Since one can compare the drag-along of the color metric with the original individuating field, we are forced to stipulate that the drag-along of the color metric represents a distinct solution to the color field equations. It is not just mathematically distinct, of course, but distinct in the sense of color metric theory. The individuating field prevents us from avoiding the hole argument by identifying all the drag-along solutions with one and the same model of color metric theory.

Clearly my use of colors in this example only serves to make the argument more "colorful." A similar argument holds for any manifold with an individuating field that is specified *independently* of any dynamical fields that obey some set of generally covariant field equations.

Returning to the metric tensor field, we can now identify the missing element in the resolution of the hole argument given above. To justify the identification of a whole equivalence class of drag-along metrics with *one* gravitational field, we must stipulate that, in general relativity, there is *no* structure on the differentiable manifold that is *both* independent of the metric tensor *and* able to serve as an individuating field—that is, to turn the points of the manifold into points of spacetime. The fact that the gravitational field equations are generally covariant does *not* suffice to allow the physical identification of a class of mathematically distinct drag-along metric fields. If such an individuating field existed, a relative dragging between metric tensor field and individuating field would be physically significant, and Einstein's hole argument would be ineluctable.

The usual way out of the hole argument seems obvious to contemporary relativists because we take it for granted that the correct mathematical setting for the general theory of relativity is a bare differentiable manifold with no additional structure that could serve as an individuating field. But this assumption places us at the *end* point of the development of the general theory, which incorporates the lesson of the hole argument, not at the *starting* point from which Einstein began his search for a relativistic theory of gravitation. (In this respect, the situation is somewhat analogous to the role of the principle of equivalence in the development of general relativity. Once one starts from a metric tensor field of Minkowski signature, there seems to be no place for the principle of equivalence in the theory—see, e.g., Synge 1964, xi—because it has already been incorporated into the starting point.)

I cannot resist a second historical comment, especially since it will be helpful later in understanding why the hole argument fails in pre-general-relativistic theories. A crucial element in the success of Einstein's approach to the electrodynamics of moving bodies that led to the special theory of relatively is his careful definition of a physical frame of reference. He defined it in terms of a network of measuring rods and a set of suitable-synchronized clocks, all at rest in an inertial system. He used such a system of rods and clocks to give physical meaning to a preferred coordinate system associated with the inertial frame of reference. In my language (admittedly anachronistic), the rods and clocks serve to establish a nondynamical individuating field for the points of Minkowski spacetime. The position of a particle with respect to the system of rods at a certain clock time, and the values of the components of a field, such as the electric or magnetic field, at a certain place and time in this network, thereby become physically significant quantities, and not just mathematical functions, as they would be in the absence of such an individuating field. Einstein ([1905] 1989) was fully aware of the importance of this step: "The theory to be developed—like every other electrodynamics—is based upon the kinematics of rigid bodies, since the assertions of any such theory concern relations between rigid bodies (coordinate systems), clocks, and electromagnetic processes. Insufficient consideration of this circumstance is at the root of the difficulties with which the electrodynamics of moving bodies currently has to contend" (p. xi).

Having painfully taught himself this lesson—the need to give physical meaning to the spatial and temporal coordinates before doing any dynamics, or as I would put it, the need to introduce an individuating field for spacetime that is independent of any dynamical theory—Einstein had to unlearn it just as painfully in order to arrive at the final version of the general theory of relativity.

A suitable modified version of the lesson remains true, however: the need for an individuating field for spacetime before *physically interpreting* any dynamical theory. The novel feature of the general theory is that this individuating field itself depends on the gravitational field, so that no spatiotemporal framework for the theory exists before the gravitational field equations are solved.

General Covariance

After this historical excursion, I now state what I take to be the principle of general covariance in general relativity: It consists of the following three assertions: (1) There are no individuating fields in spacetime that are independent of the metric tensor field; (2) the metric field determines both the chronogeometrical and inertio-gravitational structures of spacetime; and (3) The metric field obeys a set of generally covariant field equations.

What's in a name? Does it matter how we define general covariance? It depends what you are trying to make of it. Few people would disagree about the inclusion of points (2) and (3) in a definition of general covariance; but, as we have seen in the previous section, (2) and (3) alone do not suffice to avert the impact of the hole argument. So if one wants, as I do, to use the term "general covariance" to describe what it is that enables general relativity to avert the hole argument, (1) must be included in its definition.

How do the points of spacetime get individuated in general relativity? In general, a metric tensor field that satisfies the gravitational field equations will have no spacetime symmetries. In that case, the metric tensor field itself plays the role of an individuating field in regions of a pure gravitational field (such as the holes considered in the last section). More precisely, four invariants of the Riemann (or curvature) tensor constructed from the metric tensor field play this role.[14] If a particular metric tensor field does have some symmetry group, the values of the four invariants will be the same at all points on an orbit of the symmetry group, so that additional individuating elements have to be introduced to distinguish between such points. (For example, the preferred parameters of the symmetry group, one of which is associated with each of the Killing vector fields that generates the symmetry group, may be used.) However chosen, such additional elements cannot be independent of the metric tensor field since the latter serves to define the orbits in question.

If one takes seriously the point of view that, in the general theory of relativity, until a metric tensor field is specified there is no spacetime, then even the topology of the differentiable manifold associated with a model of the theory cannot be specified a priori, but must be chosen so that it is compatible with the metric of that model.[15] By this point it should be clear why I maintain that, in general relativity, when one starts with a model of the theory and abstracts the metric tensor field one has thereby passed outside the realm of physics. (Later, I will discuss why this is so for the case of pregeneral-relativistic theories.)

It is hard for me to see how this viewpoint about general-relativistic spacetimes could be made compatible with most versions of spacetime absolutism or substantivalism (see note 2). Indeed, most versions seem to require the possible existence of at least *some* spacetime structures (if only the points of spacetime) without *any* additional physical structures, such as a gravitational field, being corequisite. But as we have seen, in general relativity there is not even a global topological space, let alone any individuation of its points as points of spacetime, before the specification of a metric tensor field. This is its chronogeometrical role. In its other role, this metric tensor field also serves to generate all the gravitational field structures (in particular, the affine connection and the Riemann tensor). So, in the general theory of relativity, there is no way to have a model with a spacetime structure of any sort that is devoid of a gravitational field.

Several philosophers of science have argued that the general theory of relativity actually supports spacetime substantivalism (if not separate spatial and temporal substantivalisms) since it allows solutions consisting of nothing but a differentiable manifold with metric tensor field and no other fields present (empty spacetimes). This claim, however, ignores the second role of the metric tensor field; if it is there chronogeometrically, it inescapably generates all the gravitational field structures. Perhaps the culprit here is the words "empty spacetime." An empty spacetime could also be called a pure gravitational field, and it seems to me that the gravitational field is just as real a physical field as any other. To ignore its reality in the philosophy of spacetime is just as perilous as to ignore it in everyday life.

On the other hand, in his defense of relationalism Adolf Grünbaum (1977) contends that the gravitational role of the metric tensor field in empty spacetimes is *not* sufficient to endow this tensor with its chronogeometrical properties. If *no* individuating field were present, this would be true. But, as noted, in the generic case, the four nonvanishing invariants of the Riemann tensor constitute such an individuating field, one that is constructed exclusively from the metric tensor field and its derivatives. These four invariants can be used as coordinates, and the components of the metric tensor field at each point expressed as functions of these invariants coordinates (see Komar 1958). This procedure endows these components with the requisite chronogeometrical properties, for one can specify in terms of these invariants just *where* and *when* a clock or measuring rod should be placed to measure any desired component of the metric tensor field in the invariant coordinate system. (The presence of symmetries introduces some complications, but the procedure can be modified to take account of these.) So I maintain that even in the case of a pure gravitational field, one cannot introduce the metric tensor field in its gravitational role without its also playing its chronogeometrical role. Since, as I have emphasized, this entire construction is not independent of the metric-cum-gravitational field, it seems entirely compatible with what I would call a relational theory of the chronogeometrical structure.

The Earman-Norton Argument

Earman and Norton (1987) purport to apply the hole argument against spacetime substantivalism not only to general relativity, but to essentially *any* physical theory, that is, all that include a spacetime structure.[16] They note that all relevant physical theories can be formulated using a four-dimensional differentiable manifold, on which various tensor fields (or other geometrical object fields—see note 9) are postulated to exist. In this formulation, models of the theories they consider consist of a manifold *M* plus certain tensor or geometrical object fields on *M*, which fields are solutions to certain differential equations required by the theory (see ibid., 517). For short, I will refer to "manifold-plus" theories and models. Earman and Norton further restrict their considerations to what they call "local spacetime" (ibid.) formulations of the theories. That is, they require that all spacetime structures be subject to local (differential) field equations. They contrast such local spacetime formulations with what they call "absolute" versions of the same physical theory that introduce globally specified spacetime structures a priori. They concede; "Our present [hole] argument does not address this [global] representation of spacetime since it is commonly associated with non-local spacetime theories already beyond our compass" (ibid., 519–20).

I maintain that, for pregeneral-relativistic spacetime theories, their distinction between "local" and "global" versions is insignificant: Both versions lead to exactly the same set of models. Let us take the example they use, "special relativistic electromagnetics" (ibid., 517, 518 and 519). They formulate the spacetime portion of this theory locally by imposing the "field equation" $R^a_{bcd} = 0$ (vanishing of the Riemann tensor) on the metric tensor field g_{ab}. This is a curious sort of field equation, since locally (that is, in any finite subregion) it has only *one* solution, the Minkowski metric. So the only thing left undetermined by this "field equation" is the global topology of the manifold. Although they do not say so explicitly, let us assume that manifolds with different global topologies are possible in various models of the local version of the theory.[17] Their definition of models of "local spacetime theories" includes specification of a manifold *M* for each model.[18] So all possible models of their local version of a special-relativistic theory will fall into equivalence classes, each member of an equivalence class having the *same* manifold *M* as part of its definition. To specify a particular model in this equivalence class, one then has to choose a solution to the field equations for the remaining (nonspacetime) fields. Each such model will correspond to—indeed, be exactly the same as—a model of the global version of the same theory, in which the spacetime structure is specified globally to begin with as the Minkowski metric on the given manifold *M*, rather than obtained by "solving" a field equation that has only one solution on the given *M*. Similar comments apply to the local and global versions of any special-relativistic or Newtonian theory. If a physical theory contains absolute objects,[19] such as a fixed metric tensor field, one cannot get rid of them in a manifold-plus formulation of the theory: If they are excluded from the local formulation of the theory itself, they reappear in all the models of that local formulation. As we will see in the next section, the only way to avoid such absolute objects is to avoid manifold-plus formulations of nongenerally covariant theories.

In the light of the essential equivalence of local and global versions of nongenerally covariant theories, I choose to base my discussion of such theories on the global versions of pregeneral-relativistic manifold-plus theories. A dissatisfied localist may want to add a preliminary step: Start from the local version of such a theory on a given manifold *M*, "solve" the trivial field equations for the spacetime structure on *M*, and thus arrive at the global models involving *M*, which form my starting point.

Earman and Norton observe (as we did earlier) that, if we consider the drag-along under a diffeomorphism of *all* the fields in some model, physically we have changed nothing in the original model (see the discussion of "The Gauge Theorem," ibid., 520). They conclude, "The spacetime substantivalist holds that spacetime can exist independently of any of the things in it. . . . [S]ubstantivalists, whatever their precise flavor, will deny: *Leibniz equivalence* Diffeomorphic models represent the same physical situation" (ibid., 521).

In my language, their account acknowledges that the points of a bare manifold do not carry inherent physical individuation in *any* physical theory, and asserts that these points *should* carry such individuation in a substantivalist account of spacetime.

A major weakness of their discussion is that Earman and Norton do not consider the role of individuating fields (see the previous section "Individuating Fields"), a concept that is distinct from that of absolute objects (see note 20), which they do discuss. Hence, they cannot make the distinction between nondynamical individuating fields, which are a type of absolute object, and dynamical individuating fields (such as the metric in general relativity), which are not. When individuating fields are introduced, previously physically indistinguishable models often become distinct. For example, on Earman and Norton's definition of a "special relativistic electromagnetic" model, there is only one generic model of the Coulomb field generated by a given charge. The field of such a charge at *one* place and the field of a similar charge at *another* place are indistinguishable, as are the fields of the charge at rest and a similar one in uniform motion. The distinctions between such fields cannot be formulated in their definition of a model, which certainly makes life in the laboratory rather difficult! With the aid of spatial and temporal individuating fields, such distinctions are easily made, as Einstein realized (see the previously given citation from [1905] 1989). The scope of Leibniz equivalence, understood as the identity of indiscernibles, is considerably restricted when individuating fields are introduced into the models of a physical theory (see the next section).

Moreover, their approach leads Earman and Norton to a definition of general covariance that, if not wrong—for one cannot say that a noncontradictory, nonempty definition is wrong—is at any rate physically trivial. I will turn to this question in the final section.

First I will examine a logically prior question that goes to the heart of all arguments based on manifold-plus formulations. *Must* all theories with a spacetime structure be formulated on differentiable manifolds? I will argue that pregeneral-relativistic theories need not be—indeed, that they are properly formulated on mathematical structures that do not admit the diffeomorphism group. Hence, all arguments based on diffeomorphisms, such as the hole argument, are basically irrelevant to such theories. Even if one insists on manifold-plus formulations, the hole argument still fails for these theories when it is properly formulated.

Who Needs Differentiable Manifolds?

Presumably as a result of a preoccupation with general relativity, many philosophers of space and time seem to take for granted that the primary formulation of any physical theory involving spacetime structures should be based on a four-dimensional differentiable manifold, on which these structures (as well as other physical fields) are then imposed. "The advent of general relativity has made most compelling the identification of the bare manifold with spacetime" (Earman and Norton 1987, 518). But the spacetime structures of a physical theory have their own inherent symmetry group. In our earlier discussion of spacetime structures based on the manifold-plus approach, we defined this symmetry group as the group of diffeomorphisms that leaves the spacetime structures invariant. However, the symmetry group itself is the fundamental thing, and this symmetry group determines the mathematical space (physically speaking, the spacetime) that is most naturally associated with the physical theory. This is the lesson of Felix Klein's ([1872] 1893) approach to geometry, embodied in the famous Erlangen program (for a historical discussion, see Torretti 1978, 137–42). To give a simple example, the requirement that geometrical figures be invariant under the group of two-dimensional translations and rotations, and similar under dilations defines plane Euclidean geometry; the space that is naturally associated with this group is the Euclidean plane. We could, of course, adopt a manifold-plus approach: Begin with a two-dimensional manifold with the topology of R^2, introduce a positive-definite metric tensor field with vanishing curvature on the manifold, and then restrict the diffeomorphism group of the manifold to the symmetry group of this metric tensor. But, to do plane Euclidean geometry, this elaborate procedure is not needed; the result is still the same old plane Euclidean geometry, no more and no less.[20]

The spacetime structures of both Newtonian and special relativistic theories (with the exception mentioned in note 10), have symmetry groups (the inhomogeneous Galilei and Poincaré groups, respectively) that are subgroups of the inhomogeneous group of linear affine transformations. This group is naturally associated with a four-dimensional affine space.[21] (Physically, the law of inertia common to both Newtonian and special-relativistic spacetimes leads to this common affine structure.) Starting with the inhomogeneous affine group, one can distinguish between Newtonian and special-relativistic spacetimes by suitable choice of a preferred subgroup of the homogeneous affine group. In the Newtonian case, foliation of spacetime by the absolute time and introduction of a Euclidean metric on the leaves of the foliation—both done in such a way as to be compatible with the affine structure—single out to the homogeneous Galilei group as the preferred subgroup. In the special-relativistic case, the four-dimensional interval, again introduced in a way compatible with the affine structure, leads to the homogeneous Lorentz group as the preferred subgroup. (For detailed discussions see, e.g., Torretti 1983, sec. 1.6, and Kopczynski and Trautman 1992, 26–33, for Newtonian spacetimes; and Torretti 1983, sec. 4.2, and Kopczynski and Trautman 1992, 52–57, for special-relativistic spacetimes.)

Next, individuating fields can be introduced. In both cases, one can introduce a preferred tetrad into the spacetime at any point, the three spacelike vectors being a mathematical representation of a set of orthogonal unit measuring rods. In the case of Newtonian theory, the fourth (timelike) vector is chosen to connect two hyperplanes of absolute time differing by a unit of absolute time; in the special relativistic case it can be chosen orthogonal to the spacelike triad and represents a unit of clock time for a clock at rest in the triad frame. These tetrad vectors can be parallel transported from the initial point to any other point in spacetime, setting up a unique tetrad field in spacetime, with its associated coordinate system. (Since Newtonian and special-relativistic spacetimes are affine flat, parallel transport in each is independent of path.)

This construction serves to fix an inertial coordinate system or frame of reference in Einstein's language, an inertial individuating field in mine. Of course, many alternate ways are possible to set up such an individuating field, such as those involving the use of test particles, and light rays, to mention only two other possibilities. (Grünbaum 1977, 345, refers to these as "a family of concordant external metrical standards." For additional comments on this question, see Stachel 1973; 1983, 255–56.) The important point for present purposes is that the spacetime structures, as well as the individuating field mapped out with the help of any of these methods, are *independent* of any dynamical fields that are subsequently introduced in the spacetime.[22]

Such an account of pregeneral-relativistic theories, which makes no mention of differentiable manifolds, is entirely respectable. In one way, indeed, it is *more* respectable than one that does have such recourse, because it introduces *less* physically redundant mathematical structures, as I will now explain.

First, let me explain what I mean by "physically redundant." A current trend among some philosophers of science is toward what I will call "the fetishism of mathematics."[23] By this I mean the tendency to assume that all the mathematical elements introduced in the formalization of a physical theory must necessarily correspond to something meaningful in the physical theory and, even more, in the world that the physical theory purports to help us understand. However, in physics one should adopt the attitude toward mathematics that Humpty Dumpty adopted toward language: "When *I* use a word, it means just what I choose it to mean—neither more nor less" (Carroll [1872] n.d., 214).

Hertz's ([1894] 1956) comments about the "images" [*Bilder*] used in setting up physical theories are most apposite to the relation of the mathematical structures to the physical content of these theories. (See the "Introduction" to Hertz [1894] 1956; he compares such "images" to "models" [*Modellen*], a term that would more likely be used today.) He stresses that our choice of such images is guided by the condition that "the necessary consequents of the images in thought are always the images of the necessary consequences in nature of the things pictured" (ibid., 1). Since these images are not determined "uniquely" [*eindeutig*] by this requirement, further conditions help in their choice. He defines one image as "more distinct" [*deutlicher*] than another if it "pictures more of the essential relations of the object" (ibid., 2) and adds, "Of two images of equal distinctness the more appropriate is the one which contains, in addition to the essential characteristics, the smaller number of superfluous or empty relations [*überflüssiger oder leerer Beziehungen*]—the simpler [*einfacher*] of the two" (ibid., Hertz emphasizes that "empty relations cannot altogether be avoided" [ibid.]). Paraphrasing Hertz's criterion of simplicity, it suggests choosing the model that minimizes the number of consequences of the model that are not images of any consequences of the thing being modeled.

Applying this criterion to the choice of mathematical structures used in models of a physical theory, we may ask whether the manifold-plus models that Earman and Norton (1987) employ are always the simplest in Hertz's sense. The symmetry group of a differentiable manifold is the group of diffeomorphisms of that manifold; so only a theory that has the diffeomorphism group as the symmetry group of its spacetime structures requires a differentiable manifold for its most appropriate (in Hertz's sense) models. As we have seen, Newtonian and special-relativistic theories have much smaller spacetime symmetry groups, so manifold-plus models of these theories introduce "empty or superfluous" elements into these models. It is no accident that these theories managed to thrive for so long (in the case of Newtonian theories, for over two centuries) before the mathematical concept of a differentiable manifold was even formulated.

One might object to my argument against manifold-plus models on the ground that the differentiable structure is "more fundamental" than the affine structure, but that is not the case. The differentiable structure of an affine space can be constructed from its affine structure, just as the affine structure can be constructed from that of a differentiable manifold with a flat affine connection. The same goes for the addition structures required to turn a suitable affine space into a Galilean or special-relativistic spacetime (see Torretti 1983, sec. 4.2, for an explicit demonstration of this for Minkowski spacetime). So one cannot claim that the concept of differentiable manifold is more fundamental than the concept of an affine space.

Suppose, in spite of this argument, one *did* regard a differentiable manifold as somehow mathematically more fundamental than an affine space. Our concern is with physics, in particular the adequacy of mathematical structures to model physical relations. It is a question, then, to use Hertz's language, of giving preference to the mathematical image or model having the lesser number of consequences that are not consequences of the physical relations being modeled.

One can certainly choose to start with manifold-plus models, but Earman and Norton go much further, attaching physical significance to the manifold, "We view the manifolds *M* of the models as representing spacetime . . . we look upon the bare manifold . . . as spacetime" (1987, 518–19). They offer no argument for interpreting the *mathematical* points of the manifold as having any *physical* (spatiotemporal) properties *before* the introduction of other structures on the manifold—that is, before the "plus" has been added in the manifold-plus models.[24] As we have seen, the hole argument speaks formidably *against* such an identification for the case of general relativity. In the case of pregeneral-relativistic theories, I have given an argument, which I consider to be equally formidable, that the differentiable manifold introduces redundant elements into the formulation of the theory, such as the diffeomorphisms not belonging to its symmetry group. The points of the bare manifold on which these diffeomorphisms act are also redundant, in the sense that *no* physical significance can be attached to them once the spacetime structures are abstracted from the model.

Earman and Norton presumably would argue that an affine space formulation of a special-relativistic theory is inferior to theirs because the former is global, while they prefer local manifold-plus formulations (see the previous section). But we have seen that, since the spacetime structure is not subject to dynamical equations in Newtonian or special-relativistic theories, it must be specified globally before one can formulate (let alone solve) any concrete dynamical problem. Hence, the global nature of the affine-space formulations is not a drawback[25] The manifold-plus formulations of these theories are also subject to the condition that the *global topology* of *M* must be specified beforehand for any given model; the local conditions used to specify the spacetime structure(s) on these manifolds (such as the vanishing of the Riemann tensor for special-relativistic theories) are irrelevant.[26]

Formulations of Newtonian and special-relativistic theories based on an affine space, which I henceforth refer to as affine-space-plus formulations, are not subject to the hole argument. The allowed symmetry transformations of these theories (spatial and temporal translations, spatial rotations, and Galilei or Lorentz boosts for Newtonian or special-relativistic theories, respectively) are "rigid": If such a transformation reduces to the identity in an arbitrarily small region of the spacetime, it reduces to the identity everywhere. (Technically, this is because the symmetry groups in question are finite-parameter Lie groups.)

It is no accident that Newton and his successors, as well as Einstein and other (special) relativists before 1913, lived in blissful ignorance of the hole argument and any possible implications it might have for determinism. For the theories they were concerned with, Leibniz equivalence is restricted to equivalence under the "rigid" (finite-parameter Lie) group of symmetry transformations of the theory. If individuating fields are included in a model of the theory, the group is even further restricted to the identity transformation.

As noted, the natural symmetry group of a bare differentiable manifold is the group of all its diffeomorphisms. If we want to restrict the symmetry group of spacetime to the inhomogeneous Galilei or Poincaré group in a manifold-plus formulation, then after introducing the appropriate chronogeometrical and affine structures on the manifold, we must restrict ourselves to the subgroup of diffeomorphisms that leaves these spacetime structures invariant. By Klein's principle (see note 20), the symmetry group of the resulting physical theory is the same as if we had never introduced the full group of diffeomorphisms of the manifold in the first place.

Of course, was noted, in the manifold-plus formulation, we can consider the drag-along of *all* physical fields—both the nondynamical fields defining the spacetime structure (including any such individuating fields) and the dynamical fields subject to field equations—under an arbitrary diffeomorphism. Since they differ only in the redundant, nonphysical elements, no physical distinction can be made between the original (redundant) model and the drag-along (equally redundant) model. Recalling the color space example, this is similar to the case in which we dragged along both the color names and the color metric: In both cases, there is no relative dragging of metric and individuating fields. The difference is that, in the color space model, there are no redundant elements.[27] That is why the hole argument succeeds in color space but fails in models of Newtonian or special relativistic theories, even in the manifold-plus formulations.

Trivial General Covariance

Earman and Norton (1987) call "general covariance" what they refer to as "The Gauge Theorem" (see p. 520): If a set of fields (including all spacetime structures) is a manifold-plus model of any physical theory, then the drag-along of all these fields under any diffeomorphism provide an equivalent model. One cannot quarrel with this observation (it seems pretentious to call it a theorem), and it is pointless to quarrel over a name; but this does seem a rather trivial application of the term "general covariance." Many years ago, Peter Bergmann discussed a similar concept in the passive, coordinate transformation language, calling it "weak covariance" in view of its "fundamentally trivial nature" (Bergmann 1957, 11). I will refer to this concept as "trivial general covariance," in contrast to what I earlier called general covariance, which Bergmann calls "strong covariance."

The reason why Earman and Norton's concept of general covariance is trivial should be clear by now: By considering the drag-along of *all* physical fields over a bare manifold, we have passed beyond the real of physics. As explained earlier, contrary to the assertion of Earman and Norton,[28] the points of a bare manifold have no physical properties. So it seems obvious that dragging *everything* with a diffeomorphism in some manifold-plus model to create a "new" model changes *nothing* physically in that model. We first discussed this point for color space models, and then for models of Newtonian and special-relativistic theories. In the latter case, we emphasized the redundancy of manifold-plus models of such theories: The existence of an equivalence class under diffeomorphisms of such mathematical models that corresponds to a single physical model is a symptom of the redundancy of the manifold in such formulations. In the case of generally covariant theories (in my sense of the term), it is not a symptom of such redundancy but a consequence of the absence of any physical distinction between the points of such manifolds before metric tensor (or other) fields are specified, that is, the absence of metric-independent individuating fields.

In view of these considerations, it would be quixotic for a spacetime substantivalist to try to identify the points of a bare manifold with points of physical spacetime in a Newtonian or special-relativistic theory, as Earman and Norton suggest.[29] If the substantivalist is to look for the points of spacetime anywhere in these theories, it is in the natural, nonredundant spacetimes of the global versions of these theories. We need not go into whether a claim for such absolute spacetime structures in these theories is justified. The only conclusion I wish to draw is that the hole argument does not settle the issue.

NOTES

This article is based on a talk given at the Boston Colloquium for the Philosophy of Science, 22 January 1991. I am much indebted to Professor Roberto Torretti for critical comments on the manuscript of an earlier version, which have resulted in several important improvements.

1. Although he uses the language of coordinate transformations, substantially the same position about the significance of diffeomorphism invariance in general relativity has long been advocated by Peter Bergmann. See, for example, Bergmann (1957, 9–12; 1961).

2. Although a distinction might be made between "absolutism" and "substantivalism," I will use both words as roughly synonymous.

3. The use of the terminology "space" and "points" in connection with an abstract topological manifold does not imply that the manifold need have anything to do with what is ordinarily implied by these terms—namely, a physical space of three dimensions and its constituent points.

4. I will not discuss unified field theories, some of which employ spaces with five or more dimensions.

5. Bergman's discussion of general covariance, cited in note 1 and discussed in the last section of this article, which is phrased in the coordinate transformation language, does not suffer from such confusion.

6. Not all physical theories need include such structures. Thermodynamics, for example, may be considered as an abstract physical theory relating the equilibrium values of certain state variables characterizing a macroscopic system, and fixing the sequence in which such equilibrium states can occur. The question of the relationship of such equilibrium states and their sequential order to particular spatial and temporal events need not be considered as part of thermodynamics.

7. By the use of the word "characterizing" I am purposely begging the question of whether the behavior of rods and clocks is constitutive of the chronogeometrical structures, or merely indicative of these pre-existing structures.

8. I am simplifying a much more complicated story, which should include references to conformal and projective structures. For a much fuller account of these structures in Newtonian as well as special- and general-relativistic contexts, see Ehlers (1973).

9. The affine structure is not a tensor field, but an example of a more general structure called a geometrical object field. (Points of the manifold are another example.) As in the case of tensors, if the components of a geometrical object field are known in one chart, and a coordinate transformation to a new chart is given, the components in the new chart are determined; but the law of transformation between the components in the two charts is more complicated than the tensor law. The drag-along of such fields can also be defined (see Schouten 1954).

10. Henceforth, I will not add this qualification when alluding to pregeneral-relativistic theories. In a four-dimensional formulation of Newtonian gravitation theory, the chronogeometrical structure (consisting of the absolute time and the three-dimensional Euclidean spatial metrics) may be specified independently of the inertio-gravitational field. The latter is represented mathematically by an affine structure that is compatible with the chronogeometrical structure. But, in contrast to the situation in other physical theories, this structure not completely determined by the compatibility conditions. This freedom is sufficient to allow for incorporation of what is usually referred to as "the gravitational force" into the affine structure. (For an exposition of the four-dimensional version of Newtonian gravitation theory that highlights these structures and their relations, see Ehlers 1973.) To anticipate the upcoming discussion, in this case individuating fields based on the chronogeometrical structure, and hence independent of the inertio-gravitational structure, may be introduced.

11. This circumstance is responsible for many of the problems that arise in attempts to set up a quantum theory of gravity. For a recent discussion of this issue, emphasizing the role of the hole argument, see Rovelli (1991a,b).

12. Of course, Riemann is referring to the "space of the perceptively given color-qualities of colored light," which has three dimensions, and not to "the manifold of physical colors [that] has infinitely many dimension," as Weyl ([1927] 1949, 70) puts it.

13. Schrödinger, among others, did actually discuss a colormetric (see Stachel 1986a, 336–37, n. 21, for references). But the actual utility of such a colormetric for the theory colors is, of course, irrelevant to the point I am making.

14. An invariant of a tensor field is a quantity built out of the components of the tensor field and their derivatives that has two properties: (1) Its formal ex-

pression as a function of the components of the tensor field and their derivatives is the same in all coordinate systems; and (2) its numerical value at any point is also the same in every coordinate system. The Riemann tensor and its contractions are the only tensor fields that can be constructed from the metric tensor field and its derivatives of first and second order. In general, fourteen invariants can be formed algebraically from the components of the Riemann tensor, ten of which vanish at all points in a region of pure gravitational field (see Geheniau and Debever 1956). If the spacetime possesses no symmetries, the remaining four nonvanishing invariants take on a different set of values at each point of the hole, and can therefore be used (at least in some finite local region) to characterize distinct points of the gravitational field; that is, in my terminology they form an individuating field (see Komar 1958).

15. At one time, I espoused the view that, while the metric tensor field was relational, the manifold should be considered an absolute element of the spacetime structure of general relativity (see Stachel 1974). While it is certainly possible to adopt this viewpoint about a particular model of a general-relativistic spacetime, it now seems more in keeping with the nature of the general theory, in which different models can have maximal extensions on topologically different manifolds, to regard both the manifold and the metric tensor field as relational structures. See Stachel (1987, 1991) for more discussion of this point.

16. They do not consider theories, such as thermodynamics (see the Introduction), which do not include such a structure. Since such theories do involve a manifold of state variables, it would be amusing to extend their argument to such cases, and develop an argument against "pressure-volume-temperature substantivalism," for example.

17. I cannot understand their statement, "So we take special relativity to have models $\langle M, g_{ab}, R^a_{bcd}\rangle$ where g_{ab} can be any of many possible Minkowski metrics definable on M, which satisfy the field equations $R^a_{bcd} = 0$" (1987, 518). Do they mean to imply that these models are *inequivalent?* If so, then there would have to be some individuating field on M, independent of the metric, to account for this inequivalence, and they would have forfeited the game to the substantivalists. Perhaps the basis of this difficulty lies in an ambivalence in their treatment of the physical significance of manifolds. On the one hand, they state, "We view the manifolds M of the models as representing spacetime . . . we look upon the bare manifold . . . as spacetime" (ibid., 518–19). On the other hand, Leibniz equivalence—"Diffeomorphic models represent the same physical situation" (ibid., 522)—implies that the points of the bare manifold M do not represent *anything* physically. It seems impossible to maintain both of these positions concurrently. For a way out of this dilemma, based on the use of fiber bundle representations of physical theories, see Stachel (1986b).

18. They do not consider the question of compatibility between manifold and their "field equations" for the spacetime structures. But, to continue with their example, the Minkowski metric is well known to be only globally compatible with certain manifolds (see, e.g., Wolf 1967).

19. In the attempt to make clear the distinction between truly generally covariant theories and trivially generally covariant theories, James L. Anderson introduced the distinction between absolute and dynamical objects, that is,

between nondynamical fields on a manifold, specified once-and-for-all, and dynamical fields subject to field equations (see Anderson 1967, 83–84, 339–40). This distinction is not the same as that between individuating and nonindividuating fields. As can be seen from the discussions of individuating fields in special and general relativity in the text, individuating fields may be absolute or dynamical. Nor do absolute spacetime structures always serve to individuate the points of spacetime, as the discussion of the Minkowski metric in special relativity shows.

20. This is an example of what Schouten calls "the principle of F. Klein: If in any space with a group G_1 the subgroup G_2 is introduced, consisting of all the transformations of G_1 which leave a figure F_1 invariant, then the geometry of a figure F_2 with respect to G_2 is identical with the geometry of the set of figures F_1, F_2 with respect to G_1" (1954,65).

21. Trautman made this point clear in contrasting special and general relativity: "Now, any physical theory of space and time has a starting point as a certain [mathematical] category. Its objects are models of space-time in that theory and the morphisms are mappings preserving the structure inherent in the theory and based on physical hypotheses. For example, in special relativity the category in question is that of affine spaces Aff, whereas in general relativity it is that of differentiable manifolds Diff" (Trautman 1970, pp. 32–33). Torretti 1983, section 4.2, and Kopczynski and Trautman 1992, 26–33, contain discussions of affine spaces and the group of affine transformations in the context of their application to spacetime theories.

22. The privileged role played by rods and clocks in this procedure, and the question of whether a distinction of principole must be made between rods and clocks and other physical systems, is irrelevant for the issues under discussion. Einstein observed that, while such a distinction was provisionally necessary in the current state of physics, any *complete* physical theory should *construct* entities with the popstulated properties of rods and clocks as solutions to its equations. See Stachel (1989, 77–78; 1983, 256–58) for further discussion, with citations from Einstein. For discussion of a similar issue in quantum theory, the postulated properties of the measuring apparatus, see Stachel (1986a, 279–80).

23. I draw this terminology from Marx, of course, who, in his critique of political economy, spoke of the fetishism of commodities, of capital, and so on. He in turn drew the term from the critique of religion: "In order, therefore, to find an analogy, we must have recourse to the foggy realm of the religious world. Here the products of the human brain appear endowed with a life of their own, autonomous figures that enter into relations both with each other and with humankind. . . . I call this fetishism." (Marx and Engels [1867] 1974, 86; my translation).

24. Indeed, their own words elsewhere in the paper seem to contradict this assertion, "[E]very spacetime theory with which we deal . . . postulate[s] that there are always fields at every point in spacetime. That is, they agree that there cannot be unoccupied spacetime events" (Earman and Norton 1987, 521).

25. As pointed out earlier, if one should choose to allow more than one topology of spacetime in connection with these theories, the manifold-plus models fall into equivalence classes with respect to these topologies, each member of an

equivalence class having the same (nondynamical) spacetime topology. The models in each equivalence class could then be reformulated without reference to a differentiable manifold. No such elimination of differentiable manifolds is possible in general-relativistic theories, of course.

26. As indicated earlier, when the spacetime structure is subject to dynamical equations, as in the case of general relativity, the appropriate global topology cannot be specified before the field equations are solved. This constitutes a serious defect of manifold-plus formulations of general relativity, as pointed out in Stachel (1987).

27. If we made the color space model more realistic, there actually would be such redundant elements since the additive properties of colors give color space a linear structure.

28. "The advent of general relativity has made most compelling the identification of the bare manifold with spacetime" (Earman and Norton, 1987, 518). It is hard for me to square this assertion with Earman and Norton's Gauge Theorem, their formulation of general covariance, which seems to imply the opposite conclusion about the bare manifold (see note 17).

29. Earman (1989) refers to "manifold substantivalism" (p. 126). Earman recognizes that manifold substantivalism is vitiated by the hole argument based on what he calls general covariance and I call trivial general covariance, but calls manifold substantivalism "the only form of substantivalism presently in the offing" (ibid., 180).

REFERENCES

Anderson, J. L. 1967. *Principles of Relativity Physics*. New York: Academic Press.

Auslander, L., and R. E. Mackenzie. 1977. *Introduction to Differentiable Manifolds*. New York: Dover.

Bergmann, P. G. 1957. "Topics in the Theory of General Relativity." In *Brandeis University Summer Institute of Theoretical Physics 1957*. N.p., pp. 1–44.

———. 1961. "Observables in General Relativity." *Reviews of Modern Physics 33:* 510–14.

Carroll, L. [Dodgson, C. L.] [1872] n.d. "Through the Looking-Glass and what Alice Found There." In *The Complete Works of Lewis Carroll*. New York: Modern Library, pp. 141–272.

Earman, J. 1989. *World Enough and Space-Time: Absolute versus Relational Theories of Space and Time*. Cambridge, Mass.: MIT Press.

Earman, J., and J. Norton. 1987. "What Price Spacetime Substantivalism? The Hole Story." *British Journal for the Philosophy of Science 38:* 515–25.

Ehlers, J. 1973. "Survey of General Relativity Theory." In W. Israel, ed., *Relativity, Astrophysics and Cosmology*. Dordrecht: Reidel, pp. 1–125.

Einstein, A. 1914. "Die Formale Grundlage der allgemeinen Relativitätstheorie." *Königlich Preussischen Akademie der Wissenschaften* (Berlin). *Sitzungsberichte:* 1030–85.

———. [1905] 1989. "Zur Elektrodynamik bewegter Körper." In J. Stachel, ed., *The Collected Papers of Albert Einstein.* Vol. 2, *The Swiss Years: Writings 1900–1909.* (Originally published in *Annalen der Physik 17:* 891–921). Princeton, Princeton University Press, pp. 276–306.

Einstein, A., and M. Grossmann. 1913. "Entwurf einer verallgemeinerten Relativitätstheorie und einer Theorie der Gravitation." With an addendum. *Zeitschrift für Mathematik und Physik 62:* 225–61. (Originally published as a seperatum without the addendum. Leipzig: B. G. Teubner.)

Geheniau, J., and R. Debever. 1956. "Les quatorze invariants de courbure de l'espace riemannien a quatre dimensions." In A. Mercier and M. Kervaire, eds., *Jubilee of Relativity Theory, Bern, July 11–16, 1955. Proceedings. Helvetica Physica Acta, Suppl. 4.* Basel: Birkhäuser Verlag, pp. 101–04.

Grünbaum, A. (1977). "Absolute and Relational Theories of Space and Space-Time." In J. S. Earman, C. N. Glymour, and J. J. Stachel, eds., *Minnesota Studies in the Philosophy of Science.* Vol. 8, *Foundations of Space-Time Theories.* Minneapolis: University of Minnesota Press, pp. 303–73.

Hertz, H. [1894] 1956. *The Principles of Mechanics Presented in a New Form.* (Originally published as *Gesammelte Werke.* Vol. 3, *Die Prinzipien der Mechanik in neuem Zusammenhang Dargestellt.* Leipzig: J. A. Barth.) New York: Dover.

Klein, F. [1872] 1893. "Vergleichende Bertrachtungen über neuere geometrische Forschungen." Reprinted in *Mathematische Annalen 43:* 63–100.

Komar, A. 1958. "Construction of a Complete Set of Independent Observables in the General Theory of Relativity." *Physical Review 111:* 1182–87.

Kopczynski, W., and A. Trautman. 1992. *Spacetime and Gravitation.* Chichester: Wiley.

Logunov, A., Y. M. Loskutov, and A. Mestvirishvili. 1988. "The Relativistic Theory of Gravitation and Its Consequences." *Soviet Physics—Uspekhi 31* (7): 581–96.

Logunov, A., and A. Mestvirishvili. 1989. *The Relativistic Theory of Gravitation.* Moscow: Mir Publications.

Marx, K., and F. Engels. [1867] 1974. *Werke.* Vol. 23. (Originally published as *Das Kapital/Kritik der politischen ökonomie.* Vol. 1, *Buch 1: Der Produktionsprozeß des Kapitals.*) Berlin: Dietz Verlag.

Norton, J. 1984. "How Einstein Found His Field Equations." Reprinted in D. Howard and J. Stachel, eds., *Einstein Studies.* Vol. 1, *Einstein and the History of General Relativity.* (Originally published in *Historical Studies in the Physical Sciences 14:* 253–316.) Basel: Birkhäuser, pp. 101–59.

Riemann. B. 1919. *Über die Hypothesen, welche der Geometrie zu Grunde liegen.* Edited by H. Weyl. Berlin: Springer.

———. [1929] 1953. *The Collected Works of Bernhard Riemann.* 2d ed. Reprint. Edited by H. Weber. (English translation by H. S. White, *A Source Book in Mathematics.* Edited by D. E. Smith. New York: McGraw-Hill, pp. 411–25.) New York: Dover.

Rovelli, C. 1991a. "Quantum Reference Systems." *Classical and Quantum Gravity 8:* 317–31.

———. 1991b. "What is Observable in Classical and Quantum Gravity?" *Classical and Quantum Gravity 8:* 297–316.
Russell, B. [1897] 1956. *An Essay on the Foundations of Geometry.* Reprint. New York: Dover.
Schouten, J. A. 1954. *Ricci Calculus. An Introduction to Tensor Analysis and Its Geometrical Applications.* 2d ed. Berlin: Springer.
Stachel, J. J. 1973. "Space-Time Problems." *Science 180:* 292–93.
———. 1974. "The Rise and Fall of Geometrodynamics." In K. F. Schaffner and R. S. Cohen, eds., *Boston Studies in the Philosophy of Science.* Vol. 20, *PSA 1972—Proceedings of the 1972 Biennial Meeting of the Philosophy of Science Association.* Boston: Reidel, pp. 31–54.
———. 1983. "Special Relativity from Measuring Rods." In R. S. Cohen and L. Laudan, eds., *Physics, Philosophy and Psychoanalysis: Essays in Honor of Adolf Grünbaum.* Dordrecht: Reidel, pp. 255–72.
———. 1986a. "Do Quanta Need a New Logic?" In R. Colodny, ed., *From Quarks to Quasars: Philosophical Problems of Modern Physics.* Pittsburgh: Pittsburgh University Press, pp. 229–347.
———. 1986b. "What a Physicist Can Learn from the Discovery of General Relativity." In R. Ruffini, ed., *Proceedings of the Fourth Marcel Grossmann Meeting on General Relativity.* Amsterdam: Elsevier, pp. 1857–62.
———. 1987. "How Einstein Discovered General Relativity: A Historical Tale with Some Contemporary Morals." In M. A. H. MacCallum, ed., *General Relativity and Gravitation: Proceedings of the 11th International Conference on General Relativity and Gravitation.* Cambridge, England: Cambridge University Press, pp. 200–08.
———. 1989. "Einstein's Search for General Covariance, 1912–1915." In D. Howard and J. Stachel, eds., *Einstein Studies.* Vol. 1, *Einstein and the History of General Relativity.* Basel: Birkhäuser, pp. 63–100.
———. 1991. "Einstein and Quantum Mechanics." In A. Ashtekar, and J. Stachel, eds., *Conceptual Problems of Quantum Gravity: Proceedings of the 1988 Osgood Hill Conference.* Basel: Birkhäuser, pp. 13–42.
Synge, J. L. 1964. *Relativity: The General Theory.* Amsterdam: North-Holland.
Torretti, R. 1978. *Philosophy of Geometry from Riemann to Poincaré.* Dordrecht: Reidel.
———. 1983. *Relativity and Geometry.* Oxford: Pergamon Press.
Trautman, A. 1970. "Fibre Bundles Associated with Space-Time." *Reports on Mathematical Physics 1:* 1–34.
Weyl, H. [1927] 1949. *Philosophy of Mathematics and Natural Science.* Revised and augmented English edition. Translated by O. Helmer. (Originally published as *Philosophie der Mathematik und Naturwissenschaft.* Hanbuch der Philosophie. München: Oldenberg.) Princeton: Princeton University Press.
Wolf, J. A. 1967. *Spaces of Constant Curvature.* New York: McGraw-Hill.
Zel'dovich, Y. B., and L. P. Grishchuk 1988. "The General Theory of Relativity is Correct!" *Soviet Physics—Uspekhi 31*(7): 666–71.

Scientific Rationality and Methodology

7

Sciences and Pseudosciences

An Attempt at a New Form of Demarcation

Robert E. Butts
Department of Philosophy, The University of Western Ontario

The Demise of the Demarcation Problem

In an article appearing in an earlier *Festschrift* honoring Adolf Grünbaum (Laudan 1983), Larry Laudan argued persuasively that the attempt to distinguish science from pseudoscience by employment of one or another of the proposed demarcation criteria is now dead. What he had specifically in mind is that demarcation criteria employing verificationist and falsificationist theories of meaning have failed to accomplish the required goal.[1] However, he obviously retained confidence in what he called "our intuitive distinction between the scientific and the non-scientific" (ibid., 124) and admitted that some future successful attempt to mark the difference could not be ruled out. The intuition referred to provides a rich catalog of pseudosciences, "flat Earthers, biblical creationists, proponents of laetrile or orgone boxes, Uri Geller devotees, Bermuda Triangulators, circle squarers, Lysenkoists, charioteers of the gods, *perpetuum mobile* builders, Big Foot searchers, Loch Nessians, faith healers, polywater dabblers, Rosicrucians, the-world-is-about-to-enders, primal screamers, water diviners, magicians, and astrologers" (ibid., 121). Evidently, these assorted pseudosciences make "crank claim[s] which make ascertainably false assertions" (ibid.). One might reasonably add: *if literally construed.*[2]

The concept of literal construal of claims to knowledge will play an important role in my argument for a new kind of demarcation criterion. But first, I want to review some ways of distinguishing between science and pseudoscience that do not in any obvious ways invite exhumation of the corpses Laudan has rightly buried. What continues to nag at me is Laudan's adoption of a centuries-old intuition. Many of us have this same intuition. I am convinced that in some way we can express its message in fairly precise terms. Although we may regard as inadequate certain forms of verificationist and falsificationist *criteria* of demarcation, the *problem* of demarcation seems to remain.

I will briefly review four philosophical strategies that have some merit as attempts to distinguish science from nonscience. All four of the strategies have a bearing on the new strategy I will propose.

Forms of Reliable Knowing, Science, and Mental Aberration: The Kantian Strategy

Kant's well-known critique of classical substantive metaphysics employs an explicit form of demarcation between science and nonscience of the following kind (see Butts 1986b for a more detailed account). First, Kant's analysis of human cognition seeks to locate those species-universal mental structures that define reliable knowing. These structures are the now familiar forms of space and time and the categories functioning as a priori, hence necessary, forms of the possibility of knowledge. Kant argues that this complex cognitive machinery, applying as it does only to items given in sensation, effectively distinguishes between reliable knowledge (compliant with the cognitive structures) and pseudoknowledge (such as that alleged to be knowable in classical metaphysics). Furthermore, it was Kant's view that the premier instantiation of the formalism of spacetime and the categories is given by Newtonian physics: *The* most reliable form of human knowing is mathematical physics (see Friedman 1986 and Butts 1986a, c).

At various stages in his career, acceptance of this conclusion meant for Kant that biology can only be construed as a research strategy pursued in the interests of obtaining good taxonomy, not a full science (the position of *Critique of Judgment*), and that chemistry is not a science but an artful classification scheme (the position of *Critique* of Pure Reason and *Metaphysical Foundations of Natural Science*).[3] It also meant that substantive metaphysics has no reliable epistemic credentials, and that some claims to knowledge of that which lies beyond experience are indistinguishable from claims characteristic of mentally deranged persons (Kant's judgment on the claims of Swedenborg to possession of ways of access to a world of spirits). It was also part of Kant's strategy to propose that various forms of mental activity that deviate from, ignore, or distort application of the epistemic formalism can be classified as maladaptive mental disorders. Any such forms of maladaptive behavior will produce claims that cannot be justified as scientific (see Butts 1986b, 282–318).

In summary, here is Kant's strategy: identify the norms of reliable knowing, show that these norms are instantiated by the best science of the day, then classify pseudosciences as aberrations of these norms (metaphysics and other claims to transcendent knowledge), which in some cases involve genuine mental illnesses. There is much to recommend Kant's strategy, but unfortunately in its Kantian form it cannot be sustained. Surely we cannot require that the norms of reliable knowing must be instantiated by any science of a particular period if what we take science to be is the *content* of a particular successful theory. This defect has led many to look for distinguishing marks of science in its *methods* and what have come to be called its *virtues*. Before turning attention to such strategies, it must also be remarked that the Kantian form of demarcation can only be applied with profit after cognitive science and psychopathology have joined forces. In the end, the Kantian strategy cannot be successfully accomplished philosophically; it needs to be itself a part of science. This consequence should be applauded, for it cautions us to look for demarcation and the definition of cognitive normalcy, if they are to be found at all, not in a priori philosophical arguments, but in the results of combined theoretical and empirical investigation. The failure of the Kantian program as a philosophical account of demarcation itself points the way to some empirically ascertainable distinguishing marks of the sciences.

Prediction and Control: The Rescher Strategy

Many philosophers and scientists have directed our attention away from the specific content of sciences, and toward the expressed goals of scientific inquiry. The goals are many: accuracy of description, mathematical precision in the formulation of theory, retrodiction, and taxonomical adequacy. Two goals, however, are thought to occupy prominent positions largely unique to science: accuracy in prediction, and success in control of the environment. Nicholas Rescher (1970) presents a particularly clear and persuasive argument to the conclusion that what he calls "explanatory frameworks" (pp. 135–45), including the frameworks of the sciences as well as, for example, numerology, astrology, and black magic, can each be criticized *en bloc,* but only a posteriori. The sciences win in the competition, not because of superior content, but because of pragmatic success in prediction and control that is unmatched by other explanatory frameworks (see Rescher 1977).

Part of Rescher's strategy is clearly correct. We have learned that metaphysics (and Kant's transcendental arguments) provide no a priori grounds for favoring certain sciences over alleged nonsciences, or, alternatively stated, for preferring a definition of cognitive normalcy over claims to the possibility of heightened modes of awareness, or altered states of consciousness. Furthermore, the failed empiricist effort to locate a demarcation criterion that would single out science as the sole dispenser of cognitively significant sentences further confirms the absence of a priori grounds that would allow a decision between explanatory frameworks.

It seems unlikely, however, that one can dismiss astrology, appeals to narrative discourse, or fundamentalist creationism, or psychoanalysis, solely because these explanatory frameworks fail as accurate systems of prediction leading to at least partial control of aspects of the environment. It all depends on what is at stake. Explanatory frameworks operate to further chosen objectives. If we know what these objectives are, we can show, by admittedly a posteriori means, that those ends cannot be realized in ways promised by the given explanatory framework. Pseudosciences are shown to be defective in explanatory potency by discovering that they possess impoverished means for realizing *aims they themselves articulate.*

In different ways, the failure of the Kantian strategy and the partial success of the Rescher strategy point in the same direction. We must look for demarcation by paying attention to what kinds of things conceptual frameworks are attempting to accomplish by saying what they do.

Science and "Cognitive Accountability": The Grünbaum Strategy

Adolf Grünbaum's many writings on the epistemological status of psychoanalysis (see, for example, Grünbaum 1980, 1983, 1984) offer no new theory of demarcation. However, in other papers, his work does seek to illuminate the kinds of problems that must be faced by those who would deal with the demarcation problem. In several papers (see, for example, Grünbaum, 1989a), he has argued vigorously against Popper's claim that psychoanalysis is a pseudoscience because it is not falsifiable. Popper's conclusion is of course based on his acceptance of the falsificationist criterion of demarcation. What Grünbaum shows is that a thoroughgoing *inductivist* study of psychoanalysis reveals that some of its hypotheses are indeed falsifiable, and falsified, thus refuting Popper's claim that Freud's theories are not falsifiable, and at the same time casting doubt on the merits of the falsificationist criterion of demarcation.

Grünbaum's (1989b, 1990) study of the hermeneuticians' attempts to free Freud's theories of what Freud himself insisted is the *natural* scientific status of those theories also has an indirect bearing on the problem of demarcation. Jaspers, Habermas and Ricoeur think of the clinical situation in psychoanalysis as one eventuating in hermeneutical insight into kinships of meaning, where such affinities are also regarded as *causal*. In this way, the hermeneuticians seek to free Freud's theories from the corpus of natural sciences so that they can take their place among the human sciences. Grünbaum demonstrates, again by employment of inductivist methods of analysis of causal connection, that connections between meanings, even when very strong, do not suffice of themselves to establish causal connections. Thus the hermeneuticians' efforts to show that psychoanalysis is a pseudo *natural* science are unavailing. Such efforts, in effect, simply reinstate long-standing epistemological problems associated with the human sciences, for example, epistemic claims on behalf of empathy or *Verstehen*.

Grünbaum's contribution to the solution of the demarcation problem consists in his invitation to learn just how difficult it is to establish the scientific credentials of a theory. He does this, not by offering a criterion of demarcation, but by demonstrating that psychoanalysis, however brilliantly conceived, is *inadequate* science. His arguments are designed to show, not that psychoanalysis is a pseudoscience, but that the scientific credentials of the theory are deeply flawed by mistakes in logic, epistemological and ontological sleights of hand, and failures to credit disconfirmations when they occurred. Grünbaum (1990) refers to "the sort of cognitive accountability that is featured by the natural sciences" (p. 562) (and which the hermeneuticians seek to bypass by freeing the human sciences of the constraints normally imposed on research in the natural sciences).

The Grünbaum strategy thus adheres to my suggestion that in examining any theory (scientific or pseudoscientific), the articulated aims of the theorizing must be taken seriously into account. What are the articulated aims of Freudian science (for Grünbaum, *all* science)? What is the cognitive accountability featured by the natural sciences? The answer will surely go something like this: A theory in a natural science should have testable consequences, because, before all else, an empirical theory must be confirmed or disconfirmed by observations; a theory should be logically rigorous; a theory should be simple. There are other such "virtues" of theories in the natural sciences (Hempel 1983 calls them "*desiderata*"). It may also be true that not every theory accepted by natural scientists possesses all of the virtues. How many the theory must possess, and how basically involved they are in a given theory, are perhaps variable. In general, however, cognitive accountability as it is envisaged by Grünbaum involves essentially the accessibility of theories to rigorous and detailed criticism based on canons of inductive and deductive logic.

Marks of a Good Hypothesis: The Hempel Strategy

Hempel (Hempel 1983, 87–88) lists *desiderata* of scientific hypotheses, suggesting that they are imprecise constraints on scientific theory choice, "Among them are the following: a theory should yield precise, preferably quantitative, predictions; it should be accurate in the sense that testable consequences derivable from it should be in good agreement with the results of experimental tests; it should be consistent both internally and with currently accepted theories in neighboring fields; it should have broad scope; it should predict phenomena that are novel in the sense of not having been known or taken into account when the theory was formulated; it should be simple; it should be fruitful."

Hempel concedes that these desiderata are vague, and mentions that in earlier writings he had argued that they are arational (not irrational) constraints on theory choice. However, he thinks that his analysis should lead us to conclude that there is something left over that can count as objective traits of competing theories, for example, "the considerable role that precise and rule-governed reasoning does play in the critical appraisal of competing theories, which requires among other things a rigorous derivation of experimental implications and the performance of experimental tests that have to meet appropriate standards. But even the considerations adduced in appraising the satisfaction of the vaguer desiderata are typically perceived, it seems to me, as expressing not just individual taste or preference, but objective, if only roughly specified, traits of competing theories" (ibid., 95).

Thus Hempel agrees with Grünbaum that there are standards of cognitive accountability, and provides an outline of how these standards apply. In doing so, he seems to me to be providing grounds for distinguishing between sciences and those nonsciences appearing on Laudan's list. Apart from the desiderata, Hempel perceives in science some "quite generally acknowledged norms" (ibid., 93) for example, the requirement that scientific reasoning must comply with standards of deductive logic and the interdiction against logical inconsistencies; norms governing methods of measurement and the testing of statistical hypotheses. Such relatively precise norms, together with the desiderata, in effect define scientific appraisal of theories.

Bas van Fraassen (1980, 87–89) chooses to call the Hempel desiderata "virtues" of theories, and argues that the virtues have a pragmatic, not an epistemic, status. Consistency, empirical adequacy, and empirical strength are designated as epistemic traits of scientific theories. But, argues van Fraassen, in accepting a theory, we are not just believing it, we are also committing ourselves to certain wagers: that the research program that generated the theory is worth pursuing, that all relevant phenomena will be captured by the theory, and the like. Epistemic features of a theory (Hempel's "norms") have to do with a theory's relationship to the world; pragmatic virtues of a theory have to do with what we as human knowers want the theory to achieve over and above our belief that it is empirically adequate.

What We Learn from the Four Strategies

A bit of summing up is in order. The strategy of Kant, although it may be the most desirable one, *if achievable,* must wait upon further developments in the sciences before it can be sustained. As a philosophical program it cannot establish what it sets out to conclude: that there is just one set of species-universal conditions on the basis of which we can arrive at reliable knowing, defined by methodologies appropriate to a science regarded as fixed and finished.

As I have suggested, the strategy of Rescher has some value. It invites us to direct attention away from the content of sciences and their competitors and to features of method. In addition, this strategy, as we have seen, replaces a priori demarcation criteria with a posteriori evaluations of success. However, we are now in a position to see that accuracy in prediction and control of the environment are desiderata or virtues of theories, and must be taken into account in the full appraisal of competing theories, but that they cannot by themselves distinguish between sciences and nonsciences. If we leave aside nonepistemic goals scientific or nonscientific theories might serve—for example, improvement of various aspects of human and other animal life—and direct our attention solely to epistemic goals (those whose realization employs, in the sciences, the norms Hempel mentions), and construe these by reference to empirical adequacy, empirical support, and consistency, accuracy in prediction and success in control are clearly not decisive factors in marking off science from nonscience.

What the strategies of Grünbaum and Hempel (and of van Fraassen) enjoin is attention to the norms of cognitive appraisal that must be actually operative in determining a theory's adequacy as a set of claims *about the world.* Here, the constraints on theory are the epistemic ones of deductive consistency (and of compliance with norms of nonfallacious ordinary reasoning) and empirical *support.* Therefore, for purposes of fair comparison of theories thought to be scientific with those thought not to be scientific, we need to emphasize that, other nonepistemic things being equal, the aim of all such theories is *to tell the truth about some features of the world.* What is to be investigated, then, are the credentials a given theory has offered on behalf of its entitlement to be taken seriously as a set of empirical claims. In other words, we need to know what a given theory can offer in the way of support for its claims to empirical knowledge. In what follows, I will show a way of making a fairly precise distinction between those theories with acceptable credentials, and those lacking them.

A New Strategy of Demarcation

MAN is the Interpreter of Nature, Science the right interpretation
—**William Whewell,** *Novum Organon Renovatum*

Man interprets nature in many ways, by means of science, poetry, walks in the woods, sailing on dangerous seas, telling age-old stories, seeking spirits in trees and clouds, suns and pestilences. I am interested here only in those interpretations that are cognitive claims, or texts. Texts in this sense are of course also subject to interpretation. What is to be avoided is thinking of nature as a text. Nature is interpreted, but is not a text. Symptoms, microscope slides containing slices of tissue, animal footprints, all require interpretation, but are not texts.[4] If one thinks of all the kinds of texts there are, the problem of learning how to interpret them arises: Given a text of one sort or another, what are the permissible interpretations? Goodman has of course raised this question in a number of his writings (see Goodman and Elgin 1988, part 2, sec. 3). I think we should begin to take it seriously.

Goodman seems to me correct in thinking that texts are defined syntactically, and that interpretation of meaning is a matter of semantics. Texts are normally strings of inscriptions or vocal sounds. There are exceptions. Some texts are pictorial, and only context will distinguish text from design. What the old grammarians used to call *hypotyposis* or *subjectio sub adspectum,* referring to representational exhibition of meaning with sensuous clarity, encompassed both metaphor and pictorial image, both poetic symbol and visible sign. Thus I will want to include mandalas and international road signs as texts.[5] We can then generate a long list of different kinds of texts, and we can begin to wonder how many good or right interpretations are possible for each kind. A sentence describing chemical and morphological features of amethyst samples appears to be correctly interpretable in only one way, or in very few permissible ways, whereas

Shakespeare's *Hamlet* may have a number of right interpretations. The Hamlet of Olivier and the Hamlet of Zeffirelli are both acceptable readings of Shakespeare's text.

These considerations lead us to the idea of *a range of permissible interpretations* of a given text, and to the equally suggestive idea that kinds of texts can be distinguished from one another by *the size of that range*. Think then of a scale of texts ranging from those for which only one interpretation is permissible to those for which any number of interpretations are permissible. At one end of that scale will be, perhaps among others, those texts stating claims in formal systems of mathematics and logic, where the texts have only one permissible interpretation, just because no semantics is involved. To have only one permissible interpretation is equivalent, I think, to requiring no interpretation at all. International road signs will be at this extreme end of the scale, as will texts stating standardized rules of a game, say, chess.[6] At the other extreme end of the scale, what one might call the limit of interpretation, is nonsense or absurdity, that textual gibberish for which any interpretation at all will do. I take it that allowing any interpretation at all is equivalent to having no permissible interpretations. So we might say that gibberish is "beyond interpretation." At other places along the scale will be all of the other kinds of texts, each with a different range of permissible interpretations. Our question now becomes as follows: Where on this scale do we place the various texts of the natural sciences, and where, the texts of the various enterprises thought intuitively to be classed as pseudosciences? Since it is clearly in the interests of all parties to this dispute to have their range of textual interpretations as close to 1 as possible, a collateral question is as follows: How, in the cases of those texts where truth-telling is a central goal, does one limit the range of required hemeneutic insight, the range of permissible interpretations? We will see that the answers to both questions relate also to what we can reasonably mean by support for textual claims of various sorts.

We need to limit ourselves to those texts seeking to tell the truth about the world. They might seek to do other things as well, but for the purposes of the present exercise employment of rhetoric ingredient in a given text will be ignored. We are interested, then, only in those texts which, literally construed, can be either true or false. Notice that I say "*can be*" either true or false. I think it is totally counterproductive to argue for a distinction between sciences and pseudosciences on the basis of acceptance of the form of scientific realism which teaches that to accept a scientific theory is to believe that it is true. The pseudoscientist is always in a position to counter this view with the correct observation that no texts stating views about the world have ever been conclusively established as true, hence none are ever believable because true. This is the thin edge of the wedge actually employed by many who challenge the epistemic credentials of the sciences: Since no view has ever been conclusively established as true, mine is as good as yours, and on altogether proper epistemic grounds. Because the pseudoscientist is always in a position to exploit the logical contingency of generalizations over items of experience, van Fraassen's antircalism seems to be the best philosophical standpoint from which to argue with the pseudoscientists. Van Fraassen holds that although the language of science must be literally construed, and hence that such language can be either true or false, acceptance of a theory only amounts to believing the theory is empirically adequate, and possesses certain virtues (1980, 10–12).

In thus suggesting that van Fraassen's antirealism is a better foil to pseudoscience than is realism, I am not suggesting that this constitutes a telling reason for accepting van Fraassen's constructive empiricism. What I want to emphasize is that if the difference between science and nonscience is to be made out fairly, those of us who accept the intuitive difference had better approach the problems with a very parsimonious ontology and philosophy of theory acceptance. On one point there seems to be no fundamental disagreement: what Hempel takes to be the core norms of natural science inquiry (and which supply the basis for that cognitive accountability Grünbaum thinks to be definitive of natural science) is entirely compatible with van Fraassen's idea that the central features of science have to do with empirical adequacy, empirical support, and consistency. *At least this much* would be wanted by both realists and antirealists, and constitutes enough for us to go on.

Let us then think of the literal construal of a text as accepting that what it says can be either true or false, and as accepting nothing stronger than that. We need this idea to apply both to scientific texts and to those thought to be pseudoscientific, for claims from both quarters seek at a minimum to be empirically adequate, empirically supported, and consistent. Further, we need to endorse van Fraassen's statement that "to insist on a literal construal of the language of science is to rule out the construal of a theory as a metaphor or simile, or as intelligible only after it is 'demythologized' or subjected to some other sort of 'translation' that does not preserve logical form" (1980, 11).

It is important to note that the only form of translation allowed of sentences in organic chemistry, astrology, and creation science is translation into another natural language. For purposes of fair comparison, all contenders must state texts in one of the natural languages, which can be only translatable into any of the others.[7] There is no hidden meaning lying behind or beyond literally construed language in the sense required here. All of this granted, where do the various literally construed texts fit on the scale of texts? What is it that gives the texts of the natural sciences and those of the alleged pseudosciences different places?

My thesis is that, all things considered, the texts of the natural sciences are closer to 1 on the scale of texts with ranges of permissible interpretations than are the texts of the pseudosciences. This is to say that the texts of the natural sciences permit only one or very few interpretations, whereas those of the pseudosciences can be fit to many more legitimate interpretations. To see why this is so it is only necessary to return attention to Hempel's norms and Grünbaum's principles of cognitive accountability and to the idea of support for literally construed texts.

Science and Stability of Meaning

Hempel's desiderata and Grünbaum's implicit principles of cognitive accountability are derived mainly from work in the physical sciences, where the literally construed claims are stated with great care, are sometimes quantitative, and can even be very dangerous in consequences if not understood and applied with precision. Such claims, furthermore, require for their support experimental consequences that require exact measurements of distances and times. The language in experimental physics, chemistry and other natural sciences is thus as decisively referential (fixed in meaning) as it is possible to be, and interpretations of that language, if they are not to be taken to be completely frivolous, approach 1.[8] All properly formulated scientific discourse has a range of 1, or very near 1. This is so because such discourse must at some point make contact with public, shared, controlled observation, and because of the stability of meaning of its technical terms. This constitutes its support.[9]

We need not go to the highly sophisticated reaches of chemistry and physics, and to their applications in practical affairs, in order to understand the phenomenon of stability of meaning in the sciences.[10] Consider, for example, one of Pasteur's experiments designed to help our understanding of the impossibility of spontaneous generation. Pasteur dissolved a crystalline salt of ammonia, crystalline sugar, and phosphates from the calcination of yeast in pure distilled water, and then sowed into the liquid a few spores of *Penicillium* or some mold.[11] The spores germinated within two or three days, filling the surface of the liquid with flakes of mycelium, where they produced healthy vegetation. Pasteur's report continues, "The precaution of employing an acid salt of ammonia hinders the development of Infusoria, which would soon stop the progress of the little plant through absorbing the oxygen of the air, without which the mould cannot thrive. All the carbon of the plant is obtained from the sugar, which is slowly used up, all its nitrogen from the ammonia, its mineral matters from the phosphates. In this matter of the assimilation of nitrogen and phosphates there is thus a complete analogy between ferments, moulds, and plants of complicated organization" (1959, 228).

Pasteur next reports that if any of the dissolved principles are omitted, the growth of the fungi is arrested. The conclusion is that fungi do not obtain carbon in the same way as do phanerogamic plants, "They do not decompose carbonic acid; they do not set free oxygen. The absorption of oxygen and the setting free of carbonic acid; are on the contrary the necessary and permanent activities of their life. These facts give us precise ideas on the mode of nutrition of fungi . . ." (ibid.). This is precise enough to enable brewers, wine and vinegar makers and dairy farmers to introduce controls on their procedures based on scientific knowledge. But that is only part of the point. The larger part is that it is not possible to understand the elements of this simple inductive experiment on the basis of several interpretations. Too much precision and exact knowledge exists at every point.

What is being investigated depends upon well-understood aspects of chemistry and biology: the nature of distilled water, of calcinated yeast, of sugar, of tartrate of ammonia, the required quantities of these chemicals, the classification of plants with organs of reproduction as phanerogamic, the behavior of Infusoria, and the kind of nutrition involved in such plants. All such matters are expressed in language whose meaning is fixed; there is no room for gratuitous interpretation. There is room for disagreement, but that is another matter. Notice also that I have deliberately employed a translation of Pasteur's paper, first published in *Annales de Chimie et de Physique* in 1862. The translation does not "demythologize," it simply transfers exact language from one natural language into another. For purposes of exact science, all natural languages are one language, construed literally.

One final point about the experiment is that what Pasteur wrote can be *misunderstood*. Furthermore, Pasteur could have set out to deliberately *deceive* us. I see no insurmountable problems here. With respect to texts whose range of interpretations approaches 1, misunderstanding can be corrected, perhaps not easily, but surely in time. A misunderstanding is, in the end, a wrong interpretation, and wrong interpretations are not permissible interpretations. Any text can set out to deceive us, even in the formal sciences. Here again, the detection of deception, although in some cases it may be very difficult to realize, is always in principle realizable for those texts whose range of interpretation approaches 1. The possibility of fraud does arise in the work of natural scientists; it can never arise in the work of poets.[12]

What about astrology, creationism, and other alleged pseudosciences, can their texts be literally construed and empirically supported, establishing that the ranges of interpretation are close to 1?

Astrology and Its Support

Astrological texts are charts based on calculations of positions of heavenly bodies at various times, which relate such positions causally to events in people's lives. These results are often stated in sentences that assess trends. One can find such sentences in most newspapers published in the world today. Some condemn astrology because it does not accurately predict precisely events that will determine the course of someone's life. But careful astrologers make more cautious claims, "Astrologers are not magicians. They do not deal in specific events and must always take care that no client is injured through taking their advice too literally" (Parker and Parker 1979, 154); and:

> Serious astrologers do not "predict" events; they assess trends. When mentioning potential hazards or beneficial periods, the astrologer must always

remember to put these matters carefully to his client. He should never state: "You will have an accident on the twenty-first of the month," but he can say that: "On the twenty-first of the month you are likely to be somewhat prone to risks, so take additional care when handling hot dishes or sharp knives." Of course, in this way astrologers do lay themselves open to the criticism of using generalizations. It must be made clear to a client that progressions[13] are rather like weather forecasts, and should be used in much the same way. (Ibid.)

A host of questions arise. If astrological results cannot predict, how do we determine the success of astrological claims? If the advice given by astrologers is general, how can it avoid vagueness and loss of clarity? If the advice does not speak to specific events, are we not invited to interpret it in very many ways—that is, to what I take it to mean for my life, to what you take it to mean for yours, and so on? Are not publishing the results of theoretical and experimental work and advising clients two entirely *and importantly* different matters? Does the casting of charts have any valid relationship to the advice given a client? What, after all, is the relationship between the calculations and the general advice? That advice could, for all we know, eventuate from many other kinds of sources. Is it really possible, given these questions, to construe astrological advice literally?

Consider the following item from *The Globe and Mail* (1991):

Having a Hussein Week
Saddam Hussein's birthday is April 28. Here is some of Jeane Dixon's advice to Taurus last week:
Jan. 15 (UN's Tuesday deadline): Indulging your whims could prove costly today. Think about the image you want to present.
Jan. 16 (U.S. planes hit Baghdad, including its communications centre): Spending time alone will help you extricate yourself from a difficult situation.
Jan. 17 (Iraqi missiles strike Israel): Do what must be done at home. You can easily catch up with neglected tasks.
Jan. 18 (Jordanian parliament brands United States the Great Satan): A close relationship intensifies; two minds think as one. You sense what will happen next. Now is the time to make important connections.

I suppose *The Globe and Mail* thought this piece to be humorous; if each bit of advice is construed literally, then it would seem that Saddam has failed to heed the warnings. However, even if construed literally, each piece of advice is permissibly subject to different interpretations since the advice is general and does not speak to Saddam or any other individual. (Another question is if the advice is this general, over what instances does it generalize? the life charts of everyone born under the sign of Taurus? the character traits mythologically associated with each sign of the Zodiac? The charts and the traits are themselves open to endless interpretation.) Saddam need not have thought that failure to meet the UN deadline was whimsical; he might have wanted to project exactly the image of one who acts in defiance of the deadline. Perhaps he thought that remaining in his bunker was exactly the way to prevent his own destruction. Perhaps he thought that sending missiles to Israel was exactly the way to look after the home front.

How, then, does the astrologer attempt to limit the range of permissible interpretations of the claims involved in the advice? One way is to insist that the kind of advice given in Jeane Dixon's newspaper columns is meant only as a very loose guide. To make the advice more precise, the astrologer must look at an individual's birth chart. We have seen that even this more personally tailored advice must remain nonspecific, so it seems that no gain in restriction of interpretations is forthcoming from individualized treatment. At another level, the astrologer seeks support, not for the generalized advice, but for the entire speculative enterprise of astrology, at which point the Rescher strategy can be invoked to show that no accuracy of prediction or success in control is forthcoming from astrology. While this may be true, it is important, I think, to look briefly at what at least some astrologers present as support for their activity.

In the absence of an account of how positions of heavenly bodies can be causally related to either specific or general advice, astrologers look for support for their speculations elsewhere. A common move amounts to an appeal to authority: Aquinas's acceptance of astrology (so long as necromancy is not included) as a science, and Kepler's casting of astrological charts. Parker and Parker (1979) appeals to Jung's interest in astrology as "the first important stage in the slow revival of scientific interest in the topic [the causal relationship between planetary movements and events in human life] which we are witnessing today" (p. 59). They continue, "What is important . . . is the fact that a man recognized throughout the world as one of the greatest and most provocative thinkers of the century should have considered the data of astrology sufficiently interesting in principle not only to experiment with it, but also to include it as an integral part of his own cosmological theory" (ibid.).

If we drop the vacuous appeal to authority, what is left to support astrology and provide warrant for limiting interpretations of its claims? I think the obvious answer is "nothing." On the question of causal influence of planetary motions on human action, some astrologers point out that various forms of action at a distance are allowed by scientists, that the biological clock is an influence akin to astrological causality, and the like (ibid., 50–57). Such appeals simply complicate the speculations, lending unsubstantiated additional claims to the already heavy baggage of interpretation involved in astrology. As if this were not enough, astrologers often seek to find support for their speculations by appealing to the long history of widespread interest in their "science," and to such seemingly unrelated and highly symbolic matters as the pyramids, Stonehenge as an ancient cosmic computer, the celestial bestiary, the fact that traditionally signs of the Zodiac have been related to various parts of the human body, the close connections between mythology and astrology, even to Culpeper's herbal remedies! (For details, see Parker and Parker.)

It thus appears that, in order to take the texts of astrology to be well supported, we are driven further and further away from considerations of empirical adequacy, and closer and closer to the need for acceptance of a complete, and thoroughly questionable, metaphysical system. That system is built up of the kinds of appeals just listed, all of which boil down to being appeals based on tradition, history, community of interests, authority, collection of apparently meaningfully connected symbols (perhaps even Jung's archetypes of the collective unconscious). The consequence is that astrology cannot have its texts restricted to 1 or a few interpretations. Indeed, astrology appears as the pseudoscience *par excellence* because it is the "science" of interpretation *par excellence*. Since the range of permissible hermeneutic insight required to understand astrological claims is so large, the kind of support it seeks cannot place it close to the sciences, whose texts have a narrow range of permissible interpretations that approaches 1.

Creationism and Its Support

The same is true of creationism, although for very different reasons. Astrology simply requires acceptance of too much nonscience in order to be considered a science. The difficulty with creationism, as an alternative to scientific biology, especially the theory of evolution, is one of a multiplicity of permissible *translations*. What I will now put forward is not new, but I hope the new context I am endeavoring to supply will justify saying it once more. In brief, creationism accepts the account of the creation of life forms found in the book of Genesis of the Bible. God created everything in the heavens and on earth in six days, starting at a time. Creationists disagree about the year in which God began his work (as well they might; it is difficult to imagine how years could be counted *before* God created the heavens!). Because creationists believe this account is a true theory, they of course want it to be taught in biology classes, as a theory in competition with the theory of evolution.

If the book of Genesis is literally construed, how are we to understand the support for its claims? At the end of the first day of creation, there were God, and light and darkness. How long was that day? We are not told how God measured time. We cannot proceed with the attempt to confirm the claim that on the first day God made light and darkness. Furthermore, God's spirit was moving on the waters when he created light and darkness. Are the waters created or not? Presumably not. Again, an incoherent claim. Obviously, literal construal of the book of Genesis, or of any book of the Bible, is impossible. We could at this point simply conclude that the text of the Bible is literary, not literal, and hence is of course subject to differing interpretations. But this would not be fair to the creationists, who are also fundamentalists, and who want to teach Genesis alongside Darwin because they think that what Genesis says is not just to be literally construed, but is literally true.

The problem for the creationist lies elsewhere. To which book of Genesis do we appeal? the original Hebrew version? Luther's translation? Numerous versions of the Bible are available: the 1611 King James version, the 1979–1982 new King James version, the Revised Standard version, the New Revised Standard version (noted for its attribution of gender-free language to God), the New English Bible version, the Revised English Bible (in an ecumenical effort, "Let us make man in our image" becomes "Let us make human beings in our image"—creating a hermaphroditic deity, now we are getting somewhere!), the New International version (the best-selling version in the United States), the New Jewish version, the new American Bible (the most widely recognized Roman Catholic version) and more. We have seen that in the case of scientific languages literally construed, translation from one language into another is relatively unproblematic because of the stability of meaning of technical terms and the absence of metaphor. This does not make scientific texts fun to read, but surely makes them easy to translate.

So we see that the problem for the creationist is twofold, if the desired limitation of the range of permissible interpretations is to be accomplished. First, it must be shown that the text of Genesis can be literally construed, so that it is at least in principle capable of empirical adequacy. As suggested, a literal construal of the text of Genesis turns its sentences into either falsehoods or logical incoherencies. Second, it must be shown either that one version of the Bible is the only correct one, or that all versions are intertranslatable without loss of logical form or of literal reference. How can either alternative be rationally sustained? It cannot.

There is a final move open to the creationist. For it is possible to argue that there are sciences and sciences, and that creationism is a human, not a natural, science. The support for construing any version of the Bible literally comes not from the stability of reference of technical terms and from the absence of metaphor, but from facts having to do with the actions of human beings—again, from history, community, tradition, authority, and shared faith. Restriction of the range of hermeneutic insight required for interpreting biblical texts comes not from the collection of more confirming instances (corroboration), but from the conversion of more witnesses providing sympathetic testimony (vindication). If one seeks to justify the scientific status of creationism along this line, it would appear that creationism, like the human sciences generally, needs to provide adequate justification for the new kinds of evidential appeal. It cannot make good its claims to equal treatment on natural scientific grounds, but must appeal to what until further notice must be regarded as nonscientific forms of evidence.

I hope to have shown that at least two systems of speculation thought generally to be pseudosciences are indeed pseudosciences, and that the reason for this is that their ranges of permissible interpretations are either large, or can only be restricted by appealing to kinds of support which are scientifically unavailing. On my scale of ranges of permissible interpretations, where do the other pseudosciences Laudan lists find themselves? None of them have ranges that approach 1. Some—Rosicrucians, the-world-is-about-to-enders, probably charioteers of the gods—have ranges at the other limit; their texts are gibberish. Some may even fall off the scale and approach insanity: flat earthers, circle squarers, *perpetuum mobile* builders. Still others, literally construed, can be either confirmed or disconfirmed. Nessie might one day be found and captured, Big Foot might one day be bagged. To date, alleged sightings of both creatures seem doubtful. So-called anomalous phenomena like the two monsters and UFOs may be worthy of scientific investigation, if we think society can bear the cost. Primal screamers have only themselves to blame.

NOTES

I mention in the text some of those who have helped to shape my thinking about science. The contribution of Adolf Grünbaum is simply incomparable, and is here acknowledged with warm thanks. I have also learned about the science–pseudoscience divide from my conversations and correspondence with Andrew Lugg (see Lugg 1987, and forthcoming).

1. The employment of theories of meaning to help in drawing the line between science and nonscience now seems a lost cause. The effort was to show that the formal and empirical sciences contain sentences that are cognitively significant, whereas the sentences of nonsciences are in some respect non*sense*. However, "sense" is a notoriously ambiguous word. The attempts at demarcation along such lines (whose ancestor is the Kantian strategy to be discussed in the next section) could never answer the rejoinder of Nietzsche:

> It is no different with the faith with which so many materialistic scientists rest content nowadays, the faith in a world that is supposed to have its equivalent and its measure in human thought and human variations—a 'world of truth' that can be mastered completely and forever with the aid of our square little reason. What? Do we really want to permit existence to be degraded for us like this—reduced to a mere exercise for a calculator and an indoor diversion for mathematicians? Above all, one should not wish to divest existence of its *rich ambiguity (seines vieldeutigen Characters)*: that is a dictate of good taste, gentlemen, the taste of reverence for everything that lies beyond your horizon. That the only justifiable interpretation of the world should be one in which *you* are justified because one can continue to work and to research scientifically in *your* sense (you really mean, mechanistically?)—an interpretation that permits counting, calculating, weighing, seeing, and touching, and nothing more—that is a crudity and naiveté, assuming that it is not a mental illness, an idiocy.
>
> Would it not be rather probable that, conversely, precisely the most superficial and external aspect of existence—what is most apparent, its skin and sensualization—would be grasped first—and might even be the only thing that allowed itself to be grasped? A 'scientific' interpretation of the world, as you understand it, might therefore still be one of the *most stupid* of all possible interpretations of the world, meaning that it would be one of the poorest in meaning. (Nietzsche 1974, sec. 373)

2. Over the centuries, the list has not changed that much. Here is one from Reginald Scot (1665), "the Knavery of *Juglers, Conjurers, Charmers, Soothsay-*

ers, Figure-Casters [astrologers], *Dreamers, Alchymists* and *Philterers* [sellers of erotic love potions] . . . " (p. i).

3. In the *Opus postumum* Kant altered this assessment of the cognitive status of chemistry, and tried to give a justification of scientific status of the work of Lavoisier.

4. Anything that can be counted as a system of symbols subject to interpretation is a text. This list will be very long, but will surely contain the following: proofs of theorems, reports of results of observation or experiment, narratives (histories, news releases, novels, travelogues, stories told in psychoanalytic clinical situations), police reports, lawyers' arguments, judges' decisions, poems, clinical observations, mandalas, cryptograms, messages from another world (as in seances, contact with extraterrestrials), pictograms (cave and rock paintings), works of visual art, plays, operas, musical scores, astrological charts and advice, records of divine revelations, ceremonies (Christmas, the Festival of Lights), tribal lore transmitted verbally, the writings of philosophers, metal coins, political speeches, sermons, academic lectures, street and highway signs.

5. Although again context is important. In the neighborhood of my birthplace in Lancaster County, Pennsylvania many Amish towns have charming names. The signs read "Paradise," "Intercourse," "Blue Ball," "Bird-in-Hand." Undergraduates at the University of Pennsylvania in Philadelphia are fond of stealing these signs for use as decorations in their dormitory rooms. What were signs thus become trophies, and are not to be taken literally.

6. I am aware that one can be deceived by some road signs, especially those marking miles or kilometers. One summer, a student of mine worked for the Department of Highways; he took delight in placing mileage markers at random, rather than at precisely determined intervals. Max Black used to say that "it isn't a language unless you can lie in it." It behooves all of us to face all texts with a measure of skepticism. Even mathematical texts can be misprinted. No important epistemological point turns on these admissions.

7. I must admit that sentences in scientific reports are more readily translatable into other natural languages partly because much of the terminology is standardized, and the mathematics employed is normally semantically transparent. Such ease in moving from one language to another in science is a stroke of good fortune not also available to the pseudosciences. Translations of texts in astrology, creationism and the like might preserve logical form, but they inevitably introduce ambiguities and nuances not present in the source language. To which version of the Bible should creationists direct us? I will return to this problem later.

8. A pragmatic consideration here gives weight to Rescher's emphasis upon accurate prediction: If some experiments are not carried out with the greatest attention to all details, the experimenters are at peril. What is lost if an astrological prediction turns out to be false? Perhaps the client may murder the astrologer. That is quite different from the death of a careless experimenter.

9. It might be argued that stability of meaning of technical terms is assured by the *community* of scientists, and that similar communities of astrologers and creationists and others create their own assured stability for the meaning of their technical terms. I would not deny that fraternal conditions play some role in fix-

ing meaning. What one needs to consider here is the nature of the conditions of entry into different fraternal groups. Rigorous education and satisfaction of high intellectual standards form part of those conditions for scientists. No such training and accomplishment are required of young astrologers; all that is needed for entry into the fundamentalist community is an unwavering faith in the literal correctness of biblical claims. Indeed, further study of such conditions of entry into intellectual and quasi-intellectual communities might highlight still more ways in which the sciences and the pseudosciences differ fundamentally.

10. I am emphasizing features of the discourse of experimental and observational natural science, because I am dealing only with those texts that seek to tell truths about the world. Such an emphasis, however, should not disguise the fact that within *theoretical* natural science, interpretations of theoretical constructs are themselves part of the internal dialectic of physical science. Thus, for example, the concept of gravity in classical Newtonian physics can be interpreted as action at a distance, or as involving electromagnetic fields. So far as the mathematics and the empirical support of the classical theory are concerned, these different interpretations are equivalent, and play only heuristic roles. Like Kant's ideas of reason, they have no empirical truth-telling content. I am grateful to my colleague Robert DiSalle for calling these matters to my attention.

11. The liquid consisted of 100 parts pure water, 10 parts sugar, 0.2 to 0.5 of ammonia, and 0.1 of ash of yeast.

12. However, notice David Hume's amazing attempt to rescue some literal sense for poetry, "Poets themselves, tho' liars by profession, always endeavor to give an air of truth to their fictions; and where that is totally neglected, their performances, however ingenious, will never be able to afford much pleasure" (1951, 121).

13. Birth charts are "progressed" by, for example, taking the positions of the planets on the first day after birth to relate to the conditions in the first year after birth, the positions in the fifth day after birth, to the fifth year after birth, and so on for all days after the birth date. These progressions are employed to assess trends in the life of a person.

REFERENCES

Butts, R. E. 1986a. "Introduction: Kant's Quest for a Method for Metaphysics." In R. E. Butts, ed., *Kant's Philosophy of Physical Science: Metaphysische Anfangsgründe der Naturwissenschaft, 1786–1986*. Dordrecht: Reidel, pp. 1–24.

———. 1986b. *Kant and the Double Government Methodology: Supersensibility and Method in Kant's Philosophy of Science*. 2d ed. Dordrecht: Reidel.

———. 1986c. "The Methodological Structure of Kant's Metaphysics of Science." In R. E. Butts, ed., *Kant's Philosophy of Physical Science: Metaphysische Anfangsgründe der Naturwissenschaft, 1786–1986*. Dordrecht: Reidel, pp. 163–99.

Friedman, M. 1986. "The Metaphysical Foundations of Newtonian Science." In R. E. Butts, ed., *Kant's Philosophy of Physical Science: Metaphysische Anfangsgründe der Naturwissenschaft, 1786–1986*. Dordrecht: Reidel, pp. 25–60.

Goodman, N., and C. Z. Elgin. 1988. *Reconceptions in Philosophy and Other Arts and Sciences*. Indianapolis: Hackett.

Grünbaum, A. 1980. "Epistemological Liabilities of the Clinical Appraisal of Psychoanalytic Theory." *Noûs* 14:307–85.

———. 1983. "Logical Foundations of Psychoanalytic Theory." *Erkenntnis 19:*109–52.

———. 1984. *The Foundations of Psychoanalysis: A Philosophical Critique*. Berkeley and Los Angeles: University of California Press.

———. 1989a. "The Degeneration of Popper's Theory of Demarcation." *Epistemologica 12:*235–60.

———. 1989b. "Why Thematic Kinships between Events Do Not Attest Their Causal Linkage." In J. R. Brown and J. Mittelstrass, eds., *An Intimate Relation: Studies in History and Philosophy of Science Presented to Robert E. Butts on His 60th Birthday. Boston Studies in Philosophy of Science*. Dordrecht: Kluwer, pp. 477–94.

———. 1990. " 'Meaning' Connections and Causal Connections in the Human Sciences: The Poverty of Hermeneutic Philosophy." *Journal of the American Psychoanalytic Association 38:559–77*.

Hempel, C. G. 1983. "Valuation and Objectivity in Science." In R. S. Cohen, and L. Laudan, eds., *Physics, Philosophy and Psychoanalysis: Essays in Honor of Adolf Grünbaum*. Dordrecht: Reidel, pp. 73–100.

Hume, D. 1951. *A Treatise of Human Nature*. Reprinted from the original edition in three volumes. Edited by L. A. Selby-Bigge. Oxford: Clarendon Press.

Laudan, L. 1983. "The Demise of the Demarcation Problem." In R. S. Cohen, and L. Laudan, eds., *Physics, Philosophy and Psychoanalysis: Essays in Honor of Adolf Grünbaum*. Dordrecht: Reidel, pp. 111–27.

Lugg, A. 1987. "Bunkham, Flim-flam and Quackery." *Dialectica 41:*221–30.

———. Forthcoming. "Stuff and Nonsense: Pseudoscience as Madness."

Nietzsche, F. 1974. *The Gay Science*. Translated by W. Kaufmann. New York: Vintage Books.

Parker, D., and J. Parker. 1979. *The Compleat Astrologer*. New York: Bantam Books.

Pasteur, M. L. 1959. "Memoir on the Organized Corpuscles which Exist in the Atmosphere." In W. C. Dampier, and M. Dampier, eds., *Readings in the Literature of Science*. New York: Harper & Brothers, pp. 217–29.

Rescher, N. 1970. *Scientific Explanation*. New York: The Free Press.

———. 1977. *Methodological Pragmatism: A Systems-theoretic Approach to the Theory of Knowledge*. New York: New York University Press.

Scot, R. 1665. *The Discovery of Witchcraft*. London: n.p.

van Fraassen, B. C. 1980. *The Scientific Image*. Oxford: Clarendon Press.

8

The End of Epistemology?

Paul Feyerabend
University of California, Berkeley, and Federal Institute of Technology, Zürich, Switzerland

Epistemology arose in the West in connection with the "ancient battle between philosophy and poetry" (Plato *Republic* 607b6f). In this battle a small group of social critics, soon called philosophers, opposed traditional customs, attitudes, and ways of reasoning. Continuing the Ionian trend of unification which looked for a single substance behind the manifold events of this world, Parmenides declared Being to be the most fundamental substance of all. What can be said about Being? That it is and that not-Being is not. Since the only alternative to Being is not-Being, Parmenides inferred that, basically, the world never changes and that it is the same everywhere (compare the modern assumption that fundamental laws are valid everywhere and at all times, even under the extreme conditions of the first few moments of the Big Bang). Using later terminology we can say that according to Parmenides, change, though supported by experience and tradition (*e'thos poly'peiron*—Diels and Kranz 1957, B7,3) is unreal, that statements about it are mere opinion, that knowledge is produced by reasoning and that it is restricted to the statement "Being Is" (*estin*—ibid. B7,7). This was the first and, at the same time, the most concise theory of knowledge (*estin* was also the first explicit conversation principle—it asserted the conversation of Being).

The ancient history of the subject had two branches, one abstract, the other political (in a wide sense of the word). The branches were closely related. Protagoras's "Man is the measure of all things" (Plato

Theaetetus 152a2f) implied that the laws of a city which were usually ascribed to a god or to wise ancestors could be judged and changed by ordinary human beings. Laws were necessary for order; whoever broke them not once, but repeatedly, had to be "put to death as a plague on the city" (Plato, *Protagoras* 320ff); but they were not sacrosanct. New conditions and new wishes required new laws and new forms of knowledge. This "relativism" had little in common with modern movements of the same name. It involved experts and explained how an appeal to them could be reconciled with the man-measure doctrine (ibid. 166dlff). Plato, on the other hand, wanted greater stability. Plato and Protagoras also dabbled in practical politics, the one in Sicily, the other, according to some reports, by devising a constitution for Thurii, a Panhellenic settlement in Southern Italy. Thus an early debate between "realism" and "relativism" had major political implications.

This becomes especially clear in the case of Aristotle. Aristotle criticized Parmenides (and some features of the theory of ideas) in two ways. He analyzed the arguments, and he ridiculed the results by choosing common sense and the physics of his time as measures. The objects of physics and common sense have parts and change; common sense, in addition, sustains life in the city. It would therefore be preposterous ("next door to lunacy" Aristotle, *de gen. et corr.* 325al8ff) to regard either as "unreal." Real, the criticism implies, is what plays an important role in our research and in the kind of life we want to lead and epistemological standards must be adapted to both.

Some recent developments in the philosophy of science are the result of a first tentative, and then a more resolute, application of this principle. Without any reference to Aristotle and often without being aware of his ideas, writers such as Husserl, T. S. Eliot,[1] Wittgenstein and J. L. Austin have described the abyss that separates philosophical theories from the way we think and act in this world while historians are starting to suspect that philosophy and scientific practice are equally distinct. This does not mean that science is "irrational." There exist arguments for any move that is being made. But whatever universality we find comes from the actions of scientists who, having immersed themselves in various problem situations and using their experience, cunning and other forms of "tacit knowledge" now happen to treat different situations as if they were the same: Universal standards of scientific knowledge are contingent features of scientific practice (which may change from one research episode to the next), not transhistorical (and necessary) agencies.[2] A theory of knowledge involving such agencies is not only dead, but it never had a chance of coming to life in our world.[3] We must not be misled by the fact that philosophical ideas had a long and distinguished career and entered many enterprises; they were effective because they worked like other historical forces, that is, in a thoroughly unphilosophical way.[4] To illustrate this situation I will now discuss a recent rescue mission for old-style epistemologies: Larry Laudan's *Science and Relativism* (1991).

The book contains a debate between four characters—a relativist, a pragmatist, a realist and a positivist. Relativism is the main target, pragmatism the philosophy closest to Laudan's heart. The topic chosen by the philosophers is the nature and growth of scientific *knowledge*. What is actually being discussed is high theory, to use a term from high energy physics (HEP). This leaves out (in an incomplete list): botany, zoology, areas of science with even looser forms of classification (Hubble's classification of galaxies, Kinsman's classification of ocean waves, the Hertzsprung-Russell diagram, and so on), geography, ecology, descriptive meteorology—in short, all those subjects which accumulate and order shapes without using more abstract concepts. It also leaves out phenomenology (again in the sense of HEP), experimental design, practically oriented approaches (Noll's mechanics) and models (for example, the various models used to calculate lunar perturbations). Some models (the two-dimensional models in electrostatics, hydrodynamics and especially in quantum field theory) are unrealistic from the start, but they are easy to calculate, lead to interesting problems and their production may turn into a veritable industry. (Here science more than matches philosophy.) Is the industry useless? unscientific? a mere game? It almost seems so, until two-dimensional and even one-dimensional structures turn up in fashionable enterprises (string theory). The systematic collection of measurements and observations has given rise to special disciplines with standards of their own and with a considerable sense of self-importance. Positional astronomy is an example, "During the century and a half following 1755 when Bradley made the first accurate determinations of star positions, this class of observations constituted the main activity of most of the observatories of the world. It consumed the energies of considerably more than half the working force of astronomers during the nineteenth century" (Wilson 1960, 124).[5] Observations are not always presented in a pure form. They may be processed by low-level assumptions and theories may fail because they conflict with the assumptions, not with the observations themselves (see Feynman 1984). Then there are the catalogs of celestial objects such as the Messier catalog or the Cambridge catalogs of radio astronomy. Some arguments do not use statements, or tables; they use diagrams, computer models, even movies. It was a film and a chart that convinced James Peebles of the existence of cosmic bubbles (Regis 1989).[6] On the level of theory, scientists use approximations, special postulates designed to simplify speculation (such as the perfect cosmological principle), far-reaching ad hoc hypotheses (such as Dirac's ocean of occupied states) and other tricks to bring coherence into an otherwise untidy mass of material. No doubt Laudan believes that these idiosyncrasies of the particular are transitory wrinkles, that they are always smoothed out and that the details I mentioned are just steps on the way. The trouble is that little research was done in the past to check this belief—it was simply taken for granted that science is an orderly hierarchy of theories and facts of different generality with logical nets in between—and that more recent inquiries (done by Hacking, van Fraassen, Primas, Cartwright, and the sociology of science) have thrown serious doubt upon it. Whatever unit arises in the course of Laudan's debate is therefore a unity of faith, neglect, ignorance or double-talk, not a unity produced by a more careful analysis. (The same applies to the uniform divine meaning postulated by biblical fundamentalists.)

All the participants in Laudan's debate make general statements about this chimera. Epistemology, says the pragmatist, did not vanish with the demise of certainty. There "remain important epistemic questions" (Laudan 1991, 136) such as how best to construct theories, when to regard a theory as well supported, when to prefer it to a rival.

These are important questions indeed. They faced Einstein when he wanted to combine the insights of special relativity with the principle of equivalence; they faced scientists later on when the Brans-Dicke suggestions led to difficult experiments and a proliferation of theories of gravitation; they accompany every step of scientific research. However, can they be answered by a person who looks at science from the outside; is unaware of its divergent ingredients; lacks the mathematical skills, the judgment and especially the "tacit knowledge" which define an area of inquiry and which are unavailable to those not actively participating in the enterprise? The older epistemologists for whom such details were accidental modifications of a coherent structure that could be detached from science and judged independently said yes. Laudan agrees. True, he rejects "perfect-being epistemolog[ies]" (ibid., 105f). He wants to evaluate scientific standards empirically, using "any of the forms of reasoning appropriate to scientific research" (ibid., 99); science itself is supposed to tell us which standards are acceptable and which are not. But as I just pointed out the science Laudan discusses is a caricature, not the real thing, and the "forms of reasoning" he refers to are those of a logician dealing with stable entities of well-defined content, not those of a scientist facing incomplete and incoherent material. This difference vitiates the whole approach.

Our theories, says the pragmatist, "are worthy of acceptance . . . because they work. . . . [T]hey confer abilities on us—abilities to control, predict and manipulate nature" (ibid., 106–107). Furthermore, "our rules are worthy of acceptance, because they have shown themselves to be able consistency to pick out theories which work with a pretty high degree of reliability" (ibid., 106).

The quoted passages suggest the following scenario: There are rules (standards, norms), and there are theories. The rules are let loose on the theories and encouraged to make their favorite choice. Some rules choose losers—they are discarded. Others "consistently . . . pick out theories which work"—they are invited into the Epistemological Hall of Fame.

The scenario is most unrealistic. First, because many theories which have surprising successes in one area fail in others. Take Newton's mechanics, "The true aim of celestial mechanics is not to calculate ephemerides, for here one can always be content with short term predictions, but to find out if Newton's laws are capable of explaining all phenomena" (Poincaré 1892, 1). Poincaré was not speaking as a falsificationist, nor did he have particular discrepancies in mind; he meant the fact that the then customary series developments, though satisfactory for short times, diverged for longer periods. They could not be used to probe the overall correctness of Newton's theory or to attack problems such as the problem of the stability of the planetary system. This problem existed already at Newton's time (the great inequality of Jupiter and Saturn), it was inaccessible to Newton's geometrical procedure, seemed to resist analytic methods until Laplace considered higher approximations (and found the inequality to have a period of 929 years—Wilson 1985), returned with the realization that Laplace's series diverged, was aggravated by the proof (Bruns 1887) that no quantitative methods other than series developments could solve the problem and was then again attacked, but on the basis of an entirely new (topological) approach. Less subtle, though not less complex, difficulties, all in the nonrelativistic domain, lasted well into the nineteenth century (Grant 1852, Grosser 1962). They suggested, at least to some writers, that the law of gravitation might have to be modified. Later in the century the very foundations of the theory seemed to fade away. Treatises written by Hertz, Mach, Boltzmann and others are veritable rescue attempts. What can we say about this confusing mixture of failures and spectacular successes? Shall we say that mechanics worked? all the time? some of the time? never? That depends on standards which decide when a theory is supposed to be acceptable, which is what Laudan wants to find by his procedure.

The next point is that high theory and phenomenological laws often seem to be successful in the same area, but they are constructed in very different ways. An honest pragmatist will, of course, prefer phenomenology and engineering approaches which (usually, but not always) fit the facts much better than the theories allegedly backing them (and which have to go through approximations, ad hoc adaptations, questionable assumptions, "by-hand" insertions of relevant magnitudes, and so on to reach the facts).[7] They "work" better, according to Laudan's own criterion; yet, all his examples come from high theory.

Another point I already touched upon is that theories which "work" do not enter the world ready-made and that their success is not independent of the treatment they receive. Theoris start from modest beginnings; they go through a complicated process of growth, suffer some encouragement and numerous defeats and need different incentives and/or correctives at different stages of their development. Each incentive (corrective) can be formulated as a rule, each rule contributes to the survival and eventual success of the theory, each of them "picks out" the right candidate which means that Laudan's Hall of Fame will contain mutually incompatible standards—unless standards are tied to the situations in which they are used. But then a general account of when a particular standard is to be applied becomes as impossible as a general account of the conditions of historical events. All we can say is who did what in which circumstances, what resulted, and we may try to remember the actions for future reference just as politicians learn the tricks of their predecessors to get out of sticky situations.

An interesting example of this feature of scientific change is the long-term history of atomism. Atomism faced theoretical as well as empirical difficulties, in antiquity, when Aristotle formulated a subtle and effective criticism in the Middle Ages when it was often discredited by its connection with hedonism or atheism as well as in the nineteenth century when well-established facts and laws in chemistry and thermodynamics seemed to provide decisive refuting evidence. Its survival depended on a variety of moves each of which "picked out" and thus preserved a view that eventually worked "with a pretty high degree of reliability". One may try to restore uniformity by choosing the "last" rule of this sequence—suggested by the final and totally successful product—and ascribing the rest to a "context of discovery." The trouble is that for a strict judge the context of discovery now covers the entire history of atomism, the period of its general acceptance included while a more lenient judge may already stop at Lucretius. Again we arrive at different and mutually incompatible standards tied to different ways of judging procedures, ideas, and theories.

The early history of matrix mechanics adds details to this general story. Considering the empirical problems of the theory, Max Born (1926) wrote:

> [W]e simply cannot talk of the simultaneous existence of various states if we don't want to abandon the simple, natural interpretation of Wilson's cloud chamber tracks and of similar processes. . . . Of course, one can assume that these are very complex processes which will eventually emerge, at the end of long chains of mathematical consequences . . . but that means destroying all bridges behind oneself, which establish a connection with the nourishing homeland, observation, while advancing into as yet unchartered domains. (Quoted from Bonk 1991, 70; my translation)

Confronted with objections of this kind Heisenberg remembered Einstein's remark that it is the theory which decides what can be observed and what is being observed:

> From here it was easy to turn around our question and not to ask 'How can I represent in quantum mechanics this orbit of an electron in a cloud chamber?' but rather to ask 'Is it not true that always only such situations occur in nature, even in cloud chambers, which are to be described by the mathematical formalism of quantum mechanics?' By turning around I had to investigate what can be described in this formalism; and then it was very easily seen . . . that one could not describe at the same time the exact position and the exact velocity of an electron. (1990, 110)

Born formulated a rule (take evidence at its face value) that had "picked out" numerous successful theories. Heisenberg used a principle that implied the negation of the rule: Look at the evidence in terms of the theory you are developing and rephrase or restrict it accordingly. This principle, too, played an important role in the selection of successful theories. An example is Einstein's theory of Brownian motion.[8] Neither Born's rule nor Heisenberg's principle can claim to be universally valid. The first would have hindered Copernicus, Newton, Maxwell, Einstein, Heisenberg and numerous other scientists while the second, injudiciously applied, might forever postpone scientific reform. What is important is that we cannot regulate the proper application of either. There is no universally valid "super-rule" which would tell us when to apply the one and to subdue the other. The reason is that theory, evidence, and rules may form inseparable parts of a promising research program and that research programs containing different rules often have comparable results.[9]

Still another example, discussed in Feyerabend (1988, chaps. 6–14), is the so-called Copernican Revolution. In 1610, the year Galileo published his *Sidereus Nuncius* ([1610] 1989), the Copernican worldview had achieved some plausibility, but it also suffered from many difficulties. A hardheaded empiricist—this includes many modern scientists—would have discarded it. What were the rules which furthered its growth, eventually led to its triumph and thus can be said to have "picked out" the right candidate? This is a difficult and complex historical question further complicated by the fact that the thing that triumphed was not what Copernicus had originally proposed. I conclude that the attempt to justify rules and principles pragmatically cannot lead us back to the abstract epistemologies Laudan seems to favor.

Let us now take a look at the standards Laudan is willing to defend. According to the pragmatist, highly confirmed predictive schemes are to be preferred to theories which make some kind of sense (they conform to plausible symmetry principles or metaphysical views) but lack empirical support. "Our allegiance" (Laudan 1991, 30), he says, should be to the former; the latter are "unacceptable" (ibid.). He overlooks that "unacceptable" views can and often did discredit their "acceptable" rivals. This was done either by a cunning *transfer* of evidence from the accepted view to its "unacceptable" rival (for example, see Copernicus's—[1543] 1976, book 1—discussion of the movement of the earth, imitated and elaborated by Galileo [1632] 1967) or by an *analysis* which, starting with an "unacceptable" point of view showed the spuriousness of the evidence against it (see Einstein's analysis of Exner's measurements of Brownian motion and Heisenberg's analysis of particle tracks); alternatively "unacceptable" views were rescued by a logical examination that was *motivated* by a strong belief in their validity (see Planck's analysis of Kaufmann's experiments—Zahar 1989, chap. 6). Other reactions which, given the complexity of the matter, make good sense are as follows: refusing to take evidence at face value (many examples exist, among them the initial treatment of Olbers's paradox—see Jaki 1969, chaps. 7–9), or declaring it to be incomplete (again many examples exist, including Darwin's attitude towards the Cambrian explosion of life [Gould 1989, 56f]. In all these cases "allegiance" went to empirical underdogs, used them in ways denied success by almost all the participants in Laudan's dialogue (ibid., 84), and led to major scientific advances. It may be true that "if an approach is bogged down, and failing to produce interesting new results, [scientists] quickly vote with their feet" (ibid., 156); opportunists exist in all professions. But discovery often came from those who opposed the fashions of the time. Of course, even the most metaphysically inclined scientists hope eventually to confirm their conjectures by novel observations and striking experiments; but to achieve *this* aim they must first suspend their allegiance to a *given* harmony between theory and fact.

Arguing further along this line we are led to suspect that scientific research knows no universal boundary conditions or standards whether of a conventional, aprioristic, or empirical kind but uses and invents rules according to circumstance without regarding the selection as a separate "epistemic" act and often without realizing that an important choice is being made.[10] For example, the arrival of large-scale experimental equipment in high energy physics changed the older (and empirically selectable) demand that experiments must be repeatable, but without any explicit "epistemic" debate ("tacit" knowledge and corresponding "tacit" decisions play a large role in processes of this kind).

None of the participants of the dialogue, the relativist included, seems to be aware of this feature of scientific practice. Trying to rise to what they think is a more "philosophical" level, they engage in a debate where the relativist turns historical facts (defeated theories *occasionally* stage a triumphant comeback; discarded views *occasionally* become powerful critics of their successful rivals) into abstract principles ("any theory is a s good as any other"—Laudan 1991, 55, 76), while his opponents think that having thrown doubt upon the principles they have also disposed of the facts (ibid., 84 and passim).

Altogether the tendency to judge science by examining the logical properties of simple logical models (such as the Duhem-Quine argument) gives the debate an air of unreality. Take the assumption that, given some body of evidence, there exist "indefinitely many—arguable infinitely many" (ibid., 49) mutually incompatible theories which are compatible with it. The relativist takes the assumption for granted. But how can he do that? Given a finite universe and a finite lifetime for the human race there can only be a finite number of sentences—so, has relativism become Platonistic? Besides, Platonic infinities which are not trivial (infinitely many possible values of a constant within the range of error) and which satisfy certain minimal conditions have to be argued for, they cannot simply be assumed. Are there infinitely many prime numbers? It needs a (simple) proof to decide the question. Yet even if it could be shown that there are Platonic infinities we still could not infer that theory selection is "arbitrary" (ibid., 45) that "all rival hypotheses are on the same footing" (ibid., 54) or that social factors (as opposed to good reasons) play a "key role in explaining the doxastic life of scientists" (ibid., 157). A general may not know the infinitely many ways in which angels can win a battle, but this makes his decisions arbitrary, or one decision as good as another only in the eyes of a god contemptuous of human affairs. In *this* world (and here I repeat the argument which Aristotle used against Parmenides)—which is the world where scientists try to understand nature and philosophers scientists—we have restricted resources both as regards the theories which are being offered for choice, and the ways of choosing and doing science means operating within these resources. Inferring arbitrariness from the limited nature of our resources also means making a comparison (between the resources and a world they can never reach) which is itself beyond the resources and, therefore, "arbitrary." Strangely enough it is the relativist of the dialogue who produces metaphysical romances of this kind.

Turning now to the role of social conditions we can at once admit that they have an effect, though not in the ways envisaged by Laudan's relativist (and some social scientists). To start with, pluralistic societies contain many trends and a wide variety of reactions to them—just look at the many different forms of theater, literature, and the arts, some of which are wildly critical of prominent tendencies. Facing this plurality scientists have a choice; they are no longer at the mercy of a single type of pressure. Besides, ideas taken over from "society" are never left unchanged. Aristotle mobilized common sense against the views of Parmenides and Plato. He *consciously* tried to preserve it—he was not simply overwhelmed—but he also *revised* it, using the achievements of those he opposed. The same is true of the sciences. Even scientists who have fallen for some powerful social fashion are influenced by a second fashion, namely, the situation in their own field. Both sociologists and epistemologists have rather simpleminded views of the elements that enter scientific change.

Do we need a special subject, a philosophy, to explain "the success of science" (Laudan 1990, 167)? No, we do not. First, there is no unified and coherent entity, "science," which can be said to be successful (see the earlier discussion on Laudan 1990). Second, it is not "science" which is successful—some so-called sciences are a pretty sorry sight[11]—particular assumptions, theories, and procedures are. We may admit that some of the ideas which have influenced scientists were once defended by philosophers and may even have been invented by them. But scientists then used them in their own way, without paying much attention to the philosophical background. Thus Luria prefers "predictions that will be strongly supported or sharply rejected by a clear-cut experimental step" (1985, 115); he shows little enthusiasm for a theoretical science that is "loaded with weak inferences" (speculations about the Ozone hole and cosmology are his examples) and he reports that Fermi, for this very reason, was somewhat cool toward the general theory of relativity. A science in the style of Luria will differ considerably from a science in the style of, say, Hawking. Gauquelin (1983) assembled impressive evidence for astrological character correlations, yet few psychologists (Eysenck being a notable exception) have shown a readiness to even consider his conclusions. The defenders of supergravity do not object to tests but so far they have defined success in rather general terms: the theory gives us gravitation; chirality is excluded by point-particle theories, not by string theories; and there exist versions that seem to be anomaly-free. Feynman is not at all pleased; he wants a more direct relation between a theory and the facts (Davies 1988, 194). All the parties just mentioned are scientists, they are even empiricists, they all favor experiments, but empiricism has a different meaning for each one of them. One might say that epistemological principles become effective within the sciences only by losing their (perhaps unambiguous) philosophical content and acquiring a (highly ambiguous) scientific meaning. Third, most so-called successes can be explained by purely scientific means. Why could Newton's theory explain so many features of planetary motion? Because space is curved and Newton's approach was a good approximation to that scenario.

Do we need philosophers to make science acceptable to other cultures or to defend its "objective" or intercultural "validity"? Again the answer is no. There is not one entity, "science," which proceeds in a uniform way, is invariably successful, and should therefore be transferred as a whole. But other "cultures" also lack the unity ascribed to them by philosophers, this time by relativists. By now government officials almost everywhere wear a tie and a business suit; they welcome the investment of foreign capital and the technology, especially weapons technology that goes with it. The airports of large cities, the hotels and the downtown shopping areas become more and more similar to each other. Does this mean that the ideology that is often associated with Western Civilization (secular humanism, an objectivist view of knowledge) is "objectively valid"? Certainly not. The only "objectivity" behind the phenomena I just mentioned is money and the lust for power. Besides, many politicians and intellectuals of the so-called third world retain their "superstitions" and fall back on them in times of trouble. And even they are only a small minority. Populations in rural areas perpetuate older habits, partly out of inertia, partly by conscious effort, frequently against tremendous odds.

City slums with mixed populations grow in close proximity to high class neighborhoods in New York, Brazil and elsewhere. What is the "culture" there? Is there a culture, or just a concentration of despair? For Laudan "primitive societies" contain reasonable people who choose Western experts to improve their lot. There were of course such choices, but they often had adverse effects. One example is the so-called "Green Revolution"; it disrupted social relations and created scarcity for people who had lived well before. Another example is the transfer of modern medicine. It makes no sense for an underdeveloped country to put a supermodern hospital in the middle of a jungle (Brazil is a recent example). The people most in need of treatment cannot afford it and would not be helped if they received it as a gift: Having been cured they return to their homes which lack, for instance, sewage and clean water and to their own often unhealthy habits. Does an epistemological evaluation of "modern medicine" produce better advice? Certainly not. To start with, modern medicine, like science, has many strands, some of them in conflict with each other. Second, there exist only few comparative evaluations of special procedures of scientific medicine and corresponding third world practices. Even the role of medicine in the first world is not altogether clear. Reporting on his years as a student of medicine (in the early 1930s), Lewis Thomas writes, "It gradually dawned on us that we didn't know much that was really useful, that we could do nothing to change the course of the great majority of diseases we were so busy analyzing, that medicine for all its facade as a learned profession was in real life a profoundly ignorant occupation" (1984, 29), while Thomas McKeown has considerably reduced the roles of medicine in the reduction of mortality in the eighteenth and nineteenth centuries (1979). At any rate, those for whom "modern medicine" is a universal panacea cannot meet the *specific* needs which arise in *specific* areas among *specific* populations. And why not? Because what is required here is a detailed knowledge of particulars of life style, location, desires, past afflictions combined with relevant information about nutrition, hygiene and so on. A large part of this information depends on personal contact; general talk about the "objective validity of science" simply misses the point. To me assertions such as "science teaches us about reality" or "reality is relative to a culture" are metaphysical romances of doubtful merit. They have to be replaced by a piecemeal approach which relies on local measures and whose generalities arise, like the generalities of science, from the experiences that were gained in their course.

Epistemology is not the only discipline trying to control an activity that can stand on its own. Every area of human endeavor is surrounded by ideas which may be useful when joined with practice but impoverish our resources when imposed unchanged. Brecht had a beautiful and very rational theory of theatrical action but his plays either fell flat on their faces—which they did when they were as objective (*verfremdet*) as the theory demanded—or they moved the audience, in which case the theory was left far behind. The solution is not to abolish generalities altogether, but to bring them in close contact with the topics they are about. More concretely, it does not make sense to have confident practitioners advised by voices from a Platonic heaven or to have producers worrying about box office receipts advised by aestheticians who never even heard of *Dallas, Dynasty,* or *LA Law.* What we need are thoughtful scientists (artists, playwrights, priests, or politicians) who are experts in the twin art of modifying what is general by tying it to particulars and of explaining what is particular in general terms; in other words, what we need is a marriage of ("philosophical") universals and ("empirical") particulars. Now the interesting thing is that the marriage has already occurred and that it was consummated long ago. The artworks which surround us, the scientific theories which resemble them in so many ways, even the most trivial scientific facts and laws come from a close interaction between far-reaching fantasies and a material (marble, colors, opinions, sounds, experimental results) which resisted and modified them in many ways. Philosophers, intent on remaining in control (remember again Plato's talk about the "ancient battle between philosophy and poetry") have always tried and are still trying to prove that we are dealing with a master-slave relation where wise masters (epistemologists, for example) pushing around capable but rather ignorant slaves (scientists, artists, politicians) have succeeded in changing the world. The world has changed, and considerably so, since the arrival of Homo sapiens. But the change was not as simple as imagined by the lovers of abstraction; philosophers played a role in it but only to the extent that they replaced their ambitions by more pedestrian ways of gaining influence. And yet there are still people who mistake these ways for transhistorical patterns of order.

NOTES

1.See Perl (1989). Eliot's comments have the advantage of being concise, striking, and direct while the philosophers' are much more long-winded.

2. This applies also to the "constraints" which some recent historians, Galison and Rudwick among them, have introduced to distinguish their ideas from a relativism where having a worldview, no matter how silly, and carrying out the relevant procedures already guarantees success. Views do not disappear because somebody wants them to or because they run into history-independent "facts" or "principles"; they disappear because "resistance *emerges* in the real time of practice" (Pickering 1990, 6). Does this mean that philosophical speculation and scientific research are now replaced by an appeal to history? Not at all! A description of what has been achieved in the course of a complex and expensive research project certainly differs from the processes described; for example, it is much poorer in content. Furthermore, it is not denied that successful scientific theories may contain the idea of history- and observer-independent processes (this is how Einstein interpreted his own realism; see Schilpp 1949). What is being denied is that research is controlled by agencies (such as "facts" and Platonic "standards") which guide it without becoming part of it. This would be worse than revelation; it would be unconscious revelation. Gorgias's saying that "Being is unknown unless it *appears* in opinion" (fragment B 26, my emphasis) or, generalizing, that Being does not affect our opinions unless it participates in our lives, is as valid today as it was at the time of the sophists.

3. Interestingly enough theologians (such as Duns Scotus and William Ockam) and scientists (such as Newton) who emphasized the infinite power and the unfathomable will of God long ago came to a similar conclusion: The standards and the laws that are supported by such a being have only a contingent and transitory authority.

4. According to Eliot "the theorist's claim 'to step out of history' is specious because the objects of philosophy are 'only the reflex of his [the philosopher's] abstraction upon the current of history' " (Perl 1989, 72).

5. Being preoccupied with their work (Grosser 1962, 102, 110) and occasionally contemptuous of theory (ibid., 92), the observers of star positions were in no hurry to look for the planet that had been predicted by Leverrier and Adams and used unsuitable methods when they started (ibid., 107).

6. Regis's book (1989) contains many examples of the differences in outlook, method, and style that separate research programs, and of the tensions created by them.

7. The general problem is explained with interesting examples by Cartwright (1983, esp. essay 6, sec. 1.2). Any explanation of large-scale phenomena such as capillarity, elasticity, the behavior of complex electric circuits, the onset of turbulence, even of the apparently simple fact that macroscopic objects do not collapse or explode provides further relevant material. But the inverse case—an improvement of phenomena following a determined application of basic principles—also occurs. One example, already mentioned, is celestial mechanics. For

another example, see Birkhoff (1950). Most interesting is Maxwell's calculation of the viscosity of gases. For Maxwell this was an exercise in theoretical mechanics, an extension of his work on the rings of Saturn. Neither he nor his contemporaries believed the outcome—that viscosity remains constant over a wide range of density—and there was contrary evidence. Yet more precise measurements confirmed the prediction. Maxwell's calculations are reproduced in Maxwell (1965, 377 and passim).

8. See Einstein ([1907] 1956, 67), "Since an observer operating with definite means of observation in a definite manner can never perceive the actual path transversed in an arbitrarily small time, a certain mean velocity will always appear to him as an instantaneous velocity. But it is clear that the velocity ascertained thus corresponds to no objective property of the motion under investigation."

9. Ancestors of Born's rule belonged to the caloric theory and the phenomenological branch of thermodynamics while Einstein's principle was the moving force behind the kinetic theory. The discussion and the defense of the two alternatives goes back to antiquity, especially to the opposition between Plato (*Republic* 528d6ff and the *Timaeus*) and Aristotle (*de coelo,* esp. 293al5ff).

10. In his autobiography, Dyson (1979) describes the different approaches Bethe and Feynman used in calculating the behavior of electrons, "Hans [Bethe] was using the old cookbook quantum mechanics that Dick [Feynman] could not understand. Dick was using his own private quantum mechanics that nobody else could understand. They were getting the same answers whenever they calculated the same problems. And Dick could calculate a world of things Hans could not" (p. 54). Feynman, according to Dyson, did not use equations; he had a physical picture and he made up the rules along the way, "Until the rules were codified and made mathematically precise, I could not call it a theory" (ibid.). Quite so. My point is that scientific practice, even if it leads to theories of great precision, resists codification. Besides, Feynman's rules continue to be valid in particle physics while quantum electrodynamics, the theory in which Bethe based his calculations, does not.

11. Praying, to use an example from Laudan (1990, 169), may not shine when compared with celestial mechanics, but it surely holds its own vis-à-vis certain parts of economics or vis-à-vis a psychology based on psychoanalytic principles.

REFERENCES

Birkhoff, G. 1950. *Hydrodynamics.* Princeton: Princeton University Press.

Bonk, T. 1991. *Atomismus und Komplementaritaet.* Ph.D. dissertation, Federal Institute of Technology, Zürich, Switzerland.

Born, M. 1926. "Das Adiabatenprinzip in der Quantenmechanik." *Zeitschrift für Physik 40:* 167–92.

Bruns, . 1887. [*Acta Mathematica 11:* 25–96.]

Cartwright, N. 1983. *How the Laws of Physics Lie.* Oxford: Clarendon Press.

Copernicus, N. [1543] 1976. *On the Revolutions of the Heavenly Spheres*. Translated by A. M. Duncan. New York: Barnes & Noble.

Davies, P. C. W., and J. R. Brown, eds. 1988. *Superstrings: A Theory of Everything?* Cambridge, England: Cambridge University Press.

Diels, H., and W. Kranz, eds. 1957. *Die Fragmente der Vorsokratiker*, vol. 2, 6th ed. Berlin: Weidmann.

Dyson, F. J. 1979. *Disturbing the Universe*. New York: Harper & Row.

Einstein, A. 1956. *Investigations on the Theory of the Brownian Movement*. New York: Dover.

Feyerabend, P. K. 1988. *Against Method*. London: Verso.

Feynman, R. P. 1984. *Surely You Are Joking Mr. Feynman*. New York: Norton.

Galileo, G. [1632] 1967. *Dialogue Concerning the Two Chief World Systems*. 2d ed. Translated with revised notes by S. Drake. Berkeley and Los Angeles: University of California Press.

———. [1610] 1989. *The Sidereal Messenger*. Translated with an introduction, conclusion and notes by A. van Helden. (Originally published as *Siderus Nuncius*). Chicago: University of Chicago Press.

Gauquelin, M. 1983. *The Truth About Astrology*. Oxford: Blackwell.

Gould, S. J. 1989. *Wonderful Life*. New York: Norton.

Grant, R. 1952. *History of Physical Astronomy from the Earliest Ages to the Middle of the Nineteenth Century*. London: Baldwin.

Grosser, M. 1962. *The Discovery of Neptune*. Cambridge, Mass.: Harvard University Press.

Heisenberg, W. 1990. "Theory Criticism and a Philosophy." In A. Salam, ed., *Unification of Fundamental Forces*. Cambridge, England: Cambridge University Press, p. 110.

Jaki, S. L. 1969. *The Paradox of Olbers' Paradox: A Case History of Scientific Thought*. New York: Herder & Herder.

Laudan, L. 1990. *Science and Relativism: Some Key Controversies in the Philosophy of Science*. Chicago: University of Chicago Press.

Luria, S. E. 1985. *A Slot Machine, A Broken Test Tube: An Autobiography*. New York: Harper & Row.

Maxwell, J. C. 1965. *The Scientific Papers of James Clerk Maxwell*. Edited by W. D. Niven. New York: Dover.

McKeown, T. 1979. *The Role of Medicine: Dream, Mirage, or Nemesis?* Oxford: Blackwell.

Perl, J. M. 1989. *Skepticism and Modern Enmity: Before and After Eliot*. Baltimore: The Johns Hopkins University Press.

Pickering, A. 1990. "Beyond Constraint." Unpublished manuscript.

Poincaré, H. 1957. *Les Méthodes Nouvelles de la Mécanique Céleste*, New York: Dover.

Regis, E. 1989. *Who Got Einstein's Office? Eccentricity and Genius at the Institute for Advanced Study*. Reading, Mass.: Addison-Wesley.

Schilpp, P. A., ed. 1949. *Albert Einstein: Philosopher-Scientist*. La Salle, Ill.: Open Court.

Thomas, L. 1984. *The Youngest Science: Notes of a Medicine-Watcher.* New York: Viking Press.
Wilson, C. 1985. "The Great Inequality of Jupiter and Saturn: From Kepler to Laplace." *Archive for History of Exact Sciences 33:* 15–290.
Wilson, R. E. 1960. "A Great Catalogue of Stellar Positions and Proper Motions." In H. Shapley, ed., *Source Book in Astronomy 1900–1950.* Cambridge, Mass.: Harvard University Press, pp. 124–29.
Zahar, E. 1989. *Einstein's Revolution: A Study in Heuristic.* La Salle, Ill.: Open Court.

9

Seven Theses on Thought Experiments

Paul Humphreys
Department of Philosophy, University of Virginia

Those of us who work in philosophy of science tend to come across "thought experiments" on both sides of the street. We have famous cases from the remote and recent history of science, cases so familiar that their names alone identify them: Heisenberg's gamma ray microscope, the (original) EPR experiment, Newton's rotating bucket experiment, Maxwell's demon, Schrödinger's cat, and so on. Philosophers have used what looks like a similar technique: Reichenbach's differential and universal forces, the Turing test, Searle's Chinese Room, Hobbes's state of nature, Strawson's auditory universe, and so on. (This connection has been explored by Gerald Massey 1990. See also Wilkes 1988 for an extended discussion of the limitations on thought experiments in the philosophy of mind.)

In the small literature on the topic, the use of thought experiments in science has often been considered puzzling. The puzzles are generally taken to be a problem for empiricists, and in its simplest form, the conundrum is this: How can an imaginary rather than a real experiment relying on hypothetical data justifiably result in the rejection or acceptance of a scientific theory? Alternatively, if the thought experiment relies on existing, familiar data, how can the mere examination in thought of that data produce new knowledge? In his article "A Function for Thought Experiments" ([1964] 1977),[1] Thomas Kuhn claims that there is an "easy" and "plausible" (p. 242) answer to this puzzle. This answer, which Kuhn claims is closely related to philosophical tradition, holds that thought experiments reveal an inconsistency in the conceptual apparatus used by the scientist, and that the new understanding which results involves the elimination of the inconsistency rather than affecting our understanding of the world from which the data come (ibid.).[2]

By means of an extensive analysis of a single example due to Galileo, Kuhn argues that this traditional view is wrong. Broadly, Kuhn's strategy is two-pronged. He argues that psychological as well as logical considerations are crucial to the evaluation of Galileo's experiment, and he argues that in the context of the example he discusses, definitions are treated as if they functioned as laws of nature. This strategy leads, by an ingenious and creative set of arguments, to a cluster of what are by now familiar Kuhnian conclusions. Principal among these are the following:

C1. In addition to the claim that definitions function in a way very similar to laws of nature, one should not require concepts to apply to any situation that might arise in any possible world, but should require only that they be "unequivocally applicable in every situation which we expect ever to encounter" (ibid., 254). Thus "from thought experiments most people learn about their concepts and the world together" (ibid., 253).

C2. Thought experiments do not reveal conceptual confusions, but rather lead to a conceptual reorganization. Because the thought experiments necessarily rely only on existing empirical data, and the thought experiment must involve a "normal situation," the conceptual reorganization that results is revolutionary: "The data requisite for revolution have existed before at the fringe of scientific consciousness, the emergence of crisis brings them to the center of attention, and the revolutionary reconceptualization permits them to be seen in a new way" (ibid., 263); "A crisis induced by the failure of expectation and followed by revolution is at the heart of the thought experiment situations we have been reviewing" (ibid).

These conclusions should have a familiar ring to them; it is how Kuhn gets to them that is novel.

Kuhn also asks three questions about thought experiments:

Q1. What constraints should be put on what counts as a genuine thought experiment? Not any logically possible situation qualifies as a thought experiment, so what further restrictions are needed? He elaborates this as follows: In what sense and to what extent must the situation be one that nature could present or in fact has presented?

Q2. The second question relies on an assumption: "If we have to do with a real thought experiment, the empirical data on which it rests must have been both well-known and generally accepted before the experiment was even conceived" (ibid., 241). I will question later whether this is true, but Kuhn then asks, "How, then, relying exclusively on familiar data, can a thought experiment lead to new knowledge or to a new understanding of nature?" (ibid.).

Q3. What kind of new knowledge is produced in a thought experiment?

The argumentative structure of Kuhn's article is complex and nonlinear, but the easiest way to trace a route through it is to identify "thought" with "conceptual" and "experiment" with "data." Kuhn then identifies a conceptual scheme with the definitions and laws adopted by a discussant. This in turn allows him to argue that in many cases definitions function as laws which are revised in the light of empirical data. In so doing, he can argue to the position that concept and empirical content cannot be cleanly separated.[3] Moreover, finally, by interpreting the "conceptual" at another point in psychological terms, and allowing "empirical" to be identified with "perceptual," Kuhn would like us to arrive at another, psychologistic, conclusion; namely, that because antagonists in thought experiments have exactly the same empirical data available to them, and no new data are introduced that were not already familiar to them, the thought experiment plays much the same role as does an anomaly—it moves the antagonists toward the eventual conceptual reorganization by virtue of a psychological shift.

My first thesis, to which section I is devoted, is:

THESIS ONE. *Kuhn has misrepresented the structure of Galileo's thought experiment and, consequently, the conclusions he reaches cannot be supported by his analysis of thought experiments.*

I

No single thought experiment can possibly represent the variety of uses to which such devices have been put, and Kuhn recognizes this. He does claim, however, that Galileo's experiment bears "an obvious resemblance to certain of the thought experiments which proved effective in the twentieth century reformulation of physics and hence that the example is typical of an important class" (Kuhn [1964] 1977, 241). This claim is wrong, but more of that in section II.

The thought experiment discussed by Kuhn is described at the very beginning of Galileo (1962). (I will rely on the same Stillman Drake translation to which Kuhn refers.)

The setup consists of an inclined plane (see figure 9.1) which is "exquisitely polished and hard, upon which descends a perfectly round ball of some very hard substance" (Galileo 1962, 23). We are to consider a ball falling freely along the perpendicular *CB* and to first agree that the ball rolling down the plane would have an impetus at *A* equal to that of the ball at *B*, where the impetus is identified with the degree of velocity. Why are they equal? Sagredo's response is, "I surely believe it would. . . . [T]he impetus of each should be equally sufficient to carry it back to the same height" (ibid.), this claim being justified by an earlier appeal to experimental facts about pendulums and syphons.[4]

Next, a key admission is made by Sagredo that, prima facie, it seems "necessary that the motion by the perpendicular CB should be faster than by the inclined plane CA" (ibid.). That admission immediately raises a problem about how to reconcile this claim with the earlier one that the two balls have the same speeds at *A* and at *B*. As Sagredo says, "These two propositions seem contradictory" (ibid., 24). This intermediate stage presents an apparent contradiction, but the central argument is still to come.

To help clarify matters, Simplicio offers a definition of "faster than": One body is faster than another if and only if "one pass[es] over a greater space than the other in the same time or travel[s] an equal space in less time" (ibid.) and equality of speed is defined as "passing equal space in equal times" (ibid.). That this is a definition is acknowledged by Simplicio immediately after providing it.[5] But then, by choosing different parts of the inclined plane as reference trajectories, it is easy to demonstrate, using the definition provided by

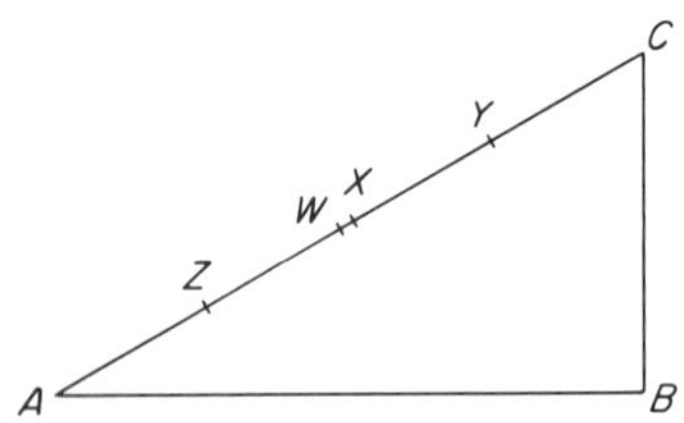

Figure 9.1

Simplicio, that the ball on the inclined plane moves slower than (from *C* to *X*), as fast as (from *Y* to *Z*), and faster than (from *W* to *A*) the vertically falling ball. Here we have an explicit contradiction and one that necessitates some change in the Aristotelian conceptual scheme.[6]

Kuhn's preliminary conclusion, which is different from Galileo's own immediate conclusion, is that the definition of "faster than" provided by Simplicio has to be abandoned in favor of a definition of instantaneous speed and "that conceptual reform is what Galileo's thought experiment helped to teach" ([1964] 1977, 251). If the analysis of the thought experiment were left there, we should still be within the traditional, easy view of thought experiments, but Kuhn goes beyond this by arguing, as I mentioned, that this reform, or revision, involves not a definition as traditionally conceived, but one that functions, or is treated, just like a law of nature.

To help us see this, Kuhn suggests we consider a world in which all motions occur at uniform speed. He writes, "What would we say if we found a scientist in this imaginary world employing the Aristotelian concept of [average] speed? Not, I think, that he was confused. Nothing could go wrong with his science or his logic because of his application of the concept" (ibid., 255). He then goes on to claim, "[G]iven our own broader experience and our correspondingly richer conceptual apparatus, we would likely say that, consciously or unconsciously, he had embodied in his concept of speed his expectation that only uniform motions would occur in his world. We would, that is, conclude that his concept functioned in part as a law of nature" (ibid.). Here the "correspondingly" gives the game away—concepts are acquired from experience and linked, via the device of definitions, with one's empirically induced expectations.

Recall that the assumption behind Kuhn's second question was that Galileo's thought experiment and all like it involve only facts familiar from everyday experience. He requires this constraint because (with one puzzling exception),[7] he endorses a recognizable but extreme empiricist line on concept acquisition. This account of concept acquisition by abstraction from everyday experience is crucial to Kuhn's argument because it allows him to shift the locus of discussion from logical considerations to psychological, and hence to reach both of the principal conclusions previously described.[8] For earlier in his article, Kuhn discusses some work by Piaget on children's ability to judge which of two objects is the faster, in which the children apparently use two distinct concepts of "faster than," one in which "perceptual blurriness" is central, that is, when one body is directly perceived to be moving faster than another, and another concept in which "goal-reaching criteria" are used, that is, whichever body reaches a preassigned goal first is assessed as the faster of the two.

Kuhn claims throughout his article that Aristotle also had two distinct concepts of "faster than," one based on "perceptual blurriness," which is related in some more or less loose way to our modern concept of instantaneous speed, and a more explicitly formulated concept closely related to our notion of average speed. The former concept, involving "perceptual blurriness," was on the periphery of the Aristotelian conceptual scheme and was not actively employed until Galileo's thought experiment necessitated moving it to a position that was central to the Galilean conceptual scheme. Both of these concepts, according to Kuhn, incorporate our expectations of how objects in our world behave. Taken together, they contain an implicit contradiction. When the thought experiment made this contradiction explicit, our expectations about the world had to alter and hence "from thought experiments most people learn about their concepts and the world together" (Kuhn [1964] 1977, 253).

Whether or not this is true in general, to claim this about the participants in Galileo's dialogue on the basis of their having committed themselves to an implicit contradiction between two perceptually acquired concepts is to misrepresent the structure of Galileo's argument. For a careful look at Galileo's actual reasoning shows that neither the concept of "perceptual blurriness," nor the perceptual criteria by means of which Kuhn links the Aristotelian conceptual framework to the psychological experiments of Piaget play a role in the central argument that accompanies Galileo's scenario, but at most appear only in the preliminary setting up of the puzzle via the intermediate apparent contradiction. To see this, let us return to Sagredo's initial admission that the vertically falling ball moves faster than the ball on the inclined plane. He is led to this by means of a judgement that on the horizontal plane, the ball would be stationary and then, presumably by an interpolation process, agrees that since the inclined plane is intermediate in slope between the horizontal and the vertical, the ball on the incline will move more slowly than on the perpendicular. This judgement could plausibly be construed as implicitly appealing to familiar observational facts that employ either perceptual blurriness or goal-reaching criteria, even though none of the participants in the dialogue refers to such criteria explicitly.

Whereas the "perceptual blurriness" or goal-reaching criteria may play a rhetorical role in the preliminary judgements, the definition of average speed is explicitly appealed to in the main argument, and it is from that explicit definition alone that the principal contradiction arises, not from a contradiction between the two criteria for "faster than" (the average speed and the "perceptual blurriness") that Kuhn has attributed to Aristotle. For when Salviati directly asks Simplicio, "[W]hy do you think the velocity of the body falling along CB should be greater than the velocity of that descending by CA" (Galileo 1962, 24), Simplicio replies "because in the time it takes one body to pass all CB, the other body will pass a part less than CB in CA" (ibid.). This response appeals directly and only to the first part of Simplicio's already stated explicit definition of "faster than," and relies on neither of the perceptual criteria. Moreover, the conclusion that is drawn from examination of the definition of average speed is that, in Galileo's own words, "You see, then, that the proposition 'Motion along the perpendicular is faster than that along the incline' is true not universally, but only for motions which begin from the initial point: that is the point of rest. Without this restriction, the proposition would be defective, that its very contrary might be true, namely that motion is faster along the incline than along the perpendicular" (ibid., 25).

Thus whatever the immediate reaction to Galileo's presentation of the scenario may be by the reader or by Sagredo, the reasoned response to it by the Aristotelian participant, which is the only one that plays a serious role in the dialogue, relies exclusively on an explicit definition that concerns the concept of average speed. This means that the Galilean thought experiment can be, and in fact is, presented entirely in terms of the objective definition of average speed and our psychological expectations need play no role in the description or analysis of this thought experiment.

Kuhn's background argumentative strategy here, which is to link Piaget's experiments and Aristotelian concepts of speed, might well seem less unusual than it did when Kuhn's article first appeared, because there is now an extensive body of research that has been interpreted as showing that some of our untutored reactions to simple physical situations resemble the responses an Aristotelian physicist might have given in those situations. But the evidence in this area is conflicting, for recent psychological investigations seem not to support the simple abstractionist account of judgements about motions to which Kuhn appeals. In Kaiser and Proffitt (1987), for example, the authors show that experimental subjects have an ability to make accurate judgements about dynamical invariants in both frictional and nonfrictional systems, and the authors conclude, "Thus the human observer has a well-defined set of implicit criteria by which to judge whether or not a collision is a natural, momentum-conserving event. These criteria appear to be based on underlying dynamic invariants such that perceptual competence is robust to dynamic properties of the system and is not constrained by our terrestrial experience" (p. 279). We should, of course, be cautious about the use of such findings, and in particular wary of transferring them to participants in seventeenth-century disputes where the entrenched physics was very different from that taught to late twentieth-century subjects, but at the very least, some current psychological evidence seems to suggest that as a general account of concept acquisition, we need not be as tightly constrained by worldly experience as Kuhn claims, and hence that his first claim (C1) is also mistaken.

There is a further reason why it is a mistake to apply psychological criteria to the analysis of Galileo's (or any other) thought experiment. For an immediate response based in perceptual judgements and a response grounded in theory are two separable responses that require very different treatments.

If someone asks me the speed of the runner who has just passed me, and I say, "He's doing a 6:50 pace," I am relying on perceptual comparisons of relative speed based on my having been passed by countless runners over the years, and knowing my pace and some of theirs. (Top 10,000 meter runners know their own pace on a track to within a few seconds per mile—wearing watches is a functionless innovation sustained by fees from watch manufacturers.) The runner's speed could also be calculated from systematic theory, simple or complex, (by means, say, of oxygen consumption calibrated for his physique, condition, ambient temperature, humidity, and wind speed), but my criteria are not based on that theory, whereas those of an exercise physiologist will be. So the fact that one *might* bring to bear on the thought experiment perceptual criteria that are not part of Aristotelian or Galilean theory is irrelevant. The only considerations that should be included in the analysis are those in terms of which the main argument of Galileo's thought experiment is actually given. These are manifestly not perceptual and can be conducted in terms of logical analysis alone once the definition has been made explicit and the empirical facts agreed upon.

Thus far our analysis would allow us to stay within a version of what Kuhn calls the "immensely plausible" view on thought experiments. But he has another reason for wanting to reject that view. Kuhn ([1964] 1977, 258) asserts that Aristotle acted as if he lived in a world where all motions were uniform, and that it was only when an example was discovered (via Galileo's thought experiment) showing that the concept of average speed leads to some undesirable consequences which could arise in our world that the concept had to be abandoned. In particular, it is crucial for Kuhn that the thought experiment concerns a situation that could arise in our world. So there are very tight constraints on thought experiments to be used in science. Not only must they be descriptive of our world, but of those parts of it with which we are already familiar. The reason for this is that since definitions for Kuhn have a function similar to that of laws and theories, we should therefore not demand of them that they be applicable in any possible scenario, but only that they "be unequivocally applicable in every situation which we expect ever to encounter" (ibid., 254).

To see that this construal of the relations between definitions and laws is wrong, we simply need to separate two elementary properties of definitions, properties that ought to be commonplace, but which are confused by Kuhn. (By "definition" here I mean nothing more than the most basic kind—explicit definitions that provide necessary and sufficient conditions for the definiendum.)[9] What Kuhn has confused are the necessary coextensiveness of the definiens and definiendum, and the contingency of the extension possessed by the definiendum (and hence the definiens) in specific worlds. Definitions are extension-fixing devices in the sense that the extensions picked out by the definiens and definiendum must be coextensive, or the offered equivalence would not be an equivalence of the correct type qua definitional equivalence. But how many things satisfy the definiendum in any given world varies from world to world and in the case of those concepts that are not universal or vacuous by virtue of their form, we need empirical knowledge to discover just how large the extension is in our world. This is not to deny, of course, that finding out what that extension is can be very difficult, and that the issue of whether the definiendum is the appropriate one for the scientific purpose at hand is an even more difficult thing to ascertain, but this confusion between the two basic features of definitions is, I think, a key reason why Kuhn succeeds in blurring the distinction between the conceptual and the factual. The correct conclusion to draw from this particular aspect of the thought experiment is simply that the extension of the concept of average speed is too narrow to serve the purposes of kinematical models of our world.

II

If scientific thought experiments do not have the function attributed to them by Kuhn, what is their role in science? Among the suggestions made in recent literature on the subject are (1) the view that they have a critical, heuristic, or apologetic function (Popper 1959, Appendix *xi), the first two of these being legitimate for Popper, the last illegitimate; (2) the view that they essentially involve abstractions and idealizations and provide us with a priori knowledge (Koyré 1960); (3) the view that they are arguments (Norton 1991); (4) the view that they provide synthetic a priori knowledge (Brown 1986); and (5) the view that they have a similar function in both science and philosophy (Massey 1990). As is often the case, many of these answers contain part of the truth, but I do not think that any of them have it quite right.

Let us start with terminology. I dislike the term "thought experiment." Ernst Mach introduced the term "Gedankenexperimente," but its English translation gives the immediate impression that such things are (a) a kind of experiment, and (b) experiments conducted in the realm of the mental, rather than in the sphere of the physical, both of which are not only wrong, as we will see, but also tend to make one think that because they are pale, mentalistic analogies of real experiments, they lack the solid empirical content of real experiments. This construal not only locates the problems for empiricism in the wrong place, but inclines one toward the view that there is some essential psychologistic component of thought experiments.

There is a tendency to present thought experiments in the secondary literature as though their effect is immediate and unmediated by argument. (I should say immediately that Kuhn himself is not guilty of this.) One can see the temptation of this view because of the appealing power of the very best thought experiments. It is true that some heuristic devices which appear to be similar to thought experiments do seem to have this instant effect. Here is a nice example. Consider an overweight monk who starts at 6 A.M. to climb a mountain path to the monastery at the top. He stops frequently for breath, walks at an erratic speed, and arrives at 6 P.M. The next day he leaves the top at 6 A.M. naturally walking much faster and less erratically, and arrives at the bottom much earlier than 6 P.M. Is there any point on the path which the monk occupies at the same time on both days? Suppose you try to solve this problem mathematically. It not only seems complex, but possibly underdetermined by the given information. To solve it, merely consider a hypothetical second monk, a replica of the first, starting from the bottom on the second day at 6 A.M. Their paths must cross somewhere and so the answer is, transparently, "yes." This kind of case tends to lead people to say that one can "just see" the solution to the problem immediately the hypothetical scenario is introduced. But this example is not a thought experiment; it is a well-posed problem for which a solution heuristic has been given.

There are two requirements for a problem to be well posed. First, the set of conditions given by the problem must be both necessary and sufficient to determine a unique solution. That is, the conditions are complete in the sense that any introduction of auxiliary information is illegitimate and unnecessary and the conditions are nonredundant in that no superfluous information is contained in them. This latter requirement is often violated in minor ways to make the presentation of the problem livelier, but serious violations of the nonredundancy requirements are legitimate grounds for complaint. In the monk problem, I mentioned that he was overweight. You correctly ignored that. Second, the conditions are given and fixed—we cannot challenge or change them. One may, on occasion, ask for clarifications of the conditions, but unless the poser of the problem has made an error, challenges are a sign that the questioner has not understood the problem.

Now compare the monk problem, which is well posed, with a different thought experiment due to Galileo, the thought experiment that is commonly taken to destroy the Aristotelian theory that heavier bodies fall faster than less heavy bodies (see Galileo 1974, 66ff. and Koyré 1960). This thought experiment suggests we consider two such bodies, one heavier than the other, first separately, and then conceptually or actually joined. In the case where they are joined, the heavier body will supposedly be slowed by the lighter, hence the overall speed of the composite body will be less than that of the heavier of the two, yet the composite body consisting of the joined pair, being heavier than both, should, according to the Aristotelians, fall faster than both. Apparently, this presents a contradiction. Of this thought experiment, which serves as Popper's sole example of their critical use, Popper says that it is "one of the simplest and most ingenious arguments in the history of rational thought about our universe" (1959, 442). Ingenious, yes; simple, no, for Galileo spends over twenty pages answering both obvious and subtle Aristotelian objections to the assumptions underlying his thought experiment, arguments that are essential for the thought experiment to conclusively establish that the Aristotelian theory is untenable.

Consider also the Galilean example discussed by Kuhn. Despite the familiarity of the experimental setup, it obviously involves a number of idealizations upon which both sides of the dispute have to agree, and Galileo takes six pages of dialogue to fully present the experiment. In this case, the extrapolations from ordinary experience would seem to be unproblematical, at least in the sense that the conclusion of the thought experiment is robust under deviations from these idealizations; but the idea that thought experiments involve inferences from familiar premises to self-evidently true conclusions would be incorrect, for on the very next page after the conclusion of this thought experiment, Galileo introduces another thought experiment which involves his famous error leading to the theory of circular inertia—an error which results from taking an idealized horizontal plane on the Earth as the correct representation of a noninclined plane. Moreover, Galileo frequently has to be very careful to postulate his idealizations in the form of limiting processes, particularly where the limit would involve a vacuum, in order for the idealization to be acceptable to an Aristotelian. Furthermore, the Aristotelian and Galilean physicists must share a methodological assumption that this method of idealization is scientifically acceptable. (I leave aside here the issue of whether early seventeenth-century Aristotelians would have granted this.)

Thus, Kuhn's assertion that "[g]ranting that every successful thought experiment embodies in its design some prior information about the world, that information is not itself at issue in the experiment" ([1964] 1977, 241) deeply misrepresents the dialectical function of many thought experiments, which is exactly to explore, probe, and often modify the assumptions on which the scenario is constructed. For much more than uncontroversial observational data and definitions are involved in the empirical data on which a thought experiment rests. The multitude of auxiliary arguments that are involved in the development and establishment of a thought experiment are frequently designed to convince a listener that he ought to treat certain theories or phenomena which are needed in the main argument as facts, when prior to those arguments he was ignorant of or skeptical about them, or to modify and revise the preconceived "information" that is brought to bear on the thought experiment. If this is so, then Kuhn's second question (Q2): "How then, relying exclusively upon familiar data, can a thought experiment lead to new knowledge or to a new understanding of nature?" is based on a false assumption. Much of the new knowledge and understanding arises from an examination of the supposedly familiar and unquestioned "facts" involved in the scenario. In this sense thought experiments differ from well-posed problems, and this process of examination frequently results in the recognition that the conditions underlying the thought experiment are either unnecessary or insufficient to reach the principal conclusion. This process should be familiar to us all, for it is that recounted in detail for mathematical progress by Imre Lakatos (1978, chap. 5, passim; and chap. 4, 45), whose account itself owes much to Georg Polya's investigations into the heuristics of mathematical problem solving.

This fact is all the more evident the more we move into the realm of highly theoretical twentieth-century thought experiments. Heisenberg's ([1927] 1983) original presentation of the gamma ray microscope thought experiment, for example, appealed to (1) the Compton effect (directly); (2) the formula for the resolving power of a microscope $\Delta x \sim \lambda/\sin \alpha$, where λ is the wavelength, Δx is the resolution, and α is the angular aperture (indirectly); and (3) diffraction involving de Broglie waves (directly). All of these assumptions are theoretical in form, and at least the second involved significant extrapolation beyond what had been empirically established at the time.

To take another example, much of Bohr's (1935) reply to the original presentation of the EPR paradox is devoted to the issue of whether the experimental arrangement could, even in principle, provide information about conjugate variables in the way that EPR had claimed, given the complementarity principle.

These points, which are drawn from examples in the scientific realm, apply even more clearly to thought experiments in philosophy, such as the resolution of the conflict between dominance principles and maximized expected utility principles revealed by Newcomb's Problem which led to a supplementation of the original problem by scenarios with an explicit causal structure on the one hand, and to a supplementation of Bayesian decision theories by various principles of ratifiability, or knowledge of one's own epistemic state, on the other.

Thus, while thought experiments involve arguments, those arguments are directed not only at the principal conclusion but at an exploratory analysis of the conditions on which the thought experiment is predicated. This exploratory function answers Kuhn's second question, (Q2)—what kind of new knowledge is produced in thought experiments—by exposing the assumption on which this question is based (that all the data are familiar and unquestioned) as false. Kuhn would have us believe that with the exception of problematical definitions, thought experiments are like well-posed problems. They are not; the increased understanding and knowledge is produced by refinement of the scenario underlying the thought experiment. Moreover, this refinement, since it is itself based on explicit argumentation, is a rational process with a logical, mathematical, and empirical basis, and is not at all psychologistic in form.

What, then, of the comparison with real experiments? I want to suggest here that although almost every writer on the subject has assumed that there is some clear parallel between thought experiments and real experiments, this is true only in a very special sense. The function of real experiments that is simulated by thought experiments is the isolation of those features of the world that are represented in a theoretical model and the approximation, as closely as possible, of the idealizations that are employed therein. This function of real experiments, which is to narrow the gap between theory and the world as it naturally presents itself to us, is nowadays almost exclusively guided by theory, and it indicates that thought experiments lie much closer to theory than to the world, or even its experimental refinements. Indeed, since (as Koyré 1960 and Körner 1980, for example, have pointed out) the concepts employed in thought experiments are designed to perfectly match those in theory, thought experiments lie squarely within the realm of theoretical entities rather than in the empirical realm inhabited by real experiments, or in the realm of the mental as suggested by psychologistic accounts.

Suppose we compare thought experiments with a more recent methodological innovation, that of computer simulations (see Humphreys 1992). Then very clear parallels emerge. First, even when computer simulations use fundamental theory to generate their models, as in the so-called ab initio methods of molecular chemistry, the boundary conditions, parameter values, approximations, and idealizations all require adjustment to bring them into conformity with existing empirical data. (This is even more obvious when phenomenological models are used.) That is, a good deal of simulation methodology involves theory refinement, and the "back and forth" method between theoretical model and output plays a significant role in the development of a clearly articulated thought experiment. Second, numerical experiments that run via simulations seem to have exactly the same kind of problems associated with them as do those that are traditionally considered to affect thought experiments. If we run a Monte Carlo simulation by generating random numbers from a specified probability distribution, and use those numbers in the simulation instead of empirically generated data, how can this possibly provide new knowledge? The answer is the same: It allows us to explore properties of the theoretical model lying behind the simulation such as its robustness under changes in the idealizations and parameter values, changes in confirmation level under alterations in the quantity of data available or grid resolution, and so on. Behind this exploration is the assumption that these data could have been generated empirically, but given that we know the form of the empirical distribution from which they come, then the numerical experiment acts as if it were a real experiment.

Third, simulations often deliberately alter the parameters involved in law schemata to produce laws that are not descriptive of our world. Thus, one can simulate the trajectory of a planet under an inverse 1.99 law, rather than an inverse square law, to explore the effects this would have on the stability of *N*-body orbital motion, a process which is clearly the same as that involved in a thought experiment, and indicates that there is no need to restrict thought experiments to worlds that satisfy the same laws as ours. Indeed, since we often do not know exactly what those laws are, to impose this restriction would be to place an absurd constraint on our exploratory investigations.

If these parallels hold, then this is further evidence of the truth of the thesis I suggested earlier, that both the "thought" and the "experiment" in "thought experiment" are nondescriptive; thought experiments are abstract entities that can be implemented on different kinds of systems, as long as the mathematical model is clearly articulated, and the so-called experiment involves theoretical rather than practical manipulation and refinement.

We may sum up the conclusions of this section in the form of six further theses:

THESIS TWO. *"Thought experiments" are neither essentially psychological nor particularly experimental in the traditional empirical sense. Rather, they are explorations and refinements of theoretical models.*

THESIS THREE. *The new knowledge that they provide involves increased understanding of the conditions under which the model holds. The ideal is to approach the conditions that underlie a well-posed mathematical problem, but that ideal is never reached.*

THESIS FOUR. *Thought experiments involve a process of rational assessment rather than providing immediate insight into a problem area.*

THESIS FIVE. *The model that best fits this process is one something like Lakatos's account of mathematical progress, applied to scientific models.*

THESIS SIX. *The data on which a thought experiment is based need not be well known; in cases where they appear to be, that appearance is often misleading. Nor need thought experiments be about what we expect. In fact the reverse is usually true.*

THESIS SEVEN. *The more the thought experiment is grounded in an explicitly formulated theoretical model, and the closer it is to a well-posed problem, the better it is. This is why so many philosophical thought experiments are useless; they are so far away from being a well-posed problem that no clear constraints on the solution are available.*

III

Nicholas Rescher has raised an important question about the view I have presented here:

> Is thought experimentation as Professor Humphreys understands it really more than a heuristic device for bringing into clearer view the implications of our theoretical models? Clearly actual experimentation is indispensable in science: If we do not monitor nature's doings interactively, we cannot hook our theories up with the real world whose workings they are supposed to help us better understand. But what about thought experimentation? Is it simply a matter of heuristic and expository aid? Or is it somehow essential—does it make a contribution to scientific understanding in whose absence the project of science would be irreparably impaired through an impoverishment that other intellectual resources cannot compensate for. Is thought experimentation a matter of convenience or of necessity within the scientific enterprise? And—accordingly—is thought experimentation something that properly belongs to the teaching (or learning) of science, or is it an essential component of science itself? (Personal communication)

This is a question that all accounts of thought experiments in science should have to answer, and within the account I have given earlier, I believe it has a straightforward answer. Thought experimentation, construed as the elaboration and refinement of theoretical models, is an essential component of scientific progress, but one that involves ineliminable heuristic elements. (For details see Humphreys 1991.) The important point to bring out here is that the heuristic aspects of model development frequently involve decisions about what are appropriate idealizations and approximations to use, and that these heuristics act in many cases as discovery devices, and are usually highly problem specific. These kinds of decisions pose, of course, serious problems for empiricism, though of a different kind than those traditional problems with which I began this essay.

IV

What, finally, does all this have to do with Professor Grünbaum's views? Quite a lot in fact, but I will restrict myself here to making one programmatical claim, which if correct, would be of some importance. By programmatical, I mean that I am prepared to defend its broad outline, but not here its detailed consequences.

Professor Grünbaum's writings have, as we might expect, made extensive use of thought experiments, both philosophical and scientific in kind. In addition to some of those famous examples already mentioned earlier, he has discussed the universal nocturnal expansion of the universe (1964), stationary light waves (1973, 373), Einstein's lightning bolt experiment (ibid., 359–67), a model pendulum universe (ibid., 200–06), a quasi-Newtonian universe (ibid., 689–96), the geometry of the rotating disk in special relativity (Grünbaum and Janis 1977), and, of course, Zeno's paradoxes and supertasks (e.g., Grünbaum 1967). As is well known, a recurring theme throughout Grünbaum's career has also been the exploration of deficiencies in Popperian methodology, and the consequences that correcting those deficiencies have for methodological issues in the philosophy of science, including those issues that arise from the Duhem-Quine hypothesis. (See, for example, Grünbaum 1976a, b, c, d and 1989.)

It is well known that there are complex and deep relationships between Kuhn's general position, the Duhem-Quine thesis, and the difficulties into which even sophisticated versions of falsificationism ran. During the 1960s, Grünbaum published a series of attacks on a version of the Duhem-Quine thesis that he called the "D-thesis." The first of these papers was published in 1960, and by 1969 the D-thesis had been refined into two subtheses, of which the second states, "For all potential empirical findings O′ [which are incompatible with an observationally testable consequence O entailed by the conjunction of H, the hypothesis under test, and A, the collection of auxiliary assumptions], there exists at least one suitably revised set of auxiliary assumptions A′ such that the conjunction of H with A′ *can be held to be true and explains* O′" (Grünbaum [1971] 1973, 587; emphasis in the original). Let us look at the last three words of this lengthy thesis, "and explains O′." Grünbaum denies any intent of completely capturing Duhem's own views, which is just as well, for these three words, when interpreted in certain ways, represent a radically different position from the one usually associated with Duhem and Quine, and one that provides crucial constraints on our implementation of the thesis. To put this in historical perspective, when Grünbaum's article was written in 1969, theories of explanation were still largely identified with logical models of the form that Popper and Hempel had defended, and with the single exception of the truth of the premises, the conditions of adequacy were purely logical in form. Thus, to insist in 1969 that a revised theoretical framework must explain O' was, for most philosophers, to stay within an approach where predictions and explanations were treated as identical in logical structure.[10] (Many specious arguments raised against Hempel's own version of the symmetry thesis are refuted in Grünbaum [1962] 1973). So when Quine in "Two Dogmas" asserted "As an empiricist I continue to think of the conceptual scheme of sciences as a tool, ultimately, for predicting future experience in the light of past experience," adding that the requirement that the new theoretical framework should explain as well as predict was no real additional burden. But the criteria for what counts as an adequate explanation have changed drastically since 1969 (which incidentally was the year immediately preceding the first publication of Wesley Salmon's seminal article on the statistical-relevance account of explanation, which began the move away from construing explanations as arguments). This is not the place to make partisan claims for what constitutes the right theory of explanation, but I do want to suggest that under the construal of explanations as picking out the correct causal structure responsible for the phenomenon O', much tighter constraints will be placed on what counts as an acceptable theoretical revision than under the logical models. In fact, because one of the themes of Grünbaum's attacks is that we need some way of making the Duhem-Quine view nontrivial,[11] the additional causal constraints seem to be an interesting candidate for that job in that the logical reshuffling of hypotheses to preserve H will by no means ensure that a causal explanation has been given of O', let alone an adequate one.

Although it has not been put this way, the attacks on Putnam's model-theoretic argument against realism (by causal realists such as Boyd, Field, and Glymour) and Putnam's almost willful refusal to engage the issue, by reiterating "it's just more theory," are a reflection of this approach in a related area. Quine's requirement that only predictive ability serve as the criterion of a satisfactory revision is too weak. Imposing the additional requirement that the revised background hypotheses allow a causal explanation of the falsifying phenomenon will not, of course, reduce the revisions available to a unique choice, but it does severely undercut the radical slackness in various holisms that seems to allow the undue degree of subjectivity in theory choice which concerns defenders of scientific rationality. And it was never part of the concept of scientific rationality that theory choice had to be uniquely determined, only that the choice be made by means of objective, rule-governed criteria. The correct formulation of the constraints on revisability is by no means straightforward if only because asserting that there exists a revision which explains O' would be to commit oneself to every falsifying instance having an explanation, which is not a doctrine of universal explainability, but close enough that we should not adopt the simple formulation. Still, in the context of contemporary advances in accounts of explanation, Grünbaum's insight, that theory revisions should not only predict but explain falsifying phenomena, looks to be one of profound importance and to offer a powerful, if partial, response to at least some of the theses propounded by Kuhn and by Quine.

NOTES

I should like to thank Adolf Grünbaum and Nicholas Rescher for helpful discussions and correspondence in connection with this essay.

1. The only detailed appraisal of Kuhn's paper of which I am aware is Cargile (1987), which exposes some serious deficiencies in Kuhn's interpretation of Aristotle. It also contains some penetrating criticisms of Kuhn's conflation of laws and definitions.

2. Kuhn does not specify anyone who has actually held this traditional view, which he rejects, and so it is a little hard to say whether the position that I will defend here falls under the scope of that view.

3. Attacks on this separability are often construed as also being attacks on the a priori—a posteriori division, and vice versa. For arguments concerning the first of these, see Putnam ([1976] 1983), where he notes Quine's explicit reference in Two Dogmas to evidence from the history of science that supposedly supports the lack of unrevisable statements, which Putnam rather hastily identifies with a priori truths. Thus, if Kuhn's arguments against the concept-content distinction were valid, one can see how this strategy would affect some other interpretations of thought experiments, such as the one espoused by Alexandre Koyré, Kuhn's mentor. In Koyré (1960), Koyré made a couple of rather incautious remarks about the a priori nature of imaginary experiments (most bluntly "Good physics is made a priori" [p. 245]), although his principal contributions in that article can be separated from this a priorism. A more recent, although much less plausible, defense of a priorism in thought experiments can be found in Brown (1986).

4. This fact could be established by means of an ingenious thought experiment described by Ernst Mach (1942, 162–63). Suppose one held that the ball rolling along the plane had a greater speed at *A* than did the ball falling at *B*. Then the speed at *A* will carry that ball to a greater vertical height than will the speed of the ball at *B*. So by arranging a contiguous series of inclined planes where the

ball rolls down the incline and rises up the vertical, any ball could be raised to an arbitrary height by gravity alone. This, one agrees, is physically impossible. A parallel argument, mutatis mutandis, will show that the ball at *A* cannot have a lower speed than the ball at *B*. One has to accept something else for this argument to be valid, which Mach overlooks. This is Galileo's theoretical assumption that all bodies are heavy, rather than the Aristotelian division into light and heavy bodies.

5. Sagredo generalizes this latter definition in a way that need not concern us here.

6. Galileo's conclusion then is that "a heavy body departing from rest passes in descent through all the gradations of slowness antecedent to whatever degree of velocity it acquires" (1962, 23) with the ulterior motive of shortly demonstrating the (incorrect) theory of circular inertial motion.

7. The puzzling exception occurs in Kuhn's discussion of the experiments by Piaget. He claims, "Presented with an animated cartoon showing [two cars, one of which has perceptually lower speed yet reaches the goal first], the child would reach the same conclusions about his concepts, even though nature were governed by the law that [perceptually?] faster bodies always reach the goal first" (Kuhn [1964] 1977, 246).

8. Aristotle himself might have agreed with Kuhn on this issue of concept acquisition. "[T]he form of man is always found in flesh and bones and parts of this kind; are these then also parts of the form and the formula? No they are matter, but because man is not found also in other matters we are unable to perform the abstraction" (*Metaphysics* 1036b).

9. By excluding implicit definitions, inductive definitions, and the like, I am not, I think, shifting the discussion away from what Galileo and the Aristotelians were concerned with, for Kuhn himself agrees that Aristotle provides an explicit definition of "faster than," "The quicker of two things traverses a greater magnitude in an equal time, an equal magnitude in less time, and a greater magnitude in less time" ([1964] 1977, 246).

10. This remark should not be interpreted as claiming that Grünbaum himself subscribed to this account of explanation, but simply that because the thesis D2 is couched in terms of entailments, any explanatory constraints that conform to the form of the thesis given here would at that time have been naturally implemented within the deductivist models then in general use.

11. Quine (1976) admits (in a letter to Grünbaum concerning the latter's attacks on the D-thesis) that the thesis is trivial as Quine has used it, and that he employed it simply as one argument against the separability of meaning and fact. I owe this reference to Adolf Grünbaum.

REFERENCES

Bohr, N. 1935. "Can Quantum Mechanical Description of Physical Reality be Considered Complete?" *Physical Review 48:* 696–702.

Brown, J. 1986. "Thought Experiments since the Scientific Revolution." *International Studies in the Philosophy of Science 1:* 1–15.

Cargile, J. 1987. "Definitions and Counterexamples." *Philosophy 62:* 179–93.
Galileo, G. 1962. *Dialogue Concerning the Two Chief World Systems—Ptolemaic and Copernican.* Translated by S. Drake. Berkeley and Los Angeles: University of California Press.
———. 1974. *Two New Sciences.* Translated by S. Drake. Madison: University of Wisconsin Press.
Grünbaum, A. 1964. "Is a Universal Nocturnal Expansion Falsifiable or Physically Vacuous?" *Philosophical Studies 15:* 71–79.
———. 1967. *Modern Science and Zeno's Paradoxes.* Middletown, Conn.: Wesleyan University Press.
———. [1962] 1973. "The Asymmetry of Retrodictability and Predictability, the Compossibility of Explanation of the Past and Prediction of the Future, and Mechanism vs. Teleology." In Grünbaum, pp. 281–313.
———. [1971] 1973. "Can We Ascertain the Falsity of a Scientific Hypothesis?" In Grünbaum, pp. 569–629.
———. 1973. *Philosophical Problems of Space and Time.* 2d ed. Dordrecht: Reidel.
———. 1976a. "*Ad Hoc* Auxiliary Hypotheses and Falsificationism." *The British Journal for the Philosophy of Science 27:* 329–62.
———. 1976b. "Can a Theory Answer More Questions Than One of Its Rivals?" *The British Journal for the Philosophy of Science 27:* 1–23.
———. 1976c. "Is Falsifiability the Touchstone of Scientific Rationality? Karl Popper versus Inductivism." In R. S. Cohen, ed., *Essays in Memory of Imre Lakatos.* Dordrecht: Reidel, pp. 213–52.
———. 1976d. "Is the Method of Bold Conjectives and Attempted Refutations *Justifiably* the Method of Science?" *The British Journal for the Philosophy of Science 27:* 105–36.
———. 1989. "The Degeneration of Popper's Theory of Demarcation." In F. D'Agostino, and I. Jarvie, eds., *Freedom and Rationality: Essays in Memory of John Watkins.* Dordrecht: Kluwer, pp. 141–61.
Grünbaum, A., and A. Janis. 1977. "The Geometry of the Rotating Disk in the Special Theory of Relativity." *Synthese 34:* 281–99.
Heisenberg, W. [1927] 1983. "The Physical Content of Quantum Kinematics and Mechanics." Translated in J. A. Wheeler, and W. H. Zurek, eds., *Quantum Theory and Measurement.* Princeton: Princeton University Press, pp. 62–84. (Originally published as "Über den anschaulichen Inhalt der quantentheorist ischen Kinematik und Mechanik." Zeitschrift fur Physik 43: 172–98.)
Humphreys, P. 1991. "Computer Simulation Methods in the Physical Sciences." Unpublished research report on NSF Grant #DIR-8911393.
———. 1992. "Computer Simulations." In A. Fine et al., eds., *PSA 1990.* Vol. 2. East Lansing: Philosophy of Science Association, pp. 497–506.
Kaiser, M., and D. Proffitt. 1987. "Observer Sensitivity to Dynamic Anomalies in Collisions." *Perception and Psychophysics 42:* 275–80.
Körner, S. 1980. "Science and the Organization of Belief." In D. H. Mellor, ed., *Science, Belief and Behaviour: Essays in Honour of R. B. Braithwaite.* Cambridge, England: Cambridge University Press, pp. 43–59.

Koyré, A. 1960. "Galileo's Treatise 'De Motu Gravium': The Use and Abuse of Imaginary Experiment." *Revue d'Histoire des Sciences 13:* 197–245.

Kuhn, T. [1964] 1977. "A Function for Thought Experiments." Reprinted in *The Essential Tension*. Chicago: University of Chicago Press, pp. 240–65.

Lakatos, I. 1978. *Philosophical Papers*. Vol. 2, *Mathematics, Science, and Epistemology*. Edited by J. Worrall and G. Currie. Cambridge, England: Cambridge University Press.

Mach, E. 1942. *The Science of Mechanics*. 9th ed. La Salle, Il: Open Court.

Massey, G. 1990. "Gedankenexperimente: Where Science and Philosophy Meet." Unpublished manuscript.

Norton, J. 1991. "Thought Experiments in Einstein's Work." In T. Horowitz and G. Massey eds., *Thought Experiments in Science and Philosophy*. Savage, MD.: Rowman & Littlefield, pp. 129–48.

Popper, K. 1959. *The Logic of Scientific Discovery*. New York: Basic Books.

Putnam, H. [1976] 1983. " 'Two Dogmas' Revisited." Reprinted in *Philosophical Papers*. Vol. 3, *Realism and Reason*. Cambridge, England: Cambridge University Press.

Quine, W. 1953. "Two Dogmas of Empiricism." In *From a Logical Point of View*. Cambridge, Mass.: Harvard University Press.

———. 1976. "A Comment on Grünbaum's Claim." In S. Harding, ed., *Can Theories be Refuted?* Dordrecht: Reidel, p. 132.

Wilkes, K. 1988. *Real People*. Oxford: Clarendon Press.

10

On the Alleged Temporal Anisotropy of Explanation

A Letter to Professor Adolf Grünbaum from His Friend and Colleague

Wesley C. Salmon
Department of Philosophy, University of Pittsburgh

June 1991

Dear Adi,

Following a venerable philosophical tradition, I am taking this occasion to address a profound problem in an open letter. Not long ago you asked me to tell you why—given the time-symmetry of most of the fundamental laws of nature—particular events cannot (accordingly to my lights) be explained by appeal to facts that obtain subsequent to their occurrences, but only by reference to those that obtained antecedently. This is, as I say, a profound question, and it has not been suitably addressed by many of those who want to exclude explanations in terms of subsequent facts. The following is my attempt to give that problem the serious attention it deserves. Unfortunately, the fact that a question is profound does not guarantee that the answer will be also; nevertheless, with my deepest respect, here is my best effort.

The Question

Can a particular event be explained by citing subsequent[1] conditions and events—along with appropriate laws—or do only antecedent conditions and events have explanatory import? For example, on 29 May 1919 there occurred an event of major importance to the history of science, namely, a total eclipse of the sun; the Isle of Principe in the Gulf of Guinea, West Africa, lay in its path of totality. Here observations were made to test Einstein's prediction of the bending of light passing near the limb of the sun. We know, of course, that a solar eclipse occurs at any given place if that place happens to be in the shadow cast by the moon. But what about the particular eclipse at Principe? Why did it occur at that particular place and time?

According to Sir Arthur Eddington's report ([1920] 1959, 113–14), the Astronomer Royal called his attention to the forthcoming eclipse in March of 1917. Using observational data concerning the relative positions and motions of the earth, sun, and moon, and applying the laws of motion, he was able to deduce the alignment of those bodies at a particular time on 29 May 1919. If, subsequent to the eclipse, someone were to ask why it happened there and then, the data and the derivation of the Astronomer Royal could be offered in answer.

One reason for picking this example is that the explanation just mentioned conforms perfectly to the deductive-nomological (D-N) model of explanation, first articulated with precision and in detail by Carl G. Hempel and Paul Oppenheim ([1948] 1965) in their epoch-making essay. Another reason is that it admirably exemplifies the *explanation-prediction symmetry thesis* propounded in that essay (ibid., sec. 3). Precisely the same deductive argument that furnishes a prediction of the eclipse prior to its occurrence provides an explanation of its occurrence after the fact.

This particular example also serves to illustrate the opening question of this letter. Since the laws of motion are time-symmetric, the Astronomer Royal could have made observations of the sun-moon-earth system two years later—in 1921—and he could have deduced that the total eclipse had occurred on Principe on 29 May 1919. Can this latter deduction qualify as an explanation of the eclipse? A great many philosophers, myself included, would reject this suggestion. With Laplace we might say, "We ought to regard the present state of the universe as the effect of its antecedent state and as the cause of the state that is to follow" ([1820] 1951). Since it is generally agreed, we might continue, that causes can explain their effects, but effects cannot explain their causes, only the derivation of the occurrence of the eclipse from antecedent conditions can qualify as an explanation. The inference from subsequent conditions can qualify only as a retrodiction—not as an explanation as well. Nevertheless, two of the most profound philosophers of science of the twentieth century, you and Peter Hempel,[2] have disagreed with this doctrine. Given the superb philosophical credentials of the two of you, we should take a close look at your point of view.[3]

A Bit of History

The classic Hempel-Oppenheim ([1948] 1965) paper, a fountainhead from which much of the subsequent philosophical literature on scientific explanation has flowed, amazingly attracted virtually no attention for a full decade after its original publication. Then, following that lapse, a flurry of sharply critical papers appeared. One major focus of attention was the explanation-prediction symmetry thesis. Your paper, "Temporally Asymmetric Principles, Parity Between Explanation and Prediction, and Mechanism versus Teleology," in which you endeavored to clarify and defend the symmetry thesis, was published in 1962. You showed convincingly that many of the criticisms were based on misunderstandings, and you sought to correct the then current misinterpretations by such authors as N. R. Hanson (1959), Nicholas Rescher (1958), and Michael Scriven (1959).

According to Hempel and Oppenheim, a D-N explanation of a particular fact is a valid deductive argument. Its conclusion (the explanandum) states the fact to be explained; its premises (the explanans) present the explanatory facts. At least one law-statement must occur essentially among the premises, and the premises must be true. Hempel and Oppenheim never claimed that all legitimate scientific explanations fit the D-N model; on the contrary, they explicitly asserted that the sciences include acceptable inductive or statistical explanations as well, even though they did not attempt to provide a model for those of the latter sort ([1948] 1965, 250–51). That project, undertaken in Hempel (1962a), yielded what later came to be called "the inductive-statistical (I-S) model of explanation."

Hempel's symmetry thesis has two parts. The first says that every scientific explanation could, under suitable circumstances, serve as a prediction. This part *seems* clearly to hold for D-N explanations. If the explanatory facts had been at our disposal before the occurrence of the fact to be explained (the explanandum) we would have been able to predict that fact, for we would have been in possession of true premises from which it follows deductively. It is worth noting that

Peter intended the symmetry thesis to apply to I-S as well as D-N explanations.[4] In Hempel (1962b, 10, 14), he maintained that either type of explanation of a particular event is an argument to the effect that the event to be explained was *to be expected* by virtue of the explanatory facts. This clearly implies the applicability of the first part of the symmetry thesis. In Hempel (1965, 367–68), he elevated this half of the symmetry thesis to a general condition of adequacy for any account of scientific explanation.

The second part of the symmetry thesis says that every legitimate scientific prediction could, under suitable circumstances, serve as an explanation. This obviously does not imply that every scientific prediction could serve as a D-N explanation. Many scientific predictions are probabilistic. According to the symmetry thesis, such predictions could, under suitable circumstances, serve as I-S explanations. As Israel Scheffler (1957) pointed out, however, a prediction is merely a statement about the future. As such, a prediction could not be an explanation, for an explanation, according to Peter, is an argument. The most that could be maintained is that legitimate scientific predictions are *the conclusions of* arguments that conform to the schemas of D-N or I-S explanation. This is the position I take you to have adopted (1962, 157–58) in defending the symmetry thesis.

One major problem introduced in the Hempel-Oppenheim essay concerns the use of the term "antecedent." Just before listing the four general conditions of adequacy for deductive explanations, the authors remark that "the explanans falls into two subclasses; one of these contains certain sentences $C_1, C_2, \ldots, C_k$ which state specific *antecedent* conditions; the other is a set of sentences $L_1, L_2, \ldots, L_r$ which represent general laws" ([1948] 1965, 247; emphasis added). Two pages later, when they offer a schema for deductive explanation, they label the *C*s "statements of *antecedent* conditions" (ibid., 249; emphasis added). One gets the impression that the *C*s are supposed to refer to facts that obtain temporally prior to the explanandum-event *E*. When, however, we examine the conditions of adequacy, which are set forth between the two foregoing characterizations of the *C*s, we see that no mention is made of temporal priority. One reason for *this* omission may be that the conditions of adequacy are intended to apply to explanations of laws as well as explanations of particular facts. In explanations of laws no premises are needed other than the *L*s; consequently, the general conditions of adequacy do not make mention of the "antecedent conditions."

When we pass from their preliminary conditions of adequacy to their precise explication of deductive explanation, however, we learn that the analysis "will be restricted to the explanation of particular events, i.e., to the case where the explanandum, *E*, is a singular sentence" (ibid., 273). A footnote inserted at this point explains why an analysis of explanations of laws cannot be furnished. Given this restriction to explanations of particular events, it becomes necessary to include in the explanans a singular sentence *C* corresponding to the "antecedent conditions." In the formal explication, however, there is no requirement that *C* describe conditions that obtain prior to the occurrence of *E*.

At this point it is tempting to suppose that the omission of any condition on the temporal relationship between the *C*s and *E* is a mere oversight—one that can easily be repaired. This supposition would, however, be unsound, for when Peter later addresses this issue directly, he denies that the *C*s should be so restricted. After offering putative examples of explanations that cite "antecedent conditions" contemporaneous with or later than the explanandum, he comments:

> Any uneasiness at explaining an event by reference to factors that include later occurrences might spring from the idea that explanations of the more familiar sort . . . seem to exhibit the explanandum event as having been brought about by earlier occurrences; whereas no event can be said to have been brought about by factors some of which were not even realized at the time of its occurrence. Perhaps this idea also seems to cast doubt upon purported explanations by reference to simultaneous circumstances. But, while such considerations may well make our earlier examples of explanation, and all causal explanations, seem more natural or plausible, it is not clear what precise construal could be given to the notion of factors "bringing about" a given event, and what reason there would be for denying the status of explanation to all accounts invoking occurrences that temporally succeed the event to be explained. (Hempel 1965, 353–54)

He thus deliberately rejects the notion that antecedent conditions should be temporally antecedent to the events that are to be explained.

In your 1962 essay, you introduced the technical term "H-explanation"—standing for "Hempelian explanation"—and defined it in a way that allows subsequent conditions to H-explain a given occurrence. Although this may seem strange from the standpoint of standard usage, we must realize that you were offering a characterization that is not only faithful to the *letter* of the 1948 paper, but also faithful to the *spirit* of Hempel, as is shown by Peter's explicit statement of 1965.[5] When, as a consequence, you defined H-prediction—standing for "Hempelian prediction"—in a way that includes what we would normally consider retrodiction, you were only being faithful to the Hempel-Oppenheim assertion that "whatever will be said in this article concerning the logical characteristics of explanation or prediction will be applicable to either, even if only one of them should be mentioned" ([1948] 1965, 249). This is, of course, simply an explicit statement of the symmetry thesis.

What, then, is the difference between H-explanation and H-prediction? As you defined these terms, it coincides precisely with the following characterization by Hempel and Oppenheim, "The difference between the two [explanation and prediction] is of a pragmatic character. If E is given, i.e. if we know that the phenomenon described by E has occurred, and a suitable set of statements C_1, C_2, . . ., C_k, L_1, L_2, . . . , L_r is provided afterwards, we speak of an explanation of the phenomenon in question. If the latter statements are given and E is derived prior to the occurrence of the phenomenon it describes, we speak of a prediction" (ibid., 249). Following the Hempel-Oppenheim text exactly, you offered figure 10.1:

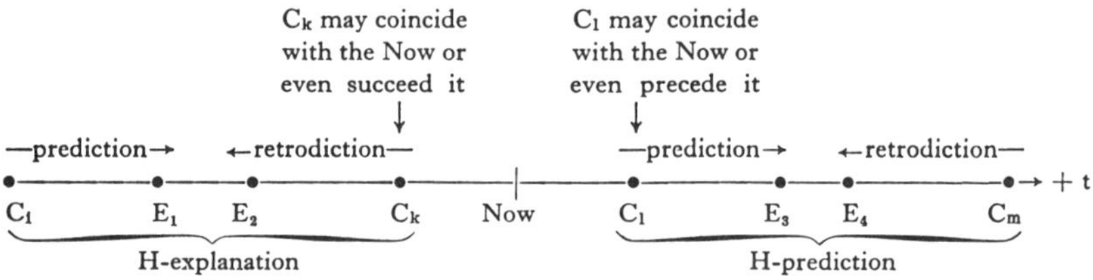

Figure 10.1 Reproduced from Adolf Grünbaum, "Temporally-Asymmetric Principles, Parity Between Explanation and Prediction, and Mechanism versus Teleology," *Philosophy of Science 29:* 156, by permission of the Philosophy of Science Association and Adolf Grünbaum.

On the one hand, as we see, whether a given event E is H-explained or H-predicted depends solely on whether E is before or after the "now" of the person deriving E from a given set of conditions C_i (in conjunction with suitable laws). It does *not* depend on whether the Cs precede or follow E. On the other hand, as we also see, if the Cs precede E we have an ordinary prediction (without the H prefix); the temporal relationship to the "now" does not matter.

Similarly, we have an (ordinary) retrodiction if the Cs follow *E*.[6] Thus, an H-prediction can be a retrodiction. If, for example, the Astronomer Royal had, in 1917, established the state that the sun-moon-earth system would assume in 1921, and had (still in 1917) derived the occurrence of the 1919 eclipse from the 1921 conditions, his inference would have been both a retrodiction and an H-prediction.

Although the foregoing terminological proposals may seem a bit odd from the standpoint of ordinary usage, they are perfectly legitimate as stipulative definitions, and they do reflect the theses propounded by Hempel and Oppenheim in 1948. In addition, they served you well in your efforts to clarify the Hempel-Oppenheim position, and to expose misunderstandings of the symmetry thesis. For this we owe you a large debt of gratitude.

The Cheap Answer

In Salmon ([1989] 1990) I offered what I now recognize as an idiosyncratic and biased historical sketch of philosophical discussions of scientific explanation in the four decades beginning with the Hempel-Oppenheim paper.[7] Pointing out that many criticisms of the "received view," which included both the D-N and I-S models of explanation, had been articulated in terms of standard counterexamples, I employed the example of the eclipse to call attention to the failure of the Hempelian account to stipulate that the so-called antecedent conditions must be temporally prior to the explanandum-event. This counterexample then served as the basis for a negative answer to the question posed at the beginning of this letter. For a large majority of philosophers, I think, this type of argument is compelling. Our intuitions rebel at the supposition that bona fide explanations could violate that constraint. This strategy amounted to taking the easy way out of the problem.

Hempel's Argument

You and Peter have approached the intuitive appeal to counterexamples in different ways. He is willing to fight fire with fire—to pit counterexample against counterexample:

> It might . . . be argued that sometimes a particular event can be satisfactorily explained by reference to subsequent occurrences. Consider, for example, a

beam of light that travels from a point *A* in one optical medium to a point *B* in another, which borders upon the first along a plane. Then, according to Fermat's principle of least time, the beam will follow a path that makes the traveling time from *A* to *B* a minimum as compared with alternative paths available. Which path this is will depend on the refractive indices of the two media; we will assume that these are given. Suppose now that the path from *A* to *B* determined by Fermat's principle passes through an intermediate point *C*. Then this fact may be said to be D-N explainable by means of Fermat's law in conjunction with the relevant data concerning the optical media and the information that the light traveled from *A* to *B*. But its "arrival at *B*," which thus serves as one of the explanatory factors, occurs only after the event to be explained, namely, the beam's passing through *C*. (1965, 353)

The answer I would offer to Peter's example is based classically on the ability of the wave theory of light to explain the phenomena of geometrical optics (including Fermat's principle). A rather similar account can be given in quantum electrodynamics.

According to the wave theory, light emitted from *A* in the general direction of *B* passes through the interface between the two optical media in a large region surrounding *C*. *After the waves have traversed this interface,* destructive interference cancels out those portions of the waves not passing through *C*, yielding the appearance of a ray that travels from *A* through *C* to *B*. Peter is certainly correct in stating that a D-N explanation of the passage of the ray through *C* can be constructed (just as he has shown) using the arrival of light at *B* as an "antecedent condition," but this fact reveals a deficiency of the D-N model. The *causal explanation* offered by the wave theory is far more satisfactory because it exhibits the actual mechanisms involved in the production of the fact to be explained. And in *that* explanation we *do not* explain any event in terms of subsequent facts.

In chapter 2 of his book, *QED: The Strange Theory of Light and Matter* (1985), Richard Feynman uses precisely the same example as Peter did to show how quantum electrodynamics deals with interactions between light and matter. Treating light as composed of particles (photons), he exploits the wave-particle duality to show how destructive interference yields the same result as Fermat's principle does. As in the classical explanation in terms of the wave theory of light, Feynman emphasizes the fact that photons pass through the interface between the two media in all possible regions—not just at *C*—and he offers convincing experimental evidence to support that claim. In the introductory chapter he asserts, moreover, that QED explains all phenomena involving interactions between light and matter, and that its predictions are the most precise predictions provided by any current physical theory.

One might be tempted to suppose that Peter chose an unfortunate example, and that a more apt example might be found by looking at other applications of variational principles such as Fermat's principle. That would be incorrect, I think. Max Born has argued in general that wherever such principles *seem* to offer an explanation of some physical fact in terms of subsequent conditions there exists an explanation in terms of straightforward efficient causes. With regard to variational formulations in classical mechanics he wrote, "From now on the world is a mechanism, ruled by strict deterministic laws. Given the initial state, all further development can be predicted from the differential equations of mechanics. The minimum principles are not due to nature's parsimony but to human economy of thinking, as Mach said; the integral of action condenses a set of differential equations into one simple expression" ([1956] 1969, 124; see also pp. 55–79).

In the past I have argued that noncausal explanations such as Peter offered in this quoted passage are illegitimate. I no longer hold that view (Salmon [1989] 1990, sec. 5.1). It seems to me that explanations that involve unification, in the sense elaborated by Michael Friedman and Philip Kitcher, provide a type of understanding that complements, rather than excludes, causal explanation. The use of variational principles does unify an enormous variety of physical phenomena, but it does not provide a suitable basis for claiming that particular events can be explained in terms of subsequent conditions.[8]

Grünbaum's Argument

You have been much less sympathetic than Peter to the appeal to commonsense example and intuition:

> But, to my mind, the philosophical task before us is *not* the ascertainment of how the *words* "explain" and "predict" are used, even assuming that there is enough consistency and precision in their usage to make this lexicographic task feasible. And hence the verdict on the correctness of Hempel's symmetry thesis cannot be made to depend on whether it holds for what is taken to be the actual or ordinary usage of these terms. Instead, in this context I conceive

the philosophical task to be both the elucidation and examination of the provision of scientific understanding of an *explanandum* by an *explanans* as encountered in actual scientific theory. (1962, 158)

In what way, you seem to ask, is the sort of use typically made of putative counterexamples different from a mere appeal to ordinary usage? In reference to another of the famous counterexamples—Sylvain Bromberger's flagpole—you retorted, "But is it not true after all that ordinary usage countenances the use of the term "explanation" only in cases employing causal antecedents and laws of succession in the *explanans?* To this I say: this *terminological* fact is as unavailing here as it is philosophically unedifying" (ibid., 168). It is now time to attempt a philosophically deeper answer to our main question.

Where to Look for an Answer

It was never my intention to appeal to untutored usage to support criticisms of the "received view," but I do believe that sound philosophical explication must involve what Rudolf Carnap called "clarification of the explicandum"—a process he exemplified superbly in *Logical Foundations of Probability* (1950, chaps. 1, 2, and 4). This process often involves consideration of examples. Neglect of such preliminary clarification might result in an explication of something different from what we set out to explicate. In my opinion, most of the literature on scientific explanation in the 1950s and 1960s was seriously deficient in this respect.

The clearest clue you have given us by way of clarification, it seems to me, is found in a previously quoted passage, namely, "the elucidation and examination of the provision of scientific understanding of an *explanandum* by an *explanans* as encountered in actual scientific theory" (1962, 158). Since the term "scientific understanding" is far from clear and unambiguous, we still have some work to do. The best approach, I think, is to look at some examples from the history of science that seem to be universally agreed *by scientists and philosophers of science* to constitute bona fide explanations, at least with respect to the body of scientific knowledge available at the time. For example, the Newtonian synthesis has been hailed for its explanation of many different types of phenomena—planetary motions, comets, and tides to mention just a few. When Edmund Halley ([1687] 1947, xiv) wrote, in his "Ode to Newton":

> Now we know
> The sharply veering ways of comets, once
> A source of dread, nor longer do we quail
> Beneath appearances of bearded stars,

he seemed to be saying that we need not fear comets because now, thanks to Newton, we understand them.

One point to be noted immediately is that in many cases, as the foregoing examples illustrate, scientific understanding results when a general regularity is explained by derivation from even more general laws. This is particularly true of situations in which a number of apparently independent regularities are unified by subsumption under a unified theory. This feature of scientific understanding was persuasively elucidated by Michael Friedman (1974), who referred to the explananda as *phenomena*. As he emphasized, such phenomena are not particular facts, but rather, general regularities. Since these regularities are not localized in time, the symmetry thesis does not apply. The question of whether "antecedent conditions" precede or succeed the explanandum-event does not arise because the explanandum is not an event. The explanans does not contain any Cs.

This point applies in a wide variety of cases. The Einstein-Smoluchowski theory of molecular bombardment explains the phenomenon of Brownian movement. Rutherford's planetary model of the atom enabled us to understand large angle scattering of alpha particles, but did not help us to understand the stability of atoms. The kinetic theory of gases explains Avogadro's law. It would be easy, but pointless, to go on collecting examples of this sort. We must conclude, I believe, in agreement with Friedman, that theoretical unification, as exemplified by the Newtonian synthesis, does produce one type of scientific understanding, but this kind of understanding sheds no light whatsoever on the temporal relations between explanatory facts and the explanandum. The examples just cited have no bearing on the model of scientific explanation offered by Hempel and Oppenheim. As just noted, they offered no precise explication of explanations of laws—only explanations of particulars.

Does science provide understanding of particular facts or occurrences? One way to a positive answer might be to look at particular experiments that have been performed. Does science provide understanding of these? Consider the Michelson-Morley experiment. Its negative outcome baffled Michelson, but Lorentz proposed a contraction hypothesis to explain it. This explanation was later judged unsatisfactory and was replaced by the special theory of relativity. By postulating the constancy of the speed of light, this theory explains why, in general, attempts to ascertain the speed of any system relative to the luminiferous ether will have a null result. Or, consider the Davisson-Germer experiment. Its result was first explained by the wave-particle duality of matter and later by quantum theory, from which one can deduce that material particles will, under certain circumstances, exhibit interference phenomena. In thinking about either one of these experiments, however, it is misleading to suppose (in spite of the singular grammatical form) that a single experiment is involved. Each encompasses a large series of experiments which established, among other things, that the results are reproducible. Hence, what is explained is not a particular fact but, rather, a general fact about experiments of particular types. Again, we must question whether bona fide explanations of particulars are involved. If not, these examples are not genuinely pertinent to our main question.

If these observations regarding scientific understanding are sound, then, apparently, theoretical science casts little light on the symmetry thesis. To find pertinent scientific material for our discussion we may have to abandon theoretical for applied science. In this domain, I believe, clear examples of scientific explanation result in genuine scientific understanding of particulars. The most obvious examples involve catastrophe, death, and destruction; in such contexts we are clearly concerned with particular cases.

Consider an example that has been widely discussed for approximately a decade, namely, the extinction of dinosaurs. Although it may have occupied a considerable stretch of space and time, it is a unique event. At a site near Gubbio, Italy, where the cretaceous-tertiary (K-T) boundary is exposed, Walter Alvarez examined a thin layer of clay and found it to be extraordinarily rich in iridium. With his father Luis Alvarez he proposed the hypothesis that this iridium anomaly could be explained in terms of a catastrophic collision of an asteroid or comet with the earth. This hypothesis has subsequently been rather strongly substantiated by additional evidence. They further hypothesized that the dire ecological consequences of this event resulted in the extinction of many living species, explaining the extinction, among others, of dinosaurs. This explanatory hypothesis has gained many adherents, though it is by no means as universally accepted as is the explanation of the iridium anomaly.

Although there are differences of opinion among experts regarding the acceptability of various proffered explanations of the iridium anomaly and of the extinction of dinosaurs, all seem clearly agreed that the search is for causal explanation in terms of temporally antecedent conditions. A convincing explanation will have to spell out the causal mechanisms whereby the explanandum is realized. I do not regard this claim as an appeal to the linguistic usage of scientists; it is a characterization of the practice of the experts in the field. It would obviously be easy to give many more examples of explanations of particular events—such as the Challenger shuttle disaster—in which the causal nature of the explanation is patent, but in examples of that sort it is possible to argue that a causal explanation is sought for the practical end of knowing how to prevent such things from occurring in the future, and not primarily for the sake of intellectual understanding. In such cases the issue of human control seems important, and indeed, is important. The examples of the iridium anomaly and the dinosaur extinction show, however, that there are cases in which causal explanations are universally regarded as being of the appropriate type even where the possibility of human control is out of the question. Scientific understanding is the principal goal.

The Right Answer

In his posthumous book, *The Direction of Time* (1956), Hans Reichenbach addressed various problems that you have dealt with at length under the heading of *temporal anisotropy.* Although you disagree with him on many fundamental points, I believe you are in accord on the following:

1. Our universe, at least in the present epoch, possesses an objective temporal anisotropy. Irreversible physical processes constitute a basis for this temporal anisotropy.

2. The fundamental laws of nature are temporally symmetric—leaving aside a few esoteric cases such as the law governing decay of the K^0-meson, which could hardly have a significant bearing on the pervasive temporal anisotropy of our world.

3. The temporal anisotropy of our world is *de facto* not *de jure.* The pervasive temporal anisotropy—what Reichenbach called "the direction of time"—is based upon pervasive matters of fact, not upon temporally asymmetric physical laws.

4. Although the laws are, within the limits just stated, time-symmetric, there are in nature *de facto* irreversible processes.

All of these seem to be sound.

Now I take it that when you raise the question about the temporal anisotropy of explanation you are prepared to distinguish two fundamentally different types of cases. In cases of the first type we are dealing with irreversible processes. Suppose, for example, that I associate with someone who has a cold, that as a result I contract a cold and lose my voice, and that in consequence I am unable to deliver a lecture to my class.[9] No one, I think, would be tempted to explain my cold by appealing to the subsequent condition of not being able to give my lecture. In nature, not being able to give a lecture does not lead causally to catching a cold. In contexts of this sort the temporal anisotropy of explanation seems to hold. The most high-powered medical research scientist would agree, I presume, with the ordinary person on the street. This is not an appeal to ordinary usage or untutored common sense.

In cases of the second type we have reversible processes; the eclipse example is a paradigm. Given the laws of mechanics, as well as what we know about general conditions in the universe, it appears to be entirely possible for there to be a solar system such as ours, but with the directions of motions of the earth about the sun and of the moon about the earth reversed. Leaving aside the extremely small frictional effects, the solar system constitutes a set of reversible processes.

When we consider the anisotropy of time, we try to take a global—or at least a large-scale—point of view. Given the facts about entropy and branch systems so clearly articulated by you and Reichenbach, we say that the temporal anisotropy is pervasive. It applies just as much to physical systems involving only reversible processes as to those that involve irreversible processes. We distinguish, consequently, earlier and later where lunar and planetary motions are concerned just as we do with regard to ice cubes melting in glasses of ginger ale. We do not say that temporal anisotropy prevails only in irreversible processes but not in reversible processes. To adopt a different policy with respect to time would play havoc with our most fundamental physical theories.

You and Reichenbach also agree on a causal theory of time, and again, I think you are right. Time and causality go hand in hand. The anisotropy of time is deeply connected to the anisotropy of causality. Causes come before their effects, not after them. Now, if one agrees that causality is an indispensable component of scientific explanations *of particular events,* it is natural to suppose that the anisotropy of time and causality would be reflected in an anisotropy of scientific explanation.

Reichenbach (1956, sec. 18), in "Cause and Effect: Producing and Recording", offers a number of extremely suggestive hints concerning the relationships among time, causality, and explanation. In earlier sections he had discussed time direction in terms of thermodynamics and microstatistics; he then attempted to apply similar considerations at the macrolevel:

> There exists an essential difference between microprocesses and macroprocesses. The former possess a natural shuffling mechanism given by the collisions of the molecules. The latter often do not possess any natural shuffling mechanisms . . . in other processes, the natural shuffling mechanism is so very slow that, at a given moment, the system remains practically unchanged. . . . This distinctive feature leads to peculiar consequences for macrostatistics: states of high order can here be preserved for a long time and can be observed conveniently. This is the reason why macrostatistics supplies what we call *records* and why, at the same time, it presents us with the key to the understanding of *causal explanation.* (Ibid., 149–50)

Reichenbach illustrates these considerations by means of the example of human footprints in the sand. Having defined order and disorder for macrosystems, he regards the sand with footprints as more highly ordered than smooth sand, and the footprints in the sand as records of a person having passed that way in the not too distant past.[10] He continues, "In addition to the clarification of the nature of records, the example of the footprints also helps us to analyze the meaning of *causal explanation.* Explanation in terms of causes is required when we meet with an isolated system displaying a state of order which in the history of the system is very improbable. We then assume that the system was not isolated at earlier times: explanation presents order in the present as a consequence of interaction in the past" (ibid., 151). He then offers an explication of cause and effect, "*The cause is the interaction at the lower end of the branch run through by an isolated system which displays order; and the state of order is the effect*" (ibid.). At this point in the text the editor (at my suggestion) inserted a footnote explaining that this explication is too narrow, but that an extension can be found a few pages later, "But in the sense of a transfer from relationships holding for irreversible processes in branch systems, the use of the word 'cause' is legitimate in application to macroprocesses governed entirely by the laws of mechanics" (ibid., 156).

As I said, Reichenbach offers hints about causality and explanation, but not a well worked-out theory. Nevertheless, he argues that causal explanation always refers to the past (ibid., 152). And his subsequent discussion of the principle of the common cause (ibid., sec. 19) reinforces the point. We explain improbable coincidences in terms of common causes, not in terms of common effects. The common cause temporally precedes the coincidence it is invoked to explain. Apparently, according to Reichenbach, anisotropy of time, anisotropy of causality, and anisotropy of explanation all go together, and are completely compatible with the time-symmetric character of the fundamental laws of nature. I think that this view is correct.

It is especially important to emphasize that the temporal anisotropy of explanation has nothing to do with the reversibility or irreversibility of the physical processes involved in the situation. The fact that there might be or might have been a solar system just about the same as ours, but with rotations reversed, is, to my mind, irrelevant to the fact that *in our solar system* the 1919 solar eclipse was a result of conditions in its past but not those in its future.

A Possible Bone of Contention

In this context, let us recall Peter's remarks, quoted earlier, concerning temporally antecedent causes and "bringing about." In contrast to explanations invoking temporally subsequent conditions, "explanations of the more familiar sort . . . seem to exhibit the explanandum event as having been brought about by earlier occurrences; whereas no event can be said to have been brought about by factors some of which were not even realized at the time of its occurrence" (1965, 353). He responds to this consideration by remarking that "it is not clear what precise construal could be given to the notion of factors 'bringing about' a given event . . ." (ibid.). We could offer Reichenbach's account of *producing* as one way of furnishing a reasonably precise construal of that notion. His discussion of causal explanation could be offered as an answer to Peter's further query as to "what reason there would be for denying the status of explanation to all accounts invoking occurrences that temporally succeed the event to be explained" (ibid., 353–54).

Although you and I have not discussed precisely this point, it occurs to me that you may feel intellectually ill at ease with the notions of *bringing about* and *producing*. Such terms may smack too much of the idea of *temporal becoming*, a concept of which you have been highly critical (e.g., Grünbaum 1967, chap. 1). For example, when Peter wrote (but did not endorse) the claim that "no event can be said to have been brought about by factors some of which were not even realized at the time of its occurrence," he was expressing a thesis about explanation that depends upon temporal becoming. Take away the notion of becoming and we can say that the universe consists of events located at various spacetime points or regions. All are (timelessly) equally real. Some of them are earlier than a given occurrence that is to be explained and some of them are later. The question is whether any of those located at times later than the explanandum can have explanatory import with respect to it. On your view there could be no question about what is or is not "realized" at a given time; events do not come into being and gain reality or pass out of being and lose reality. To be real is to have a location at some place and time in the world.

A major point of disagreement between you and Reichenbach on time has centered on his thesis regarding the existence of an objective present—the "now." I find your arguments compelling; I have no desire to defend Reichenbach's view on this matter, and I would not appeal to the argument for explanatory anisotropy mentioned in the beginning of this section. It seems to me, nevertheless, that Reichenbach's most important insights on causality and explanation can stand without reliance on his claims about temporal becoming. For example, you and I can use such Reichenbachian concepts as branch systems and causal interactions without invoking temporal becoming in any way.[11] Moreover, Reichenbach's characterization of a conjunctive fork, which plays a key role in his theories of temporal anisotropy and causal explanations, involves only probability relations that do not involve temporal becoming. In reading Reichenbach's text one must, to be sure, exercise care in not allowing becoming to creep in, but I have no doubt that it can be done.

If the analysis offered in this letter is correct, then the *alleged* temporal anisotropy of scientific explanation is also the *actual* temporal anisotropy of scientific explanation. In any case, this represents my serious and sincere effort to answer the question you posed—the one that constitutes the point of departure for this letter.

With deep affection and admiration,

Wes

cc: Peter Hempel

NOTES

1. Since the question to which this paper is devoted concerns explanation in terms of subsequent facts, the question concerning the explanatory status of simultaneous conditions and events will not be addressed.

2. Carl G. Hempel is known by colleagues and friends as Peter.

3. Because this is an *open* letter, I will review some aspects of the main question with which you are obviously completely conversant.

4. The footnotes in your (1962) seem to indicate that you were at least to some degree aware of the content of Hempel (1962a) prior to the publication of both of these papers.

5. I resist the temptation of attempting to explain your 1962 publication by reference to Peter's 1965 publication; I attribute it, rather, to your fundamental understanding, just prior to the time of writing his 1962 essay, of his position.

6. The term "H-retrodiction" is not defined; there is no need for it.

7. I was somewhat aware of this fact at the time, for I wrote, "We have arrived, finally, at the conclusion of our saga of four decades. It has been more the story of a personal odyssey than an unbiased history. . . . My decisions about what to discuss and what to omit are, without a doubt, idiosyncratic, and I apologize to the authors of such works for my neglect" Salmon ([1989] 1990, 180). Clearly you deserve such apologies, and I hereby offer them.

8. I am here using the term "phenomenon," as it is used in Friedman (1974), to designate a general fact rather than a particular occurrence.

9. This example is chosen to mirror the temporal relations in the eclipse example: Exposure to cold → contracting of cold → inability to lecture *correspond to* conditions in 1917 → conditions in 1919 (eclipse) → conditions in 1921.

10. Application of the explanation-prediction symmetry thesis for inductive-statistical explanations would lead to disastrous results if no temporal constraints were imposed, for then any event could be explained by a record of it. For example, relative widths of tree rings of logs found in an archaeological excavation could explain why a serious drought occurred in north-central Arizona at the end of the fourteenth century, for there is a strong (law-based) inductive inference from the dendrochronological data to the earlier climatic conditions. Likewise, without a temporal constraint, the iridium anomaly at the

K-T boundary could explain why a massive body collided with Earth about 65 million years ago. Without appealing to matters of usage, we can say that such pseudoexplanations would not enhance our scientific understanding of the events in question.

11. In Salmon (1984, 171) I characterize causal interactions roughly as intersections of processes (which can be represented by intersecting world-lines in spacetime diagrams) in which each process possesses some characteristic after the intersection that it did not possess prior to the intersection.

REFERENCES

Born, M. [1956] 1969. *Physics in My Generation.* 2d ed. Revised. New York: Springer-Verlag.

Carnap, R. 1950. *Logical Foundations of Probability.* Chicago: University of Chicago Press.

Eddington, A. [1920] 1959. *Space, Time and Gravitation.* Reprint. New York: Harper & Brothers.

Feynman, R. P. 1985. *QED: The Strange Theory of Light and Matter.* Princeton: Princeton University Press.

Friedman, M. 1974. "Explanation and Scientific Understanding." *Journal of Philosophy 71:* 5–19.

Grünbaum, A. 1962. "Temporally Asymmetric Principles, Parity Between Explanation and Prediction, and Mechanism versus Teleology." *Philosophy of Science 29:* 146–70. (Substantially the same article, under the same title, was published in B. H. Baumrin, ed., [1963], *Philosophy of Science: The Delaware Seminar.* Vol. 1. New York: Wiley, pp. 57–96, and in H. E. Kyburg, Jr., and E. Nagel, eds., [1963], *Induction: Some Current Issues.* Middletown, Conn.: Wesleyan University Press, pp. 114–45.)

———. 1967. *Modern Science and Zeno's Paradoxes.* Middletown, Conn.: Wesleyan University Press.

Halley, E. [1687] 1947. "The Ode Dedicated to Newton by Edmund Halley." Translated by L. J. Richardson, in F. Cajori, *Sir Isaac Newton's "Mathematical Principles of Natural Philosophy" and his "System of the World."* Berkeley and Los Angeles: University of California Press, p. xiv.

Hanson, N. R. 1959. "On the Symmetry Between Explanation and Prediction." *Philosophical Review 68:* 349–58.

Hempel, C. G. 1962a. "Deductive-Nomological vs. Statistical Explanation." In H. Feigl and G. Maxwell, eds., *Minnesota Studies in the Philosophy of Science.* Vol. 3, *Scientific Explanation, Space, and Time.* Minneapolis: University of Minnesota Press, pp. 98–169.

———. 1962b. "Explanation in Science and in History." In R. G. Colodny, ed., *Frontiers of Science and Philosophy.* Pittsburgh: University of Pittsburgh Press, pp. 7–33.

———. 1965. *Aspects of Scientific Explanation.* New York: The Free Press.

Hempel, C., and P. Oppenheim. [1948] 1965. "Studies in the Logic of Explanation." Reprinted, with an added Postscript, in Hempel, pp. 245–95. (Originally published in *Philosophy of Science 15:* 135–75.)

Laplace, P. S. [1820] 1951. *Philosophical Essay on Probabilities.* Translated by F. W. Truscott and F. L. Emory. New York: Dover.

Reichenbach, H. 1956. *The Direction of Time.* Berkeley and Los Angeles: University of California Press.

Rescher, N. 1958. "On Prediction and Explanation." *British Journal for the Philosophy of Science 8:* 281–90.

Salmon, W. C. 1984. *Scientific Explanation and the Causal Structure of the World.* Princeton: Princeton University Press.

———. [1989] 1990. *Four Decades of Scientific Explanation.* Reprinted, with a new Preface and identical pagination, from P. Kitcher and W. C. Salmon, eds., *Minnesota Studies in the Philosophy of Science.* Vol. 13, *Scientific Explanation.* (Minneapolis: University of Minnesota Press.) Minneapolis: University of Minnesota Press.

Scheffler, I. 1957. "Explanation, Prediction, and Abstraction." *British Journal for the Philosophy of Science 7.* 293–309.

Scriven, M. 1959. "Explanation and Prediction in Evolutionary Theory." *Science 130:* 477–82.

11

A New Theory of Reduction in Physics

Erhard Scheibe
Philosophy Department, Ruprecht-Karls-University

Science is patchwork. Even physics, perhaps the most successful part of science, did not achieve its goal in one attempt. It cannot yet be presented as one unified theory. Students of physics are introduced to various theories: mechanics of mass points, continuum mechanics, electrodynamics, thermodynamics, quantum mechanics, and so on. Moreover, there are other natural sciences besides physics: chemistry and, most importantly, biology. Sometimes it is said that by the invention of quantum mechanics chemistry could be *reduced* to physics, very much as, within physics, optics was reduced to electrodynamics or thermodynamics to statistical mechanics. Similarly, it is claimed that quantum mechanics has superseded classical mechanics and that, at least in principle, the latter could be replaced by the former and thus be eliminated from the foundations of physics. Thus, in addition to the vast number of scientific theories and disciplines, which today can hardly be comprehended by any one individual, there are also intertheoretic relationships, some of which seem to have a reducing or unifying effect. Galileo's law of free fall and Kepler's laws of planetary motion could be unified in Newton's gravitational theory. Although not always the consequence of a proper unification, these relationships show the totality of theories as being an *orderly net* rather than a randomly accumulated heap.

Do we have a satisfactory theory describing or explaining the reduction net of the natural sciences or even of physics? I am afraid we do not. The various attempts to find one always proceeded in the way usual for enterprises of this kind. First one looks for a sufficiently (if not excessively) general concept and only then, by way of specializing it, gets down to cases of application. In this manner the unity of the concept in question is given priority: Insofar as there are different kinds of reduction at all they are at once recognized as being only so many special cases of the one 'true' idea of reduction common to them all.[1] Although this is the usual *analytic* strategy in concept explication followed up in all the attempted explications of metascientific concepts well known from the positivistic tradition, in the case of theory reduction, I want to argue, it was a mistake. In the following, by turning the tables, I am going to develop what might be called a *synthetic* theory of reduction. In this theory we start from a couple of special and widely differing concepts of reduction and gain the most general case by their recursive combination. The most general reduction thus always appears as a certain combination of those elementary, pure cases from which the recursion started, and this feature seems to meet most happily the demands of the actual cases known from science. The various reductions occurring in physics are most conveniently analyzed, not by subsuming them under one general preconceived concept, but by recognizing them as so many combinations of a finite number of elementary reductions.[2]

We can at once indicate the new approach by correcting another mistake due to the received view. Philosophers of science seem to hold a conviction that theory reduction is a *binary* intertheoretic relation: One theory T is reduced to another one T'. In fact, however, a *third* element is always involved: the one that actually does the reduction. The binary relation view seems to be a consequence of the positivistic theory of reduction according to which a reduction of T to T' is essentially a logical deduction of T from T' (see Nagel 1961, chap. 11, sec. 2; and Scheibe 1989). We will see in the sequel that this is not the case without important qualifications. By these qualifications, among other things, the third element, the *vehicle* v of the reduction—as I will call it—is introduced. An explicit notation for a reduction statement would then be:

$$red_i\,(T, T'; v), \tag{11.1}$$

expressing that *theory T is i-reduced by v to theory T'*. Here the variable i indicates the various *kinds of reduction* that come into play, for example, (as we will see) "generalizing." Starting with initial reductions $i = 1, \ldots, n$, the most general reduction is generated by the following rules of recursion:

(R_1) $red_1, \ldots, red_n$ are reductions;
(R_2) if red_i and red_k are reductions then also $red_{(ik)}$ is a reduction.

Here "$red_{(ik)}$" stands for the reduction generated by first i-reducing any theory T to theory T' by means of vehicle v and then k-reducing T' to a third theory T'' by means of vehicle w. Accordingly, if $v \cdot w$ is the new vehicle, the operation of combining reductions to yield new reductions has to satisfy the condition (C) If $red_i(T,T',v)$ and $red_k(T',T'',w)$ then $red_{(ik)}(T,T'', v \cdot w)$. Evidently, from this requirement and the recursion rules, the vehicle of reduction is all important for the generation of new reductions from given ones and its usual repression is a fatal destruction of the whole concept of reduction.

In the following the synthetic theory of reductions based on rules (R_1) and (R_2) is presented mainly by way of illustration. Some general remarks, however, are in order. First, my main emphasis is on the various kinds of reduction—elementary and compound. Once reductions are indexed and an i-reduction is understood to be a (ternary) *relation*, different from the relation given by k-reduction for $k \neq i$, the emphasis is hardly necessary. Accordingly, in formulating the above rules the simpler expression "reduction" is used instead of "kind of reduction." Second, the number of elementary reductions is deliberately left open: We simply do not know what new kinds of reduction will have to be introduced in order to understand the further development of physics. Only combinations of reductions already given are a priori known to be reductions again. At the same time, the complex reductions make our general concept of reduction nontrivial *even if* the list of elementary reductions is not closed. Third, even more illusory than to know all kinds of reduction is the aim to precisely characterize any given kind in the sense that *all* its *instances* are genuine reductions of this kind. Genuineness here is a question of more or less, and often there will be degenerate cases, for instance, of self-reduction whose exclusion would be completely ad hoc. Fourth, the domain where a reduction of a given kind is defined depends on this kind, and this has to be observed in applying rule (R_2). Finally, (R_2) seems to be idempotent in the sense that $red_{(ii)}$ is always identical with red_i.

In this essay considerations are mainly restricted to the *formalisms* of physical theories, neglecting questions of interpretation and application beyond the absolutely necessary. With some formalization of first order set theory as our invariable background, the formalisms reduce to the axiom systems of theories. The reduction of physical theories thus becomes the reduction of set theoretical formulas functioning as physical axioms, and the vehicles of the reductions also become certain syntactical entities like formulas or terms. Accordingly, *so far* the status of reduction statements is essentially that of set theoretical theorems. Despite this purely syntactical background, the following presentation will occasionally use the "material mode of speech." Formulas and terms are then replaced by classes of structures and mappings between them in a fictitious set universe.[3]

In reconstructing physical axioms as set theoretical formulas we can go one step further and take them to be *species of structures* in the sense of Bourbaki (1968, chap. 4). (For details see Scheibe 1986.) By using mathematical structures for representing physical systems, a physical axiom says about a structure what we want it to say about the physical system represented in (or even: by) that structure. In this way we reconstruct a physical theory as being a *concept* of physical systems, and this nicely meets the terminology of physics according to which we speak of Newtonian, Hamiltonian, quantum mechanical, and such systems according to whether a system is submitted to Newtonian, Hamiltonian or quantum mechanics, respectively. It is then only with respect to a well-determined range of application Γ that a theory makes an empirical claim, namely, that *all* systems belonging to Γ are represented in mathematical structures satisfying the axioms of the theory.

I

After these preparations, let us now introduce reductions by *direct generalization*. In direct generalization we pass from a theory T to a theory T' more general in the sense that every physical system admitted by T is also admitted by T' but (as a rule) not vice versa. We reduce T to T' by knowing a contingent condition such that T' together with this condition leads back to T. The idea thus is to give T' a wider range of application than T. A very condensed formulation of direct generalization is

$$\Sigma' \wedge c \equiv \Sigma \tag{11.2a}$$

where Σ' and Σ are the axioms of T and T', respectively, c is the additional condition restricting T' to T, and all three statements are formulated in the same terms. The (set theoretical) equivalence $\equiv$ is, therefore, an ordinary reaxiomatization. In the terminology introduced previously, condition c is the vehicle of this kind of reduction. Systems admitted by Σ' but not by condition c are possible advances of T' over T, typical for direct generalizations. The formula

$$New \wedge Eny \equiv Kep \tag{11.2b}$$

illustrates the general situation: Kepler's three laws are reduced by generalization to Newton's gravitational theory (in its central field approximation) with the help of the condition that the energy of the planet is negative. The new possibilities introduced in this case are the unbounded paths admitted by Newton's but not by Kepler's theory and again excluded by the energy condition.[4]

Prior to any comments on our first type of reduction I will now introduce another: reductions by *refinement*. Theory T' is a refinement of T by means of P if P is a *surjective* mapping assigning to every system possible according to T' a possible system of T in such a way that for any two systems related by the mapping the description of the T'-system (according to T') is more complete (or more refined) than the description of the T-system (according to T). In this case, therefore, not *new* physical systems are the possible advances of T' over T but rather a more refined description according to T' of *each* system already allowed by T. Accordingly, T' has the same domain of application as has T. In syntactical terms we may write:

$$\begin{cases} \Sigma'(S') \wedge S = P(S') \vdash \Sigma(S) \\ \Sigma(S) \vdash \bigvee S'.\, \Sigma'(S') \wedge S = P(S'), \end{cases} \tag{11.3a}$$

for reduction by refinement where besides the axioms Σ' and Σ of T' and T, respectively, P is a term that "defines" a model S of Σ out of any given model S' of Σ'. In semantical terms, P stands for the mapping mentioned and is the vehicle of the reduction. The second line of (11.3a) makes P surjective.[5] As an example from black body radiation let us take the reduction of Stefan-Boltzmann's law SB to Planck's law PL:

$$\begin{cases} PL\ (\rho,\theta) \wedge u = \Delta(\rho) \vdash SB(u,\theta) \\ SB\ (u,\theta) \vdash \bigvee\rho.\ PL\ (\rho,\theta) \wedge u = \Delta(\rho). \end{cases} \tag{11.3b}$$

Here ρ and u are the spectral and integral energy density respectively and θ is the temperature. The refinement is evident from the defining relation Δ: Whereas in SB the radiation is only described by the integral energy density u, in PL the entire spectral distribution ρ as function of the frequency v of the radiation is part of the description.[6]

If we compare (11.3b) with the corresponding deduction in (11.2b) we can realize how little it means to characterize reductions as derivations; we have derivations in both cases. Still the effect is entirely different in the two cases according to the different nature of the additional premise—the vehicle: In the Kepler-Newton case it restricts the totality of Newtonian systems to a subset; in the other case it coarsens the description of each single system. The difference in question—in contrast to the mere derivation—can also elucidate the different kinds of *progress* that are brought about in the two cases. In reduction by generalization possible progress lies in the discovery of new physical systems allowed by the reducing theory but excluded by the reduced one. Experience has convinced us that hyperbolas are physically possible orbits in a central gravitational field. It was just a mistake to exclude them. On the other hand, in reduction by refinement nothing of this sort is involved. The more refined description of each system made available in the reducing theory can be an advance if we are provided with confirming experiences. Let us point out that both kinds of progress would mean nothing if the reductions would not allow us to recover the reduced theories in some sense. If they had been empirically adequate theories and could not be recovered we would hardly be in the position to talk of progress. It is here, in the conservative part of reduction, where the importance of deduction lies: By being deduced the reducing theories are recovered in a most literal sense.

Let us now look closely at the work done by the vehicle of a reduction if the latter is to fulfill its *explanatory role.* For generalization, (1) alternatives to the special theory T are available within the more general theory T', and (2) a condition c is distinguished under which, given T', theory T and *not* any of its alternatives is bound to hold. In this sense, we know *why T.* The explanatory function of generalization transpires most clearly from the situation where the old theory T once had been declared to be valid on a priori grounds. This, of course, not only means that no alternatives to T were known but that on the alleged evidence of those grounds no alternatives could exist. Consequently, no explanation of T apart from self-explanation could be given. But as soon as alternatives arc made available through generalization the major question becomes: Why T and not any of those alternatives? This question is answered by means of the *condition c.* A story of this kind is given by the development of geometry in the time between Kant and Einstein (to be discussed later).[7] An explanation different *in kind* is provided by reduction by refinement. Here the *terms P* take over the burden of explanation by telling us how the basic concepts of theory T depend on those of theory T' and *therefore* the axiom Σ holds if Σ' is assumed to hold.[8]

II

Before considering combinations of generalization and refinement let us address in what sense, if any, these kinds of reduction are *elementary.* If this is meant in an absolute sense, I do not know the answer. However, interesting as the answer may be, it certainly need not be given in order to establish our theory; for we are free to declare these and further reductions to be elementary in the relative sense that by fiat they are chosen to be the initial settings of the recursion (R) and thus made the building blocks of the theory. In the case before us we can distinguish *subkinds* of direct generalization and refinement so that in this sense these reductions are simply not elementary. An obvious case in point is *equivalence* as a subkind of refinement: T and T' are equivalent if they are mutual refinements of each other. This implies that the terms P and P^{-1} transforming T into T' and T' into T according to (11.3a) are inverses of each other. As reductions these equivalences may be called "degenerate" because equivalent theories can be reduced to each other. Still in physics we cannot do without them. Being changes in the conceptual basis of a theory they teach us to view one and the same physical system in different perspectives, and their application sometimes leads to unexpected insights into intertheoretic relations. A first type of *combination of two reductions,* one of which is an equivalence, may confirm this.

Reductions by direct generalizations are much too special in order to be an adequate explication of our usual understanding of generalizations. A case in point is geometry. One of the momentous generalizations in the history of mathematics was the step from Euclidean to Riemannian geometry. Its application to physics by Einstein led to the analogous generalization of Minkowskian to Lorentzian geometry of spacetime.[9] In both cases it was a step from a metrically flat to a curved manifold. And it was a genuine reduction of the special to the general case with respect to the vanishing of the Riemann tensor as the reducing subsidiary condition (the vehicle). But it was *not* a direct generalization. Direct generalization assumes—as is implicit in formulas (2)—that the theories related are formulated in the same language. This, however, was hardly the case for the present example: Before Riemann nobody thought of defining Euclidean space by means of a metric tensor field. To include this case and many others known from the history of physics in our theory we have to allow for reductions by *generalization involving equivalences.* More precisely, "generalization" is direct generalization preceded or followed by an equivalence. A Euclid-style version of Euclidean geometry has first to be equivalently transformed into its Riemann-style formulation which in turn leads to Riemannian geometry by direct generalization.

Furthermore, there is also the important case of *combining refinement with* (direct or indirect) *generalization.* This type frequently occurs as forward reduction beginning with specialization followed by coarsening. In the systematic sense a forward reduction is just the inverse of a reduction proper. There is, however, a pragmatic difference between the two. In a reduction proper we assume that the theory to be reduced was known prior to the reducing theory. Many applications of our concept exist, however, where it is the other way round. In particular specialization and coarsening as the relations inverse to generalization and refinement are frequent steps taken in physics. If, for instance, Hamiltonian mechanics is put to use, the first step is always a specialization leading to a fairly definite phase space and Hamiltonian. In a second step this theory is coarsened by the definition of some quantities for which an empirical law is derived from the equations of motion, for example, Kepler's third law from Newton's equations. Another example is the scattering theory in quantum mechanics. Here, on a very general level, in a first step (specialization) the bound states, if any, are eliminated which in a second step (coarsening) allows the definition of the *S*-operator and the derivation of its characteristic properties.

The importance of equivalences as constituents of complex reductions is *conceptual assimilation*. We have already seen this in the case of geometry in which Riemannian geometry assimilates basic concepts of Euclidean geometry. Stressing the heuristics in the process one could also say that conceptual transformation is part of the insight as to how Euclidean geometry can be fruitfully generalized. Such assimilation and transformation is brought about by equivalence. A well-known lesson that things, looking entirely different at face value, eventually turn out to be the same was taught by quantum mechanics: the equivalence of matrix and wave mechanics. An even more amazing case was Cartan's geometrization of Newton's gravitational field theory (see Künzle 1972). He and his followers could show that under suitable boundary conditions Newton's theory is equivalent to the differential geometry of a 4-dimensional flat Galilean metric together with a nonvanishing symmetric connection compatible with the metric. The equivalent of Newton's field equation is a field equation expressed with the help of the Ricci tensor of the connection. The equivalent of Newton's equations of motion are the equations for the geodesics of the connection. The proof of this equivalence was the first step in the attempt to reduce Newton's theory to Einstein's theory of general relativity. However, to obtain this reduction no combination of the elementary reductions introduced so far would suffice. Rather our list of elementary reductions has to be enlarged.[10]

III

We here meet with the weakness of the elementary reductions considered in the foregoing sections: They as well as their combinations always allow for an *exact recovery* of the axioms of the theories superseded by their historical successors, and the progress, if any, does *not* include any essential *correction* of the earlier theories. Such cases, however, rarely occur in the history of physics. As a rule a theory and its presumable successors are *rivals,* their axioms contradict each other, and the progress that is made does include a modification of the axioms of the theory superseded. Such is the case with Einstein's theory of gravitation vis à vis Newton's; it is even the case with the latter vis à vis Kepler's laws of planetary motion, and indeed we can find very simple illustrations of this kind of progress. The main difficulty in finding new types of reduction taking account of rivalry between theories is to reconcile this situation with the demand that the predecessor theory, insofar as it had been confirmed, *must* be recovered from the theory superseding it in *some* sense; otherwise, we were not entitled to speak of any progress attached to the reduction.

The essential idea to find reductions relating incompatible theories is to allow for *approximations:* Instead of an exact recovery the reduced theory has to be reproduced only approximately. Let us see what this can mean by looking at the simple case of the reduction of the ideal gas law

$$p \cdot v = R \cdot T \tag{11.4a}$$

to van der Waal's law

$$\left(p + \frac{a}{v^2}\right) \cdot (v - b) = R \cdot T\ (a,b > 0). \tag{11.4b}$$

It is obvious that these laws contradict each other in the sense that, given the constant R, there is no solution $\langle a,b,p,v,T \rangle$ of (11.4b) such that $\langle p,v,T \rangle$ is a solution of (11.4a). As to reduction we can first specialize the van der Waals's law to be a law for one sort of gas, say, oxygen only. In this case we have numerically fixed values a_o and b_o for the van der Waals's constants. The parameters left, that is, p, v and T, are then the same in the two theories, and it is evident that for

$$a_o p^{-1} v^{-2},\ b_o v^{-1} \ll 1 \tag{11.4c}$$

given values p, v, T satisfying (11.4b), though not also satisfying (11.4a), are located very close to values that *do* satisfy (11.4a). Whereas in this case of, as it might be called, *asymptotic reduction,* the two theories are built over the same conceptual material; the general van der Waals's law with variable a and b in fact includes a refinement of the description of a gas as compared with the ideal gas law. In this case the approximation is most conveniently reconstructed as an ordinary limit process: If we look at a and b in van der Waals's equation as parametrizing the solutions p, v, T then the union of these 2-dimensional hypersurfaces (in the 3-dimensional space of all $p, v, T > 0$) has the hypersurface made up of the solution of the ideal gas law as its boundary. Thus here the reduced theory appears as a *limiting case*[11] of the reducing theory for $a,b \rightarrow 0$.

The foregoing example is typical for a large class of reductions in which the two theories involved are not essentially different in their most fundamental, that is, topological or geometrical, parts but have contradicting *laws* in the proper sense of the term. This holds in particular for the difference between asymptotic reduction (I) and limiting case reduction (II). We give a parallel description of these two kinds in three parts according to the three components of their respective vehicles.[12]

(1) The first component appears as the common part Σ_o in the assumed decompositions

$$\begin{cases} \Sigma'(S_o,s) \equiv \Sigma_o(S_o) \wedge s \in M'(S_o) \\ \Sigma(S_o,s) \equiv \Sigma_o(S_o) \wedge s \in M(S_o) \end{cases} \tag{11.5a}$$

where $M'(S_o)$ and $M(S_o)$ involve typifications by the *same* scale set $\sigma(X_o)$, with $-X_o$ the principal base sets of S_o. In case (I) Σ' and Σ are the reducing and reduced axioms respectively while in case (II) Σ' is already a "refinement."

$$\begin{cases} \Sigma'(S_o,s) \equiv \bigvee S_o's'.\ \Sigma''(S_o,S_o',s') \\ \qquad\qquad\qquad \wedge\ s = q(S_o,S_o',s') \end{cases} \tag{11.5b}$$

of the proper reducing axioms Σ'', and the term q also belongs to the vehicle. Though Σ' and Σ contradict each other—see (2)—they have common partial models in the sense[13] that

$$\begin{cases} \text{for all } S_o \text{ such that } \Sigma_o(S_o) \text{ there are s, s}' \text{ such} \\ \text{that } s' \in M'(S_o) \text{ and } s \in M(S_o). \end{cases} \tag{11.5c}$$

Evidently, these assumptions allow us to define a relation between any two structures satisfying Σ' and Σ, respectively: They stand in this relation if their fragments typified according to Σ_o are identical.

The further reduction only concerns $M'(S_o)$ and $M(S_o)$ for any *given* S_o, and we drop the argument.

(2) The second component of the vehicle is a topological space $\langle M_o, t_o \rangle$ depending on and deducible from S_o by means of Σ_o such that

$$M, M' \subseteq M_o \subseteq \sigma(X_o). \qquad (11.6a)$$

Moreover, with the bar indicating topological closure,

$$\overline{M} \cap M' = M \cap \overline{M'} = \phi \qquad (11.6b)$$

in case (I) and

$$M \cap M' = \phi \qquad (11.6c)$$

in case (II) hold as expressing mutual inconsistency of the laws. In case (I) we actually assume t_o to be a uniform structure.

(3) Finally, the third component of the vehicle is a monotonously decreasing family of sets $c_\delta \subseteq M_o$ $(\delta > 0)$, independent of M' in case (I) and subsets of M' in case (II). With their help the reduction statement in case (I) is

$$\begin{array}{l}\text{for all } u \in t_o \text{ there is } c_\delta > 0 \text{ such that} \\ M' \cap c_\delta \subseteq M_u\end{array} \qquad (11.7a)$$

(where M_u is the u-neighborhood of M) and in case (II) is

$$\left\{\begin{array}{l}M = \{s \in M_o \mid s \notin M' \text{ and for all } u \in t_o, s \in u \\ \text{there is } s' \in M_o \text{ and } \delta > 0 \text{ such that } s' \in c_\delta \cap u\}.\end{array}\right. \qquad (11.7b)$$

What does this mean? Because of (11.6b) in case (I) there is no question of any solution of law M being a limiting case of solutions of M'. There are, however, common accumulation points (possibly in the infinite) in the sense that for all neighborhoods $u \in t_o$: $M_u \cap M'$, $M \cap M'_u \neq \phi$. Equation (11.7a) then says that under the special conditions of c_δ any solution of M' comes very near a solution of M. Thus we have a typical asymptotic behavior in this case.[14] By contrast, in case (II) the solutions of M *are* limiting cases of solutions of

M': It follows from (11.7b) that M is a subset of the boundary of M' in the narrow sense that $M \subseteq \bar{M}' - M'$. The function of c_δ in this case is to pick out that part of the boundary that coincides with M: In general, a law may have several limiting cases.

Obviously, from (11.7a) asymptotic reduction is an approximate variant of direct generalization. In view of (11.5b) it seems that limiting case reduction, if Σ'' is the reducing axiom, may even properly be decomposed into an exact refinement preceded by a truly approximate reduction as described in (11.7b). So why not just omit (11.5b) in (II) and take Σ' to be the reducing axiom, thus making full use of the very idea of a synthetic theory of reduction? From a purely formal point of view we would be entirely justified to do this. However, there is the difficulty that in successive reduction of axioms Σ to Σ' and Σ' to Σ'' and thus of Σ to Σ'' it may happen that, whereas Σ and Σ'' represent ordinary physical theories T and T'', respectively, no *reasonable* theory T' corresponds to Σ'. In our above example the ideal gas law (Σ) and the van der Waals's (Σ'') are reasonable physical laws. But Σ' in this case would be the statement (about p, v, and T) that there are (constants) a and b such that van der Waals's law holds. And this, though a perfectly clear statement, is not the kind of statement that we would like to call a physical law. Thus the insertion of Σ' into this reduction is only "virtual" in the sense of introducing virtual particles in a physical interaction. A solution of this problem cannot be given in the present essay.

IV

Asymptotic reduction as described before is still not general enough to cover cases in which the approximations concern only partial descriptions of the respective physical systems. Such is the case, for instance, with reduction of the Rayleigh-Jeans radiation law to Planck's law. But an appropriate generalization is easily obtained, and I will leave this case in order to address more important ones as we find them in relativity and quantum theory and microreduction. In this essay, I cannot even touch upon the deep questions that can be and have been raised in connection with intertheoretic relations in these fields. It goes without saying that my work on theory reduction was motivated by the well-known difficulties concerning the folk views on progress and unity of science as they were pointed out by

Kuhn and Feyerabend and, even earlier, by the physicists themselves (see Kuhn 1983; Feyerabend 1975, chap. 17; and Scheibe 1988b). As regards these difficulties I have nothing to add to the main *general* idea on which the present approach to reduction is founded. But I hope to show in the last two sections that the greater flexibility that this idea gives to theory reduction is an advantage in the treatment even of those difficult cases.

The main lesson taught by a comparison of the various (special and general) *relativity theories* with their nonrelativistic predecessors is the effect of equivalence transformations. As mentioned earlier, the main reduction statement was either a statement of equivalence itself or of equivalence combined with direct generalization or refinement. Now that approximate reductions are also at our disposal, new combinations with equivalence are possible. Their major effect in the case of rival theories is that conceptual assimilation allows not only the very formulation of rivalry but also the wanted approximate reduction with respect to the conceptual bases obtained. On account of their rich equivalence structure the situation is best illustrated by physical theories based on geometries. There is, for instance, a common formulation for field theories based on either Galileo or Lorentz manifolds. In it each metric is described by a triple consisting of two tensor fields g and h, roughly standing for time and space, and a real non-negative parameter λ. This may be symbolized by

$$\left\{\begin{array}{l} g_{ab}\, h^{bc} = -\lambda \delta^{c}_{a} \\ \text{and other axioms} \end{array}\right. \qquad \left\{\begin{array}{l} \lambda = 0 \text{ Galileo} \\ \lambda = c^{-2} \text{ Lorentz} \end{array}\right. \tag{11.8}$$

where it is understood that the entire difference between the theories is expressed by the different values of the parameter λ (see Ehlers 1986). The formulation signalizes that approximation as far as it is possible at all is of the limiting case type ($\lambda \to 0$).[15]

At the same time it is not to be expected that the possibility of an approximate reduction is invariant under equivalence transformations. This remark is, if not all, then at least a great deal of what can be said about the alleged difficulties with conceptual incommensurabilities. Their major source is that axioms of theories contradicting each other on a common conceptual basis allow us to define new concepts in either theory whose definition is not possible in the other. Thus, starting with the given version of field theory, the Galilean axioms allow the introduction of an absolute time while this is not possible in the Lorentz-Einstein case. It is then natural to look for an equivalent formulation of the Galilean theory using the concept of absolute time as primitive. If *this* formulation is compared with the original version of the Lorentz-Einstein theory, incommensurability of the theories becomes evident. In general, however, it amounts to no more than that contradicting theories T and T' allow for equivalent formulations T_1 and T'_1 using concepts that are not definable in T'_1 and T_1, respectively. From this I draw the conclusion that so far it is not incommensurability as such that we have to be puzzled about. It is rather particular instances of it that happen to be of a fundamental character.

However, this conclusion is not final as long as we have not taken *quantum theory* into consideration. Perhaps the most recalcitrant reduction problem within physics is the one of reducing classical mechanics to quantum mechanics. The present approach offers no solution to this problem. The following observation may throw some light on why the problem is so difficult. On the one hand, the divergence of quantum mechanics from its classical predecessor begins already at the deepest possible level of state descriptions: Quantum mechanical observables and states in their usual representation on Hilbert space seem to have almost nothing in common with their classical counterparts on phase space. On the other hand, some identification on this level was the basis of all approximate reductions considered so far. In (11.5c) we have sharpened this condition as the existence of (sufficiently many) common partial models of the two theories. In the case now before us, what would the common theory Σ_o occurring in (11.5c) be like?

Configuration space is a suitable common partial model of the two theories (see Ashtekar 1980). There are formulations of the theories in which configuration space plays just this role. But if in these formulations we look at the other concepts to be compared the situation seems hopeless. In classical mechanics the observables, for instance, are represented by real-valued functions on the cotangent bundle over configuration space. In quantum mechanics they are represented by self-adjoint operators on a Hilbert space of complex functions on that space. We are thus left with two classes of entities of completely different types with respect to configuration space, suggesting no comparison. Moreover, in this reconstruction the theories do not even contradict each other. If, on the other hand, we follow the formulation, very popular for quantum mechanics, which makes the set of observables the principal base set of the describing structure, then we easily find contradicting consequences of the two axiom systems. But in this reconstruction the two theories seem to have no common partial models including such basic entities as observables (and states).

Under fairly restrictive assumptions to be specified in a moment the difficulty may be overcome by means of the so-called Weyl-Wigner-transformation. It provides us with two real Lie-Algebras of observables having the *same* underlying vector space $\mathfrak{A}$ but two different Lie-products: one classical $[A,B]$, the other quantum mechanical$[A,B]_h$, depending on $h(=$ Planck's constant) In the classical phase space representation, $[A,B]$ is the Poisson product; in the Hilbert space representation $[A,B]_h$ is $(hi)^{-1}$ times the commutator product. The first representation of $[A,B]_h$ and the second of $\{A,B\}$ are more complicated. At any rate, the situation permits a topological comparison of the two products (as subsets of $\mathfrak{A}^3$), and it turns out that in a sense

$$\lim_{h\to 0} [A,B]_h = [A,B]. \tag{11.9}$$

In its topological features this approximation may be reconstructed as a limiting case reduction in the sense of the previous section. However, it has to be emphasized that the result (11.9) has been proven only for Hilbert-Schmidt operators (in Hilbert-space representation) and does not include as important observables as position and momentum (see Emch 1983, Pool 1966, and Baker 1958).

Thus, while in general relativity the remaining reduction problems seem to be mainly of a mathematical nature, in quantum mechanics the situation still is beset with conceptual difficulties having an ontological root. They cast their shadow also on *microreduction*. The major problem here is, of course, the explanation of the essentially classical behavior of our macroscopic environment from nonclassical, quantum theoretical premises about the microscopic constituents. Evidently this is a very specific problem whose solution will hardly follow from the idea of the recursive nature of reduction in general. However, it seems that the solution will hardly be found without that idea. At any rate we should carefully study the contributions coming from the various components of a complex reduction. It has been mentioned already for relativity theory that the definability of concepts may significantly depend on the axioms of a theory: Concepts that are not definable in a given theory may become so in one of its specializations under additional contingent assumptions. Trivial as this step may appear from a purely logical point of view, it can have amazing effects in a field where the weight of additional contingent assumptions is considerably high when compared with the weight of the original laws. Such a field is microreduction. Many, if not all, of the so-called emergent properties of a physical system may come about by such restrictions.

Equally important in the field is the admission of and even emphasis on coarsening and approximate reductions, in particular, their combination as it occurred in the limiting case reduction of the previous section. A case in point is the reduction of the Boltzmann equation to statistical mechanics with the usual collision dynamics of hard spheres. In the reducing theory (Σ'') a single system receives a microdescription essentially consisting of Boltzmann's μ-space X, the particle number n, the dynamics H_n mentioned and a (timely) trajectory $\langle q_i,p_i\rangle_{1\leq i\leq n}$ in Γ-space X^n satisfying the Hamilton equations with respect to H_n. In the theory to be reduced (Σ), the same system is macrodescribed by its 1-particle density f in X (as function of time) satisfying the Boltzmann equation. These seemingly different theories are assimilated by a coarsening assigning to every microdescription $\langle q_i,p_i\rangle_{1\leq i\leq n}$ the macrodescription

$$F_{\langle q_i,p_i\rangle_{1\leq i\leq n}}(\Delta) = \frac{1}{n}\cdot\sum_{i=1}^{n} X_\Delta\,(q_i,p_i) \qquad (11.10)$$

where Δ is any Borel set in X and X_Δ its characteristic function. This step corresponds to (11.5b) leading to the "virtual" theory Σ' of macrodescriptions (11.10) generated by any microdescription as indicated. The essential idea then is to approximate the Boltzmann density f by the measures (11.10) for large n (and vanishing diameters of the spheres). It turns out that this is not possible for single systems but only for statistical ensembles (in X^n) in the sense that for "almost all" members the approximation in question can be obtained (see Lanford 1976 and 1975, esp. secs. 6 and 7). This leads to considerable complications in the formulation of the reduction, and no reasonable generalization comparable with the one of the previous section is known. Without doubt, however, they will be found soon, and it may be noted already now that in addition to coarsening and approximation, a specialization of the probability distributions (in X^n), essential for bringing about irreversibility, enters the scene, again nicely illustrating our synthetic theory of reduction.

V

For some of the more basic unsolved problems mentioned in the last section, I will offer a provisional solution. It is characterized by the concept of *partial reduction*. It is a fact that not merely most but indeed all statements to be found in the relevant text books to the effect that Newtonian mechanics is a limiting case of relativistic mechanics, that Newton's theory of gravitation is a limiting case of Einstein's and statements of a similar stature simply are not founded on any proof of a corresponding reduction *in general* but are just referring to so many *partial* reductions that indeed can be established in the field. At the same time the present synthetic theory of reduction is well-suited for a treatment of just these partial reductions. In partial reductions several elementary reductions are put together to form a symmetric structure—a *reduction square*—thus exploiting the basic feature of our theory in a natural and elegant way.

Let us first look at an example of a *closed* reduction square in Fig. 11.1. Suppose we want to explain Kepler's 3^d law from Newton's gravitational theory. We can *either* explain the law approximately from its well-known Newtonian correction and then reduce this ex-

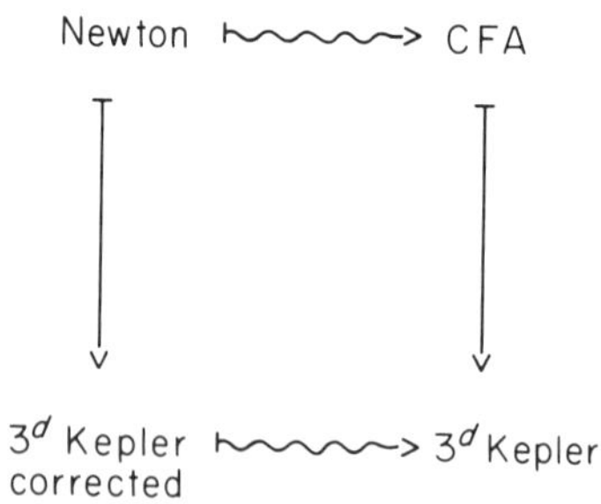

Figure 11.1

actly (indeed by a refinement followed by a generalization) to Newton's theory, *or* we can begin with an exact reduction (again a refinement and a subsequent generalization) of our law to the central field approximation of Newton's theory and explain this approximately according to its name. The procedure can be visualized in general by the square shown in figure 11.2: where the arrows point to the respective theory that is reduced. The vertical reductions ER are exact, and they correspond to each other in the sense that ER_2 imitates ER_1 as far as the new circumstances permit this. By contrast, the horizontal reductions AR are approximate and may be different. The main question answered by a closed reduction square is as follows: Given an approximate reduction AR_1 of Σ to Σ', what about corresponding reductions AR_2 of exact "consequences" θ of Σ to exact "consequences" θ' of Σ'? The example given shows that under favorite circumstances the answer to this question is in the affirmative.

Now our present interest in the reduction square does not apply to the closed but to the *open* one. A reduction square is open if we omit AR_1. This omission may be deliberate. But, of course, our interest derives from cases where we do not know whether Σ can be reduced to Σ' or where we even surmise that it cannot be reduced. It is in such cases that we must have recourse to partial reduction of Σ to Σ' in the sense that merely special cases, coarsenings, or their combinations, derived from Σ, can be reduced (approximately) to special cases, coarsenings, or combinations of such, derived from Σ'. As mentioned earlier, general relativity and quantum mechanics are cases in point. Not knowing the solution of the reduction problem in general we may, for instance, specialize Newton's and Einstein's theory to static, spherically symmetric solutions in empty space and then approxi-

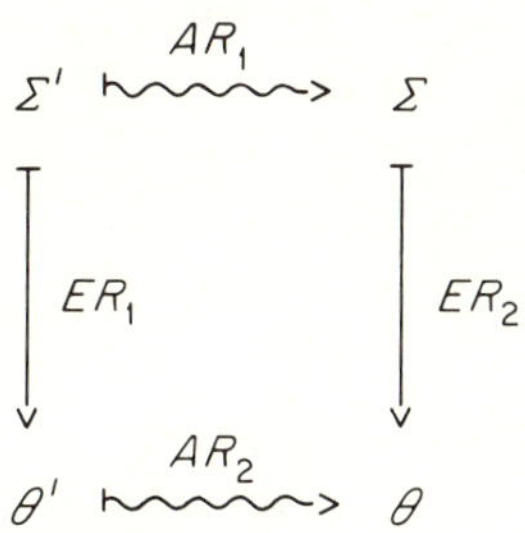

Figure 11.2

mately reduce Newton's theory thus specialized (the central field approximation) to Einstein's theory specialized correspondingly (Schwarzschild metric). Similarly, the position distribution of a classical harmonic oscillator in thermodynamic equilibrium can easily be reduced asymptotically to the corresponding quantum mechanical case, although we do not know how to handle the general case. Many, many examples of this kind could be given from the quantum and relativity domain showing the importance of partial reductions. However, the concept of partial reduction is essentially incomplete as long as one has not specified the correspondence between the two vertical reductions in the square: They have to be the same as far as possible in view of the difference between Σ and Σ'. To make this precise is no easy task in the case of an open reduction square though it is certainly easier than the reduction itself.

In this essay I have indicated a new approach to theory reduction. It was called "synthetic" because, on the basis of rules (R), the most general reduction is obtained recursively from certain initial reductions. The method of rule-generated reductions was chosen because the typical situation that we meet with in physics are reductions that are combinations of other reductions, sometimes widely differing in kind. Once one has analyzed a given reduction as the combination of, say, an asymptotic reduction followed by a direct generalization followed by equivalence it is hard to see what *other* description of the situation could be given that would be equally satisfactory as the one here suggested. Partial reductions as they were considered at the end of the essay obviously are not reductions in the sense of our notion, but their definition rests on this very notion, and again it is hard to see how else one could obtain this extension. On the other hand, there certainly are extensions of our concept in the sense that new elementary (initial) reductions are introduced. They may be useful in microreduction and also elsewhere.

NOTES

1. The "structuralist" theory of reduction is the most recent example. See Balzer et al. (1987, chaps. 6 and 7). Note, however, that in another sense the structuralists are more restrictive in their usage of the term "reduction" than is suggested in this approach (see note 7).

2. Some considerations in Ludwig (1978, chap. 8) are suggestive of a synthetic theory of reductions. A comparison with the present approach is an urgent demand.

3. It has become fashionable to replace the syntactical approach of (the earlier) Carnap by a so-called "semantical" approach (see note 1; Stegmüller 1979, secs. 1 and 2; van Fraassen 1980, chap. 3, sec. 6). Apart from the elimination of some irrelevancies I can see no essential advantage of this approach. If we come down to the special cases, it is hard to see how we can do without having to resort to the languages.

4. In (11.2b) *both* sides of the equivalence are to be understood as statements about the path of the planet.

5. The first line of (11.3a) is a deduction in the sense of Bourbaki (1968, chap. 4).

6. In (11.3b) the temperature θ is unaffected by the derivations. This case is covered by (11.3a) with partially trivial terms P.

7. Despite the evident explanatory and reductive features of generalization, it is not classified as a reduction in Balzer et al. (1987).

8. Equations (11.3a,b) involving the vehicle is here taken to be an *analytical* statement thus allowing the whole reduction statement to be analytical in character. This situation may occur even in the proper case of a reduction in which the reduced theory T is known prior to T'. The equations in question then are a real (as opposed to a nominal) definition. The case where the equations are a so-called "synthetic identity" has to be treated separately; see Nagel's (1961) treatment and more recently Sklar (1967) and Scheibe (1988a).

9. Although the geometry of general relativity is a generalization of the geometry of special relativity, the relation between general and special relativistic *physics* is more complex.

10. Conceptual assimilation can also be brought about by refinements proper if they concern the reducing theory. See the case of microreductions in section 5.

11. The expression "limiting case" is used by physicists. The earliest usage I could find is the German equivalent "Grenzfall" in Hertz ([1892] 1914, 21ff).

12. For details concerning species of structures the reader must be referred to Bourbaki (1968). We use a vector notation for the structures: S_o is a tuple of base sets and typified sets, s is a tuple of typified sets, and so on.

13. The class of physical systems admitted by a physical theory is in general much smaller than the class of corresponding structures satisfying its axioms. The further restriction is then brought about by fixing one particular structure as, for instance, S_o in (11.5a). Such is the case in the gas examples given. This qualification is suppressed in our treatment for the sake of simplicity.

14. It seems that in fact we always also have $M \cap c_\delta \subseteq M'_u$ which makes asymptotic reduction an almost symmetric affair.

15. It is still not clear what kind of approximate reduction has to be applied in reducing Newton's gravitational field theory to Einstein's, let alone Newton's n-body theory of gravitation. See Künzle (1976), Ehlers (1981), and Lottermoser (1988).

REFERENCES

Ashtekar, A. 1980. "On the Relation Between Classical and Quantum Observables." *Communications in Mathematical Physics 71:* 59–64.

Baker, G. A., Jr. 1958. "Formulation of Quantum Mechanics Based on the Quasi-Probability Distribution Induced on Phase Space." *Physical Review 109:* 2198–2206.

Balzer, W.; C. U. Moulines; and J. Sneed. 1987. *An Architectonic for Science: The Structuralist Program.* Dordrecht: Reidel.

Bourbaki, N. 1968. *Elements of Mathematics: Theory of Sets.* Paris: Hermann.

Ehlers, J. 1981. "Über den Newtonschen Grenzwert der Einsteinschen Gravitationstheorie." In J. Nitsch, J. Pharr, and E. W. Stachow, eds., *Grundlagenprobleme der modernen Physik.* Mannheim: Bibiographisches Institut, pp. 65–84.

———. 1986. "On Limit Relations Between, and Approximate Explanations of, Physical Theories." In R. B. Marcus et al., eds., *Logic, Methodology and Philosophy of Science.* Vol. 7. Amsterdam: North-Holland, pp. 387–404.

Emch, G. 1983. "Geometric Dequantization and the Correspondence Problem." *International Journal of Theoretical Physics 22:* 397–420.

Feyerabend, P. 1975. *Against Method.* London: Verso.

Hertz, H. [1892] 1914. *Untersuchungen über die Ausbreitung der elektrischen Kraft.* 3d ed. Leipzig: Barth.

Kuhn, T. 1983. "Commensurability, Comparability, Communicability." In P. D. Asquith, and T. Nickles, eds., *PSA 1982.* East Lansing: Philosophy of Science Association, pp. 669–88.

Künzle, H. P. 1972. "Galilei and Lorentz Structures on Space-Time: Comparison of the Corresponding Geometry and Physics." *Annales de Institut Henri Poincaré 17:* 337–62.

———. 1976. "Covariant Newtonian Limit of Lorentz Space-Times." *General Relativity and Gravitation 7:* 445–57.

Lanford, O. E. 1975. "The Evolution of Large Classical Systems." In J. Moser, ed., *Dynamical Systems, Theory and Applications. Batelle Seattle 1974 Rencontres.* Berlin: Springer-Verlag, pp. 1–111.

———. 1976. "On a Derivation of the Boltzmann Equation." *Astérisque 40:* 117–37.

Lottermoser, M. 1988. "Über den Newtonschen Grenzwert der Allgemeinen Relativitatstheorie und die relativistische Erweiterung Newtonscher Anfangsdaten." Ph.D. dissertation. München: Ludwig-Maximilian Universität.

Ludwig, G. 1978. *Die Grundstrukturen einer physikalischen Theorie.* Berlin: Springer-Verlag.

Nagel, E. 1961. *The Structure of Science: Problems in the Logic of Scientific Explanation.* New York: Harcourt, Brace & World.

Pool, J. C. T. 1966. "Mathematical Aspects of the Weyl Correspondence." *Journal of Mathematical Physics 7:* 66–76.

Scheibe, E. 1986. "Mathematics and Physical Axiomatization." In F. Singer-Polignac, ed., *Mérites et limites des Méthodes Logiques en Philosphie.* Paris: Librairie Philosophigue J. Vrin, pp. 251–69.

———. 1988a. "Äquivalenz und Reduktion. Zur Frage ihres empirischen Status." *Conceptus* 22: 91–105.

———. 1988b. "The Physicists' Conception of Progress." *Studies in History and Philosophy of Science* 19: 141–59.

———. 1989. "Two Types of Successor Relations Between Theories." *Zeitschrift für allgemeine Wissenschaftstheorie* 14: 68–80.

Sklar, L. 1967. "Types of Inter-theoretic Reduction." *British Journal for the Philosophy of Science* 18: 109–24.

Stegmüller, W. 1979. *The Structuralist View of Theories*. Berlin: Springer-Verlag.

van Fraassen, B. C. 1980. *The Scientific Image*. Oxford: Clarendon Press.

12

Analogy by Similarity in Hyper-Carnapian Inductive Logic

Brian Skyrms
Department of Philosophy, University of California at Irvine

Adolf Grünbaum (1976, 1978) has longstanding critical interest in Bayesian methods in the philosophy of science. In this essay I will show how Bayesian methods can be applied to provide a natural solution to a problem in Rudolf Carnap's program for inductive logic—the problem of analogy by similarity. That the inductive methods in Carnap's (1952) λ-continuum failed to give an adequate account of various kinds of analogical reasoning was pointed out by Achinstein and Putnam in a series of papers in 1963 (see Achinstein 1963a,b; Putnam [1963] 1975a, [1963] 1975b; and Carnap 1963a,b). In his posthumous "Basic System" (1980), Carnap is still struggling with the problems of analogical inference. Carnap distinguished between two kinds of analogical effect in inductive logic: that due to similarity information carried by names and that due to similarity information carried by predicates. These he called, respectively, "Analogy by Proximity" and "Analogy by Similarity."

The first kind of analogy is to be represented in a future inductive logic defined on "coordinate languages." The names are interpreted as temporal or spacetime indices. The name "Analogy by Proximity" suggests that Carnap had something like Markov chains or Markov random fields in mind. Then, to take the example of a Markov chain, the probability of an outcome on a trial may depend on the outcome of the preceding trial. This requires that Carnap's postulate of symmetry or exchangeability—here being invariance under finite permutations of trials—be dropped. The necessity of such a radical change of framework might suggest that a treatment of Markov chains along Carnapian lines would be exceedingly difficult. In fact, the classical Carnapian methods transfer smoothly and easily to this setting (Kuipers 1988, Skyrms 1991). "Analogy by Similarity" requires us to move outside the class of Carnapian methods. However, by considering methods which are, in a sense, mixtures of classical Carnapian methods, we can give a treatment of analogy by similarity which remains faithful to the spirit of Carnap's program.

Analogy by Similarity

Carnap introduces the problem of analogy by similarity by an example where the color of insects of a certain species is observed. The observer is assumed to partition the color space into nine regions forming a linear series from pure blue to pure green, with a sample of 20 individuals observed, obtaining results shown in table 12.1. Carnap considers the probabilities, on this evidence, of finding the next instance in *P*3 and the probability of finding it in *P*8. He argues that in spite of the sample containing exactly two individuals in each of these categories, the probability of finding the next individual in *P*3 should be higher than the probability of finding it in *P*8 because there are five members of the sample in the categories adjacent to *P*2 and none in the neighbors of *P*8 (see Carnap 1980, 40).

Carnap wishes to generalize this example by supposing that there is some "similarity metric" on the predicates such that the analogy influence is stronger between more similar categories. The categories need not, of course, form a linear order. To vary the example, suppose that we spin a roulette wheel of unknown bias. Carnap believes that it may not be reasonable, on the evidence of a few trials, to let the probability of a given segment on the next trial depend only on its relative frequency on preceding trials and not of the relative frequency of other segments. If segments close to it have come up often, it should be more likely than if segments far away from it have come up often. At a more general level, without putting any structure on

Table 12.1

# observed:	*3*	*6*	*2*	*4*	*3*	*0*	*0*	*2*	*0*
	*P*1	*P*2	*P*3	*P*4	*P*5	*P*6	*P*7	*P*8	*P*9

similarity, we can ask how instances not in a category can have a differential effect on the predictive probability that the next instance will be in that category.

Carnap's Continuua of Inductive Methods

Suppose that we have an exhaustive family of k mutually exclusive categories, and a sample of size N of which n are of category F. Carnap ({1950}1962) originally proposed the following inductive rule, C*, to give the probability that a new sampled individual a would be in F on the basis of the given sample evidence e:

$$pr(Fa \mid e) = \frac{1 + n}{k + N}. \quad C^*$$

On the basis of no sample evidence, each category gets equal probability of $1/k$. As the sample grows larger, the effect of the initial equiprobable assignment shrinks and the probability attaching to a category approaches the empirical average in the sample n/N. Soon Carnap (1952) shifted from this method to a class of inductive methods of which it is a member, the λ-continuum of inductive methods.

$$pr(Fa \mid e) = \frac{\lambda + n}{\lambda k + N}. \quad \lambda\text{-continuum}$$

Here again we have initial equiprobability of categories and predominance of the empirical average in the limit with the parameter $\lambda (\lambda > 0)$ controlling the rate at which the sample evidence swamps the prior probabilities. In his posthumous paper, Carnap (1980) introduced the more general λ-γ continuum:

$$pr(F_i a \mid e) = \frac{\lambda\gamma_i + n}{\lambda + N} (\Sigma_i \gamma_i = 1). \quad \lambda\text{-}\gamma \text{ continuum}$$

The new parameters $\gamma_i > 0$ allow unequal *a priori* probabilities for different categories. For Carnap there are intended to reflect different "logical widths" of the categories.

Carnap's continuua are familiar to Bayesians. Carnap is modeling the inductive problem as having the structure of throwing a die with unknown bias. Conditional on any given bias the trials are independent and identically distributed. Then the quantification of prior ignorance consists in picking a prior probability over the possible biases. The family of natural conjugate priors which are proportional to the likelihoods are the Dirichlet priors. On the evidence of a series of trials, we update by Bayes's theorem. If we use a natural conjugate prior and start with our probabilities being a Dirichlet mixture of multinomials, then the result of updating on a series of trials is again a Dirichlet mixture of multinomials; that is, we need only update the parameters of the Dirichlet distribution. Consider the predictive probabilities in such a model—that the next trial will yield a certain outcome. These will depend on the particular Dirichlet prior chosen. If we take the uniform or "flat" prior we get Carnap's C^* rule. If we take the symmetric Dirichlet priors, we get Carnap's λ-continuum. If we take all the Dirichlet priors, we get Carnap's λ-γ continuum.

Let us consider two characteristic features and two general virtues of these models. The general virtues are *computational tractability* and *Bayesian consistency*. The first virtue is evident from the rules. Bayesian consistency is not much discussed by philosophers of induction, but should be. In the statistical model of the die with independent and identically distributed trials, any way of fixing the bias parameter θ can be thought of as determining a possible chance distribution $pr\theta$. A Bayesian prior is consistent for a value of the bias parameter θ if with $pr\theta = 1$ the posterior converges (in the weak-star topology) to a point mass at θ. A Bayesian prior is *consistent* if it is consistent for all possible values of θ. Perfectly coherent priors can fail to be consistent. For example, consider a coin-tossing prior which puts probability 1/2 on θ(chance of heads) = 1/3 and probability 1/2 on $\theta = 2/3$. This prior is obviously not consistent for $\theta = 1/2$. Dirichlet priors are consistent (Diaconis and Freedman 1986). In the obvious correlative sense, the inductive methods of Carnap's λ-γ continuum are consistent. Provided that the true chances give independent identically distributed trials, there is zero chance that they will not converge to the "chance-optimal" predictions.

The first characteristic feature of the Carnapian continuua is *exchangeability* of trials, or, in Carnap's terminology, *symmetry*. The probability of a finite sequence of outcomes is invariant under permutations of trials. For example, sequences $F1,F1,F2,F3$ and

*F*1,*F*2,*F*3,*F*1 have the same probability since the second can be derived from the first by permuting the second and fourth trials and then permuting the second and third trials. Exchangeability means that order does not make a difference and the probability of an outcome sequence of fixed length is determined solely by the frequency counts of each of the outcome categories. Exchangeability is a property of independent identically distributed sequences of outcomes and of any mixture of them, not just of the Dirichlet mixtures under consideration here. The second characteristic feature is what I call *generalized sufficientness*. For the rules in Carnap's λ-γ continuum, the probability that the next trial will have outcome F_i conditional on sample evidence depends only on i (because different categories may have different values of γ) and the number of outcomes in F_i in the sample, and on the total number of trials in the sample. The distribution of the outcomes not in F_i among the other categories do not matter. Thus, as Carnap well knew, none of the methods in λ-γ continuum is capable of exhibiting analogy by similarity. We will see in the next section that in a certain sense the converse is true. Carnap's λ-γ continuum consists of just those methods which are incapable of capturing the desired analogical influence.

Exchangeability and Generalized Sufficientness

The two features noted at the end of the last section have far-reaching consequences. Suppose that we consider an infinite sequence of random variables $X_1, \ldots, X_i, \ldots$ (one for each trial of the experiment) taking any of a finite number of values $O_1, \ldots, O_k$ (the possible experimental outcomes). Suppose that this infinite sequence of random variables is *exchangeable,* that is, that for every finite subsequence of length *N*, $X_1, \ldots, X_N$ the probability:

$$Pr(X_1 = O_1 \wedge \ldots \wedge X_N = O_N)$$

is invariant under permutations of the index i (of trials). Then, by a famous theorem of de Finetti ({1937}1980), this probability has a representation as a mixture of probabilities which make the random variables independent and identically distributed. Exchangeability here gives us a mixture of multinomial probabilities, that is, that in the case of a finite number of possible outcomes, exchangeability guarantees that our probabilities correspond to modeling the situation as throwing a die with unknown bias.

This leaves open the nature of the mixing measure—of the prior probability on the bias of the die. We can characterize those priors which cannot exhibit any analogy by similarity in the resulting predictive probabilities $Pr(X_{N+1} = O_i \mid X_1, \ldots, X_N)$ as those which satisfy the *generalized sufficientness condition:*

$$Pr(X_{N+1} = O_i \mid X_1, \ldots, X_N) = f_i(n_i)$$

where n_i is the frequency count of outcomes in O_i. Suppose that the number of outcomes is finite but at least three, that the conditional probabilities in the generalized sufficientness condition are all well defined[1] and that the condition holds, and that we have exchangeability. Zabell (1982) shows that these conditions characterize the mixtures of multinomial probabilities according to Dirichlet priors (see also Stegmüller 1973, 490, and Kuipers 1978). That is, the predictive probabilities possible under these conditions are just those which fall within Carnap's λ-γ continuum of inductive methods.[2] Thus, given the background assumption of exchangeability, Carnap's λ-γ continuum consists of just those methods which cannot exhibit analogy by similarity. Can we implement analogy by similarity outside the λ-γ continuum against the background assumption of exchangeability and at the same time retain the generalized virtues of computational tractability and Bayesian consistency?

Hyper-Carnapian Methods and Analogy by Similarity

The most conservative move outside Carnap's λ-γ continuum would be to consider finite mixtures of methods that are themselves in the λ-γ continuum. One could think of this as putting a "hyperprior" probability on a finite number of metahypotheses as to the values of the λ and γ_i hyperparameters. Conditional on each metahypothesis, one calculates the predictive probabilities according to the Carnapian method specified by that metahypothesis. The probabilities of the metahypotheses are updated using Bayes's theorem. We will call these *Hyper-Carnapian* Methods.

The hierarchical gloss, however, is inessential. The model just described is mathematically equivalent to using a prior on the multinomial parameters which is not Dirichlet but rather a finite mixture of Dirichlet priors. Evidently, if the number of Carnapian methods in the mixture is not too great, the *computational tractability* of Carnapian methods is not severely compromised. Furthermore, *Bayesian consistency* is retained. Finite mixtures of Dirichlet priors are consistent (Diaconis and Freedman 1986). Furthermore, they can exhibit the kind of analogy by similarity that Carnap wished to model. We can illustrate this by means of a simple example: A wheel of fortune is divided into four quadrants: **N,E,S,W.** There are four "metahypotheses" which are initially equiprobable. Each requires updating by a different Carnapian rule as indicated in table 12.2 where n is the number of successes in N trials.

Since the hypotheses are initially equiprobable, the possible outcomes **N,E,S,W** are also initially equiprobable. Suppose that we have one trial whose outcome is **N.** Then updating the probabilities of our hypotheses by Bayes's Theorem, the probabilities of **H1,H2,H3,H4** respectively become .5,.2,.1,.2. Applying the Carnapian rule of each hypothesis and mixing with the new weights gives probabilities:

$$pr(\mathbf{N}) = 44/110$$
$$pr(\mathbf{E}) = 24/110$$
$$pr(\mathbf{S}) = 18/110$$
$$pr(\mathbf{W}) = 24/110.$$

The outcome **N** has affected the probabilities of the nonoutcomes **E,S,W** differentially even though each Carnapian rule treats them the

Table 12.2

	N	E	S	W
H1:	$\frac{5+n}{10+N}$	$\frac{2+n}{10+N}$	$\frac{1+n}{10+N}$	$\frac{2+n}{10+N}$
H2:	$\frac{2+n}{10+N}$	$\frac{5+n}{10+N}$	$\frac{2+n}{10+N}$	$\frac{1+n}{10+N}$
H3:	$\frac{1+n}{10+N}$	$\frac{2+n}{10+N}$	$\frac{5+n}{10+N}$	$\frac{2+n}{10+N}$
H4:	$\frac{2+n}{10+N}$	$\frac{1+n}{10+N}$	$\frac{2+n}{10+N}$	$\frac{5+n}{10+N}$

same. The outcome **N** has reduced the probability of the distant outcome **S** much more than that of the close outcomes **E** and **W** just as Carnap thought it should.

Two Kinds of Analogy and Their Impact on Carnap's Program

As we saw earlier, for cases in which the number of outcomes is greater than two, and the predictive conditional probabilities are all well defined, Carnap's λ-γ continuum of inductive methods was exactly characterized by two main properties: *exchangeability* and *generalized sufficientness*. In his last work on inductive logic, Carnap noted two kinds of analogy which were lacking in this continuum. He called them *analogy by proximity* and *analogy by similarity*. Each leads outside the λ-γ continuum in a different way. Analogy by proximity (for example in the case of inference regarding Markov chains) requires that one go beyond exchangeability. Analogy by similarity, on the other hand, is compatible with exchangeability but requires that one give up generalized sufficientness. In both cases, however, many of the virtues of Carnapian inductive methods can be maintained. It is, of course, also possible to explore richer models in which both types of analogy are operative.

Related Literature

There is a rich philosophical literature exploring ways to introduce analogy by similarity into a Carnapian framework (see Kuipers 1978, 1984a,b, 1988; Niinilouto 1981, 1988; Costantini 1983; Spohn 1981). The dominant technique is to make modifications at the level of the Carnapian inductive rules. Most of these suggestions are incompatible with exchangeability (see, however, Kuipers 1984b, sec. 4). There are interesting open questions regarding the Bayesian analysis of these proposals and their relation to hyper-Carnapian methods.

NOTES

I would like to thank Richard Jeffrey, Theo Kuipers, and Ilkka Niiniluoto for discussion of an earlier draft of this essay.

1. All possible finite outcome sequences have positive probability:

$$Pr(X_1 = O_{i_1}, \ldots, X_N = O_{i_N}) > 0.$$

2. W. E. Johnson (1932) originally assumed a stronger sufficiency condition which required that f be independent of i. This stronger assumption gets the symmetric Dirichlet priors, and the resulting predictive probabilities form Carnap's λ-continuum. For details see Zabell (1982).

REFERENCES

Achinstein, P. 1963a. "Confirmation Theory, Order and Periodicity." *Philosophy of Science 30:* 17–35.

———. 1963b. "Variety and Analogy in Confirmation Theory." *Philosophy of Science 30:* 207–21.

Carnap, R. 1952. *The Continuum of Inductive Methods.* Chicago: University of Chicago Press.

———. {1950} 1962. *Logical Foundations of Probability.* 2d ed. Chicago: University of Chicago Press.

———. 1963a. "Replies and Systematic Expositions." In P. A. Schilpp, ed., *The Philosophy of Rudolf Carnap.* La Salle, Ill.: Open Court, pp. 711–37.

———. 1963b. "Variety, Analogy and Periodicity in Inductive Logic." *Philosophy of Science 30:* 222–27.

———. 1980. "A Basic System of Inductive Logic, Part 2." In R. C. Jeffrey, ed., *Studies in Inductive Logic and Probability.* Vol. 2. Berkeley and Los Angeles: University of California Press, pp. 7–155.

Costantini, D. 1983. "Analogy by Similarity." *Erkenntnis 20:* 103–14.

de Finetti, B. {1937} 1980. "Foresight: Its Logical Laws, Its Subjective Sources." Translated in H. E. Kyburg, Jr., and H. Smokler, eds., *Studies in Subjective Probability.* Huntington, N.Y.: Kreiger, pp. 53–118. (Originally published as "La Prévision: ses lois logiques, ses sources subjectives." *Annales de l'Institut Henri Poincaré 7:* 1–68.)

Diaconis, P., and D. Freedman. 1986. "On the Consistency of Bayes Estimates." *Annals of Statistics 14:* 1–26.

Grünbaum, A. 1976. "Is Falsifiability the Touchstone of Scientific Rationality? Karl Popper vs. Inductivism." In R. S. Cohen, P. K. Feyerabend, and M. W. Wartofsky, eds., *Essays in Memory of Imre Lakatos.* Dordrecht: Reidel, pp. 213–52.

———. 1978. "Popper vs. Inductivism." in G. Radnitzky and G. Anderson, eds., *Progress and Rationality in Science.* Dordrecht: Reidel, pp. 117–42.

Johnson, W. E. 1932. "Probability: The Deductive and Inductive Problems." *Mind 49:* 421–23.

Kuipers, T. A. F. 1978. *Studies in Inductive Logic and Rational Expectation.* Dordrecht: Reidel.

———. 1984a. "Inductive Analogy in the Carnapian Spirit." In P. D. Asquith and P. Kitcher, eds., *PSA 1984.* Vol. 1. East Lansing: Philosophy of Science Association, pp. 157–67.

———. 1984b. "Two Types of Inductive Analogy by Similarity." *Erkenntnis 21:* 63–87.

———. 1988. "Inductive Analogy by Similarity and Proximity." In D. H. Helman, ed., *Analogical Reasoning.* Dordrecht: Reidel, pp. 299–313.

Niiniluoto, I. 1981. "Analogy and Inductive Logic." *Erkenntnis 16:* 1–34.

———. 1988. "Analogy and Similarity in Scientific Reasoning." In D. H. Helman, ed., *Analogical Reasoning.* Dordrecht: Reidel, pp. 271–98.

Putnam, H. {1963} 1975a. "Degree of Confirmation and Inductive Logic." Reprinted in H. Putnam, *Philosophical Papers.* Vol. 1, *Mathematics, Matter and Method.* Cambridge, England: Cambridge University Press, pp. 270–92.

———. {1963} 1975b. "Probability and Confirmation." Reprinted in H. Putnam, *Philosophical Papers.* Vol. 1, *Mathematics, Matter and Method.* Cambridge, England: Cambridge University Press, pp. 293–304.

Skyrms, B. 1991. "Carnapian Inductive Logic for Markov Chains." *Erkenntnis 35:* 439–60.

Spohn, W. 1981. "Analogy and Inductive Logic: A Note on Niiniluoto." *Erkenntnis 16:* 35–52.

Stegmüller, W. 1973. *Probleme und Resultäte der Wissenschaftstheorie und Analytischen Philosophie.* Band 4, *Personelle und statistische Wahrscheinlichkeit.* Berlin: Springer.

Zabell, S. L. 1982. "W. E. Johnson's 'Sufficientness' Postulate." *Annals of Statistics 10:* 1091–99.

13

Capacities and Invariance

James Woodward
Division of Humanities and Social Sciences, California Institute of Technology

A sign prominently displayed in California restaurants warns that consumption of alcoholic beverages during pregnancy "can cause" birth defects. My interest in this essay is in what claims of this sort mean. What kind of causal knowledge do they represent and what kind of evidence is relevant to their truth or falsity? Following Nancy Cartwright (1989b) I will argue that such claims are about causal capacities and that they play an important role in causal reasoning, both in the natural and in the behavioral and social sciences.[1]

My discussion will proceed as follows. In section 1, I try to distinguish the capacity notion on which my subsequent discussion will focus from two other related causal notions. Section 2 then introduces some examples which are designed both to illustrate how claims about causal capacities can be investigated empirically, and also to show how capacity claims figure in establishing other kinds of causal claims. Sections 3 and 4 then sketch a positive theory of causal capacities. In contrast to many other writers, I argue that causal capacities need not conform to deterministic or probabilistic laws and need not satisfy anything like the unanimity condition standardly imposed in probabilistic theories of causation. The underlying idea in my alternative account is that causal capacities must satisfy an invariance condition, but that this condition is consistent with a substantial degree of nonuniformity or irregularity in the behavior of capacities. While law-governed capacities are the rule in physics, experimental and nonexperimental investigations in the behavioral and social sciences often provide convincing evidence for the existence of causal capacities satisfying this invariance condition, but rarely for the existence of capacities that are law-governed.

I

It is useful to begin by distinguishing among three different kinds of causal claims. The sign described at the beginning of this essay expresses a claim about (what I will call) a *causal capacity* of alcohol consumption—about its power or ability to produce a certain kind of effect. We use a variety of locutions to make such capacity claims—for example, we say that some event or factor is a "possible cause" or "a potential cause" or "is the sort of thing that could cause" another event. As we will see, many (but not all) claims of the form "*X* causes *Y*" where *X* and *Y* are types of events (e.g., "smoking causes heart attacks") are also plausibly interpreted as claims about causal capacities. A great deal of causal knowledge, both in scientific and everyday contexts, takes the form of information about causal capacities. For example, we know or suspect that air pollution can cause lung cancer, that aspirin can relieve headaches, that cumbersome registration laws can lower voter turnout and that the collision of a large asteroid with the earth could cause mass extinction.

Claims about causal capacities should be distinguished from *singular causal claims*. These are claims about the existence of an actual causal connection between particular events, as in the claim that (1) Jones's ingestion of an aspirin on some particular occasion caused his headache to go away a short time later. If a singular causal claim is true, then the cause and effect event must actually occur. That is, if (1) is true, it must be true that Jones ingested aspirin and that his headache disappeared. By contrast, a claim that a cause has the capacity to cause some effect can be true even if the capacity is never realized. For example, according to some current theories, the detonation of large numbers of nuclear weapons has the capacity to cause widespread and extensive climatic changes ("nuclear winter"), even though no such detonation has ever occurred and (let us hope) never will. In contrast to singular causal claims, causal capacity claims to the effect that Cs can cause *E*s do not tell us on which particular occasions instances of *C* cause instances of *E*.

Both capacity and singular causal claims should be distinguished from what I will call *causal role claims.* These are claims that some causal factor or process is playing the role of producing some effect (commonly an effect whose magnitude or rate of occurrence can be given some more or less rough quantitative description) in some population. When we make such claims our interest is not, as in the case of a singular causal claim, in why some particular individual exhibits or is subject to a certain effect (why Jones's headache has disappeared), but rather with why the feature in question is at a certain level or occurs at a certain rate or takes on a certain average value in a population as a whole. Causal role claims thus always make reference to a particular population, unlike causal capacity claims which (as we will see below) are not similarly population-relative. For example, the incidence of lung cancer increased explosively in the American population in the twentieth century. The claim that much of this increase was caused by cigarette smoking (or more precisely, that some particular fraction of the total increase was due to smoking) is a claim about the causal role of smoking in producing lung cancer in the American population during this period. Like singular causal claims but unlike causal capacity claims, causal role claims require the actual occurrence of tokens of the causal factors to which they refer—if an increase in cigarette smoking is causally responsible for the increased incidence of lung cancer among twentieth-century Americans, then smoking and lung cancer must be actually occurring in the American population and individuals in this population must be caused to get lung cancer by their smoking at the rate indicated in the causal role claim. As we will see below, so-called causal modeling techniques—quantitative techniques for making causal inferences from statistical data that are widely used in epidemiology, and in the social sciences—are often directed toward establishing causal role claims, rather than claims about causal capacities or singular causal claims.

How are these three varieties of causal claims related? This question will occupy us in some detail in the remainder of this essay, but here are some preliminary suggestions by way of orienting the reader to what follows. First, I take it to be a sufficient but not a necessary condition for the truth of the claim that Cs have the capacity to cause *E*s that the singular causal claim that some particular instance of *C* has caused a particular instance of *E* is true. If some particular token of *C* has actually caused some particular token of *E*, then *C*s have the capacity to cause *E*s, but, as we have noted, *C*s can cause *E*s without any such actual token-causation occurring. Second, I take it to be a necessary but not a sufficient condition for the truth of a singular causal claim that the factor or event cited as cause have the causal capacity to bring about the event cited as effect. Jones's taking aspirin on some particular occasion cannot have caused his headache to disappear if ingestion of aspirin does not have the causal capacity to relieve headaches. (Oddly, this seemingly obvious claim is denied in several recent discussions.)[2] But even if Jones takes aspirin, and his headache disappears and aspirin has the capacity to relieve headaches, it does not follow that (1) Jones's taking aspirin caused his headache to disappear. His headache might instead have been caused to disappear by some other cause. As we will see, one establishes that a singular causal claim like (1) is true by showing that (a) Jones took aspirin and his headache was relieved, that (b) ingestion of aspirin has the capacity to cause headache relief, and (c) by ruling out or rendering implausible the claim that any other possible cause might have been effacious in relieving Jones's headache.

The truth of causal role claims also requires the truth of various causal capacity claims. A necessary condition for the truth of the claim that cigarette smoking accounts for some portion of the incidence of lung cancer in the present American population is that smoking has the causal capacity to produce lung cancer. As we will see, knowledge of causal capacities helps to tell us which variables are appropriate candidates for exclusion or inclusion within a causal model, and to ground claims about causal ordering and independence—all matters which are crucial to the reliable use of such models.

The relationships just sketched may be usefully compared with the well-known distinction between token-causation and type-causation introduced by Elliot Sober (1985) and Ellery Eells (1988) (see also Eells and Sober 1983). Claims about the existence of what Sober and Eells call token causal relationships seem to coincide with what we have called singular causal claims. Sober's and Eells' notion of type-causation is meant to explicate a claim like (2) "smoking causes heart attacks" or "smoking is a positive causal factor for heart attacks" which these authors take to be elliptical for the claim that smoking causes heart attacks in some particular population, for example, "the human population" (Eells and Sober 1983, 38).[3] As Eells and Sober see it, claims about type-causation are thus "population-relative"—smoking may cause (be a positive causal factor for) heart attacks in one population, but not in another. According to Eells and Sober, familiar probabilistic theories of causation are most plausibly construed as attempts to explicate such claims about type-causation. Also, Eells and Sober hold that at least on the most plausible interpretations of probability, claims like "smoking causes heart attacks" do not require that it be true there actually exist individuals for whom smoking causes heart attacks—it is enough that the introduction of smoking would increase the probability of heart attacks uniformly within the population of interest, in accordance with a unanimity principle which I will later discuss in more detail.

From my perspective, Eells and Sober's notion of type-causation conflates the causal capacity and causal role notions just distinguished. A claim like (2) "smoking causes heart attacks" is susceptible of (at least) two readings. On what is perhaps the most common and natural construal, in which the claim is not made in a context which suggests reference to any particular population, the claim means that smoking has the capacity to cause heart attacks. On the other hand, a claim like (2) may also be used with reference to some particular population to suggest that smoking is actually causing heart attacks within that population at a certain rate, in which case it is to be understood as a causal role claim.

A number of features which Eells and Sober assign to type-causation seem to fit either the causal role notion or the causal capacity notion but not both. For example, as I have already remarked, the causal role notion, like type-causation, is population-relative while the causal capacity notion is not similarly population-relative. The capacity of smoking to cause lung cancer does not vary across populations in the way that its causal role does. To anticipate my upcoming discussion, the capacity of smoking to cause heart attack is (at least roughly and approximately) invariant across many populations, but the rate at which smoking causes lung cancer in particular populations or how much of the incidence of lung cancer in particular populations is due to smoking—the sort of information conveyed in claims about the causal role of smoking—often does vary considerably across different populations. Moreover, while the truth of the claim that *C* has the capacity to cause *E*, like claims about type-causation, does not require that instances of *C* actually occur and cause instances of *E*, the truth of the claim that *C* is playing the role of causing *E* within a particular population *P* does require that instances of *C* and *E* occur, and be (token) causally connected within that population. Neither the causal capacity notion nor the causal role notion have all the features Eells and Sober assign to type causation.

Indeed, the very phrase "type-causation" seems to me to be potentially misleading. There are not two kinds or varieties of causal connection, one having to do with token causal connections between particulars, and the other having to do with the obtaining of causal connections between properties or types of events. Actual token causal connections between particular events, as reported in singular causal claims, are all the causal connections that exist. When we say that "smoking causes lung cancer" we are not asserting the existence of a causal relationship between the property of smoking and the property of having lung cancer. Instead we are most likely claiming that smoking has the capacity to cause heart attacks, where this implies that individual instances of smoking by particular people can actually cause (i.e., token cause) those people to develop lung cancer. Alternatively, if the causal role interpretation is intended, we are claiming that such token-causation is occurring at some level or rate among the individuals in some particular population. One of the virtues of distinguishing between the capacity and role notions is that it allows us to avoid the notion of type-causation with its Platonic overtones of causal relations between (possibly uninstantiated) properties.

Several other general features of capacity claims are worth noting at this point, before we turn to a more detailed discussion. When we learn that a capacity claim is true, we learn of course that a cause can produce a characteristic sort of effect. But, as my discussion will illustrate, there is usually much more to be learned about the behavior of a causal capacity. For example, there are usually general facts to be discovered about the conditions or circumstances in which a cause has the capacity to produce an effect, and about the conditions that prevent or interfere with the production of the effect. Similarly empirical investigation will often provide further information (although typically in a somewhat loose and gappy form) about the qualitative behavior of the cause or about the results of altering its magnitude or intensity or about the direction or sign of its effect. Thus, for example, we know not just that aspirin causes headache relief, but that an increase in dosage from one tablet to two increases the likelihood of headache relief, and not just that smoking causes lung cancer, but that groups who smoke more heavily and for a longer period of time have a higher incidence of lung cancer. In addition to this, causal capacities typically have a characteristic mechanism or modus operandi—a distinctive way or method by which they produce their effects. As we will see, signs of the operation of such mechanisms are often an important evidential constraint in establishing causal claims. For example, even in the absence of knowledge of the precise biochemical mechanism by which smoking causally contributes to lung cancer, it was clear to early researchers that if smoking caused lung cancer, the mechanism by which it produced this effect involved the inhalation of tobacco smoke into the lungs. This fact by itself has important implications. Given this mechanism, if smoking does cause lung cancer one would expect to find that the smokers who inhale tobacco smoke into their lungs will experience a higher incidence of lung cancer than smokers who are noninhalers. Similarly, one would also expect to find various sorts of organic changes or damage in the lungs of smokers who inhale. The discovery of both of these effects helped to convince early researchers that tobacco smoke was indeed carcinogenic.

When philosophers think of general causal knowledge they tend to focus on knowledge of laws or exceptionless regularities. The kind of causal knowledge about capacities which we have just been discussing is certainly general, but it often falls well short of knowledge of laws or exceptionless regularities. One of my main themes will be that it is this sort of general knowledge, rather than knowledge of laws, that we may reasonably expect to get in the behavioral and social sciences.

II

My discussion so far has been rather abstract. I turn now to some specific examples which will serve as a focus for the remainder of this essay. The first has to do with a social scientific investigation in which a somewhat surprising claim about a causal capacity was established. The example also illustrates nicely the difference between the causal capacity and causal role notions and how claims about capacities are relevant to establishing claims about causal roles. A second example, drawn from the natural sciences, then illustrates how claims about causal capacities figure in establishing singular causal claims. A final example exhibits the use of capacity claims in the causal modeling literature. These examples will be used to motivate the positive account of causal capacities to be defended in the remaining sections of this essay. Very briefly, the overall strategy of my argument will be that to understand what causal capacity claims mean we need to understand what kinds of evidence are used to establish them and how capacity claims figure in arguments for singular causal and causal role claims. When we consider the ways in which capacity claims are established and the use of such claims in causal reasoning we are led to a picture of causal capacities that is rather different from that emphasized in standard accounts.

Shanto Iyengar and Donald Kinder's recent book *News That Matters* (1987) is a study of the effects of television news on political beliefs and attitudes. It is a paradigm of an investigation into the capacity or, as the authors themselves frequently say, the "power" of television news to produce certain kinds of effects. I will focus on just one aspect of this investigation which has to do with what the authors call the "agenda-setting" capacity of television news—with the fact that "television news powerfully influences which problems viewers regard as the nation's most serious" (ibid. 1987, 4). An important part of Iyengar and Kinder's investigation into agenda-setting consists of an ingenious series of experiments. In one such series, participants were asked to fill out a questionnaire which, among other things, asked them to name the most important problems facing the country. They were then randomly assigned to various treatment groups. People in each group were led to believe that they were watching unedited versions of network newscasts, but in fact they were subjected, over a period of time, to a series of carefully edited newscasts which were systematically altered to vary the amount and nature of coverage given various national problems. For example, one treatment group received a heavy dose of new stories discussing inadequacies in United States military defenses, another group was subjected to a large number of stories dealing with economic problems, another group to stories dealing with civil rights, and so forth. For each group, stories covering other sorts of problems besides the targeted problem were systematically excluded, so that each group served as a control group for the others. At the end of the experiment each group again completed a questionnaire about the country's problems.

These experiments showed a striking agenda-setting effect. In the case of each group, the effect of the edited news was to sharply increase the importance assigned to the problem on which the news focused, in comparison with the judgement of importance assigned prior to the experiment or in other groups. For example, in the case of national defense:

> Before seeing any newscasts, participants who were randomly assigned to the defense treatment condition ranked defense sixth in relative importance, behind inflation, pollution, unemployment, energy and civil rights. After exposure to the newscasts, the same participants now believed that defense was the country's second most important problem, trailing only inflation. Among viewers in the control condition, meanwhile, the relative position of defense as a national problem did not change.
>
> Such a dramatic shift in priorities, induced by such modest and unobtrusive alteration in television news coverage, constitutes powerful confirmation of the agenda-setting hypothesis. (Ibid., 1987, 18)

As the authors go on to show, similar agenda-setting effects occur in a wide variety of other circumstances and contexts. For example, similar effects emerge in experiments of different design, in connection with different groups of subjects, and, with one exception, in connection with all the problems investigated. That is, agenda-setting effects also exist in connection with coverage of economic problems, environmental problems, and so forth, although the precise magnitude of such effects, and exactly who is affected, varies from experiment to experiment.

Iyengar and Kinder then investigate the mechanisms and characteristic behavior of this causal capacity in more detail. For example, they show that there is a gradient effect of just the sort one would expect. Increased coverage of the target problem systematically leads to a more pronounced agenda-setting effect. They also show (what is politically important) that the agenda-setting effect persists for some time after the increased coverage stops. They then investigate various hypotheses about (as we might put it) the way which agenda effect works and the conditions under which the effect is enhanced or diminished. For example, they show that contrary to what one might have expected, news stories that illustrate and personalize national problems (e.g., stories about some particular person who has lost his job) do not enhance the agenda-setting effect. They also show that the position of a story in a broadcast does affect agenda-setting: Lead stories have much more of an agenda-setting effect setting than non-lead stories. Like the basic agenda-setting effect itself, these additional features of agenda setting turn out to be fairly stable and robust; they show up in a variety of different experimental designs.

The experimental work just described strongly supports the claim that television news has the causal capacity to alter people's judgements of political importance given the right background circumstances. But for the purposes of political science, the really interesting question is what we earlier called a causal role question. Given that changes in television news coverage *can* cause agenda-setting effects, have changes in network news coverage in fact caused significant agenda-setting effects within the contemporary American population and political system? In accordance with our earlier discussion, Iyengar and Kinder regard this as a distinct question from the question of whether television news has the capacity to produce such an effect. This second question is not a question which can be attacked experimentally; instead the authors approach the problem by looking at trends in television news coverage and in public opinion over time and by using a set of linear equations designed to estimate the impact of the former variable on the latter. Here the method employed involves the use of causal modeling techniques of the sort referred to earlier. For most problems investigated, the authors do indeed find a substantial agenda-setting effect. For example, in the case of energy, they find that "for every seven news stories broadcast, public responses citing energy as one of the countries most important problems increased by about 1 percent" (ibid., 1987, 30). Moreover, various specific features of the agenda-setting effect detected experimentally also show up in public opinion surveys concerning the entire United States population. For example, the authors find a very substantial effect on public opinion due to lead stories. They are thus able to show that the effect in which they are interested—changes in public opinion—behaves in detail in just the way one would expect it to behave if in fact it was being causally influenced by changes in television news coverage.

I turn next to an example in which reasoning appealing to claims about causal capacities is used to establish a singular causal claim. The example illustrates a strategy which is very widely used in arguing for singular causal claims. I will call this the eliminativist strategy. One argues for the singular causal claim that some particular instance *c* of a causal factor of kind C caused some particular instance *e* of a factor of kind *E* by showing that (a) *C* has the capacity to cause *E* (that it is the sort of thing that could have caused *E*), and by (b) then ruling out or rendering implausible the claim that anything else with the capacity to cause events of kind *E* did in fact cause this particular occurrence *e* of *E*. There are a variety of different ways of accomplishing (b). In the case of some causal factors with the capacity to cause *e*, one may be able to show that they did not occur on the occasion in question or, even if they did occur, that they did not bear the spatiotemporal relationship to *e* that they would need to bear if they were to cause *e*. In other cases, one may be able to show that specific features of the effect *e* are such that they only could have been caused by *c* or that the evidence for the specific mechanism or modus operandi by which *c* produces the effect was present and that no such evidence is present in the case of other possible causes of *e* or that evidence was present which is inconsistent with the operation of the mechanisms by which these other causes might produce *e*. Obviously in carrying out this sort of reasoning, information about causal capacities and the mechanisms by which they produce their effects is crucial.

In 1979 Luis and Walter Alvarez put forward their celebrated theory that (1) the mass extinction at the end of the Cretaceous period was caused by the impact of a large extraterrestial object such as a comet or an asteroid. (1) is a singular causal claim and the Alvarezs reasoning in support of (1) as described in their published papers, and in a recent book by their collaborator Richard Muller, appears to be a paradigm of the eliminativist strategy. The Alverezs begin by attempting to show that the impact of an extraterrestial body like an asteroid *could* have caused the mass extinction of the dinosaurs—that it has the capacity to produce this effect. Direct experimentation is of course out of the question, and so their argument in support of this claim appeals instead to theoretical considerations and to data concerning causes thought to be relevantly similar or analogous, in their effects, to asteroid impacts. They first determine from astronomical data the largest asteroid that might reasonably have been expected to hit the earth during the period in which dinosaurs existed. They then calculate the kinetic energy that would be released from such a collision and argue, by extrapolating from data from nuclear explosions and by looking at the estimated effects of other impact craters, that such a collision would have thrown up enough dust in the atmosphere to block sunlight for a considerable period of time, thus disrupting photosynthesis (Alvarez and Alvarez, 1980). Subsequent research has suggested several other possible "killing mechanisms" associated with such an impact including extensive acid rain and widespread fires (see Alvarez et al. 1980, Muller 1988, and Alvarez and Asaro 1990).

This shows that such an impact *could* have caused the extinction of dinosaurs (and, indeed, that impacts with such catastrophic effects must have occurred, since we know that there have been suitably large extraterrestial impacts), but it does not of course show that the extinction at the end of the Cretaceous period were actually due to such an impact. To show *this,* evidence must be found that rules out other possible causes of the extinction. Alvarez and Alvarez attempt to accomplish this in a variety of different ways. Some of the alternative causes that have been proposed have the capacity to produce extinction, but fail to bear the right temporal relationship to it. For example, climatic changes or changes in sea level "take much longer to occur than did the extinction; moreover they do not seem to have coincided with the extinction" (Alvarez and Asaro 1990, 83). In other cases, their argument is that various possible causes can be ruled out because they fail to explain or are inconsistent with various detailed features of the effect, or because various kinds of evidence that ought to be present, were those causes operative, are not found. For example, paleontological investigation shows a thin layer of clay at the boundary between the Cretaceous and Tertiary periods (the KT boundary) when the mass extinctions occurred. This boundary contains unusually high concentration of iridium, which is distributed uniformly worldwide. Iridium is extremely rare in the earth's crust and in sea water, but much more abundant in extraterrestrial sources like asteroids. Alvarez and Alvarez and their collaborators argue that the presence of iridium is a distinctive "signature" showing that the cause of the extinction was extraterrestrial in origin; a suitably large extraterrestrial impact would have injected iridium dust into orbit, and it then would have settled uniformly to earth. No terrestrial source could have produced this uniform distribution. Richard Muller summarizes this part of the argument in a way that nicely illustrates how it conforms to eliminativist strategy just described, "By the end of the summer, [Alvarez] had concluded that there was only one acceptable explanation for the iridium anomaly. All other origins could be ruled out, either because they were inconsistent with some verified measurement or because they were internally inconsistent. The iridium had come from space" (1988, 51–52).

Other evidence at the KT boundary seems to support such a conclusion. For example, other chemical elements are present in this boundary in the same ratios as in meteors, but in quite different ratios than are usual in the earth's crust (Alvarez and Asaro 1990, 80). Mineral spherules and grains of shocked quartz of a sort that (Alvarez and Alvarez argue) are known to result from meteor impacts are also found at the KT boundary (ibid., 80). Finally Alvarez and Alvarez argue on the basis of paleontological evidence that the extinction itself was very abrupt rather than gradual. This is exactly what one would expect given the mechanism or modus operandi associated with an asteroid impact, since the impact would have killed many plants and animals very quickly. By contrast, Alvarez and Alvarez argue, other possible causes are inconsistent with this evidence or at least leave it unexplained. For example, probably the most frequently suggested alternative hypothesis is that the extinction was due to a massive volcanic eruption. (An enormous eruption involving the Deccan Traps is thought to have occurred at approximately the same time as the KT extinction, so it has roughly the right temporal characteristics to have caused the extinction). But while there is iridium in the interior of the earth that might have been ejected during such an eruption, it is not easy to see how the eruption could have resulted in the observed global distribution of iridium, and controversial whether such an eruption would have had the capacity to produce shocked quartz or mineral spherules of the sort found at the KT boundary. Similarly, the mechanism of killing associated with a volcanic eruption is such that it would have produced extinction over a period that is too long to be consistent with the paleontological evidence.

We find a similar strategy at work in connection with another alternative hypothesis that Alvarez and Alvarez briefly considered: that the cause of the extinction, while extraterrestrial in origin, was a nearby supernova, rather than an asteroid impact. According to generally accepted theory, a supernova explosion would have injected an isotope of Plutonium (Pu-244) in the KT clay layer. None of the other possible causes postulated to account for the extinction would have produced Plutonium. Evidence for Pu-244 in the KT layer thus would have supported the supernova hypothesis precisely because it would conform so completely to the eliminativist ideal:

> If Pu-244 could be found in the clay layer, it would prove the theory. Not just verify it, not just strengthen it, but prove it. There is no other conceivable source of Pu-244. Its presence would strengthen the supernova explanation beyond all reasonable doubt. (Muller 1988, 53)

When no Pu-244 was found, the investigators rejected the supernova theory as "dead" (ibid., 60).

This example illustrates how, in a natural scientific context, information about how causal capacities work (that is, information about the ability of a cause to produce some effect or about its characteristic modus operandi) plays a crucial role in establishing singular causal claims. Eliminativist strategies similar to that just described are very common in both the natural and the social sciences. As we will see, the ubiquity of such eliminativist strategies represents an important clue concerning the interpretation of causal capacity claims, for the use of such strategies does not require that the capacity claims to which they appeal be law-governed.

I turn now to a final example, which illustrates yet another context in which claims about causal capacities figure importantly: the causal modeling techniques referred to in section 1. For the ease of exposition, I will focus largely on the simplest and most transparent of these techniques, which is ordinary linear regression. Compressing greatly, we can think of this technique in terms of the following basic picture. We suppose that within some population of interest there is a fixed linear relationship, that is causal in character and does not merely hold accidentally, between a dependent variable Y and a set of independent variables $(X_1, \ldots, X_n)$, that is,

$$Y = B_1X_1 + B_2X_2 + \ldots + B_nX_n. \qquad (13.1)$$

The problem we face is that of making use of statistical information about the variables $[Y, X_1 \ldots X_n]$—in particular information about their variances and covariances—to infer the values of the coefficients $B_1 \ldots B_n$. However, we cannot do this without making use of extrastatistical causal information of various kinds. For example, we must make use of such information in deciding which independent variables are legitimate candidates for inclusion within the regression equation itself. The expression which serves as the estimator for the coefficients B_i is such that one can always alter the coefficient of any variable in the regression equation by the inclusion or deletion of other independent variables as long as these variables exhibit a nonzero correlation with the original variable and with the dependent variable.[4] Given any finite body of data, it is very likely that we always will be able to find such additional, correlated variables. If one is going to use a regression equation to draw determinate causal conclusions, one thus needs some sort of principled rationale that tells one that certain variables are suitable candidates for inclusion in the equation and that others are not. My contention is that knowledge of causal capacities supplies just such a rationale—that one kind of consideration that helps to justify the claim that a certain variable *X* ought to be included in a regression equation with dependent variable *Y* is that *X* has the capacity to cause *Y*, and that one sort of consideration which justifies the exclusion of variable *X* from such an equation (even if *X* is highly correlated with the dependent variable *Y*) is that *X* lacks the capacity to cause *Y*.

Here is an example to illustrate the basic idea. Edward Tufte (1974) describes an investigation into the causes of differences in automobile fatality rates among different states in the contemporary United States. Tufte regresses a variable representing automobile fatality rates on, among others, variables measuring whether a state has automobile inspections, its population density, whether it was one of the original thirteen colonies, and whether it has less than seven letters in its name. In each case he finds a statistically significant nonzero regression coefficient: states with high fatality rates tend to lack inspections and to have low population densities. They also tend not to have been one of the original thirteen colonies and to have seven or less letters in their name.

Tufte relies on what he calls "substantive judgement" to support the conclusion that the former set of variables and not the latter causally influence the fatality rate. For example, he argues that "thinly populated states have higher fatality rates compared to thickly populated states because drivers go for longer distances at higher speeds in less dense states" (ibid., 20–21). He also remarks that "while we observe many different associations between death rate and other characteristics of the state, it is our substantive judgement and not merely the observed association that tells us density and inspections might have something to do with the death rate and that the number of letters in the name of a state has nothing to do with it" (ibid., 9).

The substantive judgements on which Tufte relies are claims about causal capacities and it is to such claims—for example, that inspections have the capacity to lower the fatality rate and that the number of letters in a state's names does not—that Tufte appeals in order to justify the inclusion of certain variables in the regression equation and the exclusion of others. Although I lack the space for a detailed discussion, I believe that causal capacity claims must be relied upon at a variety of other points in the use of causal modeling techniques as well—for example, in justifying claims about causal ordering, as when we decide, when *X* and *Y* are correlated, that *X* causes *Y*, rather than that *Y* causes *X* (for further discussion, see Woodward 1988, pp. 259–60).

As the example under discussion illustrates, the capacity claims that serve as input to a causal model are often vague, gappy, imprecise and qualitative; they may say no more than that *X*s are the sorts of things that can cause *Y*s. Nonetheless, such claims are general in the sense that they are applicable to many different populations and circumstances—inspections can and presumably do reduce fatalities in many populations and in many different background circumstances. By contrast, the output represented by the regression equation itself—what we are calling a causal role claim—is much more functionally precise and quantitative, but lacks generality in the sense that it is population specific. For example, Tufte calculates that states with inspections typically have a death rate lower by around six deaths per 100,000 people than states without inspections (1974, 14). However, there is no reason to suppose that this precise quantitative relationship would be the same if the same regression equation were estimated on other populations—for example, on administrative units drawn from other parts of the world with automobile inspections (e.g., French provinces or English counties). That is to say, while it is reasonable to expect that automobile inspections will have some effect on the fatality rate in many of these other populations, there are no good grounds to expect that the exact magnitude of this effect—as reflected in the estimated regression coefficient—will be stable across these different populations. The quantitative relationship between inspections and death rate discovered by Tufte is thus a population-specific relationship that results from the combination of generic capacity information that is applicable to many populations with statistical information about variances and covariances that is restricted to a particular population.

We can find what seems to be a similar picture of the use of capacity claims and causal role claims in methodological treatises in the social sciences. For example, Christopher Achen (1982) also insists that the results of a particular regression analysis will be population specific. As Achen puts it, the researchers intent in estimating a regression equation is to describe, for example, "the effect of the Catholic vote on the Nazi rise to power, or the impact of a preschool cultural enrichment program like Head Start on poor children's success in school. Whatever the truth in such cases, one would not characterize it as a law. Neither Catholics nor impoverished youngsters would behave the same way in other times and places" (ibid., 12).

Achen contrasts this kind of population-specific causal role information with the sorts of "general theoretical statements" that he thinks are characteristic of good social theory. According to Achen, general claims—that can be expected to apply across a variety of different circumstances—in the social sciences are typically not "functionally specific" in the sense that they describe exact, quantitative relationships between variables in the way that typical physical laws, like the Newtonian inverse square law, do. Instead, Achen writes:

> A typical example [of good social theory] would be the "law of supply and demand," which says that in a laissez-faire market subject to certain conditions, a drop in supply or a rise in demand will cause a rise in price. Another instance would be the "overlapping cleavages" rule in comparative politics. Religious, political, sectional, or other divisions within a society are said to "overlap" when each of them divides the society in approximately the same way. . . . The cleavages rule asserts that, all else being equal, societies whose divisions overlap will experience more strife than those with nonoverlapping cleavages.
>
> Notice that neither of these theories specifies a functional form for its effects: prices may rise rapidly or slowly, to a great or modest height, by a linear or nonlinear path, and they may do so in different ways in different circumstances—all without falsifying the theory. Similar remarks apply to the increase in strife among states with overlapping cleavages. Causes and effects are specified, along with the direction and continuity of their relationship, but not the functional form. This kind of generality is typical of good social theory. (Ibid., 12–13)

Elsewhere Achen adds "that functionally specific laws in the sciences are sure to fail serious empirical tests. . . . [Plausible] social theories rarely say more than that, ceteris paribus, certain variables are related" (ibid., 15–16). I suggest that we regard the sorts of claims Achen has in mind—claims to the effect that certain variables are causally related, along with claims about the direction of causal relationships—as claims about the existence and behavior of causal capacities, but capacities that we do not know how to describe as conforming to laws or exceptionless regularities. Ideas like Achen's about the form that good social theory will take are an important part of motivation for the account of causal capacities I will explore in the remainder of the essay—an account that will admit the possibility of non-law-governed capacities.

III

The previous section has illustrated some issues having to do with the epistemology of claims about causal capacities. We have described some of the evidence that scientists have used to support claims about causal capacities and some of the ways in which claims about causal capacities figure in reasoning about singular causal claims and in claims about causal roles. These epistemological considerations constitute an important constraint on the interpretation of capacity claims: Any positive account of causal capacities must be such that it fits with and allows us to make sense of their use in causal reasoning in examples of the sort I have described. For example, given some proposed analysis of the notion of a causal capacity, we may ask whether the sort of evidence which is ordinarily taken to support claims about the existence of a causal capacity is evidence that the conditions specified in the analysis are satisfied—if not, this may suggest that either the evidential practices in question are misguided or that the conditions imposed in the analysis are mistaken. Similarly, given some proposed analysis of causal capacities, we may ask whether the conditions specified in the analysis are really necessary for the role that causal capacity claims play in causal reasoning—if not, we may decide that there is no principled rationale for insisting on those conditions.

In what follows, I will argue that these epistemological considerations support the conclusion that many of the standard accounts of causal capacities are too stringent. In particular, I will argue for the intelligibility of what I will call irregular capacities—capacities that are not law-governed and that do not satisfy a unanimity requirement. In many cases our evidence for the existence of capacities does not differentially support the claim that they are law-governed rather than irregular. Moreover, in many cases the role that capacities play in causal reasoning does not require us to suppose that such capacities are law-governed or unanimous.

In order to explore the question of whether all capacities are law-governed, we need some sort of working understanding of when a generalization counts as a law. While this is not the place for a detailed discussion, I will take it that laws are reasonably precise generalizations of wide scope that are counterfactual supporting in an appropriate way, or, as I will put it later, stable or invariant across some wide range of background charges. I will also assume that laws, whether deterministic or probabilistic, claim that if certain conditions are met, a result of a certain kind will invariably follow. If a law is deterministic, the result that is claimed to follow is that some nonprobabilistic outcome occurs or that some nonprobabilistic quantity or some magnitude that is not a probability takes on a certain value. If a law is probabilistic the result that is claimed to follow is that some event occurs or that some quantity takes on a certain value with a fixed probability or that these conform to some specific probability density function. For reasons that will emerge in section 4, my view is that a generalization can count as a law even if it has exceptions or holds only within a certain domain of validity and breaks down outside of this domain. The claim that a generalization makes about what invariably happens when certain conditions are met thus need not be strictly or literally true for the generalization to count as law. But generalizations which are such that they do not even purport to delimit the conditions under which some outcome will invariably follow and which do not even pretend to describe functionally specific relationships will not count as laws. In particular, I follow Achen in thinking that generalizations that merely tell us that a certain kind of event or feature can or does cause a certain kind of effect or which merely describes the typical modus operandi of a causal capacity will not qualify as laws. My intent, in other words, is to preserve the contrast Achen draws between generalizations describing functionally exact, invariable relationships that apply to a wide variety of circumstances or populations and which do count as laws, and generalizations in which "causes and effects are specified, along with direction and continuity of their relationships, but not their functional form" (Achen, 1982, 13) and which do not describe laws.

With this in mind we may distinguish three different kinds of ways in which a causal capacity might behave.[5] The capacity of *C*s to cause *E*s is *deterministic* if there exists some deterministic law linking *C*s and *E*s which, so to speak, can be formulated at the level of, or in terms of the vocabulary used to characterize, *C* and *E*.[6] For example, if the capacity of aspirin ingestion to cause headache relief is deterministic, then there must exist a deterministic law specifying further conditions such that in those conditions aspirin ingestion is invariably followed by headache relief. Many capacities described by familiar physical theories—for example, the capacities of electromagnetic fields described by Maxwell's equations—are deterministic capacities.

While some philosophers continue to believe that talk of causation is only appropriate if a capacity operates deterministically, most now also recognize the possible existence of *probabilistic* causal capacities. For our purposes, the capacity of *C*s to cause *E*s is probabilistic if there exists a probabilistic law, again formulated at the level of *C* and *E*, specifying that when *C* and perhaps other conditions obtain, *E* will follow with some probability, strictly between 0 and 1. The issue of when a generalization qualifies as a probabilistic law is a complex and vexing one that I will not try to explore in detail here. But at least two necessary conditions naturally suggest themselves. First, I assume that the notion of probability that figures in a probabilistic law must be susceptible of some sort of "objective" interpretation—presumably (since actual relative frequency interpretations of probability look like nonstarters) as a hypothetical relative frequency or propensity. Second, I assume that the probabilities that figure in the law must be stable and counterfactual supporting in the right sort of way; we must get the same probabilities for the occurrence of the events specified in the consequent of the law whenever the events specified in the antecedent recur. I take this stability requirement to mean that the probabilities specified in the law must be either time invariant (i.e., stationary) or such that, if they do vary with time, they do so in some regular and systematic way, which is specified in the probabilistic law itself. The rationale for this second requirement derives from the idea that a sufficiently large degree of irregularity and non-systematicity in the relationship between two variables should be inconsistent with the claim that their relationship is governed by a probabilistic law. If, for example, in the presence of C and background circumstances B_i, E sometimes occurs and sometimes does not but with no fixed, stable relative frequency when C and those "same" background circumstances B_i recur on different occasions, then a generalization linking just C, B_i and E will not qualify as a probabilistic law; there will be no such thing as "the" probability with which E would occur, were C to occur in the future in circumstances B_i.[7]

In many recent discussions, an additional requirement is imposed on causes that operate probabilistically. This is a so-called unanimity condition, which for our purposes can be understood as the requirement that a genuine cause will raise the probability of its effect in every possible background circumstance compatible with the occurrence of the cause. That is, C causes E only if $P(E/C.B_i > P(E/\bar{C}.B_i)$ for all background circumstances or additionally causally relevant factors B_i compatible with C and $\bar{C}$.[8] I will explore this requirement in more detail later in this section.

Can we make sense of the notion of a causal capacity that is not governed by either a deterministic or probabilistic law? I will call such capacities *irregular* capacities and will argue later that for all we know many of the causal capacities that figure both in everyday life and in the social and behavioral sciences are irregular capacities. A cause has an irregular capacity to produce some effect if it sometimes causes the effect and sometimes does not, but does not do so in accordance with a deterministic or probabilistic law. For at least some of the background circumstance B_i, no matter how finely or relevantly specified, in which the cause C produces the effect E, the frequency with which E occurs does not converge to some fixed, stable value under repetition, but rather varies in an irregular, non-law-governed way.

Here is an example to illustrate the sort of possibility I have in mind. The investigations into television news described in section 2 provide convincing evidence that television news coverage can cause agenda-setting effects, and that, furthermore, television news actually did so in both experimental and nonexperimental contexts. But the investigations certainly do not result in the discovery or exhibition of anything that looks like a deterministic law which specifies the conditions under which news coverage is always followed by some agenda-setting effect or which specifies conditions under which a given level of news coverage of some national problem is always followed by an agenda-setting effect of a certain magnitude. Nor do the investigations claim to have discovered probabilistic laws relating various levels of news coverage to stable probabilities with which various agenda-setting effects will occur. Nor is it obvious how the investigations could be interpreted as showing that there must exist unknown deterministic or probabilistic laws having this character, which "underlie" or "ground" the causal relationships that were actually discovered.

Relative to the (partial) characterizations or descriptions of background circumstances that the investigations are able to provide, then, it thus does not appear to be the case that for each such background circumstance, television news coverage is followed either always or with some stable relative frequency by outcomes in which agenda-setting occurs. If the capacity of television news coverage to cause agenda-setting is irregular then this pattern continues to hold no matter how detailed the specification of these background conditions. That is, for at least some of these various background circumstances B_i, no matter how precisely specified, there is no determinate probability that agenda-setting or a certain level of agenda setting will occur when there is increased news coverage; it is simply the case that increased news coverage sometimes causes agenda-setting and sometimes does not, and there is no more uniformity to the results of increased news coverage than this. One might make this plausible for some background context B_i by showing that under repeated experimental trials in which there is increased news coverage in B_i the frequency of agenda-setting does not converge to any stable limit, but rather varies in an irregular way, and that this instability remains under an increasingly detailed causally relevant description of B_i. It is surely not logically impossible that the relationship between increased news coverage and agenda-setting might have this character and the claim that it is intelligible to suppose that the *causal* capacity of television news coverage to cause agenda-setting might be irregular is simply the claim that the discovery that the relationship between television news and agenda-setting behaves in this way would not be tantamount to the discovery that television news does not really cause agenda-setting.

In many cases, our evidence for the existence of a causal capacity does not differentially support the claim that the capacity is deterministic or probabilistic rather than irregular. That is to say, while such evidence does not rule out the possibility that the capacity may be law-governed, it provides no particular reason to suppose that this is the case. We can provide evidence that a cause has the capacity to produce some effect, without providing any evidence that it conforms to a deterministic or probabilistic law. The investigations into television news described above illustrate this point. What these investigations do show is the agenda-setting effects associated with television news coverage occur across many different background contexts, among many different groups of people, and in connection with many political problems. Moreover, they show that various alternative explanations according to which this association is spurious, accidental or noncausal (for example, alternative explanations according to which news coverage and judgements of increased problem importance are joint effects of a common cause or the result of an accidental correlation between increased news coverage and some other factor that does cause such judgements) are implausible.[9] However, the precise character and magnitude of these effects—exactly who will be affected, and by how much, and under what circumstances or with what probability—seem to vary in an irregular and unsystematic way across different background contexts. Not everyone in the experimental groups is affected and while the investigators have identified some of the factors relevant to this, they clearly have not identified all of them. Similarly, the level of the agenda-setting effect varies from problem to problem and the investigators have no systematic theory of such variations. What is fairly stable or invariant across many (but not all) different background contexts is not some law or exceptionless regularity linking coverage and agenda-setting. Rather what is stable is (a) that television news coverage has the capacity to produce (and in many cases does produce) *some* agenda-setting effect or other, and (b) that the relationship or mechanism by which this effect is produced behaves in a qualitatively similar way across many different background contexts. (Thus it is consistently the case, across many different background circumstances, that position of a story in a news cast does matter, but whether the story "personalizes" a problem does not. Similarly, more news coverage generally means a larger agenda-setting effect and so forth.)

My claim, then, is that while the experimenters do establish that television news has the capacity to cause agenda-setting effects, their evidence does not support the claim that this capacity is law-governed, rather than irregular.[10] I think that this fact by itself puts considerable pressure on those who hold that it is part of the logic or semantics of "cause" that causes must produce their effects in accordance with some deterministic or probabilistic law or that such laws must "underlie" or ground every causal relationship. Someone who takes this view must either hold that (a) the investigations described above do not really show that television news coverage causes agenda-setting effects, since they provide no reason for thinking that the relevant laws exist, or that (b) contrary to what appears to be the case, the investigations do after all provide good evidence for the existence of such laws. It is not obvious how to develop the latter position in a convincing manner.[11]

A similar conclusion seems to hold in connection with the unanimity condition. In establishing that television news coverage produces an agenda-setting effect, Iyengar and Kinder do not establish or provide serious evidence that television news raises the probability of this effect across every background condition or even across every background condition in some particular population. For one thing, as we have seen, for some background circumstances, the relevant probabilities in the sense of stable limits of relative frequencies may not exist. For another, the experimental results described certainly allow for the possibility that, for some individuals in some permissible background circumstances, the result of increased television coverage may be to leave unchanged or even lower the probability that the subject will report that the problem is more important. The experimental results show that if one looks at the *average* or *aggregate* impact of television news on various experimental groups, there is an agenda-setting effect in many different circumstances, not that such an effect occurs in all circumstances or for all individuals. That is, what the experimental results show is that, in many circumstances and for many groups, exposure to increased coverage of some problem results in an increase in the frequency of those who regard the problem as important in comparison with a control group. This increase in frequency would occur if, for example, (a) increased coverage sometimes caused judgements of increased problem importance and never caused judgements of decreased importance or if (b) increased coverage caused judgements of the former sort more frequently than judgements of the latter sort. In both of these cases, it seems true that increased coverage has been shown to have the capacity to cause judgements of increased problem importance—after all, it does in fact produce this effect under both (a) and (b). It might of course be true in addition that exposure to increased coverage raises the probability of judgements of increased problem importance for all individuals in all background circumstances in some or all of the treatment groups, but the assumption that this is the case is certainly not necessary for the observed experimental results to occur and the observed experimental results are not evidence that this is the case. In short, the experimental results establish that television news has the capacity to cause agenda-setting, but not that this capacity operates unanimously.[12]

We should now be able to see that there is a correct idea about the testing of causal capacity claims that needs to be distinguished from the contention that such capacities must conform to a deterministic or probabilistic law or must satisfy a unanimity principle. The correct idea is this: We think that if *C*s have the capacity to cause *E*s in background conditions B_i, then if large enough collections of individuals in B_i are randomly divided into treatment groups that are exposed to *C* and to control groups that are not exposed to *C* (and if the randomization works in the sense that there is no systematic difference between the treatment and control groups in other factors besides *C* that are causally relevant to *E*), then (barring complications having to do with interaction effects and mixed capacities)[13] we should usually observe a higher frequency of *E*s in the treatment group than in the control group. The systematic appearance of this sort of result in a number of different experiments conducted with different individuals in different circumstances consistent with the holding of conditions B_i will usually constitute convincing evidence that *C*s do indeed have the capacity to cause *E*s; the failure of this effect to appear in any systematic way will (again barring the complications previously referred to) be strong evidence against the claim that *C*s can cause *E*s. It is just this sort of reasoning that Iyenger and Kinder appeal to in their experimental investigations. It should now be clear, however, that the evidence for such frequency increases is not evidence for and does not require that it be true that the capacity in question operates either deterministically or probabilistically. What the frequency increase shows is simply that *C*s sometimes cause *E*s (and thus that *C*s have the capacity to cause *E*s), but it is perfectly possible that the capacity in question may be irregular, rather than deterministic, probabilistic or unanimous.

My argument so far has been that if one looks at the evidence that is used to support claims about the existence of capacities, this evidence often fails to support the claim that such capacities are law-governed or unanimous. I want now to argue that typical uses of causal capacity claims in causal reasoning—and in particular the use of capacity claims to support singular causal claims or to support claims about causal roles—also do not require that the capacities in question be deterministic, probabilistic or unanimous.

Consider first the role that capacity claims play in establishing singular causal claims. As I have suggested, the relevant reasoning often takes an eliminative form. If one wants to show that exposure to substance C_1 (e.g., cigarette smoke) caused Jones's lung cancer, one needs to consider the other possible causes $C_2, \ldots, C_n$ of lung cancer and rule out or render implausible the possibility that any of these caused Jones's cancer. As we have seen, one might do this in a variety of ways. For example, one might show that Jones was not exposed at all to some of these possible causes (that they do not bear the spatiotemporal relationships that they need to bear to Jones if they are to cause lung cancer in him), that in other cases evidence for the characteristic modus operandi of these alternative causes is not present, and so forth. The point I wish to emphasize is that in making use of this eliminative method, we do not need to assume or to rely on the assumption that causal capacities of $C_1, C_2, \ldots, C_n$ are deterministic or probabilistic or satisfy the unanimity principle. Given the appropriate evidence, the eliminative method will work perfectly well even if the possible causes $C_1, \ldots, C_n$ produce their effects irregularly. Suppose that Jones is a heavy smoker, develops lung cancer, and that it is unlikely that he has been exposed to any other causes that can produce lung cancer. Then even if cigarette smoking produces lung cancer in an irregular, nonunanimous way, we have very good reason to think that in this particular case, it was Jones's smoking that caused his lung cancer. We reason that something must have caused Jones's cancer and, that in the absence of any other candidates, we are justified in concluding that this is one of those cases in which C_1 produced this effect. From the point of view of our reasoning about this particular case, the assumptions that smoking causes lung cancer in accordance with a law or the unanimity condition are idle wheels; these assumptions may be true, but we do not need to assume them to make sense of our reasoning about the causation of Jones's cancer.

Although I lack the space for a detailed discussion, I believe that a similar conclusion holds in connection with causal role reasoning. To briefly recapitulate my remarks in section 2, we can think of such reasoning in terms of the following basic picture. Such reasoning takes as input information about causal capacities as well as statistical information of various sorts. This information about capacities is typically not information about the truth of deterministic or probabilistic laws. Instead it is general, qualitative information to the effect that *C*s are the sorts of things that can cause *E*s—that automobile inspection are the sorts of things that can reduce traffic fatalities, and so forth. The output of such techniques, on the other hand, is much more precise and quantitative: a linear equation or set of such equations that tell us that within this particular population, changes in the level of various independent variables will produce specific changes in the aggregate level of the dependent variable, with the magnitude of the change specified by a set of fixed, numerical coefficients. It is simply an empirical fact that the values of these coefficients will often vary considerably as the same set of linear equations is estimated on different populations. For example, the coefficients in an equation which purports to estimate the deterrent effect of capital punishment on the murder rate may vary considerably depending on whether the relevant population is drawn from the United States or from some other country with the death penalty (see Leamer 1983). The death penalty may have some deterrent effect or some capacity to deter in each population, but how much of an effect it has may vary significantly from population to population. Similarly while it is reasonable to expect that television news coverage can and will produce some agenda-setting effect in many different countries with appropriate institutional arrangements (democratic elections, for example) the quantitative details of such effects—how much of an increase in judgements of importance for a given kind of problem results from, say, two additional news stories per week—probably varies considerably from country to country.

Because the information about casual capacities that serves as input to a causal model or to a causal role claim is typically loose, qualitative information rather than precise, quantitative information of a sort provided by laws of nature, the use of such information is consistent with the possibility that the capacities in question are irregular. Indeed, if we knew a law of nature specifying precisely the conditions under which various levels of television news coverage would always (across every population) be followed by an agenda-setting effect of some specified magnitude or under which this effect would follow with fixed, stable probabilities, there would be no need to estimate the coefficients in our causal model separately for each population of interest; we would already know a single stable quantitative relationship which held across all populations. The very fact that the coefficients in typical causal models are unstable across different populations tells us that neither the inputs to such models nor the sorts of causal knowledge they yield as outputs represent laws of nature. When we make use of capacity claims as input to causal models, we thus do not need to suppose (do not at any point need to rely on the assumption) that the capacities in question conform to deterministic or probabilistic laws or satisfy a unanimity principle. The case, based on such models, for the agenda-setting role of television in the U.S. population, or for the role of automobile inspections in reducing traffic fatalities in the U.S. population would be no less good (or bad) if the relevant capacities were irregular rather than law-governed.

IV

My discussion so far has attempted to make the idea of an irregular capacity intuitively plausible and to point out that, in case of many causal capacities, our evidence for the existence of such capacities is consistent with the idea that those capacities are irregular. I turn now to a sketch of a positive theory of causal capacities. The underlying idea of this positive theory is that there is a close connection between the notion of causation and (what I will call) invariance—stability of a relationship under some class of changes.[14]

I begin by introducing the basic idea of invariance and then discuss how this idea may be applied to causal capacities, including irregular capacities.

As I will understand the notion, an invariant relationship is a relationship that holds at least roughly or approximately, or in some suitably large number of cases, in the actual circumstances and that would continue to hold in a similar way under some specified class of interventions or for specified changes in background conditions. Causal and nomological relationships are invariant in the sense under discussion, with those relationships that are representable by deterministic or probabilistic laws constituting a proper subset of the more general class of invariant relationships. By contrast, the mark of a relationship that holds in the actual circumstances, but only accidentally—a relationship which is not nomological or causal—is precisely that it is not invariant: It would not continue to hold if circumstances were to change in various ways.

Here is an illustration of the underlying idea of invariance that involves a deterministic law. The ideal gas law $PV = NRT$ describes in a roughly accurate way the relationship between pressure, temperature and volume that obtains in many (although by no means all) actual samples of gas. But we think of this relationship as a law not merely because it holds in the actual circumstances, but because it would also continue to hold even if various sorts of changes or interventions were to occur in connection with these samples. Thus, the relationship $PV = NRT$ will continue to hold, at least approximately, for many samples of gas if we were to change their pressures, temperatures, and volumes in various ways, although of course this will only be true within a certain range of such changes—for example, at high temperatures and pressures, the ideal gas law will not even be approximately correct. Similarly, the relationship $PV = NRT$ would continue to hold if we were to introduce many other kinds of changes—for example, if we were to change the spatial location of the gas, or the color or shape of its container. We can think of the set of changes or interventions under which $PV = NRT$ continues to hold as constituting the domain or regime or scope of invariance of this relationship.

By contrast, a relationship is not invariant under some class of changes or interventions if the result of those changes will be to disturb the relationship so that it no longer continues to hold, even approximately or except for some limited class of exceptions. Suppose, for example, that the relationship $PV = NRT$ accurately describes the relationship between the measured values of the pressure, volume and temperature for some set of actual samples of gas, but that the result of, say, doubling the pressure of these samples would be to change the relationship between P, V and T, so that $PV = NRT$ no longer holds, even approximately. In these circumstances the relationship would be noninvariant under this change and if this failure of invariance turned out to be the general pattern, we would be justified in thinking of the relationship $PV = NRT$ as nonlawful, and only accidentally true of these particular samples of gas for which it holds.

This association of invariance with lawfulness and causation is widely reflected in scientific practice. I lack the space here for a systematic discussion, and confine myself to two very brief illustrations. First, one of the features of fundamental physical laws most emphasized by physicists themselves (but very largely ignored in traditional philosophical discussion of lawfulness) is that laws must satisfy various symmetry requirements. Such symmetry requirements are in effect invariance requirements: They amount to the demand that laws express relationships that remain invariant under certain kinds of transformations or changes. For example, to take some very simple possibilities, it is generally required that fundamental physical laws remain invariant under translations in space and time, under spatial rotations and under translation from one inertial frame to another. Second, a similar association of invariance with those generalizations which express genuine causal connections can be found in econometrics and in the literature on causal modeling. For example, it is standard in this literature to distinguish "structural equations," which express genuine causal connections from equations that merely express an "empirical association" (Fox 1984, 63 ff).[15] The distinguishing mark of a structural equation is that, in contrast to an empirical association, it expresses a relationship that will remain stable or unaltered under some specified class of interventions.

As a number of writers have emphasized, there is also a close connection between the idea that causal and nomological relationships are invariant relationships and the common idea that one mark of a causal or nomological relationship is that it can sometimes be exploited for purposes of manipulation and control in a way that a merely accidental, noncausal relationship cannot. Of course, there are many cases in which there exists a causal or nomological relationship between *A* and *B* and in which it is not possible to control or manipulate *A*. However, if such a causal or nomological relationship does exist, and one can alter or manipulate *A*, then doing this will often be a reasonable strategy for controlling or manipulating *B*. In the absence of such a causal or nomological relationship this will not be true. The close connection between causality and invariance is reflected in the fact that invariance is exactly the feature that a relationship needs to possess if we are to use it for purposes of manipulation and control. If the relationship $PV = NRT$ is invariant, then it will continue to hold for some sample of gas if, say, I increase the temperature from T_1 to T_2 or if various other changes in background conditions occur spontaneously. I can thus avail myself of the stability of this relationship to alter the pressure of the gas by, for example, altering its temperature while holding its volume constant.

As I have already intimated, it is important to understand that the claim that (1) a generalization is invariant is different from the claim that (2) it is an exceptionless generalization of wide scope. The truth of (2) is clearly not sufficient, and, more controversially, is not even necessary for the truth of (1). To begin with the question of sufficiency, it is perfectly possible for a regularity to hold over a wide spatiotemporal interval, and to be exceptionless, or nearly so and yet for a generalization describing the regularity to be such that virtually any change in initial conditions or background circumstances from those that actually obtain would render it false. Perhaps some pervasive cosmological regularities have this character.

More controversially, if a relationship is invariant, it does not follow that it must take the form of a regularity that holds universally and without exception in the actual world. Invariance, as I have characterized it, is always invariance with respect to some class of changes or conditions, and it is perfectly possible for a relationship to be invariant with respect to some class of changes and to be radically noninvariant with respect to changes outside this class. In some cases we may be able to identify the scope of invariance of a relationship in a fairly precise way. For example, as already noted, the ideal gas law breaks down at relatively high temperatures and pressures, although under other conditions it is a relatively invariant generalization. In other cases, a relationship may be invariant across some set of circumstances and fail or break down in some other set of circumstances, but we may be unable to usefully identify or characterize these circumstances in advance, at least at the level of analysis at which the relationship in question is formulated. For example, we know from statistical mechanical considerations that in a rather special set of circumstances (a set of measure zero) mechanical systems will fail to obey the second law of thermodynamics. However, for many systems to which we wish to apply the thermodynamical concepts, we are unable to ascertain in advance whether these very special circumstances obtain. While many philosophers may conclude that this shows that the so-called second law does not really state a lawful or nomological relationship, it seems to me that it rather shows that a relationship can be sufficiently invariant to qualify as lawful even though it does not take the form of a completely exceptionless regularity. My view is that it makes perfectly good sense to say that the second law is a highly invariant relationship—it holds across an extremely wide set of physical circumstances and is stable across changes in this set—even though it is not exceptionless. Indeed, in my view, this fact about invariance explains why it is reasonable to regard the second law as a genuine law even though it is not exceptionless. For similar reasons, I would hold that it is reasonable to regard the ideal gas law as a genuine law even though, when formulated as a generalization covering all possible or even all actual samples of gas, it has many exceptions.

These remarks will help, I hope, to make plausible a more general suggestion that the notion of invariance that is relevant to causation should be understood in such a way that a relationship can qualify as invariant even though it is not associated with an exceptionless regularity. Invariance has to do with stability or robustness—that is, with whether a relationship would continue to hold in some significant number of cases under circumstances and conditions different from the actual circumstances. This is simply a different feature of relationships than the feature of being associated with a regularity (or of being describable by a generalization) that holds exceptionlessly in the actual circumstances. As we will see, this decoupling of the notion of invariance from the notion of an exceptionless regularity allows us to make sense of the claim that a capacity can be causal even though it is irregular.

Put baldly, my idea is that causes that produce their effects irregularly can qualify as genuine causes despite their irregularity as long as they satisfy a suitably formulated invariance condition. This condition has been heavily foreshadowed in my earlier remarks. The underlying idea is that if *C*s have the capacity to cause *E*s, then the association between *C* and *E* must be such that it continues to hold (in the sense that, controlling for other causal factors, *E*s occur more frequently in the presence of *C*s than in their absence) across some large set of (but not necessarily all) changes in circumstances or background conditions even though the details of the association itself—its functional form and so forth—may vary in a sufficiently irregular way that it does not count as a law.[16] In other words, what must be invariant in such cases is the *capacity* of *C*s to produce *E*s rather than any exceptionless generalization linking *C* and *E*. Put differently, the idea is that a genuine causal capacity must exhibit a significant degree of context-independence; it must be such that it "does the same thing" or "acts in the same way" in the production of some characteristic effect in a variety of different environments. If an alleged cause is such that it is too context-sensitive—if any one of an extremely large number of changes in background circumstances will disrupt its apparent production of, or association with, some effect—it is no cause at all.

We have already seen how the experiments conducted by Iyengar and Kinder show that the relationship between television news coverage and agenda-setting satisfies this sort of invariance condition, although the authors make no pretense of being able to formulate a law linking these variables. Here is a second illustration, drawn from a well-known paper by Cornfield et al. (1959), which describes evidence which, as of 1957, was regarded by the authors as convincingly showing that smoking causes (i.e., has the capacity to cause) lung cancer. It is a good example of what working scientists mean when they claim that a causal agent possesses the capacity to produce a certain effect. This paper was written in the absence of a detailed knowledge of the precise biochemical mechanism by which smoking produces lung cancer and relies largely on epidemiological evidence and to a lesser degree on experimental studies of laboratory animals. Cornfield et al. make no attempt to formulate a law or exceptionless generalization describing the conditions under which smoking will be followed by lung cancer or to formulate a law specifying the probability that one will develop lung cancer given that one smokes. Instead they lay great emphasis on the stability or invariance of the relationship between smoking and lung cancer, in the sense described earlier. For example, the authors note that some association appears between smoking and lung cancer in every well-designed study on sufficiently large and representative populations with which they are familiar. There is evidence of a higher frequency of lung cancer among smokers than among nonsmokers, when potentially confounding variables are controlled for, among both men and women, among people of different genetic backgrounds, among people with quite different diets, and among people living in quite different environments and under quite different socioeconomic conditions (Cornfield et al. 1959, 181). The precise level or quantitative details of the association do vary—for example, among smokers the incidence of lung cancer is higher among those in lower socioeconomic groups—but the fact that there is some association or other is stable or robust across a wide variety or different groups and background circumstances. A similar stable association is also found among laboratory animals. In controlled experiments animals exposed to tobacco smoke and other tobacco products have a higher incidence of various kinds of cancer than unexposed controls.

Moreover, the qualitative character of the association—the mechanism or modus operandi of the cause—is also stable in just the way that one would expect if smoking were causing lung cancer. Thus, for example, heavy smokers and inhalers in all populations show a higher incidence of lung cancer than light smokers and noninhalers, groups that have smoked the longest show the highest incidence of lung cancer and the incidence of lung cancer is lower among those who have smoked and stopped than among relevantly similar groups who are still smoking. As the authors explain, the cumulative effect of such evidence is to render implausible alternative causal scenarios according to which the relationship between smoking and lung cancer is accidental or noninvariant.[17]

This example provides another illustration of the sort of invariance or stability that I take to characterize a genuine causal capacity. While the authors do not exhibit a precise deterministic or probabilistic generalization that is invariant across different circumstances or populations, they do show that the relationship between smoking and lung cancer satisfies the weaker, less demanding invariance requirement I have sought to describe. That is, what the authors show is that smoking has the power or ability to produce a characteristic effect in a wide variety of different circumstances and populations. They thus show that the capacity of smoking to cause lung cancer is a capacity that, in Nancy Cartwright's phrase, smoking "carries around with it, from situation to situation" (1989a, 196).[18] This sort of stability of effect must be present if the relationship between smoking and lung cancer is genuinely causal.

What would a world (or a domain of inquiry) be like in which there are no relationships between events which are invariant even in the loose, exception-permitting sense I have been describing? Here is one possibility. Michael Oakeshott (1966) holds that "nothing in the world of history is negative or non-contributory" (p. 208). His picture is one in which all historical events are "connected" or "contribute" to one another in a sort of seamless web of reciprocal interdependence (ibid., 207 ff). Presumably if some historical event—for example, the French Revolution—had not occurred, then everything that followed this event and the entire web of relationships in which it is embedded would have been different in some way, but, according to Oakeshott, there are no grounds for even an educated guess (and it is no part of the task of the historian to speculate) about the specific determinate way in which subsequent history would have been different under this counterfactual assumption. Oakeshott claims that in such a world of interdependence the notions of causation and causal explanation would be inapplicable. The account of causal capacities that I have sketched supports this conclusion. Oakeshott's historical world is not merely one in which historical events are not law-governed; it is a world in which historical events are not causes—they fail to have characteristic effects at all to which they are linked in a way that satisfies even the weak invariance condition I have described. On Oakeshott's view, what follows any particular historical event "depends" (if that is the right word) upon everything else that is true of the society in which the event occurs—every difference in historical context could make a difference to what follows that event. On the account I have sketched, this sort of possibility—in which the (putative) effects of a cause vary in irregular and intractable ways from context to context—is incompatible with genuine causation.[19]

Thus while the accounts of causation and causal capacity for which I have been arguing are more permissive than many traditional accounts in the sense that they tolerate causes that are not law-governed, they are far from indefinitely permissive. It is an open empirical possibility that, at some levels of analysis in some domains of inquiry, it may not be possible to construct causal theories that identify invariant relationships at all—that is, that there are no causal relationships at that level of analysis.[20]

I conclude with a final observation about the notion of invariance under discussion. We have already noted a close relationship between invariant relationships and those that can be used for purposes of manipulation and control; this in turn is connected to the point that causes are often exploitable as means for bringing about effects. One good reason for thinking that the weak, exception-permitted notion of invariance I have described, which is applicable to relationships that are not law-governed, is nonetheless an appropriate notion for explicating the notion of a causal capacity is that this notion preserves the connection between causation, manipulation and control. Even if the capacity of *C*s to produce *E*s is irregular and invariant only in the weak sense described earlier, it can still be a reasonable strategy to manipulate *C* as a way of manipulating *E*. For example, even if the agenda-setting capacity of news coverage operates irregularly, it can still be reasonable, if one wants to alter the political agenda, to try to do so by altering television news coverage, especially if no alternative, more regular cause of this effect is available. The irregularity of the relationship means, of course, that success is not guaranteed, (and that it may not even be possible to estimate the probability of success) but as long as the relationship is invariant or roughly so across different background circumstances and one has no reason to think it likely that one is in a background circumstance in which it is known that the relationship will fail to hold, it can be reasonable to try to make use of the relationship to produce the effect of interest. In order to exploit the relationship between *C* and *E* for purposes of manipulation, it must be the case that, at a minimum, the result of altering *C* will not be to automatically disturb the relationship between *C* and *E* and that the relationship between *C* and *E* also not be too readily disruptable by other kinds of background changes. The weak notion of invariance which I have tried to describe has this feature—it is the weakest kind of stability and context-independence that is consistent with the use of a relationship for purposes of manipulation and control. In my view, one of the strongest reasons for rejecting the claim that if *C*s have the capacity to cause *E*s, the relationship between *C* and *E* must be governed by a deterministic or probabilistic law, or must satisfy the unanimity requirement, is that *C*s can be used to achieve some measure of manipulation and control over *E*s even in the absence of these features. As long as manipulating *C* is a way of manipulating *E*, it is appropriate to regard the relationship between *C* and *E* as causal even if the further requirements of law-governedness or unanimity are not satisfied. It seems to me to be a virtue of the invariance-based account of causal capacities that I have sketched that it supports this commonsensical idea about the relationship between causation and manipulation.

NOTES

Adolf Grünbaum's friendship and philosophical work have been a source of support and inspiration to me for many years. Themes related to those explored in this essay concerning the analysis and testing of causal claims, particularly in the behavioral sciences, have been extensively discussed by Grünbaum, most recently in his work on Freud. It is a great pleasure to contribute this essay to a volume in Adolf Grünbaum's honor.

1. While Cartwright (1989b) has been an important stimulus to my thinking, my account of causal capacities differs in some fundamental ways.

2. Consider, for example, Elliot Sober's discussion of Debra Rosen's well-known example in which a golf ball is rolling toward the cup, is kicked by a squirrel, ricochets about, and finally (and improbably) drops in the cup. Sober holds that the squirrel's kick token-caused the ball to go into the cup, but the squirrel's kick was not merely not a positive causal factor for the ball's dropping in, but was actually a negative causal factor for this effect (1985, 407). If Sober's notion of a positive causal factor is meant to capture the preanalytic notion of a factor which has the capacity to produce an effect, this is a puzzling result; it seems obvious that the kick has the capacity to cause the ball to drop in the cup, in the sense that it is capable of causing this effect, since on Sober's own theory it did token-cause this effect. This represents an additional reason for concluding that Sober's notion of a positive causal factor does not coincide with the causal capacity notion we are trying to explicate. Similar remarks apply to Cartwright (1989a) in which she holds that pressing on the accelerator of a car causes its increased speed, but seems to deny that pressing on the accelerator has the capacity (actually she says "natural capacity") to produce this effect (p. 144). Here again, it seems to me to be highly desirable to retain the idea that if *X* token causes *Y*, then *X* has the capacity to cause *Y*. On this understanding of causal capacities, one or another of Cartwright's two claims about the accelerator must be given up.

3. As the quotation from Eells and Sober perhaps suggests, the concept of a population stands in some need of clarification. My own view is that talk of a

population only makes sense if one is able to specify what it is to sample in a random or representative way from that population. Populations must also be of definite, finite size. Thus, for example "human beings alive in 1990" describes a perfectly good population, and it is sensible to ask what the causal role of smoking is in this population. On the other hand causal claims that apply to any arbitrary individual satisfying some general background condition are not helpfully understood as referring to a particular population. Instead the set of individuals to which such claims apply is open-ended and of indefinite size. If "smoking causes cancer in human beings" is understood as applying to any arbitrary human being, or to all human beings who have lived or will ever live, it does not make reference to a particular population any more than "short circuits cause fires in the presence of oxygen" refers to some particular population of fires occurring in oxygen-laden environments.

4. For example, in the case in which there are two independent variables X_1 and X_2, the expression which serves as estimator for the regression coefficient B_1 will be a function of, among other things, the correlations between X_1 and Y, X_2 and Y, and X_1 and X_2 If we were to drop X_2 from the equation and substitute another variable X_3, the value of the coefficient B_1 would change as long as Cov $(X_3, Y) \neq$ Cov (X_2, Y) and Cov $(X_3, X_1 \neq$ Cov (X_2, X_1). For additional details, see Woodward (1988).

5. A similar three-part division of causal capacities into those that are deterministic, those that are probabilistic, and those that are neither deterministic nor probabilistic can be found in Dupré and Cartwright (1988). Dupré and Cartwright call causal capacities that are neither deterministic nor probabilistic "random" capacities. I have chosen the name "irregular capacities" instead, because (a) probabilistic capacities presumably can generate sequences of outcomes that pass the usual mathematical tests for randomness and (b) irregular capacities presumably can generate sequences of outcomes that fail to pass such tests. While Dupré and Cartwright admit the possibility of non-law-governed capacities, they say relatively little about such capacities in their subsequent discussion.

6. The intent of this characterization, according to which the deterministic law governing the capacity must be formulated at the level of C and E, is to avoid a certain kind of trivialization. It is uncontroversial that the fundamental laws of microphysics are probabilistic, and that everything in the universe is made of objects that conform to these laws. Someone might argue that it follows from this fact alone that in some relevant sense the behavior of aspirin (like everything else) is "governed" by probabilistic laws, and hence that the capacity of aspirin to cure headaches is probabilistic. This is *not* the sense of "governed by a probabilistic law" that is relevant to determining when a capacity is probabilistic. From the general microphysical facts described above, it does not follow that there exists a probabilistic law formulated in terms of predicates like "aspirin ingestion" and "headache relief." Similarly if the fundamental physical laws are deterministic, it would not follow that all capacities are deterministic capacities (see also note 7).

7. When are frequencies or probabilities sufficiently stable to figure in a probabilistic law? While I have no general account of this matter, a few further remarks may be helpful. To begin with I assume that processes, the underlying

dynamics of which are deterministic, can be characterized by probabilistic laws. Even for processes that are deterministic, the truth of an ergodic hypothesis (or some suitable analogue to such a hypothesis) is a sufficient condition for the existence of stable probabilities of a sort that can figure in probabilistic laws. Similarly, many games of chance and gambling devices produce outcomes which occur with stable probabilities, even though, for all relevant purposes, the underlying dynamics of such devices are deterministic. One also finds grounds for belief in stable frequencies in other areas of scientific investigation, for example, genetics. What I want to reject is the automatic assumption that every case in which an outcome *E* repeatedly occurs can be assimilated to these examples. That is, I reject the assumption that whenever *E* occurs, there must be some causal capacity or mechanism that is generating it in accordance with a probabilistic (or deterministic) law or endowing it with some stable probability of occurrence. In my view, the philosophical literature on probabilistic causation has been much too ready to make this assumption. There is no reason to suppose that stable probabilities exist in the case of many of the phenomena to which such theories are applied.

8. There are a variety of different formulations of the unanimity condition in the philosophical literature. I have made use of what I hope is a relatively generic version that captures the central idea (see Cartwright 1989a and Dupré 1990). The condition itself raises a number of additional issues that I lack the space to discuss. There are, however, two points on which I want to comment since they bear directly on the skepticism I have expressed about the condition. Several recent discussions seem to suggest (or at least make it natural for the reader to suppose) that imposition of the unanimity condition is necessary to avoid Simpson's paradoxlike difficulties (see Eells and Sober 1983). If this claim were correct, it would constitute a powerful argument in favor of the unanimity condition. I think, however, that the claim is mistaken. Very roughly, Simpson's paradox difficulties arise when one takes the presence of a correlation between an independent variable C_1 and a dependent variable *E* to reflect the existence of a direct causal connection between C_1 and *E* when in fact the correlation (or part of the correlation) is due to some additional factor C_2. Factor C_2 may, for example, be a common cause of both C_1 and *E* or may be a cause of *E* and merely correlated with C_1. The crucial point is that the conditions that must be satisfied to avoid this mistaken causal inference do not require that the causal factors with which one is dealing obey the unanimity condition. One can perhaps see this most readily in the case of randomized experimental designs. A properly designed randomized experiment in which the frequency of *E*s in a group treated with C_1 exceeds the frequency of *E*s in a control group not treated with C_1 can rule out the possibility that this increase is due to the operation of some further causal factor C_2, but it is not a necessary condition for the success of this experimental design that the causal relation between C_1 and *E* satisfy the unanimity principle. For example, we can use a randomized experimental design to show that aspirin causes headache relief even if aspirin, while increasing the probability of headache relief among most subjects, lowers this probability for a small subgroup in the treatment group. Here aspirin acts nonunanimously but the sort of con-

founding of cause and correlation characteristic of Simpson's paradox has been ruled out by the experimental design.

A similar conclusion emerges if one thinks about this issue within the context of causal modeling techniques. The natural analogue of the unanimity requirement within these models is a condition of some kind on the functional form employed in the model—perhaps an additivity condition or a condition to the effect that the coefficients in the model are constant across the population of interest. (This captures the idea that the cause uniformly affects all individuals within the population of interest.) The sorts of worries associated with Simpson's paradox, on the other hand, are addressed in such models by the standard conditions imposed on the distribution of the error term—for example, by the condition that the errors must not be systematically correlated with both the independent and dependent variables. These two sets of conditions (on the functional form and on the distribution of the error term) are conceptually distinct in the sense that each set of conditions can be satisfied without the other set being satisfied. Here too, it is just wrong to argue for the unanimity condition on the ground that we need it in order to avoid conflating cause and correlation.

A second point concerns the connection between unanimity and invariance. I will argue in section 4 that genuine causes must satisfy an invariance requirement—they must be capable of acting in the same way in a variety of different background circumstances. It is natural to think of the unanimity condition as one way of trying to capture or represent such an invariance requirement, the idea being that the requirement that the cause act in the same way can be cashed out in terms of the idea of the cause raising the probability of the effect across all background contexts. However, I will also argue that unanimity does not capture the sort of invariance condition appropriate to causation.

9. These alternative explanations are rendered implausible through the use of randomized experiments and through the detection of the agenda-setting effect in many experimental and nonexperimental contexts.

10. I should acknowledge that this claim rests on a substantive assumption about confirmation or evidential support. This is the assumption that confirmation is always differential—that to provide evidence for p one must provide evidence that discriminates in favor of p and against some competing claim q or that supports belief in p rather than q. Observational claims that are derivable from both p and q but do not differentially favor p against q are not evidence for either p or q. It is in this sense of evidential support, that our evidence for the existence of many causal capacities is not evidence that such capacities are law-governed (rather than irregular). While I believe that this general conception of evidential support as differential can be defended, I lack the space to do so here.

11. Some readers will undoubtedly think that my arguments for the possible existence of irregular capacities conflate epistemological and metaphysical issues. It will be said that while it is true that one can provide grounds showing that a causal capacity exists without providing grounds that it conforms to a law, this is a fact about the epistemology of causation; it does not show that there are (or even could be) capacities that do not conform to laws. I agree of course that the epistemological practices I have emphasized do not show that irregular causal ca-

pacities exist. But I do think that such practices strongly suggest there is nothing conceptually confused or unintelligible about the notion of such capacities—that it is an empirical matter whether such capacities exist. Those who take the contrary view face the difficulty that their epistemological and metaphysical views about causation fail to fit together in any natural way. Any story we tell about the truth conditions about causal claims (e.g., that such claims must always be grounded in laws) must make it understandable how the epistemic practices we have for establishing such claims sometimes succeed in providing information that shows that these truth conditions have been satisfied. If an account of the truth conditions or the correct analysis of causal claims is such that we can virtually never tell, using the epistemic practices ordinarily thought to be relevant to such claims, that those truth conditions are satisfied, then either (a) something is wrong with those epistemic practices, or (b) the account of truth conditions is, if not mistaken, at least gratuitous and unmotivated. If it is true that all causal capacities must be law governed, it is left mysterious why our ordinary epistemic procedures, like randomized experiments, for testing such claims work, since pretty clearly such practices do not provide evidence for the existence of the relevant laws.

12. The observation that randomized experiments fail to provide evidence that probabilistic causes obey the unanimity condition is also made in a very clear and forceful way in Dupré (1990). Dupré also emphasizes the importance of providing an account of the truth conditions of causal claims that fits together with our epistemic practices for testing such claims. I might also add, in connection with this portion of my discussion, that in assessing the unanimity condition it is particularly important to keep in mind the distinction between causal capacity claims and causal role claims. My concern in this essay is with unanimity as a requirement on causal capacities and my argument has been that, when understood in this way, the requirement looks pretty implausible. The discovery of some group of people whose headaches are made worse by the ingestion of aspirin does not show that aspirin lacks the capacity to cure headaches in the sense that it (can and does) cause headache relief among other groups of people. The plausibility of unanimity as a condition on causal role claims raises somewhat different issues, although here too, for reasons I have described in Woodward (1990), I think that the condition is too strong.

13. An interaction effect occurs when a cause produces a characteristic kind of effect in one background circumstance, and a very different kind of effect in another background circumstance. A factor has mixed causal capacities if, within the same background circumstances, it has both the capacity to cause some effect by one causal route and a capacity to prevent or inhibit the occurrence of the effect by some other causal route. The notion is discussed in an illuminating way in Dupré and Cartwright (1988). As emphasized by Dupré and Cartwright, the experimental strategy described may give a misleading picture of the causal facts when mixed capacities are present, or when certain kinds of interaction effects are present.

14. I claim no originality for this general suggestion. As I remark in section 4 the connection between lawfulness, causation and invariance is widely recog-

nized both in discussions of physical laws and in economics. The connection also surfaces, although in rather different forms from the one that I advocate, in a good deal of recent philosophical discussion. Skyrms (1980) and van Fraassen (1989) both emphasize the connection between lawfulness, symmetry, and invariance, but within a subjectivist framework that is very different from the realist account of causation favored here. The connection between causation and invariance is also emphasized in Humphreys (1988) and, as I have already noted, Cartwright (1989a). However, both Humphreys and Cartwright interpret invariance in a different, less permissive way than I. Finally, the robustness condition on causation imposed by Redhead (1987, 1990) is also an invariance condition, although again different from the condition I have imposed.

It is also worth underscoring that this part of my discussion is not intended as a noncircular reductive analysis of the notions of law and cause in the sense of an analysis that explains these notions in terms of ideas that do not themselves presuppose causal, nomological or modal notions. For reasons described in Woodward (1992), I do not believe that such a noncircular analysis is possible. The notion of invariance has ideas like counterfactual dependence and physical possibility built into it—it is part of the same circle of notions as "law" and "cause." My discussion is an attempt to elucidate the relationships between notions in this circle, not to explain the circle in terms of something outside of it.

15. The connection between lawfulness (and causation) and invariance under some class of interventions is implicit in early work in econometrics and the representation of causal relations by systems of linear equations carried out by writers like Koopmans (1950), Marschak (1953), Simon (1953) and Hurwicz (1962). For a more recent discussion, see, for example, Engle, Hendry and Richard (1983), especially their remarks on what they call "superexogeneity." For an excellent overview of the current state of discussion within econometrics, see Hoover (1988). The econometrics literature has also tended to stress the connection between invariance and our practical concern with manipulation and control. For further discussion of the notion of invariance as it occurs within econometrics, see Woodward (forthcoming).

16. But when is the set of background conditions across which an association holds sufficiently "large" for the association to qualify as causal? With respect to what measure do we assess largeness? It seems to me that it is a mistake to expect any sharp answers to these questions. I see a continuous gradation leading from those relationships that are sufficiently invariant to qualify as causal to those relationships that are not invariant at all. For any given relationship, there is a determinate fact of the matter about whether it is stable or invariant under some particular change; but there need be no sharp cutoff point between causal and noncausal relationships. There is no further fact about whether a relationship is causal over and above facts about the range of changes under which it is invariant.

17. As I have attempted to explain in Woodward (1992), one can help to make it plausible that a relationship is invariant by providing evidence that undercuts alternative hypotheses according to which the relationship depends upon the operation of additional factors in a way that makes it noninvariant. For example, if C_1 and E are correlated, but this correlation is simply due to the fact that C_1

happens to be correlated with C_2, the true cause of E, then the relationship between C_1 and E will be noninvariant and readily disrupted by changes in C_1 or C_2. Ruling out this sort of possibility—for example, by a randomized experiment—is one way of providing evidence that the relationship between C_1 and E is invariant.

Cornfield et al. (1959) provide a nice illustration of what this involves in connection with the suggestion, attributed to R. A. Fisher, that a disposition to smoke and a disposition to develop lung cancer are effects of some common cause, presumably a genetic factor. Cornfield et al. point out that although this suggestion accounts for the gross correlation between smoking and lung cancer, it is inadequate as an explanation of the qualitative details of the smoking-lung cancer relationship. Fisher's hypothesis does not explain why inhalers should develop lung cancer more frequently than noninhalers, or why those who smoke and stop should be less likely to develop cancer than those who continue to smoke. Of course Fisher's hypothesis could be complicated to try to explain these facts—for example, one might suppose that there are two genetic factors G_1 and G_2 and that G_1 disposes people to smoke but makes it easy for them to stop, while G_2 causes smoking and makes it hard to stop, and that G_2 is more likely to cause cancer than G_1. But such an account is implausible and ad hoc, and becomes even more so as one tries to account for the other qualitative facts previously mentioned. Consider, for example, the problem of accounting for the correlation between cancer and exposure to tobacco smoke among laboratory animals within Fisher's framework. The cumulative effect of this evidence is to rule out Fisher's hypothesis and thus to rule out one important ground for thinking that the association between smoking and lung cancer is noninvariant.

18. While I agree with Cartwright about this point, I disagree with what I take to be her view about *what* is carried around from situation to situation by capacities. At least in this essay, Cartwright seems to suggest that if econometrics is to tell us about capacities, it is the numerical values of the coefficients in the linear equations we are estimating—for example, the parameter representing price elasticity of demand in a demand equation—that must be stable from situation to situation (1989a, 195–96). However, as I have insisted, it is simply an empirical fact that such parameters are often not stable from situation to situation in the sorts of contexts in which econometric techniques are used. Thus, if capacities must satisfy the sort of stability requirement that Cartwright favors, econometrics discovers few, if any, facts about capacities. My own reaction to this nonstability of parameter values is to adopt a different conception of what must be carried around from context to context for a factor to qualify as possessing a causal capacity. On my view a factor has a causal capacity if it produces qualitatively the same effect in many situations, even if quantitative details (and parameter values) vary across those situations. Thus, an increase in price has the capacity to cause a decrease in quantity demanded if it produces this effect in many different situations, even though the precise numeral value of this decrease for a unit change in price may vary from situation to situation.

19. Needless to say, my interest in these remarks is in describing an example in which there are no invariant relationships in the sense I have sought to capture. I am not concerned with whether Oakeshott's account of historical under-

standing is correct, or with whether I have accurately rendered every feature of Oakeshott's views.

20. I hope to discuss this very compressed (and no doubt obscure) suggestion in more detail elsewhere. But as a further illustration, consider again causal modeling techniques. It seems to me likely that many of the variables employed in causal models constructed by social scientists fail to meet even the weak invariance requirement I have been describing. Consider, for example, the recent study, by Michael Timberlake and Kirk Williams, discussed by both Clark Glymour et al. (1987) and Nancy Cartwright (1989), that purports to show that foreign investment causes political repression. Quite apart from the issues about model specification and parameter estimation raised by Glymour and Cartwright, it seems to me that there is an important prior question: Is "foreign investment" even the sort of variable that one can plausibly suppose has the kind of relatively stable, context-independent effect on political repression that we have taken to be characteristic of a causal capacity? My suspicion is that the answer to this question is "no" and that a similar conclusion holds for many of the other variables invoked by causal modelers. The methodological recommendation that follows from my account of causal capacities is that it is not enough in order to justify the inclusion of a variable within a causal model, to argue that the hypothesis that the variable has a nonzero coefficient when other variables are properly controlled for passes the usual tests for statistical significance. Rather, one needs independent grounds for supposing that the variable has the capacity to causally affect the dependent variable, and this requires providing evidence that the variable produces this effect in a number of other contexts besides the population with which one is presently working. Those who use causal modeling techniques often fail to follow this methodological recommendation. This shows again that the invariance requirement that I have proposed is far from toothless.

REFERENCES

Achen, C. 1982. *Interpreting and Using Regression.* Beverly Hills: Sage Publications

Alvarez, C.; R. Alvarez; F. Asaro; and H. Michael. 1980. "Extraterrestrial Cause for the Cretaceous-Tertiary Extinction." *Science 208:*1095–1108.

Alvarez, W., and F. Asaro. 1990. "An Extraterrestrial Impact." *Scientific American 263* (4):78–84.

Cartwright, N. 1989a. "A Case Study in Realism: Why Econometrics is Committed to Capacities." In A. Fine and J. Leplin, eds., *PSA 1988.* Vol. 2. East Lansing: Philosophy of Science Association, pp. 190–97.

———. 1989b. *Nature's Capacities and Their Measurement.* Oxford: Oxford University Press.

Cornfield, J.; W. Haenszel; and E. Hammond. 1959. "Smoking and Lung Cancer: Recent Evidence and Discussion of Some Questions." *Journal of the National Cancer Institute* 22. 173–203.

Dupré, J. 1990. "Discussion: Probabilistic Causality: A Rejoiner to Ellery Eells." *Philosophy of Science 57:*690–98.
Dupré, J, and N. Cartwright. 1988. "Probability and Causality: Why Hume and Indeterminism Don't Mix." *Nous 22:*521–36.
Eells, E. 1988. "Probabilistic Causal Levels." In B. Skyrms, and W. Harper, eds., *Causation, Chance and Credence*. Dordrecht: Kluwer, pp. 109–34.
Eells, E., and E. Sober. 1983. "Probabilistic Causality and the Question of Transitivity." *Philosophy of Science 50:*35–37.
Engel, R.; D. Hendry; and F. Richard. 1983. "Exogeneity." *Econometrica 51:* 277–309.
Fox, J. 1984. *Linear Statistical Models and Related Methods*. New York: Wiley & Sons.
Glymour, C.; R. Scheiners; P. Spirtes; and K. Kelly. 1987. *Discovering Causal Structure*. Academic Press.
Hoover, K. 1988. *The New Classical Macroeconomics*. Oxford: Basil Blackwell.
Humphreys, P. 1988. *The Chances of Explanation*. Princeton: Princeton University Press.
Hurwicz, L. 1962. "On the Structural Form of Interdependent Systems." In E. Nagel, P. Suppes, and A. Tarksi, eds., *Logic, Methodology and Philosophy of Science*. Palo Alto: Stanford University Press, pp. 232–39.
Iyengar, S., and D. Kinder. 1987. *News That Matters*. Chicago: The University of Chicago Press.
Koopmans, T. C. 1950. "When is an Equation System Complete for Statistical Purposes?" In *Statistical Inference in Dynamic Economic Models*. New York: Wiley & Sons, pp. 393–409.
Leamer, E. 1983. "Let's Take the Con Out of Econometrics." *American Economic Review 73:*31–43.
Marschak, J. 1953. "Economic Measurements for Policy and Prediction." In W. C. Hood and T. C. Koopmans, eds., *Studies in Econometric Method*. Cowles Commission Monograph, no. 14. New York: Wiley & Sons, pp. 1–26.
Muller, R. 1988. *Nemesis*. New York: Weidenfeld & Nicolson.
Oakeshott, M. 1966. "Historical Continuity and Causal Analysis." In William Dray, ed., *Philosophical Analysis and History*. New York: Harper & Row, pp. 192–212.
Redhead, M. 1987. *Incompleteness, Nonlocality and Realism*. Oxford: Clarendon Press.
———. 1990. "The Nature of Reality." *British Journal for the Philosophy of Science 40:*429–41.
Simon, H. 1953. "Causal Ordering and Identifiably." In W. C. Hood and T. C. Koopmans, eds., *Studies in Econometric Method*. Cowles Commission Monograph, no. 14. New York: Wiley & Sons, pp. 49–74.
Skyrms, B. 1980. *Causal Necessity*. New Haven: Yale University Press.
Sober, E. 1985. "Two Concepts of Cause." In P. Asquith, and P. Kitcher, eds., *PSA 1984*. Vol. 2. East Lansing: The Philosophy of Science Association, pp. 405–24.

Tufte, E. 1974. *Data Analysis for Politics and Policy.* Englewood Cliffs, N.J.: Prentice-Hall.

van Fraassen, B. C. 1989. *Laws and Symmetry.* Oxford: Oxford University Press.

Woodward, J. 1988. "Understanding Regression." In A. Fine and J. Leplin, eds., *PSA 1988.* Vol. 1. East Lansing: Philosophy of Science Association, pp. 255–69.

———. 1990. "Laws and Causes." *The British Journal for the Philosophy of Science* 41:553–73.

———. 1992. "Realism About Laws." *Erkenntnis* 36:181–218.

———. Forthcoming. "Causation and Explanation in Econometrics." In D. Little, ed., *The Reliability of Economic Models.* Dordrecht: Kluwer.

14

Falsification, Rationality, and the Duhem Problem

Grünbaum versus Bayes

John Worrall
Department of Philosophy, London School of Economics

Can We Ascertain the Falsity of "Isolated" Scientific Theories Despite the Duhem Problem?

It was in a fascinating lecture of Adolf Grünbaum's that I first heard A. R. Anderson's wonderful admonition, "Keep an open mind, but not so open that your brains fall out!" Given the evidence we now have, it would surely be brain-endangering to be open-minded about any of the following theories:

T_1: *Light beams consist of tiny material particles.*
T_2: *Light beams consist of waves in an elastic solid ether.*
T_3: *The universe was created in 4004 B.C. with the earth and the various "kinds" that inhabit it very much as they are today.*
T_4: *The parting of the Red Sea, the fall of the walls of Jericho and sundry other Biblical events were caused by a series of close encounters between the earth and a massive comet that originally broke away from Jupiter.*[1]

Each of these theories is surely false and we *know* that each is false "to all scientific intents and purposes" (Grünbaum 1969, 1092). We need an account of reasonable belief that saves our brains by delivering the judgement that the only reasonable epistemic belief about these theories in the light of the evidence we now have is that they are false (or so likely to be false that it makes no difference).

The problem is that the simplest account of how we know that the above statements are false fails when subjected to a little analysis.

This simplest account is that we know that each of these statements is false because it has been directly empirically refuted. That is, we know it to be false because it has turned out to be directly logically inconsistent with experimental or observational reports that we know to be true (or can reasonably take ourselves as knowing to be true). But, even discounting skeptical doubts about observation statements, and as Duhem ([1906] 1954, part 2, chap. 4) long ago pointed out, no observational consequences in fact follow from "single," "isolated" theories of the above kinds. Such theories are therefore *never* directly inconsistent with observation sentences. The sort of structures that *do* have observational consequences are better characterized as *theoretical systems.* These systems are "built around" such single claims but also involve other theories: further "specific" assumptions—so-called auxiliary, or instrumental theories—and often "closure assumptions"—which state that nothing plays any significant role in the observable phenomenon at issue aside from the factors explicitly taken into account.

In order, for example, to test the claim that light consists of material particles, particular assumptions have to be added about the particles and about the forces acting on them in various circumstances, as well as instrumental assumptions about, perhaps, the effect of light on photographic emulsion (or on our retinas). In order to test Newton's theory of universal gravitation using astronomical data about the position of some planet, assumptions are needed which tie our telescopic observations to claims about planetary positions (including, for example, an assumption about the amount of refraction that light undergoes in passing into the earth's atmosphere), as well as assumptions about the other planets. Moreover, in the Newton case, a "closure" assumption is needed, saying in effect that only gravitational effects produced by the interaction with the sun and those planets specifically taken into account in the derivation play any nonnegligible role.

If none of the theories T_i listed above has any observational consequences of its own, then it follows that, given any set of observational or experimental results $\{e_1, \ldots, e_n\}$, there always exists a stronger theory T_i' that is (i) consistent, (ii) entails T_i, and (iii) also entails $e_1 \& e_2 \& \ldots \& e_n$.[2]

Indeed such a T_i' can be constructed trivially by, for example, conjoining T_i to an "auxiliary" A_i of the conditional form $T_i \rightarrow (e_1 \& \ldots$

$\&e_n$). However, we can easily see how to perform the trick in the case of some of the above theories without resorting to this general logical result. For example, the creationist hypothesis might *seem* straightforwardly refuted by the evidence indicating an immensely older age for the universe than the one postulated by that hypothesis or by the evidence from the fossil record, homologies and the like. But, notoriously, a creationist can always make what might be called the "Gosse dodge": God built the radioactive decay products, the fossils, the homologies and the rest into his creation, thus creating the universe *as if* it were already immensely old, and *as if* present species had evolved from earlier and now extinct species. Similarly Velikovsky's theory that there were worldwide cataclysms associated with the earth's "close encounters" with an enormous comet that had broken away from Jupiter might seem to be refuted by the lack of records of any such cataclysms in several contemporary record-keeping cultures. But, as Velikovsky himself pointed out, nothing is easier than to produce a theoretical system built on his claim about these "close encounters" that entails the observed absence of records: Simply assume that in just those cultures that have no appropriate records the people at the time suffered "collective amnesia" brought on by those very cataclysms.

Does this undeniable Duhemian fact of no direct refutation mean that, despite whatever initial inclinations we might have, epistemic fair play dictates an open mind about the falsity of the theories listed? Obviously not: It means only that the simplest account of how we come rationally to believe in the falsity of some theories fails. It means only that a purely deductive theory of rationality is hopelessly weak (or rather that a theory of rationality based only on deductive logic, together with the requirement that [sufficiently low level] observation statements, once accredited, be regarded as beyond rational doubt[3] is hopelessly weak.) The problem is surely to strengthen our theory of rationality so that it *does* deliver the judgement, for example, that although it is (deductively) logically possible to accept all the known observational evidence and still believe in special creation, it is not rational to hold this belief, and, having strengthened our theory of rationality, to supply some justification for the strengthening.

Adolf Grünbaum would agree, I think, that this is the right sort of view and the right set of problems to tackle. His work played a key role in focusing the attention of philosophers of science in the 1960s and 1970s on the Duhem problem. In an important series of papers culminating in "Can We Ascertain the Falsity of a Scientific Theory?" of 1971,[4] he accurately identified the problem and argued that, at any rate in certain special cases, "component hypotheses"—that is, hypotheses that have no observational consequences in isolation from other assumptions—can nonetheless be *ascertained* to be false "to all scientific intents and purposes." Moreover, in an addendum to the version of that paper reprinted in the second edition of his *Philosophical Problems of Space and Time* (1973), he explicitly "reject[ed] Imre Lakatos's universal falsificationist agnosticism in regard to the following contention of his: Empirical results cannot make it rationally mandatory to presume that a given scientific hypothesis is false" (p. 849). Grünbaum analyzed in detail a specific case in which it would, so he holds, indeed be "rationally mandatory" to believe in the falsity of a particular component hypothesis. I will recall the main lines of Grünbaum's analysis in the third section of this essay; and I will argue that, although explicitly aimed at a special kind of case, his analysis highlights, and indeed relies on, intuitions that can be generalized so as to underwrite a more encompassing solution of the Duhem problem.

What Adolf Grünbaum did *not* supply was any attempt to *validate* those intuitions. A large body of opinion (personalist Bayesian) holds that they cannot in fact *be* validated and therefore that we should do without them in our account of rational scientific methodology. I wish I could provide a general, plausible and justified theory that underwrites the intuitions at issue. Instead I here argue only that the task of trying to develop such a theory is unavoidable—the price of going along with the large body of opinion and rejecting these intuitions is altogether too high.

In the next section, I present the personalist Bayesian account of the Duhem problem and argue for its inadequacy. Then, in the third section, I outline Grünbaum's analysis and highlight what I take to be its chief difference from the personalist Bayesian one. Undoubtedly unsolved difficulties are involved in the Grünbaum-style approach but they are, I will argue, outweighed by those involved in the personalist Bayesian approach. As so often with basic philosophical problems, no proposed solution is bulletproof: I give some reasons for holding that your friendly neighborhood epistemological dentist would prefer you to bite the bullet with Grünbaum than to bite the Bayesian bullet.

The Personalist Bayesian Account of the Duhem Problem and Its Inadequacy

The Personalist Bayesian Account

For present purposes I take the general Duhem problem to be constituted by the fact that each of the following is true:

(i) If the deductive structure of any experimental test of a "single" theory in science is fully analyzed, then some wider system of theoretical assumptions is always found to have been invoked.

(ii) In many cases in the history of science a negative test result was dealt with by "blaming" (that is, replacing) a "peripheral" assumption rather than the "central theory under test"; in other cases a negative test result was dealt with by "modifying" rather than "rejecting" the central theory;[5] moreover, in many such cases the "defense" of the central theory seems entirely reasonable—to the extent that some such episodes result in what seem, from the intuitive point of view, to be great successes for that central theory.[6]

(iii) This possibility of blaming a "peripheral" assumption or "modifying a theory rather than rejecting it"—though always present as a deductive possibility—is by no means universally exploited in science; some results *are* taken as directly threatening the "central" theory and, should a sequence of results of this kind occur, that theory's credibility is eroded, perhaps so as to become indistinguishable from zero; again, this procedure seems intuitively to have been reasonable in some of these cases.

The problem, then, is what general rule, if any, distinguishes these two types of case and *why* (that is, for what reason do scientists take some results as affecting peripheral assumptions and others as affecting the "central" theory of a system)? Let us suppose that the theoretical system at issue can "naturally" be represented as a conjunction of the central theory T and a set of auxiliaries A.[7] Then the problem can be posed more sharply: Given that e refutes $T\&A$ but neither T nor A separately, how do scientists nonetheless localize disconfirmation between T and A, and why?

Jon Dorling (1979, 1982) has claimed that the following account, based on (personalist) Bayesian principles,[8] solves the problem. Dorling's account has been further explained and elaborated by Michael Redhead (1980) and enthusiastically endorsed by Colin Howson and Peter Urbach (1989).

Although there is *deductive* symmetry in these cases (that is, e is inconsistent neither with T nor A alone but only with their conjunction), there is, of course, no reason why, on the Bayesian approach, the effect of e on T and A should be symmetric. The Bayesian looks at the "posterior" probabilities of T and of A, compared to their prior probabilities. These "posterior" (conditional) probabilities of T and A can be expressed using Bayes's Theorem as

$$p(T/e) = p(T) \cdot (p(e/T)/p(e)), \tag{14.1}$$

$$p(A/e) = p(A) \cdot (p(e/A)/p(e)). \tag{14.2}$$

Clearly if, for example, a Bayesian "agent" judges $p(e/A)$ to be very close to $p(e)$, while she judges $p(e/T)$ to be very much less than $p(e)$, then by the above equations, the posterior probability of A will be very close to its prior, while the posterior probability of T will be significantly reduced compared to its prior. In particular, should the probability of T in the light of e be very close to 0 while its prior had been reasonably high, then T could presumably be regarded as refuted by e "at least to all intents and purposes of the scientific enterprise" (Grünbaum 1969, 1092).[9] (The Bayesian would object to this gloss, claiming that all we can properly do is assign posterior probabilities in the light of any evidence we have, and should eschew all talk of refutation. For present purposes, I am happy to regard "e to all intents and purposes refutes T" as *meaning*—or even perhaps as more perspicuously expressed by—"T has a probability greater than a half ahead of e, but close to zero probability given e.") Should, on the other hand, A's posterior probability be significantly lowered while T's is close to its prior, then this would "justify" an agent in retaining her central theory T despite the refutation of the overall system $T\&A$ by e.[10]

In order to count as a rational Bayesian, the agent's degrees of belief must be representable as genuine probabilities and this requires in particular that deductive relationships be respected: Since in the cases we are considering e is deductively inconsistent with the conjunction $T\&A$, this means that the Bayesian requires his agent to set $p(T\&A/e) = 0$.

As equation 14.1 shows, the crucial factor determining the impact of e on what the Bayesian sees as the rational credibility of T is the ratio $p(e/T)/p(e)$. (Similarly, in the case of A, the crucial factor is $p(e/A)/p(e)$.) Applying the theorem on total probability, the likelihoods $p(e/T)$ and $p(e/A)$ can be expressed as

$$p(e/T) = p(e/T\&A)p(A/T) + p(e/T\&\neg A)p(\neg A/T) \text{ and}$$
$$p(e/A) = p(e/A\&T)p(T/A) + p(e/A\&\neg T)p(\neg T/A).$$

This allows us to incorporate the constraint distinctive of the Duhem problem; for when $T\&A$ yields $\neg e$, then the first of the summands disappears in each case. Thus the Bayesian requires in such cases that, in order to count as rational, his agent's degrees of belief must satisfy

$$p(e/T) = p(e/T\&\neg A)p(\neg A/T) \text{ and}$$
$$p(e/A) = p(e/\neg T\&A)p(\neg T/A).$$

As long as his agent's degrees of belief can be represented as probabilities satisfying these last equations, and his agent "conditionalizes" (that is, as long as the agent's degree of belief in any theory after evidence e has been discovered to hold is her prior degree of belief in that theory conditional on e) then, *independently of any further details,* she acted as a Bayesian-rational agent.

Dorling develops his account via consideration of a historical case of a scientist who accepted experimental results that were inconsistent with his initial overall theoretical system. The treatment given by Howson and Urbach also focuses on a similar (though different) historical case. In fact both take a case of the opposite sort to the one I primarily concentrate on: namely, a case in which the scientist concerned regarded the negative experimental findings as posing a negligible threat to her "central" theory and as instead pointing to some deficiency in some "auxiliary" or other. They then

(a) first, claim to show that the scientist held, ahead of the relevant evidence, beliefs about the "central" and "auxiliary" theories concerned and about the likelihood of that evidence on various suppositions about the truth and falsity of those theories that are appropriately idealized as ascriptions of particular prior probabilities and (probabilistic) likelihoods which meet the above formal constraints;[11] and

(b) second, demonstrate that the probability calculus leads, once the evidence is in, from these priors and likelihoods to radically (and, allegedly, "surprisingly") asymmetric posterior probabilities for the "central" and "auxiliary" theories at issue: the posterior probability of the "central" theory being only negligibly different from its prior, while the (intially plausible) auxiliary has a posterior probability close to zero.

The claim then is that the scientist's rationale for "holding onto" her central theory despite this negative evidence has been revealed: The scientist's belief dynamics were to all intents and purposes those of an ideal Bayesian agent.

Consider the often-cited case of Prout, reanalyzed by Howson and Urbach (1989, 96–102). Prout developed the hypothesis that the atomic weight of (any pure sample of) any element is an integer multiple of the atomic weight of hydrogen. Moreover, at any rate on the usual reconstruction (as Howson and Urbach point out, not much is known about Prout's real beliefs), his belief in this hypothesis seems to have been scarcely affected (if indeed affected at all) by the evidence (which he accepted) that repeated measurements of the atomic weight of chlorine consistently produced values of 35.83 (± 0.01). Prout held that the reason for this discrepancy between observation and theory was not the falsity of his fundamental hypothesis, but instead the falsity of the assumptions about the necessary purification and measurement procedures.

Letting T be Prout's central hypothesis, A the then accepted auxiliaries about the necessary purification and measurement procedures, and e the evidence on chlorine, Howson and Urbach claim that the following are all reasonable idealizations of Prout's beliefs:

(a) $p(A/T) = p(A)$ (i.e., A and T are probabilistically independent)
(b) $p(T) = 0.9$
(c) $p(A) = 0.6$
(d) $p(e/\neg T\&A) = p(e/\neg T\&\neg A) = 1/2\ p(e/T\&\neg A)$.

They then demonstrate that, applied to these assumptions, the probability calculus entails that

$$p(T/e) = 0.878 \text{ while } p(A/e) = 0.073.$$

Thus a Bayesian agent, beginning with these priors and conditionalizing on the evidence about chlorine, would find the credibility of T

scarcely affected, while she would now regard A, which she had originally found more plausible than not ($p(A) = 0.6$), scarcely credible. (A might, then, be considered to be, so far as this agent is concerned, "refuted to all intents and purposes.") This effectively mirrors the reactions to the evidence of Prout (and the majority of his fellow early nineteenth-century chemists), and so the Bayesian claims to have exhibited the rationale of those reactions. Howson and Urbach write:

> Thus Bayes's Theorem provides a model to account for the scientific reasoning that gave rise to the Duhem problem. And the example of Prout's hypothesis, as well as others that Dorling . . . has described, show, in our view, that the Bayesian model is essentially correct. (P. 101)

The Inadequacy of the Personalist Bayesian Solution of the Duhem Problem: Two Further "Rationally Reconstructed" Case Studies

As already mentioned, it is by no means straightforward to specify exactly what counts as an acceptable rational model of some piece of historical "belief-dynamics." Not even the most committed Bayesian would, of course, claim that such accounts are straightforwardly descriptive, since no one seriously holds that the historical agents (or indeed anyone) exhibited degrees of belief that are exactly expressed by the numbers used in these Bayesian analyses. A range of criticisms could be made about at least some of the assumptions even as idealizations. (This is especially true of the likelihood assumptions (d): Did Prout and the Proutians really have *any* quantifiable beliefs about the likelihood of the chlorine evidence on the assumption that both their hypothesis and the accepted auxiliaries were false?) However, let us concede, for present purposes, that these Bayesian accounts *are* reasonable idealizations of the changes of belief that occurred in these episodes.

It does not follow that the Bayesian can supply the *rationale* for Prout's retention of his "central" theory; nor that this Bayesian account is in general the "essentially correct" solution of the Duhem problem. The chief worry here is the old one of the extent to which these accounts depend on "initial conditions"—that is, on the values ascribed to the priors and likelihoods, values that are extraneous "givens" so far as Bayesian principles are concerned. Even if it is accepted that the particular initial conditions invoked do accurately reflect the beliefs of, say, Prout and his colleagues, *different* initial conditions might lead to radically different posteriors relative to the same evidence. Hence the reasoning of a scientist who regarded the chlorine evidence as making Prout's hypothesis nigh on incredible might equally well be "modeled" by the Bayesian. An important feature of the Prout case is surely that we think of his "shifting the blame" onto the auxiliary assumptions as reasonable (or at any rate as not unreasonable). However, many cases exist in which such defenses of a favored theory carry the clear hallmark of pseudoscience. Could the Bayesian not equally well "model" such apparently *pseudo*scientific defenses of a theory?

It can be seen straightforwardly that these possibilities are indeed actualized. The obvious and major complaint against the Bayesian is *not* that he cannot adequately model trains of reasoning, like Prout's, that seem intuitively scientifically correct, but rather that he can, by invoking suitable assumptions about priors and likelihoods, just as well model pieces of reasoning that seem intuitively scientifically incorrect or even pseudoscientific.

To see just how flexible the Bayesian framework is, let us consider a further case study to match Howson and Urbach's case of Prout.

Immanuel Velikovsky held that a gigantic comet once broke away from Jupiter and orbited the earth on a series of occasions, causing a series of notable events such as the parting of the Red Sea and the apparent "standing still" of the sun and attendant fall of the walls of Jericho. Velikovsky and his followers continued to believe in that basic theory despite the observation that cultures aside from the Judaic one were keeping records at the time but recorded no cataclysms. Velikovsky explained that there is no *direct* clash between these observations of no appropriate records and his basic theory. In order to generate such a clash, at least one further assumption is needed: that scribes in the different cultures would have recorded any events on this scale. Velikovsky believed that the observations severely undermined that auxiliary and suggested that, on the contrary, the cataclysms had occurred but had induced collective amnesia in just those cultures concerning just those events. Thus these observations, on Velikovsky's view, hardly affected the credibility of his basic theory.

Formally speaking, the belief dynamics of Prout and of Velikovsky were isomorphic: Each continued to believe strongly in his "central" theory despite an inconsistency between observation and the latest theoretical system based on that theory and each blamed the inconsistency on an auxiliary assumption that was later duly amended to restore consistency with observation. Despite this formal similarity, there is a clear and strong intuitive asymmetry: Prout was surely scientifically justified and Velikovsky was not. Bayesianism fails to capture this asymmetry, for it is possible that Velikovsky acted just as much in accord with Bayesian principles as did Prout.

Velikovsky's beliefs can be easily modeled in Bayesian terms. We need only suppose, for example, that his beliefs about his core theory and the auxiliary ahead of the evidence are appropriately expressed by

(a′) $p(T) = 0.9$,
(b′) $p(A) = 0.6$,
(c′) $p(T/A) = p(T)$,

and that his beliefs about the likelihood of the evidence of no records on various suppositions about the truth and falsity of T and A are appropriately expressed by

(d′) $p(e/\neg T \& A) = p(e/\neg T \& \neg A) = 1/2\, p(e/T \& \neg A)$.

Since these are the same assumptions that Howson and Urbach make in the case of Prout, obviously the same consequences about the posteriors follow, namely that

$$p(T/e) = 0.878 \text{ while } p(A/e) = 0.073.$$ [12]

The Bayesian does not, of course, allow his agent to make post hoc adjustments to her priors in order to supply a rationale for continued adherence to a favored theory. Instead, the numbers must be plausible as approximations or idealizations of the agent's real beliefs. A historical investigation would doubtless reveal that the real story here was more complicated than my reconstruction of it; and such an investigation might reveal that the real Velikovsky, for example, is most plausibly represented, not as a consistent Bayesian agent, but as making a "conditionalization error," or perhaps a whole series of them. But the criticism here is, I believe, independent of the historical facts. Suppose that the above assumptions about priors *were* accurate reflections of Velikovsky's beliefs (or perhaps of his little known but avid supporter Veriweirdsky), then these agents' beliefs that the lack of records had negligible adverse impact on their central cometary hypothesis would be exhibited, on the Bayesian account, as the result of reasoning every bit as scientific as Prout's. The logical possibility of a "rational" Velikovsky is, I would claim, enough to impugn the theory of scientific reasoning that allows the possibility.[13]

In fact you and I and the next reasonable person would want to object in Velikovsky's case to some of the assumptions about the values of the priors and likelihoods and in particular to assumption (d′); surely, for example, the nonexistence of the records of cataclysms is, contrary to this assumption, *really* at least as likely on the supposition that there were no cataclysms but accurate scribes as it is on the assumption of cataclysms but inaccurate scribes.[14] But such protests are not, indeed *cannot* be, underwritten by the personalist Bayesian. For him there is no fact of the matter about the likelihoods and priors, but only each person's opinion.

Similar considerations apply to the sort of case that most concerns me in this essay, namely, the sort in which some generally accepted "central" theory comes to be unambiguously rejected as false. Intuitively speaking, such a rejection seems invariably to have as its rationale a long series of negative results which can be accommodated by that theory only at the cost of increasingly elaborate, increasingly ad hoc specific assumptions, together with the eventual articulation and predictive success of a rival "central" theory that contradicts it. For example, Fresnel *argued* that the material emission theory of light had become ("to all intents and purposes") incredible by the 1830s by showing *first* that a whole series of phenomena—especially interference and diffraction effects—could be fitted into the emissionist framework only by making increasingly baroque and implausible assumptions about the light particles, the forces acting on them, and the circumstances in which those phenomena were produced; and *second,* that a series of phenomena—especially circular polarization and conical refraction—were correctly predicted *in advance* by the rival theory that light consists of waves in a medium. Redhead (1980) has shown how to "model the life and death of a research programme" (p. 341) in Bayesian terms. It would be a straightforward exercise to develop such a model for any particular historical process by which belief in some initially accepted central theory was gradually eroded and hence to develop a model of the sort of argument given by Fresnel.

Once again the problem is not a paucity of "explanatory" resources, but a surfeit. The Bayesian could just as well "model" what was surely an entirely rash and premature rejection of a central theory.

Suppose, for example, that the usual story abut Flamsteed and Newton were correct.[15] Flamsteed, the Astronomer Royal of the time, and famous for the accuracy of his "observations" of planetary positions, was asked by Newton to test certain predictions about those positions based on the latter's theory of universal gravitation. Flamsteed's so-called data was in fact *calculated* from "raw data" about telescopic inclinations, clock-readings and the like using various auxiliary and instrumental assumptions. However, these assumptions were accredited by his and his predecessors' well-entrenched practices. Flamsteed accordingly gave these assumptions high credibility. Finding his results inconsistent with the Newtonian predictions, Flamsteed inferred that Newton's theory of universal gravitation was very likely to be false.

Assuming that Flamsteed did indeed reason in this way, then he made, I would claim, a scientific error—one visible at the time, not simply with hindsight. Newton soon showed, of course, that the inconsistency between his theory, the auxiliaries underlying Flamsteed's "calculations" of planetary positions and the raw data was best resolved by amending the particular auxiliary assumption about the amount of refraction light undergoes in entering the earth's atmosphere. However, independently of that analysis, it is surely clear that the belief that the blame *must* (with very high probability) lie with the "core" theory in this case was entirely premature and certainly not intuitively good scientific reasoning.

Despite the fact that Flamsteed, on this account, made what is intuitively a scientific error, he can readily be "modeled" as a Bayesian agent. Suppose we take the same numbers as in the Prout case and simply switch T and A, then the arithmetic plus the Bayesian principle of conditionalization obviously dictates a posterior degree of belief in Newton's theory of 0.073 and a scarcely reduced posterior for the auxiliaries underlying Flamsteed's own practice of 0.878. Thus Flamsteed's belief that Newton's theory was to all intents and purposes refuted by his observations would be exhibited as based on reasoning just as scientific as that by which Fresnel came, on the basis of a whole series of negative experiments, to the belief that light could not consist of material particles.

Whether or not these initial assumptions about the priors and likelihoods are accurate *descriptions* of Flamsteed's beliefs is again irrelevant; they are consistent with the Bayesian approach and, were they to have been accurate, they would, according to the Bayesian, have provided the rationale for beliefs that are intuitively scientifically erroneous. As in the Velikovsky case, most of us would want to protest, not just that we do not share those priors, but that anyone who did have such prior beliefs was *not* fully scientifically *rational.* In particular it was not rational of anyone to have such a strong prior belief in the assumptions underlying current astronomical calculations of planetary positions. But again no such judgement is available to a personalist Bayesian for whom *any* prior distribution—provided that it is a *probability* distribution—is as reasonable as any other.

Redhead (1980) has claimed that Dorling's Bayesian analysis shows how a Bayesian can "*justify* retaining T and abandoning [A]" (p. 341; emphasis added), while Howson and Urbach (1989) claim that they have shown that "Bayes's Theorem provides a model to *account* for the kind of scientific reasoning that gave rise to the Duhem problem" (p. 101; emphasis added) and that "the Bayesian model is essentially *correct*" (ibid.; emphasis added). *Assuming that a "correct"* (*or even "essentially correct"*) *solution of the Duhem problem requires some means of discriminating those decisions* (*to reject a core or auxiliary theory*) *that seem intuitively scientific from those that do not,* the above demonstration that the Bayesian cannot in fact discriminate Prout from Velikovsky nor Fresnel from Flamsteed shows that the Bayesian has not correctly solved the Duhem problem.

Bayesian Reactions to This Charge of Inadequacy

The argument that personalist Bayesianism is *too* personalist, that, because of overdependence on initial conditions, it explains nothing is, of course, familiar from the literature. It would, for example, be difficult to improve on the formulations of this criticism given by Clark Glymour (1980, 60), "Bayesian accounts of methodological truisms and of particular historical cases are of one of two kinds: either they depend on general principles restricting prior probabilities or they don't. My claim is that many of the principles proposed by the first kind of Bayesian are either implausible or incoherent, and that, for want of such principles, the explanations the second kind of Bayesians provide for particular historical cases and for truisms of methods are chimeras" (p. 68), and later, "It seems to me that we never succeed in explaining a widely shared judgment about the relevance or irrelevance of some piece of evidence merely by asserting that degrees of belief happened to be so distributed as to generate those judgments according to the Bayesian scheme" (p. 85).

I have argued only that this overdependence on priors infects, and to my mind invalidates, the Bayesian attempt to solve the Duhem problem in particular. Since the general objection is old, it is not surprising that there are longstanding Bayesian responses. These fall into two categories.

First Response: The Priors Do Not Matter That Much. It would make no sense to claim that "everything" in the Bayesian analysis depends on initial conditions. Redhead (1980) shows, for example, that so long as one assumes that the likelihoods of evidence *e*, relative to $T\&\neg A$, $\neg T\&\neg A$ and $\neg T\&A$ are the same then, in the case of a refutation of the conjunction $T\&A$, the asymmetric effect on T and A individually of the evidence (measured by his "asymmetry ratio") is independent of the absolute value of that likelihood.[16] Obviously, no conflict exists between the demonstration that the Bayesian model is "robust" in *certain* respects and the claim that it is nonetheless *over*dependent on initial conditions. This latter claim seems to be established by cases like those of Velikovsky and Flamsteed.

The results most often appealed to by Bayesians in response to the charges of "oversubjectivism" are various results about the "swamping" or "washing out" of priors. This sort of response advertizes the virtues of the subjective element in the personalist view, pointing out that many situations arise even in science in which it would be counterproductive if all reasoners had the same beliefs about the theories available to them (see Kuhn 1978, chap. 13, and Salmon 1990; see Worrall's 1990 for a discussion of Kuhn's [quasi-Bayesian] views), but then goes on to claim that this subjective element does not, in the long run, matter very much, because in many situations the subjective element is overwhelmed. In certain situations and subject only to "weak" constraints, all Bayesian agents will agree in the long run, whatever their initial priors.[17]

A general solution of the Duhem problem, however, requires an account of the impact of "*single*" pieces of evidence on individual components of a theoretical system. A satisfactory solution, as I see it, must differentiate those cases where a single piece of negative evidence *should* be taken as telling against the "central" theory of the refuted system from those cases where it need not.

Certainly the process by which a once established "central" theory in some field of science comes to be rejected is generally, as Kuhn and others have made us aware, a long drawn out attritional affair. Single pieces of evidence do not knock out "central" theories, but they may unambiguously tell against such theories in a more limited sense. Dispersion always counted against the basic idea of the wave theory of light, even though in the early nineteenth century its effect was greatly outweighed by all sorts of observational evidence which told in favor of that theory, and even though it was always possible that some development in the wave optics program might turn dispersion from a negative to a positive result (basically by producing a theory that dealt with dispersion and enjoyed independent predictive success). As things actually stood, however, dispersion was a phenomenon which, taken by itself, clearly counted against the wave theory. Similarly, while the lack of records of any suitable cataclysms in contemporary record-keeping cultures does not on its own justify the rejection of Velikovsky's cometary hypothesis, that lack does unambiguously count against that hypothesis (at any rate pending a Velikovskian theory which, unlike collective amnesia, explains the lack of records in an independently testable and independently successful way).

Long run "swamping" effects obviously cannot help the Bayesian deliver these judgements about single pieces of evidence, which indeed we already saw the Bayesian cannot deliver at all.

Despite the high formal interest of some of the results, "swamping" is, I believe, of limited general philosophical significance. Priors are only ever fully "swamped" or "washed out" in the limit, which of course we never achieve. In any real evidential situation—say the one in which we presently find ourselves concerning Darwininism versus creationism—*any* theoretic preference (even the intuitively most bizarre) can be "explained" in Bayesian terms; provided he began with a sufficiently high prior for the creationist theory and correspondingly low prior for Darwinism, a creationist can have conditionalized away on the accumulating evidence and *still* have arrived, as of right now, at an overwhelmingly high posterior for his creationism.[18] The fact that, although still overwhelmingly high, this posterior probability is somewhat lower than the prior probability he had earlier assigned his favored hypothesis is of little consolation. Nor can I see much consolation in the fact that (provided he does not call for a "new round" of bets and simply rejig his priors—to be discussed in the next section), his posterior degree of belief is destined in the long run to converge on that of his erstwhile opponent. Not only are we all dead in the long run (perhaps bored to death by clichés), but surely scientific rationality is dead if only achievable in the long run. The creationist, psi-buff and the rest surely fail to be scientifically justified *now.* Any account of scientific reasoning which fails to deliver this judgement—as the Bayesian account patently does fail—is surely too weak.

Second Response: Anything More Is Not Logic and Is Therefore "Arbitrary." The second type of Bayesian response to the charge of overdependence on priors is basically that the personalist Bayesian position is all that can be justified by logic alone and therefore all that can be justified. Bayesians from Ramsey onwards have defended their position as a natural extension of deductive logic;[19] the requirement of (probabilistic) coherence is a natural extension of that of deductive consistency. (Indeed Ramsey used the term "consistency" for the probabilistic notion, too.) It is just as illogical to regard a system of odds as fair if it does not conform to the probability calculus as it is to hold both a proposition and its negation. The principle of conditionalization is also supposed (at any rate by *some* Bayesians) to be logically or analytically justifiable. (More of this shortly.)

The imposition of *any further* constraints, and in particular any constraints on prior probabilities, would, according to this response, definitely transcend pure logic. Thus Ramsey suggested (and later Bayesians such as Dorling and Howson have explicitly asserted) that to go beyond personalist Bayesianism, to require anything *more* of an agent than that her degrees of belief be deductively *and* probabilistically consistent and that she conditionalize on accumulating evidence is at the same time to go beyond logic. Any such requirement would therefore in some way presuppose some synthetic claim about the world and our abilities to comprehend it. This raises the awkward question of how such a claim itself could have any reasoned credentials. How could a synthetic claim that was allegedly constitutive of reason itself be given a reasoned justification without getting involved in some vicious circle? Personalists, like Howson, conclude that any such further requirement on an agent's degrees of belief would ultimately be "arbitrary."[20]

The import of this argument is that inductive *logic must* treat any agent's prior distribution as simply given, as an "initial condition." To expect inductive logic to do more than this—to expect it to dictate (or even to constrain) priors—would be like expecting *deductive* logic to dictate (substantive) premises. Deductive logic in fact dictates "only" what else an agent must believe, *given* that she starts from certain premises—the only constraint on the premises, so far as deductive logic alone goes, being that they are mutually consistent; inductive logic (that is, personalist Bayesianism) dictates "only" what an agent's posterior degrees of belief must be, *given* that she starts from a certain set of prior degrees of belief, the only "logical" constraint on these being that they be probabilistically consistent.

Are the requirements imposed by personalist Bayesians "logical"? The weak point here, so far as the explicit position goes,[21] is clearly the principle of conditionalization. Even assuming that the Dutch Book (or some other) argument establishes that an agent's degrees of belief *at any given time* must, on pain of illogicality, conform to the probability calculus, there is still the question of the status of the *independent* "principle of conditionalization," governing *changes* in degrees of belief. Ahead of coming to accept some evidential statement e as known, an agent's degrees of belief are distributed in a certain way; *after* coming to know e his degrees of belief are distributed differently. Assume that the Bayesian is correct that logic dictates that both distributions be probability distributions. Does logic also dictate that $p_{new}(h) = p_{old}(h/e)$? Or is the requirement that this equation hold an extralogical constraint?

A substantial body of opinion, following Hacking and others, holds that the principle of conditionalization is a substantive, synthetic assumption. Howson and Urbach claim, however, that this body of opinion is based on a simple error:

> To see that this conclusion is incorrect . . . is very straightforward. If, as will be assumed, the background information relative to which $p_{new}(h)$ is defined differs from that to which $p_{old}(h/e)$, $p_{old}(h)$, $p_{old}(e)$, etc., are relativised only by the addition of e, the principle [of conditionalisation] follows immediately; for $p_{old}(h/e)$ is, as far as you are concerned, just what the fair betting-quotient would be on h were e to be accepted as true. Hence from the

knowledge that e is true you should infer (and it is an inference endorsed by the standard analyses of subjunctive conditionals) that the fair betting-quotient on h is equal to $p_{old}(h/e)$. But the fair betting-quotient on h after e is known is by definition $p_{new}(h)$.[22] (1989, 68)

Howson and Urbach's view is, in other words, that "$p_{old}(h/e) = r$" simply *means* that the agent would regard r as the fair betting quotient on h in the situation identical to the present one except that she has in the meanwhile come to know e. Thus if indeed all that distinguishes the new from the old situation is that e is now part of "background knowledge," then the agent would simply contradict herself if she did not set $p_{new}(h) = r$.

This seems to me unconvincing for several (no doubt interrelated) reasons. First, the argument clearly extends the range of "logic" to cover Quine's broader class of analytic statements, and there are well-known difficulties in successfully distinguishing these from synthetic statements. Second, nothing is easier than to defend *any* theory of correct scientific reasoning as "logical" if one is allowed to invoke "meanings" (i.e., "definitions") in this way. Suppose, for example, I wished to defend my own favored account of scientific confirmation as "logical." I could simply claim that it is part of the *meaning* of the phrase "h supports e" that e is an empirically correct genuine prediction of h's—that it follows from h without having been built into it. Others might respond either to this latter claim or to the Bayesian one that this is not part of the meaning for them. How could we proceed?

Third (and relatedly), no doubt even the best intentioned, knowledgeable and formally acute reasoner might *as a matter of fact* fail to conditionalize. She might, having thought deeply about the matter, assign $p_{old}(h/e) = r$ and even agree that this is to mean that, should she come to know e and nothing else changes, then her degree of belief in h will be r. Yet despite all this when it comes to it and e is discovered to hold (and nothing else of epistemic relevance *seems* to occur), it turns out that her degree of belief in h ($p_{new}(h)$) is r', where r' is different from r. Howson and Urbach and others who defend conditionalization as a logical principle would presumably have to argue that such an agent had simply been *mistaken* in thinking that for her $p_{old}(h/e) = r$: Assuming that her new probabilities are accurate reflections of her real beliefs, then unbeknown to her at the time, her earlier real belief was that $p_{old}(h/e) = r'$. But how could we know that this assumption about the *new* beliefs is itself correct? This defense of the principle of conditionalization as "logical" forces the Bayesian into a highly fallibilist view of his ascription of probabilities to agents: An agent's *real* degrees of belief may be very different from what she believes them to be!

Fourth, the appeal to "standard analyses of subjunctive conditionals" is surely inadmissible. The standard criticism of these standard analyses is precisely that, when subjected to critical scrutiny, they all rely on synthetic (metaphysical) claims rather than purely logical ones. They may, of course, be arguably none the worse for that; but, given this reliance, they may not legitimately be appealed to as part of an argument for the genuinely logical nature of the principle of conditionalization.

Since Howson and Urbach's argument seems unconvincing, and since no other argument that I am aware of does better, the claim that the principle of conditionalization is part of logic can scarcely be said to have been uncontroversially established. But even *were* conditionalization a logical requirement, would it not simply follow that "logic" (even in this extended Bayesian sense) is not strong enough to ground an adequate theory of proper scientific reasoning? The great range of intuitively unscientific reasoning that is sanctioned by a purely personalist Bayesianism certainly seems strong support for this conclusion. Given that a creationist with a present degree of belief in his creationist hypothesis as high as you like could have legitimately arrived at that degree of belief, assuming only that he began at some still more enormously high figure (a starting point with which the personalist can have no quarrel), and given that a Velikovskian can legitimately barely shift his degree of belief in his central cometary hypothesis in view of the lack of cataclysmic records, provided only that he began with the right (or rather wrong) priors and likelihoods, then, assuming that the principles of personalist Bayesianism are all that is sanctioned by logic, we need a theory of scientific rationality that goes beyond logic.

This is, in effect, the option recommended by Adolf Grünbaum. Grünbaum's analysis of the Duhem problem uses probabilities and employs Bayes's theorem, but, as I show in the next section, it implicitly advocates judgements of an objectivist kind that go beyond personalist Bayesian "logic." My account of his position will therefore provide a specific focus for the Bayesian's complaint that any theory of scientific reasoning stronger than his own must, by transcending "logic," invoke assumptions that are ultimately "arbitrary."

Grünbaum's Objectivism

Ascertaining That a "Component Hypothesis" Is False

Adolf Grünbaum wrote a series of articles on the Duhem problem in the 1960s, culminating in his "Can We Ascertain the Falsity of a Scientific Hypothesis?" of 1969 (and then 1971, which was reprinted with an addendum in 1973). Although there is much else of interest in these articles, I shall concentrate on their central methodological claims.

Grünbaum (1971) differentiates two Duhemian theses:

D_1: "No constituent hypothesis H of a wider theory can *ever* be sufficiently isolated from some set or other of auxiliary assumptions so as to be separately falsifiable observationally" (p. 87).

D_2: For any piece of evidence e which is inconsistent with a theoretical system T&A, there always exists at least one set of revised "auxiliaries" A′ such that T&A′ can be held to be true and *to explain* e.

He correctly insists that "[i]n its non-trivial form D_2 has not been demonstrated" (ibid., 88). While admitting that he has no general criterion for what is to count as nontrivial, Grünbaum is basically asserting, as his discussion makes clear, that the mere deductive possibility of retaining *T* in the face of "negative" evidence is not in itself interesting. The only interesting question so far as science is concerned is whether *T* can be retained in a *properly scientific* way (which excludes, in particular, resort to purely semantic trickery and to such logical tricks as using $T \rightarrow e$ as the revised auxiliary A'). Grünbaum takes it that the onus is on the Duhemian to produce an argument for D_2 in his nontrivial form, that is, that the onus is on the Duhemian to show, not just that there is always an A' such that $T\&A'$ entails any given *e*, but that there always exists such an A' which is *nontrivial* (I would prefer "scientifically acceptable")—an A' which, when conjoined with *T*, would *explain e* (where clearly Grünbaum assumes here that explanation requires more than mere deducibility).

The situation is, I believe, better described as follows. Duhem himself would not, I believe, have defended D_2 in Grünbaum's "nontrivial" form. Duhem's point was that the constraints imposed just by deductive logic alone and the acceptance of (low-level) observation statements left the freedom consistently to defend any "isolated" scientific theory *T* "come what may" from experimental results. This point is indeed trivial (and propounded as such). He emphasized that this freedom is not *in fact* exploited in science; instead, "good sense" tells the scientist when to hold onto *T* and when to reject it in favor of an alternative. The *challenge* is then first to articulate the principles that underlie the good sense (Duhem seems himself to have believed that there would always be something "ineffable" about "good sense" that eluded full articulation) and then to explain why those principles should be accepted as governing *correct* scientific practice. (The same point underlies Laudan's 1965 criticism of Grünbaum for attributing D_2 to Duhem.)

Looked at in this way, Grünbaum's treatment of the basic logical point as simply trivial reflects a claim that I endorse, namely, that this Duhemian challenge is to be met positively. There *are* further principles of rationality to be articulated and defended which will dictate at least when a component scientific theory can no longer be reasonably defended, that will, in other words delineate those circumstances in which we can "justify the strong presumption" that that theory is false, despite its deductive consistency with all observation reports.

Indeed Grünbaum's counterargument to D_1 consists precisely of the delineation of one such set of circumstances—circumstances in which, according to him, we *can* falsify a "component hypothesis," at least "to all intents and purposes of the scientific enterprise." His argument is constructive; he argues that it is *sometimes* possible to refute a component hypothesis by producing a case in which it would be possible. He takes as the component hypothesis at issue the following claim about the metric geometry of a given spatial surface *S:*

> *H:* "if lengths $ds = \sqrt{g_{ik}dx^i dx^k}$ are assigned to space intervals by means of *rigid* unit rods, then Euclidean geometry is the metric geometry which prevails physically on the given surface S" (Grünbaum 1971, 111).

He supposes that we have taken a set of rods of (so it seems) greatly varying chemical compositions and checked the rods for rigidity in *S* by ascertaining, first, that they are initially congruent and, second, that they maintain that congruence, whenever checked, through transport over *S*. Next, he supposes that measurements made using some of those apparently rigid rods have produced "negative" results for *H*. Suppose it has been consistently found, for example, that the ratio of circumference to diameter of circles in *S* is other than π and indeed that this ratio varies with the size of the circle. The Duhem problem arises here. *H* is not directly inconsistent with these findings but only when conjoined with the further assumption:

A: Region *S* is free from perturbing influences.

This is because, if there were perturbing influences in *S*, then apparently non-Euclidean results would be compatible with the real geometry of *S* being Euclidean.

Grünbaum argues that in these supposed circumstances we could in fact regard *A* as verified. We are imagining, remember, observational results that fall into two main parts: (a) all the initially congruent rods of apparently diverse chemical compositions remained congruent through all observed transports in *S*, and (b) measurements were made using some of those rods which were inconsistent with the joint assumption that the rods are *truly* rigid *and* that the geometry of *S* is Euclidean. Suppose first that we accept the "natural" (double) inductive generalization of part (a) of these findings; that is, we accept

C: "whatever their chemical constitution any and all rods invariably preserve their initial coincidences under transport in S" (ibid., 114).

(The question of whether *H* can legitimately be preserved by accepting the observational findings but *not* generalizing to *C* is also considered by Grünbaum, as we will see shortly.)

On Grünbaum's account, the following assumption *R* is to be taken as part of unchallenged background knowledge, common to all competing theories about the metric geometry prevailing in *S:*

R: "[T]here is a concordance in the coincidence behavior of solid rods such that no inconsistency would result from the subsequent interchangeable use of initially coinciding unit rods which remain unperturbed or rigid" (ibid., 111).

(Where a rod is defined to be rigid, so far as its use in any region *S* is concerned, just in case it is not affected in *S* by "independently designatable perturbing influences" [ibid.].)[23] Assumption R thus, in effect, asserts that for any region S, if statement A (no perturbing forces in S) holds, then so will statement C (initial coincidences of rods preserved). As Grünbaum points out, C clearly does not *deductively* yield A, even given R. Nonetheless, Grünbaum holds that C does in effect establish A (given R as a generally accepted background assumption). This is because the probability of A, given R and C, is very close to 1.

Grünbaum argues for this latter claim as follows. First, by Bayes's theorem:

$$p(A/R\&C) = \frac{p(A/R)}{p(C/R)} \cdot p(C/R\&A)$$

and so, since $p(C/R\&A) = 1$,

$$p(A/R\&C) = \frac{p(A/R)}{p(C/R)}.$$

He then argues that the right-hand side of this equation must be close to 1. The crucial step in this argument is for the claim that $p(C/R\&\neg A) \approx 0$:

> For suppose that A is false, i.e., that S *is* subject to differential forces. Then C can hold only in the *very rare* event that there *happens* to be just the right superposition. . . . Hence we are confronted here with a situation in which the Duhemian wants C to hold under conditions in which superimposed differential forces are actually operative, although there is no independent evidence for them. (Ibid., 119)

Given that $p(C/R\&\neg A) \approx 0$, it follows that $p(A/R)$ and $p(C/R)$ are almost identical. The easiest way to see this is by using the theorem on total probability:

$$p(C/R) = p(C\&A/R) + p(C\&\neg A/R).$$

But, since A entails C given that R holds, the first summand is $p(A/R)$ and the second summand is close to zero if $p(C/R\&\neg A)$ is—because

$p(C\&\neg A/R) = p(C/\neg A\&R)p(\neg A/R)$. Hence
$p(C/R) \approx p(A/R)$.

Finally, as required,

$$P(A/C\&R) = P(C/R)/P(A/R) \approx 1.$$

Since Grünbaum uses probabilities and makes specific use of Bayes's theorem, he might seem to be advocating a Bayesian view. But it is surely not a personalist view. Although he gives no explicit account of the status of his ascriptions of probabilities, Grünbaum's objectivist intentions seem patent. The probability that there are in fact differential forces in *S* but it "just happens" that they have cancelled one another out in *all* the repeated observations on different rods of apparently different chemical constitutions is not simply one that Grünbaum regards as low, though others may give it a nonnegligible value; that probability just *is* low. This superposition conjecture, Grünbaum later says, is one of those that "may try a working scientist's patience with philosophers" (ibid., 122). Without explicitly saying so, he clearly holds that this is more than simply a quirk that working scientists happen to evince; they are *right: This is the sort of conjecture which, although logically possible, is not rationally believable.* Or, more precisely, it is not rationally believable *in the absence of independent evidence.*[24]

So far, we assumed that anyone who accepted the observational findings in this case would be ready to accept the general claim C. Is there any way in which someone could accept the findings about the rods but *not C?* Grünbaum takes it that this would entail the "postulate that the great differences in chemical appearance [of the rods] belie a true crypto-identity of chemical constitution" (ibid., 121). If the apparent chemical differences between the rods were *merely* apparent and hid a real "crypto-identity," then the Duhemian "could hope to render the observed concordance among the rods unproblematic for $\neg$A *without* the superposition conjecture" (ibid.). This is because it might[25] then be consistent to assume that only one differential force was operative in *S* and that that force affected all the rods in the same way.

This conjecture deservedly gets equally short shrift from Grünbaum, "But there is no reason to think that the crypto-identity conjecture is any more probable than the incredible superposition conjecture" (ibid.). Like the latter, the former conjecture "may try a working scientist's patience with philosophers" (ibid.). Again the objectivist tenor is clear: The claim is not just that you and I who happen to be scientifically inclined, as a matter of fact, give no weight to the crypto-identity conjecture but that no one *ought* to give it any rational weight; not that you and I happen to see no reason to think this conjecture anything other than incredibly unlikely, but rather that *there is* no such reason. (This again relies on the assumption that there is no *independent* evidence for "crypto-identity.")

Grünbaum's analysis, then, differs from any sort of personalist Bayesian analysis in that the judgement that, in this particular case, $p(C/R\&\neg A)\approx 1$ is taken to be the *right* judgement, to be, in effect, rationally obligatory.

I should add that Grünbaum has elsewhere expressed explicit criticisms of, and doubts about, any sort of Bayesian approach (for example, see Grünbaum 1976). These include doubts about the whole program of using numerical-valued probabilities in confirmation theory (although he at the same time insists that any theory of rationality must include a way of *ordering* the credibilities of different theories in the light of evidence).[26] I too am unsure about the commitment to formal probabilities in this area, and sure only that *if* we are to use probabilities, they cannot be entirely subjective. Stripped of any attachment to formal probabilities, the main point that I would want to draw from Grünbaum's analysis is simply this: It is an adequacy requirement on any theory of rationality that it deliver such judgements as that *the superposition conjecture is, in the absence of independent evidence, and as an objective matter, sufficiently implausible as to be rationally incredible.*

Can Grünbaum's Analysis Be Extended to Deal with Other Aspects of the Duhem Problem?

As I already indicated, Grünbaum is careful to acknowledge that his analysis applies only to certain special cases of the Duhem problem—ones in which the necessary auxiliaries can arguably be verified. It is certainly no criticism of that analysis, therefore, if there are other cases where belief in the falsity of a particular assumption from within a theoretical system is again "rationally mandatory" but where that conclusion cannot be reached via Grünbaum's route. Indeed there are such cases, including that of Duhem's own favorite (and guiding) example.

Whatever it was rational to believe about the emission theory of light in the period when its truth or falsity was in a live issue in science (see Worrall 1990), given the evidence we have *now,* the only rational belief about its truth value is that it is false. This seems so obvious it is almost embarrassing to say: Of course, no one can deny the fallibility *in principle* of any ascription of either truth or falsity to any scientific theory, but that should not—just as Grünbaum warns—tempt us into a sort of universal agnosticism. We now *know*—to "all scientific intents and purposes"—that the emission theory is false.

The historical process by which belief in the falsity of the emission theory became "rationally mandatory" involved a whole *series* of experimental results. Each refuted a theoretical system based on the "core" emission theory, but each could be accommodated by switching to a new theoretical system based on the same "core." This new theoretical system in turn was refuted by a further experimental result, leading to the development of a third emissionist theoretical system, and so the pattern was repeated.[27] In intuitive terms, these theoretical systems became ever more elaborate and "disunified"; the specific assumptions which "flesh out" the basic particulate idea bore more and more signs of having been "cooked up" precisely to yield the results which refuted their predecessors. At the same time, *some* of these experimental results had been *predicted* in advance by theoretical systems based on the different core idea that light consists of waves in a mechanical medium. Historically the whole emissionist approach was then abandoned, while the wave program enjoyed a series of further predictive successes.

Suppose that I am correct to claim that this process made belief in the falsity of the emission theory "rationally mandatory." One reason why a Grünbaum-style account will not yield this result directly is that no single experimental result (or rather no single type of experimental result) was involved. But this is a problem of detail; the essential point in Grünbaum's analysis is that the set of auxiliaries A should have a probability of close to 1 in the light of the evidence, and there is no reason why this should always be as the result of anything naturally described as a "single type" of experiment. The main problem is that the process which led to the definitive rejection of the emission theory was not one in which the "auxiliaries"—that is, the assumptions making up the "rest of the theoretical system" aside from the "core" emission claim—were ("to all intents and purposes") observationally verified.

Duhem was certainly right that if the full deductive structure of any of the tests of the emission theory is articulated, then further assumptions aside from the basic claim that light consists of material particles are always involved. These further assumptions fall into at least two categories. The first consists of assumptions—perhaps the only ones naturally termed "auxiliaries" here—about the particular experimental set-up and about the operation of the instruments involved—including, in the case, say, of the Foucault result, assumptions underpinning the method of measuring the speed of light in different media. In the second category are the more specific assumptions about the nature of light that "flesh out" the basic idea that light consists of material particles of some sort—these assumptions might, for example, specify the mechanical properties of the particles themselves, and the forms of the forces acting on those particles in different circumstances.

Grünbaum's analysis can, I think, be applied here up to a point: The assumptions in the first category—the genuine "auxiliaries"—can be made (and historically *were* made) so probable as to be "to all intents and purposes verified." This would mean that a negative experimental result could justify the "strong presumption of falsity" of the emission theory as "fleshed out" by some particular set of second category assumptions—those specifying the light particles' masses, velocities, and rotations; those specifying the forces acting on the particles; and so on. But we cannot "zero in" any further on the "core" emission theory in direct Grünbaum-style. What eventually justified the inference that the core theory is false was nothing like the verification of the rest of the (second category) assumptions. Indeed, a Grünbaum-style analysis does not even get started here: The "emission theory as fleshed out by some particular set of more specific assumptions" cannot be represented (or cannot naturally be represented) as the conjunction of the core theory—light consists of some sort of material particles—and further assumptions (which means that a Dorling-Redhead-style Bayesian analysis also fails to apply to this case). Instead, each of those specific assumptions—that the light particles fall into groups having masses $m_1, \ldots, m_n$; that in passing from air into glass, the light particles are affected by force *F;* and so on—*entails* the basic core claim, without that claim appearing as a separate conjunct. Clearly it would be incoherent to claim that evidence verified such a claim and at the same time refuted one of its logical consequences. The intuitive story here is not one of the gradual corroboration of the specific assumptions exposing the central core idea to ever more directly negative evidence. Instead it is a story of *many different* (mutually inconsistent) sets of specific assumptions being tried and turning out to be unsatisfactory.

However, although Grünbaum's analysis does not apply directly, the argument for the falsity of theories like the "core" emission theory of light rests, I believe, on formally similar judgements of objective improbability to the one he invokes.

His argument, stripped of its probabilistic clothing, was essentially that what would have to be maintained in order to deny the auxiliary *A* in his case and therefore maintain the basic physical geometric hypothesis *H* was, given the evidence, so (objectively) implausible that it could rationally be dismissed. The defender of *H* would have to hold one of the following claims. *Either* the differential forces "just happened" to superpose so as to have no effect on the coincidence behavior of the rods in any of the multiple observations actually made with different rods in different places, *or,* despite all the apparent differences between the rods, they "just happened" to be of sufficiently similar underlying chemical structures as to have reacted in exactly the same way to the differential forces.

The argument—in this case built up over time by a whole series of experimental results of different types, rather than, as in Grünbaum's case, a single type of experiment—for the falsity of the basic claim of the emission theory is, at its crudest, this. In order to maintain the emission theory in view of all the observational evidence that accumulated in the eighteenth and early nineteenth centuries one would have to hold (a) that the reason for the sequence of failures of the specific versions of that theory lies always with the specific assumptions made, and not with the component common to all the specific versions (and this despite the fact that so many different sets of specific assumptions had been tried), and (b) the rival wave theory's predictions about hitherto unrecognized phenomena had "just happened" to be correct in every case, despite the fact that this theory is radically false (as it would have to be were the emission theory true). Just as in Grünbaum's case, what one would have to hold is surely so (*objectively*) improbable as to be rationally incredible.[28]

The Status of Judgements of Objective Improbability

Of course it is one thing to assert the need to incorporate some objective element into our theory of rationality so as to yield the judgements endorsed by Grünbaum, and another to *justify* whatever objective element one favors. How could we reply to someone who *denied* the total implausibility of, say, Grünbaum's superposition conjecture or someone who insisted that, despite the evidence, we should still have an open mind on the falsity of the corpuscular theory of light or Velikovsky's theory?

The personalist Bayesian will claim that such questions are ultimately unanswerable, which is precisely why the addition of any "objective" element to his position cannot result in a legitimate, defensible theory of rationality. Clearly, the weaker the assumptions involved in our account of proper scientific reasoning, the fewer the problems that face the attempt to justify that account; but, on the other hand, the weaker those assumptions, the greater the risk of failing to capture very firm intuitions about what counts as proper scientific reasoning and what does not.

The fully committed personalist Bayesian can always reject any intuitive judgement that his account fails to capture.[29] He may himself largely agree with Prout, Fresnel and so on while disagreeing with Velikovsky, Flamsteed and Grünbaum's superpositionist—because he emphathizes with the priors ascribed by those in the first group, while not sharing the priors of those in the second; but the Bayesian may be ready to accept that these are simply facts about himself, not judgements for which he can legitimately claim any rational cogency. There really is (once we have reflected on the matter) no *reason* to categorize Prout, Fresnel and the rest as "genuinely scientific" or "rational," and to categorize Velikovsky, Flamsteed and Grünbaum's superpositionist as "irrational" or "scientifically mistaken." So long as they were not inconsistent (in the Bayesians' extended sense) then Velikovsky, Flamsteed et al. were *just as scientific* as Prout, Fresnel et al.

My earlier discussion indicates, however, just how many and just how strong the intuitive judgements are that the fully committed Bayesian is asking us to forgo. It is easy to gain the impression from the accounts of scientific reasoning available in the literature that the Bayesian delivers the best of both worlds: Allegedly making no assumptions that go beyond logic, he nonetheless captures our intuitive notion of proper scientific reasoning. But this is an illusion based on judicious choice of examples. In the particular case of the Duhem problem, the impression is easily gained from the examples analyzed by Dorling and by Howson and Urbach that their Bayesian analyses supply the "essentially correct" solution, but only because we implicitly underwrite the "priors" and "likelihoods" assigned to the scientists concerned as (to a reasonable approximation) "correct." The illusion is exposed by cases like those of Velikovsky and Flamsteed in which the priors and likelihoods themselves seem "irrational." The Bayesian can claim to supply an "essentially correct" solution of the Duhem problem only by denying the assumption that such a solution must supply some principled distinction between those "defenses" of a "core" theory that seem intuitively to be scientifically legitimate and those that do not.

Indeed still more of these intuitive judgements about rationality need to be surrendered than so far indicated. In applying their theory to real episodes, Bayesians, so far as I can see, invariably, but implicitly, make at least two further assumptions that are no part of official doctrine. First, they invariably take as "the evidence" what you and I and the next scientifically inclined person take as *really* the evidence: that the telescope's inclination was $\Theta \pm \epsilon°$, or that the stuff in the bottle weighed *g* grams. But the pure Bayesian ought, of course, to be personalist here too: *not* taking an "objective" view of the evidence, but instead taking as "evidence for the agent" any statement to which that agent comes to assign probability 1. So, should an agent come to assign probability 1 to the statement that "the needle in the meter points to 117" when you and I can see that it points to 4 (give or take a bit), then the Bayesian must solemnly record this fact and can have no quarrel with that agent—so long, of course, as his overall distribution of degrees of belief is indeed a probability distribution. Now this particular agent is not likely to be around to trouble us for very long, but while he is, it seems difficult indeed not to regard him as somehow irrational, as not properly scientific. Even this judgement is beyond the reach of the truly personalist Bayesian.

The second implicit assumption that the Bayesian seems invariably to make when applying his analysis to real cases is that the agent will "call for a new round of bets" (or for a new assignment of fair odds), that is, insist that background knowledge has changed, *only* when there is (intuitively) good reason to do so: standardly when some initially uncertain evidence is established (precisely when the principle of conditionalization comes into play) or when some relevant new theory is proposed. It is generally assumed that the agent will not simply and "capriciously" begin "to see things differently," that is, he will not simply, and in an unmotivated way, change what he sees as background knowledge and hence reassign all his probabilities. But again the judgements implicit in the terms "good reason" and "capriciously" are not really available to the Bayesian. The personalist seems committed to the view that agent *A*'s background knowledge at time *t* just is whatever it is for that agent, so that if, as a matter of fact, *A*'s conception of his background knowledge at *t′* is different, then the personalist must simply record this fact and has no justification for inquiring into any "reasons" for the shift. But this means that even more bizarre belief-dynamics are (implicitly) sanctioned by the personalist.

Suppose that a creationist, for example, assigns, against background *B*, a very high probability to his favored hypothesis and a correspondingly low one to the Darwinian hypothesis; he proceeds correctly to conditionalize on accumulating evidence and finds this initial asymmetry being eroded—the posteriors of these two hypotheses are getting ever closer; he then, however, says that all of a sudden he sees his fundamental framework in a different light and accordingly has reassigned priors relative to this new state of his background "knowledge" *B′*. It turns out that, against background *B′*, the creationist hypothesis has an extremely high prior (higher than the prior relative to his initial background knowledge) and that Darwinism has a correspondingly low (even lower) prior.

Again I should emphasize that the personalist does not condone the plucking of figures out of the air in order to defend some favored theory; the agent's probabilities at all times are supposed to reflect his *genuine* degrees of belief. And I suppose most of us would suspect that, in the situation I have envisaged, the creationist agent was *professing* new beliefs (for purposes of an extraepistemic kind) rather than really possessing them. But suppose (as before, the logical possibility is surely enough to embarrass the personalist) we strap our agent, before and after, into a polygraph or give him, before and after, an infallible truth drug, and it turns out that he really did have a change of "mind set" between *t* and *t′* and that his professed beliefs throughout were his real beliefs. The personalist Bayesian could have no reason then to regard this agent's belief-dynamics as anything other than perfectly scientifically proper, no doubt somewhat eccentric, speaking purely empirically,[30] but *evaluatively* on a par with the belief-dynamics of Fresnel or Einstein, for example. But, as Russell said of the man who believed the world is a poached egg, this creationist agent's beliefs are surely to be regarded as irrational on more than just the grounds that he is in a minority.

As before, the Bayesian could reply that I have simply shown that these intuitions about the "irrationality" of the creationist in my story and of the man who "misreads" the meter are to be rejected as invalid. The Bayesian can always defend his position against any judgements about particular cases in this way. But I suppose all but the most committed will find these last cases, if not the earlier ones, just too much to accept. If our theory of correct scientific reasoning were restricted to the principles involved in personalist Bayesianism, then the consequences we should have to swallow—among them the possibility of a properly scientific flat-earther, creationist, psi-buff and the rest—elicit the gagging-reflex much too strongly to tolerate.

It might be different if the personalist, by restricting herself in this way, really did avoid entirely the sort of difficulty she herself brings against any attempt to strengthen her theory. But, even supposing that she succeeded in showing that her principles were logical (and so supposing in particular that her suspect argument concerning the principle of conditionalization in fact worked), this would not save her from the determined skeptic, from a potential infinite regress and therefore from needing ultimately to defend her position in "dogmatic," "arbitrary" fashion. After all, it is precisely the point of Lewis Carroll's (1895) famous dialogue between Achilles and the Tortoise that even *deductive* logic (specifically, even the rule of modus ponens) cannot be justified in a noncircular fashion. Someone (or even some tortoise) who does not accept that modus ponens is truth-preserving cannot be presented with any argument that it *is* truth-preserving, which does not itself presuppose modus ponens. A similar result applies in the case of rationality theory. As, for example, Reichenbach and Popper both saw clearly, *no* theory of rationality can itself be fully defended in a rational way.[31]

Of course, this should not be taken as condoning the adoption of any old principle as constitutive of reason, but it does mean that the choice is not, as the personalist here presents it, one between an ultimately unjustified rationality theory that captures widespread intuitions about proper reasoning and proper reaction to evidence, on the one hand, and an unquestionable "logic," on the other. This is *not* the choice even if the "logic" option really is restricted to the bare principles of personalist Bayesianism. If the personalist were to make her own position more palatable by explicitly adding, for instance, the requirement that a properly scientific agent at least take as the evidence the *real*, objective evidence, then, although her position now captures more intuitive judgements about proper scientific reasoning, it is even clearer that it cannot count as unquestionable logic. For, as several centuries of philosophy have shown, there is no answer to the skeptic about observation statements except, in the end, to dig in one's heels. If it is permissible to be "dogmatic" about (sufficiently low-level) observation statements, if acceptance even of these is ultimately "arbitrary," then the principle has already been breached that no nonlogical ultimately "unjustified" assumptions can be allowed in our account of scientific rationality.

John Earman (1992) has recently defended Bayesianism in a limited sort of way. As I understand his position, it is that Bayesianism is the only well worked-out general theory of confirmation that we have available and that gives the right results in at least a goodish range of cases (for example, according to Earman, in the case of the notorious ravens). In that situation, to give blanket epistemic superiority to the sorts of intuitive judgements about particular cases that I have invoked would be a mistake. Instead we should move into a state of agnosticism concerning those judgements, pending the development of an equally formally adequate and more encompassing theory of confirmation.

Clearly it can be no bad thing to re-examine these intuitive judgements to see if they are firm. Some might, on reflection, seem less clear-cut than they did initially, but *not,* I suggest, those concerning the falsity of the theories with which I began nor those discussed in this section. Of course a formally articulated general theory is the aim, but given that logic, even when supplemented by the requirement to accept (low-level) observation statements, yields an impoverished account of scientific rationality, there seems to me no way to avoid reliance, in building that general theory, on those intuitive judgements that are firm. For one thing, what underwrites Earman's judgement that Bayesian implications about some cases (for example, the ravens) are "right"? Just as I would have refused to allow Descartes to push me into a state of even temporary agnosticism concerning judgements like "the meter needle points to 4 (more or less)," so I refuse to allow Earman to push me into a state of even temporary agnosticism concerning judgements like "it is not scientifically acceptable in view of the evidence we have now to believe that the universe was created largely as it presently is in 4004 B.C." Some intuitive judgements about proper scientific reasoning that personalist Bayesianism fails to underwrite are firm, and so we can tell *now* that there is a wider, more encompassing account of proper scientific reasoning than that supplied by the personalist Bayesian. Doubtless, consistency with the probability calculus is a necessary condition for "rationality," but it is not sufficient. Taken alone, it leaves us with minds too open for the good of our brains. Let's go with Grünbaum, not with Bayes.[32]

NOTES

An earlier version of this paper was delivered at the Special Colloquium for Adolf Grünbaum held in Pittsburgh in October 1990 and I thank the organizers Jerry Massey, Al Janis and Nick Rescher for the invitation to what was indeed a special occasion. I had several very helpful discussions after my talk and would like to thank Clark Glymour, Wes Salmon and, especially, John Earman. (John Earman kindly allowed me to read a preprinted draft of his book *Bayes or Bust?* [1992]. Wes Salmon kindly redrew my attention to his paper [1976] which raises many interesting concerns that would certainly need to be treated in a fuller account of the virtues and vices of the Bayesian approach.)

My friends Colin Howson and Peter Urbach have been as forceful as ever in their criticisms; my paper is substantially modified and, I believe, improved on account of the helpful points they made. An early draft was read and helpfully criticized by Elie Zahar.

I especially thank Adolf Grünbaum for his generous comments on my Pittsburgh talk and for his encouragement over many years.

1. Of course, nothing is special about these particular theories: They simply happened to come off the top of my head. Equally clearly while all are surely false, some are more interestingly false than others: There were times when the evidence gave us some reason to believe T_1 and subsequently T_2, the same cannot be said for either T_3 or T_4.

2. I assume here, of course, that T_i is itself consistent, and that so is the conjunction of the e_j's.

3. I believe that this *is* a requirement on rationality (see Worrall 1978, 1980), though this is presently an unfashionable thesis.

4. Grünbaum (1971) is a revised version of Grünbaum (1969), which in turn draws on (and revises) Grünbaum (1960 and 1966).

5. The idea of "modifying a theory rather than rejecting it" hints at an important possibility but makes no strict sense. If a theory is modified, it is modified into something else—standardly a theory inconsistent with its predecessor. Sometimes, however, that successor theory has enough in common with its predecessor—they share the same "core" component(s)—to count as a "modified version" of it. Thus, for example, Fresnel "modified" the wave theory of light rather than "reject it" in the light of the evidence that oppositely polarized light beams cannot be made to exhibit fringes through interference. What this means is that he rejected one theory—"the" wave theory together with the specific assumption that the waves are longitudinal—in favor of another theory inconsistent with the first, saying that the waves are transverse. Because these two theories, while mutually inconsistent, share the "core" assumption that light is *some sort of* wave in a medium, it seems natural to refer to the second as a modification of the first.

6. One example is the much analyzed case of the discovery of Neptune, which was one of the great successes for Newton's theory—see, for example, Grünbaum (1971).

7. Dorling, Howson and Urbach, and Grünbaum all make this supposition; as we will see later, though, it is not representative of all cases in which the Duhem problem arises.

8. There are of course various "objective" versions of Bayesianism (see, e.g., Salmon 1970 as well, e.g., as Abner Shimony's 1970 "tempered personalist" Bayesianism), but the personalist form is so much in the ascendancy that it is standard to use the term "Bayesian" to mean *personalist* Bayesian. I will sometimes follow this practice in what follows.

9. In the revised version (Grünbaum 1971), he replaced his claim that component hypotheses can be falsified "to all intents and purposes" with the claim that a "strong presumption of falsity" (p. 126) can be established for such hypotheses.

10. According to Redhead (1980, 341), "Dorling has demonstrated an asymmetry in the effect of the refutation on the posterior probabilities of a hard core (T) and auxiliary hypothesis (H) . . . which would justify retaining T and abandoning H."

11. By no means are all Bayesians committed to the idea that any real person (as opposed to their acknowledgely somewhat idealized "agents") has degrees of belief that are real-number valued. So, for example, Howson and Urbach state, "There is nothing in the theory we have put forward [i.e., personalist Bayesianism] which asserts that people actually do have point-valued degrees of belief. That theory is quite compatible with people's beliefs being as vague as you like: it merely states that if they were to be point-valued, and consistent, then they would formally be probabilities" (1989, 70). Howson and Urbach in particular make no claim that the probabilities they assign to Prout and his fellow chemists are straightforwardly descriptive "[T]hese numbers and those we . . . assign to other probabilities are intended chiefly to illustrate how Bayes's Theorem resolves Duhem's problem" (ibid., 98), although they immediately

add "nevertheless we believe them to be sufficiently accurate to throw light on the progress of Prout's hypothesis" (ibid.). Of course, what counts as an acceptable idealization, as opposed to an accurate description, is an awkward question.

12. Urbach and Howson (1989, 106–07) mention the Velikovsky case in another context—that of why scientists have a negative attitude toward some ad hoc explanations. Their Bayesian "explanation" of the unacceptability of Velikovsky's collective amnesia manoeuvre seems to be simply that scientists will have found Velikovsky's revised auxiliary "incredible." That is, that (most) scientists will have given it an extremely low prior. But *which* scientists? (Clearly not Velikovsky himself.) And is not the "incredibility" of that auxiliary precisely the problem rather than a solution of it? That is, is it not the job of the methodologist to *explain* the incredibility of such hypotheses, given the evidential circumstances (explain it as "rational," as scientifically justified), rather than taking it as an empirical datum about the attitudes of (most) scientists?

13. I am not, of course, claiming that this lack of records alone refuted the central Velikovskian claim—not even "to all intents and purposes." Moreover, I do not deny that the Velikovskian system with collective amnsia is, in a certain sense, better than the system without it: With collective amnesia the system is at any rate clearly consistent with the facts of no suitable records in the cultures at issue. However, I *do* claim that this result of no suitable records should have had, when considered in isolation, a distinctly negative impact on the standing of the central cometary hypothesis (where this negative impact could have been compensated for by the positive impact of *other* evidence).

14. No doubt you and I and the next reasonable person would feel unhappy about Velikovsky's (or Veriweirdsky's) priors, feeling that they did not *really* represent the probabilities of the theories concerned ahead of the relevant evidence. In fact, claims that certain priors and likelihoods are objectively correct often creep into Bayesian accounts, despite being officially banned. This is especially true of Dorling's historical reconstructions, which, largely for this reason, seem to me very insightful. For example, Dorling (1979) remarks at one point, "It is essential to the quantitative conclusions which I shall obtain that e was *a priori* unlikely in the sense that there was no *plausible* rival to T available which predicted e" (p. 182; emphasis added). Similarly, his values for $p(e/T\&\neg A)$ and $p(e/\neg T\&\neg A)$ are given justifications in which he considers, but dismisses, the suggestion that the second "*ought*" (ibid., 183) to have a value higher than the one he gives it; decides that there is in fact "no *reason*" (ibid.) for this; and concludes, "so it seems to me *correct* to take both these conditional probabilities as the same to a sufficient approximation for our purposes" (ibid.; emphasis added).

15. For the same reason as before, the historical facts are not really important: It is enough that *had* anyone reasoned in the way that I will take Flamsteed to have reasoned then, while his reasoning would have been fully in accord with Bayesian principles, it would not have been properly scientific reasoning from the intuitive point of view.

16. Nonetheless, the assumption that these likelihoods are all equal is itself an extraneous, unexplained given, and, as Redhead himself emphasizes, the asym-

metric effect *is* crucially dependent on the priors of *T* and *A*. (Redhead also demonstrates that, relative to the assumptions he makes and in particular the assumption of a considerably higher prior for *T* than for *A*, the asymmetric effect of a refutation on *T* and *A* is not diminished by a previous series of successful predictions.)

17. Edwards, Lindman and Savage, for example, write:

> If observations are precise, in a certain sense, relative to the prior distribution on which they bear, then the form and properties of the prior distribution have negligible influence on the posterior distribution. . . . [T]he untrammelled subjectivity of opinion about a parameter ceases to apply as soon as much data become available. More generally, two people with widely divergent prior opinions but reasonably open minds will be forced into arbitrarily close agreement about future observations by a sufficient amount of data. (1963, 201)

Again notice the recommendation implicit in the phrase "reasonably open mind" which is officially unavailable to the personalist.

18. The erosion of the creationist's belief in his theory, on the Bayesian account will be slowed down by a few (intuitively bizarre) likelihood judgements. Even if he accepts what we would regard as plausible likelihoods, it is still true that for any given (finite!) evidence, and any given present posterior probability for creationism, there is a probability which the creationist "agent" could have assigned to his creationist hypothesis ahead of that evidence which makes his present posterior probability Bayesian-correct.

19. See Ramsey (1931, 182), "We find, therefore, that a precise account of the nature of partial belief reveals that the laws of probability are laws of consistency, an extension to partial beliefs of formal logic, the logic of consistency. They do not depend for their meaning on any degree of belief in a proposition being uniquely determined as the rational one; they merely distinguish those sets of beliefs which obey them as consistent ones."

20. See Howson and Urbach (1989, 55), "[A]ny a priori probability distribution, be it an equiprobability distribution over the state-descriptions of some language or other, or some other distribution, is going to be arbitrary. For this reason we do not regard people who try to evaluate the probabilities of hypotheses relative to data as doing exercises in a genuine logic . . . for we take logic to be essentially noncommittal on matters of fact." The argument that anything more than personalist Bayesianism is "arbitrary" has been forcefully pressed on me on a number of occasions by Howson. Although I attribute it to "the" Bayesian, it seems by no means to be shared by *all* Bayesians.

21. As I argue in the next section, there are further assumptions to which the Bayesian seems to commit himself *implicitly* in applying his model to particular episodes—assumptions which clearly cannot be justified by logic alone.

22. I have changed the terminology to bring it into line with my own. Hacking (1967) argued against the logical status of the conditionalization principle. For a more recent argument along the same lines, see van Fraassen 1989.

23. Grünbaum argues that if *R* were in fact denied then the "physical significance" of *H* would have been changed; denying *R* in the attempt to save *H* would be self-defeating. (It would not be *H* that was being saved). "R must be

assumed by the Duhemian if his claim that S is Euclidean is to have the physical significance which is asserted by his hypothesis H" (Grünbaum 1971, 115).

24. If, for example, an explanation were proposed of why such a superposition should always occur whenever we made a measurement and if that explanation made testable predictions in some other area (clearly it could not, by assumption, be testable by direct measurements of the kind under discussion) and these predictions were surprisingly confirmed, then the situation would be radically altered. Grünbaum holds that, were the "superposition conjecture" to be understood in such a way that it *precludes* such independent evidence in principle (instead of such evidence being possible in principle but simply lacking in fact), and in particular if the conjecture were proposed without any attempt to specify the physical sources of the allegedly superposing differential forces, then that conjecture would be without any significance. Anyone who proposed the superposition conjecture in this form would, he holds, in effect be admitting that the geometry of *S* is not Euclidean (1971, 199, n. 68).

25. Grünbaum notes that *in fact*, "[I]t is not obvious that the Duhemian could make this assertion in this context with impunity: even in the presence of only one kind of differential force, there may be rods of one and the same chemical constitution which exhibit the hysteresis behavior of ferromagnetic materials in the sense that their coincidences at a given place will depend on their paths of transport . . . " (1971, 121, n. 71).

26. For example, in an (unpublished) reply to Imre Lakatos, he clarifies his account (1971) of how a component hypothesis may be falsified by explicitly *denying* that he was "predicating the strong presumption of the falsity of H on the availability of an accepted inductive logic which can assign a *numerical* measure to the strength of the corroboration of the auxiliary A and to the strength of the refutation of H." He states, "I emphatically do *not* rest my case on the availability of *numerical* degrees of confirmation or refutation" (Grünbaum n.d., 12). (As we saw, this would not on its own separate him from all Bayesians.) On the other hand, he continues:

> I would not know how we could *ever* distinguish a sane belief from an utterly insane one unless we made the assumption that at least for some hypotheses, corroboration is susceptible of degrees in the qualitative or ordinal sense of more or less. For example, I do fervently hope that Lakatos will agree that the hypothesis "Napoleon Bonaparte is dead" is immensely better confirmed in this ordinal sense than the contrary hypothesis that unbeknownst to everyone else, *I* am actually Napoleon and am therefore now 202 years old! (Ibid.)

27. This is a rational reconstruction of the history. Of course no one either in science or in any other field in which argument and inference are important ever in fact spells out their arguments in full detail. Moreover, there is the complication in science that the defenders of a theory may be more or less committed to a full-blooded realist version of it. Following Newton's lead, most eighteenth- and early nineteenth-century optical scientists regarded themselves as committed only to "parts" of light rather than particles—but these parts were regarded as acting *very much* like particles (e.g., they consisted of discrete chunks and moved along rectilinear paths whenever left to themselves). These scientists would also

have regarded themselves as searching for the correct laws of interaction between their "parts" of light and "gross matter," rather than making definite assertions about these which then were refuted and replaced by something more sophisticated. Despite these and other complicating historical facts, I regard the account in the text as the "right" rational reconstruction—but this claim requires, I admit, a detailed defense.

28. Indeed the argument for the falsity of a "central" theory (or part of that argument) is sometimes even closer in form to the one pointed to by Grünbaum. An example is the argument which built up against the idea, central to classical physics, of a fixed ether frame. Again in crude form the argument was this. It turned out that, in a whole series of instances, no effects of the earth's motion through the ether were observed in circumstances in which they might have been expected. In each case a classical explanation could be given of how some compensating mechanism had just happened to compensate for the effects of our motion through the ether. But as these instances accumulate, it becomes more and more improbable that the earth should indeed be in motion through the ether and so many compensating factors just happen to conspire so as precisely to mask the effects that that motion would otherwise have.

29. He can also, as Howson forcefully pointed out to me, deny that Bayesianism has anything to do with "rationality." There are really two Bayesian positions here: atheist and agnostic positions. The atheist version positively asserts that there is no more to proper scientific reasoning than is captured by "logic," where logic is personalist Bayesianism; the atheist deems all talk of a scientific "rationality" that goes beyond what is sanctioned by his principles as misguided because any *non*logical principles of rationality would themselves need a justification and this opens up a justificatory infinite regress. The agnostic position, on the other hand, allows that some stronger principles of rationality *might* have some sort of justification, though not the justification that they are required by logic (since personalist Bayesianism exhausts what can count as logic). Nothing I say in this essay affects the weaker "agnostic" position which simply amounts to the claim that it is a *necessary* (though not perhaps sufficient) condition for the rationality of one's degrees of belief that they conform to the probability calculus. Far from criticizing this latter claim, I believe it to be correct. My criticisms are directed entirely against the stronger "atheist" view which *identifies* rationality with consistency with the probability calculus.

30. Of course, he will appear "eccentric" only within scientific circles, not within the population more generally. Like those who believe that philosophy of science can be fully "naturalized," Bayesians foster the illusion that their very weak account of scientific reasoning does not clash with strong normative intuition simply by concentrating exclusively on the "good guys."

31. See Popper:

The rationalist attitude is characterized by the importance that it attaches to argument and experience. But neither logical argument nor experience can establish the rationalist attitude; for only those who are ready to consider argument or experience, and who have therefore adopted this attitude already, will be impressed by them. . . . We have to conclude from this that no rational argument will have a rational effect on a man who does not want to adopt a rational attitude. . . . But this means that whoever adopts the rationalist attitude does so because he has adopted consciously or unconsciously, some proposal, or decision, or belief . . . an adoption which may be called irrational ["non-" or "a-rational" would surely have been more accurate]. (1945, 230–31)

32. I really mean, "Let's go with Grünbaum, not with the *personalist* Bayesian." As Earman (1992) shows, Bayes himself was not a Bayesian.

REFERENCES

Carroll, L. 1895. "What the Tortoise Said to Achilles." *Mind 14:* 278–80.

Dorling, J. 1979. "Bayesian Personalism, the Methodology of Research Programmes, and Duhem's Problem." *Studies in the History and Philosophy of Science 10:* 177–87.

———. 1982. "Further illustrations of the Bayesian Solution of Duhem's Problem." Unpublished manuscript.

Duhem, P. [1906] 1954. *The Aim and Structure of Physical Theory.* Translated by P. P. Wiener. Princeton: Princeton University Press.

Earman, J. 1992. *Bayes or Bust? A Critical Examination of Bayesian Confirmation Theory.* Cambridge, Mass.: MIT Press.

Edwards, W.; H. Lindman; and L. J. Savage. 1963. "Bayesian Statistical Inference for Psychological Research." *Psychological Review 70:* 193–242.

Glymour, C. 1980. *Theory and Evidence.* Princeton: Princeton University Press.

Grünbaum, A. 1960. "The Duhemian Argument." *Philosophy of Science 32* 75–79.

———. 1966. "The Falsifiability of a Component of a Theoretical System." In P. Feyerabend, and G. Maxwell, eds., *Mind, Matter and Method: Essays in Honor of Herbert Feigl.* Minneapolis: University of Minnesota Press, pp. 273–305.

———. 1969. "Can We Ascertain the Falsity of a Scientific Hypothesis?" *Studium Generale 22:* 1061–93.

———. 1971. "Can We Ascertain the Falsity of a Scientific Hypothesis?" In E. Nagel, S. Bromberger, and A. Grünbaum, eds., *Observation and Theory in Science.* Baltimore: The Johns Hopkins University Press, pp. 69–129.

———. 1973. *Philosophical Problems of Space and Time.* 2d ed. Enlarged. Dordrecht: Reidel.

———. 1976. "Is Falsifiability the Touchstone of Scientific Rationality? Karl Popper versus Inductivism." In R. S. Cohen, P. K. Feyerabend, and M. Wartofsky, eds., *Essays in Memory of Imre Lakatos.* Dordrecht: Reidel, pp. 213–52.

———. N.d. "Falsifiability and Rationality." Unpublished manuscript.

Hacking, I. 1967. "Slightly More Realistic Personal Probabilities." *Philosophy of Science 34:* 311–25.

Howson, C., and P. Urbach. 1989. *Scientific Reasoning: The Bayesian Approach.* La Salle, Ill.: Open Court.

Kuhn, T. S. 1978. *The Essential Tension.* Chicago: Chicago University Press.
Laudan, L. 1965. "On the Impossibility of Crucial Falsifying Experiments: Grünbaum on 'The Duhemian Argument.' " *Philosophy of Science 32:* 295–99.
Popper, K. R. 1945. *The Open Society and Its Enemies.* Vol. 2. London: Routledge & Kegan Paul.
Ramsey, F. P. 1931. *The Foundations of Mathematics and Other Logical Essays.* London: Routledge & Kegan Paul.
Redhead, M. 1980. "A Bayesian Reconstruction of the Methodology of Scientific Research Programmes." *Studies in the History and Philosophy of Science 11:* 341–47.
Salmon, W. C. 1970. "Bayes's Theorem and the History of Science." In R. H. Stuewer, ed., *Historical and Philosophical Perspectives of Science.* Minneapolis: University of Minnesota Press pp. 68–86.
———. 1976. "Confirmation and Relevance." In G. Maxwell, and R. M. Anderson, eds., *Induction, Probability and Confirmation.* Minneapolis: University of Minnesota Press, pp. 3–36.
———. 1990. "Rationality and Objectivity in Science or Tom Kuhn Meets Tom Bayes." In C. W. Savage, ed., *Scientific Theories.* Minneapolis: University of Minnesota Press, pp. 175–204.
Shimony, A. 1970. "Scientific Inference." In R. G. Colodny, ed., *Pittsburgh Studies in the Philosophy of Science.* Vol. 4. Pittsburgh: University of Pittsburgh Press, pp. 79–172.
van Fraassen, B. C. 1989. *Laws and Symmetry.* Oxford: Oxford University Press.
Worrall, J. 1978. "Is the Empirical Content of a Theory Dependent on Its Rivals?" In I. Niiniluoto, and R. Toumela, eds., *The Logic and Epistemology of Scientific Change.* Amsterdam: North-Holland, pp. 298–310.
———. 1980. "Facts and Feyerabend." In H. P. Duerr, eds., *Versuchungen: Aufsätze zur Philosophie Paul Feyerabend.* Zweiter Band. Frankfurt: Suhrkamp Verlag, pp. 298–319.
———. 1990. "Scientific Revolutions and Scientific Rationality: The Case of the 'Elderly Holdout.' " In C. W. Savage, ed., *Scientific Theories.* Minneapolis: University of Minnesota Press, pp. 319–54.

Philosophy of Psychiatry

15

The Dynamics of Theory Change in Psychoanalysis

Morris Eagle
Ontario Institute for Studies in Education

Adolf Grünbaum's book, *The Foundations of Psychoanalysis: A Philosophical Critique* (1984), has had an enormous impact on philosophers, psychoanalysts, and others interested in the status of psychoanalytic theory. One of the main tasks of that book was to rigorously examine the empirical and logical foundations of certain basic Freudian claims such as the ubiquitous occurrence of repression and its purported central pathogenic role in the development and maintenance of neurotic symptoms. In Eagle (1986a), I have tried to show that current psychoanalytic developments, in particular, object relations theory and self psychology—the two predominant, current theoretical developments in psychoanalysis—do not escape the criticisms that Grünbaum has directed against Freudian theory. In that paper, I tried to demonstrate that self psychology is based on even weaker epistemological grounds than Freudian theory.

In this essay, I continue my examination of so-called contemporary psychoanalytic theory. The main question I will address is how theoretical changes occur in psychoanalysis. If analysts (as they have claimed) no longer think in accord with core Freudian propositions, how *do* they think and how did they get to think the way they do today? On the basis of what kinds of considerations did many analysts come to adopt the point of view that is referred to as contemporary psychoanalytic theory? In addressing these questions, I am assuming that one may gain insight into the nature of a discipline by understanding how theoretical changes occur in that discipline.

Before dealing with the question of how contemporary psychoanalytic theory developed, we must first clarify just what contemporary psychoanalytic theory is and just how contemporary analysts do think. It is not at all clear that there is a uniform body of thought analogous to the main corpus of Freudian theory that can be called contemporary psychoanalytic theory. In the last forty or fifty years there have been three major theoretical developments in psychoanalysis: ego psychology, object relations theory, and self psychology. If contemporary psychoanalytic theory is anything, it is one of these three or some combination, integrative or otherwise, of the three. In the following sections, I will present briefly some core tenets of these three theoretical approaches and will examine the basis of their development.

Ego Psychology

I begin with those theoretical developments known as ego psychology. In contrast to object relations theory and self psychology, ego psychology consists mainly in corrections and extensions of Freudian theory and in making explicit what was, in large part, already implicit in that theory. The main tenets of ego psychology are as follows: (1) Contrary to Freud's claim that thinking (and by implication, other ego functions) develop primarily because of the pressure of instinctual drive gratification, ego psychologists acknowledge that, in large part at least, the development of a wide variety of ego functions—such as thinking, motility, memory and perception—unfold on the basis of an innate genetic program. Hartmann (1958, 1964) refers to this as the *primary autonomy* of ego functions. (2) Certain behaviors that developed in the course of gratifying instinctual needs and dealing with intrapsychic conflict assume a *secondary autonomy*. For example, a child may use the defense of intellectualization as a means of dealing with conflict. However, in the course of development, intellectual activities acquire their own autonomous status, their own adaptive value, and are engaged in for the autonomous pleasures they provide and the adaptive functions they serve. Hence, a good deal of behavior may be autonomous in the sense of being "conflict-free."

What Hartmann tried to accomplish with his concept of primary autonomy is make room in psychoanalytic theory for the facts of biological maturation, facts of which developmental psychologists and biologists, for example, were fully aware, but which the logic of Freudian instinct theory seemed to exclude. In other words, when Hartmann wrote that in an "average expectable environment," practically all children will develop, within a particular age range and in particular sequences, certain motor and cognitive skills and capacities independent of drive vicissitudes, he was not presenting a new discovery—the facts of the matter were long available—but was rather attempting to correct an obvious deficiency of Freudian instinct theory.

In a certain sense, then, the formulations of ego psychology reflect a degree of responsiveness and accommodation to empirical findings and phenomena which are derived, however, from outside the psychoanalytic context. Thus, the facts of biological maturation are not distinctively psychoanalytic and, of course, were uncovered completely independent of the psychoanalytic context. To correct and accommodate one's theory in the light of some basic and obvious data that the theory had overlooked and had even contradicted represents, I suppose, a kind of progress. But it is a limited kind of progress. And, more important in the present context, the theoretical changes that constitute ego psychology do not entail any essential modifications in Freudian theory. In particular, the primary issues addressed by Grünbaum—for example, the role of repression in the development and maintenance of neurosis, the so-called tally argument, and the limitations of free association—are not altered in any essential way by the theoretical developments of ego psychology.

So, should analysts, in response to Grünbaum's critique of Freudian theory, say "we do not think *that* way anymore, we think along the lines of ego psychology," the response would be somewhat tangential insofar as the formulations of ego psychology do not really vitiate or even directly bear on Grünbaum's criticisms. As noted, the formulations known as ego psychology have to do mainly with such metapsychological issues as whether ego functions develop entirely under the pressure of drive gratification or can unfold "autonomously" on the basis of genetic programming.

As far as the issue of the dynamics of theoretical change in psychoanalysis is concerned, the developments of ego psychology were not occasioned by new data from the psychoanalytic situation—analysts do not do longitudinal studies on the development of ego functions. Rather, as I have tried to show, they were prompted by a recognition of certain obvious deficiencies of Freudian theory, in particular, its failure to take account of the obvious facts of biological maturation as they apply to ego functions.

Object Relations Theory

I turn now to the theoretical developments known as object relations theory. Object relations theory has had an enormous influence on contemporary psychoanalytic theorizing and the thinking of someone who views himself or herself as a contemporary analyst is likely to reflect this influence. What is object relations theory and how did it come to play the role that it does in contemporary psychoanalytic thinking?

Given space limitations, one can only describe the bare essentials of object relations theory (for a fuller account see Eagle 1987, and Greenberg and Mitchell 1983). Object relations theory or the so-called English school of psychoanalysis has its most immediate origins in the work of Melanie Klein (1948, 1975) and in the elaboration and selective revision of that work by Fairbairn (1952), Guntrip (1969) and Winnicott (1958). Fairbairn presents the most systematic account of object relations theory and it is mainly his writings that will be emphasized.

In traditional Freudian theory, the external object is defined as "the thing in regard to which the instinct achieves its aim" (Freud [1915] 1957, 122). The idea that the primary function of objects is instinctual gratification is also seen in Freud's ([1914] 1957, 87) characterization of infant-mother attachment as an "anaclitic" relationship. That is, the relationship "leans upon" the mother's role in gratifying the instinctual hunger drive. Still other expressions of the secondary and derived status assigned to objects and object relations in Freudian theory are seen in Freud's related ideas that we only reluctantly come to develop an interest in objects and that were it not for the fact that objects happen to be necessary for satisfaction of the hunger drive, we would never develop either a conception of or an interest in objects (Freud [1900] 1953).[1]

Object relations theory is partly characterized by its rejection of this view and by its insistence that an interest in objects and object relations are not derived from or secondary to instinctual drive gratification, but rather are primary, autonomous, and have an inborn basis. This claim and the concurrent rejection of Freudian instinct theory is stated most succinctly in Fairbairn's (1952) dictum that "libido is object-seeking, not pleasure seeking" (p. 82) and in Balint and Balint's (1965) concept of "primary object love."

In Freudian theory, the primary "task" of the individual is the achievement of drive gratification without undue guilt and anxiety. This claim is based on Freud's core assumption that the primary danger facing the individual is the build up of ego-damaging excessive excitation resulting from failure to adequately discharge instinctual impulses. This foundational assumption helps one understand the central pathogenic role assigned to repression in Freudian theory and the formulation of neurotic symptoms as "compromise formations" that permit disguised and partial instinctual discharge and that thereby represent an "escape valve" and a kind of "solution" to an irresolvable intrapsychic conflict between unconscious sexual and aggressive wishes and defenses against these wishes.

Object relations theory entails a rejection of all of these (as well as other) key features of Freudian theory. In object relations theory, the primary "tasks" facing the individual have little to do with instinctual gratification, but rather concern the development of an independent and differentiated sense of self, a differentiated sense of the object and a well functioning capacity to relate to others, both in external reality and intrapsychically in one's inner world, in a manner that will sustain one's sense of self. According to object relations theory, we are all born with a natural propensity to relate to objects. One way of tersely summarizing object relations theory is to say that the core concerns of that theory are the fate and vicissitudes of that inborn propensity. Psychological development, the conception of psychopathology, and the nature of mental functioning are all understood very differently in object relations theory than in Freudian theory.

There are also critical differences in conceptions of treatment between Freudian theory and object relations theory. In the former, the primary goals of treatment are to "make the unconscious conscious" and to replace id with ego ("where id was, there should ego be"). Another way to put this is to say that the primary "curative" factors are the lifting of repression and the achievement of insight. In object relations theory, the primary goals of treatment are to "exorcise" the "bad" internalized object and to replace a "bad" object with a "good" object, and accordingly, the primary "curative" factor is the therapist as "good" object. That is, a benevolent therapeutic relationship is the central efficacious factor in successful treatment.[2]

Side by side with the given differences between Freudian theory and object relations theory, there are also important areas of similarity and agreement that make the latter recognizable as a psychoanalytic theory. Foremost is a common emphasis on the role of repression in psychopathology (a point noted by Grünbaum 1984), although each theory employs a somewhat different conception of repression.

Fairbairn (1952) refers to the "libidinal ego," the "antilibidinal ego" (and the "central ego") as basic components of the personality. The term "libidinal ego" refers generally to one's early object seeking yearnings for nurturance and care and the term "antilibidinal ego" (also referred to by Fairbairn as the "internal saboteur") refers to one's condemnation and contempt for one's own libidinal longings. Fairbairn suggests that under the impact of early experiences of deprivation and frustration, one represses the libidinal ego. One can understand this to mean that because they are associated with pain and rejection, one *represses* one's "infantile" longings to be cared for and nurtured. Despite Fairbairn's rejection of the role of instinctual drives and the Freudian concept of the id, this description of the libidinal ego-antilibidinal ego conflict is similar to the Freudian id-ego model (a similarity that Fairbairn acknowledges)—save that it is object-seeking longings that are repressed rather than sexual and aggressive wishes emanating from id drives. Hence, it is inaccurate to claim, as do some critics of Grünbaum, that the concept of repression and the repression etiology of psychopathology are no longer of consequence in object relations theory. Repression does not play precisely the same role in object relations theory that it does in the Freudian theory of psychopathology, but it does continue to play an essential role.[3]

It does appear as if the lifting of repression is no longer a central issue in the object relations theory of *treatment* partly because of Fairbairn's (and even more so, Guntrip's 1969) emphasis on the therapist's provision of "a real good object" (1952, 348), that is, an emotionally corrective benevolent relationship. But the logic of the object relations theory of treatment suggests that the issue is somewhat more complicated. Thus, Fairbairn's (1952) emphasis on the need to extirpate the internalized "bad" object in treatment surely includes making conscious what was hitherto unconscious, that is, bringing to awareness and making more accessible both one's repressed early libidinal longings and the ways in which one has internalized attitudes of rejection, condemnation, and contempt for these longings. Although a "good object situation" may facilitate this whole process, the process itself certainly seems to include the lifting of repression and the making conscious what was previously unconscious.[4]

It is possible, I suppose, to construe object relations theory differently than I have and to emphasize aspects of it that have little to do with repression. Hence, those analysts who view themselves as object relations theorists can perhaps legitimately say, with regard to Grünbaum's criticisms of Freudian theory, "we no longer think *that* way." Thus, Grünbaum's disturbing criticisms of Freud's repression theory would be less disturbing and even obviated to the extent that repression (or, at least, Freud's version of repression) is no longer held to play a central role in at least certain forms of psychopathology. A number of critical questions and considerations, however, immediately arise.

In what way would object relations theory continue to be a *psychoanalytic* theory? If object relations theory rejected *all* central features of Freudian theory and could legitimately claim the status of *contemporary* psychoanalytic theory, one clear implication is that the essence of Freudian theory has been virtually discarded by contemporary analysts. But, since what is known as a psychoanalytic theory has always been understood as, in large part at least, Freudian theory, in what way is a theory that rejects most or all the key features of Freudian theory an updated *psychoanalytic* theory? It would be more accurately described as a replacement for or an alternative to psychoanalytic theory.

I turn now to the question of how object relations theory came to play its influential role in the psychoanalytic community. A basic assumption of object relations theory is that lives are made or broken, not as a function of the vicissitudes of sex and aggression, but as a function of the nature of early object relations. According to object relations theory, if early object relations (read "early mothering") are characterized by love and caring—what Winnicott (1958) refers to as "good enough" mothering—the risks of pathological development are minimized and the likelihood of healthy development is maximized. Whatever the details and the frequent convoluted metapsychology of object relations theory, one of its central claims and its basic point of departure from Freudian theory can be stated very simply: Psychological development, including pathology versus health, is not a function of the vicissitudes of drive gratification, but of the overall quality of early object relations and their impact on one's sense of self and one's later object relations. When one states this central claim of object relations theory as such, it becomes apparent that it is essentially identical to a primary contention of the neo-Freudian theorists, especially Sullivan (e.g., 1953) and Horney (e.g., 1945). They, too, have maintained that the etiology of pathology versus health lies not in the complexities, conflicts and traumas surrounding sex and aggression (including the kind of oedipal conflict resolution achieved), but rather in the quality of early interpersonal relations. Consistent with object relations theory, they further maintained that the primary *expressions* of pathology in adulthood were found in the pattern of one's interpersonal relations rather than simply in the appearance of discrete ego-alien symptoms or in the interrelationships of different components of the "mental apparatus."

Clearly, then, at least some core claims of object relations theory were already articulated by at least some neo-Freudians in the 1930s and 1940s. Thus, as far as certain key aspects of object relations theory are concerned, it is not that object relations theory is essentially new or more "contemporary" than, say, the formulations of Sullivan or Horney. It is, rather, that object relations theory "caught on" more recently and more readily became part of American mainstream psychoanalysis.

Although object relations theory has gained enormous influence in North American psychoanalysis mainly in the last 15 to 20 years and is viewed by many as a central part of "contemporary" psychoanalysis, the fact is that the writings of Fairbairn and Winnicott, the most important and most influential proponents of object relations theory, were essentially contemporaneous with the writings of Sullivan and Horney. Thus, object relations theory is really not any more "contemporary" than the neo-Freudian theories of Sullivan and Horney.

One way of understanding "contemporary" psychoanalysis, particularly the popularity of object relations theory, is to recognize that many neo-Freudian cogent criticisms of Freudian theory, particularly of Freudian instinct theory (which, at the time they were made, were dismissed as "superficial" and as a defensive retreat from accepting the overriding importance of sexual and aggressive instincts and impulses) have essentially been assimilated into so-called contemporary psychoanalysis—*without explicit acknowledgement that these earlier criticisms were valid or that they have, in fact, been assimilated.* Instead, these "contemporary" developments are sometimes presented as an *evolution of Freudian theory.* Hence, analysts who view themselves as object relations theorists can shield themselves from the fact that many of the central formulations of Freudian theory have essentially been discarded and overthrown. They can continue to believe that like other cumulative and orderly disciplines, psychoanalytic theory shows a steady course of growth and expansion and can continue to use terms of which they are very fond, such as "breakthrough" and "the expanded scope of psychoanalysis."

That object relations theory is viewed as a more contemporary *psychoanalytic* theory has more to do with what can broadly be referred to as sociological and political factors than with the content and substance of the theory. That is, in contrast to Sullivan and Horney and their followers, object relations theorists did not openly break with established psychoanalytic training institutes and did not set up their own competing institutes. Hence, their formulations could be more readily accepted and assimilated by many analysts who viewed themselves as essentially belonging to the mainstream (rather than to the "revisionist" or heretical) psychoanalytic tradition. Analysts who were sympathetic to object relations theory did not need to join new maverick institutions; they could remain as loyal members in the established mainstream institutes. They could even form or become part of a more "progressive" bloc within these institutes and enjoy the heightened self-regard derived from viewing themselves as more progressive and more contemporary than the really orthodox Freudians.

Apart from these sociological and political considerations, what is the evidential empirical base for the central claims of object relations theory? In considering this question, let us make certain distinctions, the most important of which is the distinction between, on the one hand, object relations theory's broad conception of human nature and its claims regarding the primary basis for infant-mother attachment as well as for intimate interpersonal relations in general and, on the other hand, its particular conception of the nature and etiology of psychopathology, and its rather convoluted metapsychological formulations regarding the structure of the personality.

With regard to broad conception of human nature (see Eagle 1987), the empirical evidence on infant development overwhelmingly supports the claims of object relations theory to the effect that "an interest in objects . . . is an autonomous inborn natural propensity which appears at birth or shortly after birth and is not dependent on the vicissitudes of instinctual gratification" (p. 87).

As for infant-mother attachment, Harlow's (1958) classic study definitively showed that the infant monkey's attachment to mother is not a secondary derivative of the latter's role in the gratification of so-called primary drives (i.e., hunger, thirst), but is based on the autonomous need for what Harlow called "contact comfort" (p. 677). Also, evidence from a variety of animal studies shows that despite the adequacy of nutrition and other forms of physical care, the interruption of physical contact between infant and mother, in particular of tactile stimulation, leads to abnormal development (e.g., Shanberg and Kuhn 1980). Finally, the body of work generated by Bowlby's (1969, 1973, 1980) seminal attachment theory attests to the existence of an inborn and autonomous attachment system in animals and humans.

At the level of general conception of human nature then, the available evidence from a variety of sources seems to support object relations theory's claim that we are inherently "object-seeking" creatures. However, as noted earlier, what is known as object relations theory is also made up of a number of specific formulations having to do with the nature and etiology of psychopathology, the nature of mental functioning, and prescriptions regarding treatment. These components have had the greatest influence on contemporary psychoanalysis and it is mainly these components that analysts have in mind when they refer to object relations theory. These components of object relations theory include a complex and convoluted metapsychology in which concepts and hypotheses such as "splits in the ego" and "internalized objects" play a prominent role. There is no greater empirical evidence adduced for these hypotheses than for Freud's central hypotheses. And there is no indication that the kinds of evidential and epistemological issues raised by Grünbaum are somehow better met by these object relational hypotheses than by Freudian hypotheses.

Indeed, for some of the object relations hypotheses and concepts it is not clear what would constitute supportive evidence. For example, it is not clear what evidence one could cite that would support Fairbairn's (1952) hypothesis that the development of "internalized objects" (resulting from early traumatic experiences), which, in turn, lead to "splits in the ego," is at the heart of psychopathology. Largely, this is so because concepts such as "internalized objects" and "splits in the ego" are murky and vague. What would constitute evidence for the existence of "internalized objects"? for "splits in the ego"? Further, even if one were clear about what these concepts mean empirically, how, short of longitudinal studies, can one adduce evidence for the causal chain of traumatic experiences to internalized objects to splits in the ego to development of psychopathology?

Object relations theory (and, as we will see, self psychology theory), no less than Freudian theory, has to address the issue of the warrant for etiological conclusions based entirely on the productions of adult patients in treatment. The mere shift from repression as a pathogen to early traumatic deprivation and frustration that presumably produce splits in the ego does not attenuate the magnitude of the epistemological problems involved. It is difficult to understand on what basis advocates of object relations theory (and those who think of object relations theory as representative of "contemporary" psychoanalysis) believe that that theory somehow addresses or bypasses those of Grünbaum's criticisms of psychoanalytic theory that deal with its etiological claims.

One may want to argue that the evidential support available for its broad conception of human nature lends some general degree of plausibility to object relations etiological theory of psychopathology. But surely, evidence attesting to the inborn basis for an interest in objects and for an autonomous attachment system hardly carries any direct implications for the validity of the etiological claims that psychopathology is a consequence of early traumatic experiences that, in turn, engender the internalization of "bad" objects and the production of "splits in the ego."

Let me return once again to the historical and sociological question of the popularity of object relations theory within the psychoanalytic community. Although, as noted earlier, the writings of the object relations theorists and many of the neo-Freudians were roughly contemporaneous, the former reached North America (that is, was widely read by analysts) a bit later (for example, Guntrip's book, *Schizoid Phenomena, Object Relations, and the Self,* which did much to bring object relations theory to the attention of the North American psychoanalytic community, was only published in 1969). By the time the object relations literature became known in America, it was becoming more and more apparent—both to analysts and others interested in psychoanalysis—that many central and framework formulations of Freudian theory were simply untenable. For example, with regard to one of Freud's most basic underlying assumptions, it was clear that the central nervous system was *not* so constructed that its primary "function" was to rid itself of excitation and that it required drive as a source of energy in order to function. Holt's (1962) devastating critique of Freud's concept of psychic energy had already been published. Infant research was burgeoning and it was clear that the Freudian conception of the nature of the infant was inadequate and often simply wrong. Harlow's (1958) very important study on "contact comfort" had been published and Bowlby's (1969) seminal work on attachment theory was getting known. All these developments, plus others I have not mentioned, were making it increasingly clear that certain core formulations of Freudian theory were untenable and, therefore, that the brilliant, complex, and elegant structure known as Freudian theory simply could not stand unmodified and, indeed, was in great danger of toppling. Something had to be done.[5]

Object relations theory "arrived" in North America around the time that the developments I have described were occurring. Hence, formulations and critiques that in the past were labeled as "revisionist" and that were often arrogantly dismissed by confident Freudians were now taken far more seriously. And because, as noted earlier, object relations theorists were politically acceptable, their criticisms of Freudian theory could more readily be assimilated. It should be noted again that during this process of "modernizing" of psychoanalytic theory there was little or no explicit acknowledgement that, in effect, it constituted at least a substantial dismantling of Freudian theory or that many earlier neo-Freudian criticisms had been assimilated.

Mitchell (1979) has noted an additional characteristic of the dynamics of theoretical change in psychoanalysis, namely, the "discovery" of a new form of psychopathology which presumably requires a new theoretical formulation. Thus, object relations theory was presumably especially appropriate to so-called schizoid pathology and was initially formulated in order to account for such pathology. Similarly, self psychology was initially formulated in order to account for so-called narcissistic personality disorders (see Eagle 1987). Certainly in the case of self psychology, the clear implication in the early literature was that while Freudian theory was applicable to the classical neuroses, the newer theory—self psychology—was applicable to the "new" pathology, namely, narcissistic personality disorders (Kohut 1971). According to Mitchell, the allocation of each theory to its own domain and the argument that the newly "discovered" or newly formulated pathology required its own theory were initially politically acceptable ways of introducing new "revisionist" theories. (Be that as it may, this more ecumenical position was soon replaced with the more ambitious claim that the newer theory more accurately accounted for *all* pathology. Once having established a foothold of acceptance, it could broaden its claims).

Kohut (1977) has suggested that the nature of psychopathology has changed and has maintained that while "structural conflicts" may have been the modal pathology of Freud's era, and adequately accounted for by Freudian theory, narcissistic personality disorders are the modal pathology of our time and are best accounted for by self psychology theory. The whole question of whether the nature of psychopathology has changed with time is an interesting one that cannot be discussed here at great length. However, some clear implications of Kohut's position are worth noting. One such implication is that contrary to what had always been assumed and claimed, psychoanalytic theories, including Freudian theory, are not properly understood as universal and historical statements about human nature and mental functioning, but are to be viewed as "local" and limited theories that will vary with historical period and type of pathology. Thus, if Kohut is correct, it follows that self psychology theory, too, will be replaced if and when the nature of pathology changes.

The issue of the changing nature of pathology is even more complex. For diagnostic categories such as narcissistic personality disorders, self-defects, and schizoid conditions are not transparent, but are rather theory-saturated. As described in Eagle (1987), recently two case books were published, one by a traditional Freudian from the New York Psychoanalytic Institute (Firestein 1978) and one by a follower of Kohut (Goldberg 1978). Although the clinical phenomena presented in both casebooks are relatively similar, the Firestein casebook is replete with references to oedipal conflicts, with not a word about self-defects, while the Goldberg casebook has nothing to say about oedipal conflicts and finds self-defects at the core in virtually every case. Hence, whether psychopathology itself has changed or *the way in which psychopathology has been conceptualized* has changed, or some combination of the two, is unclear. If "contemporary" psychoanalytic formulations are, to a significant extent, a matter of how psychopathology is conceptualized, we are back to the original question of why object relations theory and self psychology theory have become so predominant in the contemporary psychoanalytic community.

Theoretical changes in psychoanalysis, to a considerable extent at least, likely reflect broad social and economic developments and changes in the culture at large. For example, as is suggested in Cushman (1990), self psychology may be as much a *reflection* of what the author refers to as the "empty self" that has emerged in post-World War Two industrial society as an explanatory account of that empty self. That is, by putting the self at the center of its theory and self-defects at the center of its conception of psychopathology, self psychology is giving voice to the widespread *feeling* of an empty and impoverished self. This "resonance" between the theory and the phenomenology of many people's experiences probably accounts, in large part, for the evocative power and appeal that the theory currently holds for many. But such "resonance" is not to be confused with a deep explanatory account that is buttressed by systematic and reliable evidence.

A similar point can be made with regard to object relations theory. The theory "resonates" with the current conviction (although even in our era it is not a universally held conviction—see Storr 1988) that deep and intimate interpersonal relations are, both developmentally and in adult psychological functioning, at the core of the personality and of psychopathology (just as, perhaps, Freud's theory "resonated" with the Victorian middle class and presexual liberation conviction that conflicted and repressed sexual wishes and impulses were at the core of personality and of psychopathology). However, here too such "resonance" is not equivalent to a powerful explanatory theory that is adequately supported by reliable evidence.

Self Psychology Theory

Although I have already referred to self psychology a number of times in the course of the discussion, I have not dealt with it as a separate development. The same set of questions that were directed to object relations theory can be addressed to self psychology. What is self psychology? How did it come to occupy the place that it does in contemporary psychoanalysis? What is the evidential basis for the major hypotheses of self psychology? And finally, does self psychology, as presumably an expression of contemporary psychoanalysis, escape the kinds of criticisms that Grünbaum has directed against Freudian theory? I will briefly deal with each question.

What is self psychology? Current psychoanalytic self psychology is mainly a product of the work of Kohut (1971, 1977, 1984) and his followers. Kohut's main emphasis has been on the developmental achievement of a cohesive self. He has argued that this aspect or dimension of psychological growth, which he refers to as a narcissistic line of development, is a central one and should be considered apart from the traditional emphasis on psychosexual sequences or even ego development. According to Kohut, the eventual development of a cohesive self requires the presence of a mothering figure who will provide empathic mirroring and the later availability of a parental figure permitting idealization. At an early stage of development, the mirroring and idealized figures function as "self-objects" rather than fully separate others. As long as union is maintained with these self-objects, the child feels powerful and secure rather than powerless and empty. People who show deficiencies in the development of self-cohesiveness—presumably because of failure of the mother to provide mirroring and empathy and failure of either parent to meet the child's need for an idealized self-object—require the presence of self-objects in order to maintain some feeling of self-cohesiveness and in order to avoid the experience of "disintegration anxiety." Such people, described by Kohut and his followers as narcissistic personality disorders, relate to others, not as fully separate others or as objects for drive gratification, but as self-objects necessary for self-cohesiveness.

As in Eagle (1986b), self psychology radically departs from Freudian theory in a number of critical ways, some examples of which follow:

1. There is a clear shift from the traditional *conflict* model of psychopathology to a psychology of self-*defects*. According to Kohut (1984), self psychology understands pathology in terms of "faulty structures responsible for [the] faulty functioning" (p. 86).

2. Accompanying the shift of emphasis from conflict to defects and deficits is an associated shift from wishes to needs. Kohut argues that viewing pathology and its expressions in the transference in terms of infantile sexual and aggressive wishes entails the imposition on the patient of an "adult morality" and tends to be censorious and disapproving, while an understanding of the patient in terms of needs that are impelled by self-defects tends to be more empathic and therapeutic.

3. As far as the therapeutic process is concerned, there has been a shift involving a de-emphasis on interpretation and insight and an increasing emphasis on the role of empathic understanding and the patient-therapist relationship. Kohut (1984) explicitly argued that insight is not curative and that the analyst's empathic understanding and other associated elements (such as "optimal failures" and acknowledging one's failures and errors) are the curative ingredients in psychoanalytic treatment.

In effect, the main therapeutic vehicle in self psychology is the provision in treatment of traumatically unmet needs for empathic mirroring and understanding. The meeting of these hitherto unmet early needs permits the "resumption of developmental growth."

Implied in the given examples are some very basic differences between Freudian theory and self psychology, including the latter's rejection of the idea that conflict about and repression of sexual and aggressive wishes are at the heart of psychopathology. Also, Kohut's self psychology emphasizes in its etiological theory, not unconscious fantasies, conflicts, and meanings, but *the effect of (purported) actual environmental events,* namely, parental traumatic failures in provision of empathic mirroring and opportunities for idealization.

In view of its rejection of virtually every core tenet of traditional psychoanalytic theory, in what way(s) can self psychology theory be legitimately viewed as a *psychoanalytic* theory? Kohut's (1977, 302) response to this question is to argue that any approach that focuses on "complex mental states" and obtains its data through empathic introspection should be viewed as a psychoanalytic approach. I leave it to the reader to judge the adequacy of this reply. However, this description hardly distinguishes self psychology from other *nonpsychoanalytic* approaches to human behavior and to treatment (e.g., Carl Rogers's approach).

I think one must, in part at least, look to sociological and political factors in order to understand why a theory that rejects core features of traditional psychoanalytic theory and that is as radically different from it as self psychology is viewed as a *psychoanalytic theory rather than a nonpsychoanalytic alternative to or replacement for psychoanalytic theory.* As was already noted with regard to object relations theory, here, too, a significant factor lies in the fact that self psychology analysts remained within mainstream psychoanalytic training institutes and did not set out to establish their own institutes.

An additional important reason that both self psychology and object relations theory are viewed as psychoanalytic despite their radical departures from traditional theory *lies in the contemporary tendency to implicitly define psychoanalysis solely in terms of treatment setting and treatment techniques.* Thus, because patients come to treatment a set number of times per week, use the couch, free associate and form transferences, and because therapists interpret these free associations, often in terms of the patient's transference, the approach is seen as psychoanalytic. This is perhaps appropriate. But what also happens is that the theory that, so to speak, "surrounds" the treatment approach is also viewed as psychoanalytic—even if that theory rejects or ignores virtually every core feature of what has been understood as psychoanalytic. The recognizable psychoanalytic nature of the treatment techniques automatically endows the theory with a psychoanalytic stamp, whatever the content and nature of that theory. Or, as I strongly suspect, many people in the psychoanalytic community are not especially interested in theory—*any* theory—and are comfortable with limiting their conception of what is psychoanalytic to treatment techniques. Thus, if it includes transference, resistance, free association, the couch, interpretation, and so on, it is psychoanalytic, whatever the nature of the accompanying etiological theory and theory of personality development and mental functioning.

Finally, I want to consider two questions about self psychology that were also raised with regard to object relations theory: (1) What is the nature of the evidence for self psychology's etiological theory as well as its approach to treatment? and (2) Does self psychology theory successfully escape the kinds of criticisms that Grünbaum has directed to Freudian theory?

With regard to the nature of the evidence for the etiological theory of self psychology, there is little one need say. As remarkable as it may seem, there is simply no reliable evidence for the etiological claims that so-called narcissistic personality disorders were subject in their infancy and childhood to traumatic unfulfillment of needs for empathic mirroring and idealization; that such purported traumas are causally related to the production of self-defects; or that the experience of empathic mirroring and opportunities for idealization are universal needs which, if not adequately met, will generate narcissistic pathology. About the only "evidence" that Kohut and his followers provide in support of these given hypotheses is the (purported) spontaneous development of mirroring and idealizing transferences of adult patients in treatment. Of course, such "evidence" is highly problematic on a number of counts. Even if one reliably found that such transferences regularly developed, this would hardly constitute convincing evidence that (1) they represent attempts to meet infantile needs; (2) that such purported needs were traumatically unmet in infancy and childhood; and (3) that even if unmet, they are causally implicated in current pathology.

As for evidence supporting the self psychology approach to treatment, there simply is none beyond selective case reports from advocates of self psychology theory and Kohut's (1979) report of his experiences in the "two analyses of Mr. Z." According to Kohut, in his first analysis of Mr. Z, when Kohut was still an adherent of classical theory and the classical approach to treatment, he interpreted Mr. Z's feelings of entitlement as his clinging to infantile wishes, a therapeutic stance which, according to Kohut, tended to be judgmental and censorious. Although Mr. Z showed some improvement in his first analysis, it was a limited improvement and according to Kohut, the analysis did not get to the core of Mr. Z's problems. In the second analysis of Mr. Z, because Kohut had begun to question his previous adherence to classical theory and treatment approach, he responded to and understood Mr. Z's feelings of entitlement as the product of his self-defects and lack of self-cohesiveness. This, according to Kohut, *more empathic* therapeutic response and understanding permitted deeper and more extensive therapeutic gains.

That those experiences played a pivotal personal role in Kohut's approach to treatment is understandable. But that they should have also played a pivotal evidential role in the establishment and appeal of self psychology, including its entire baggage of etiological theory, is, of course, another matter. In my view, Kohut's report of the two analyses of Mr. Z can be instructive in helping to account for the wide appeal of self psychology. As Holzman (1976) has observed, many classical analysts had, in effect, caricatured the "rule" of analytic neutrality so that it came to be equated with aloofness, coldness, and stodginess. By contrast, Kohut's explicit emphasis on empathic understanding and normal human warmth came as a breath of fresh air. It permitted analytic therapists, particularly young therapists, to heed their commonsensical and natural feelings and intuitions. As observed in Eagle (1986b), it seems remarkable that anyone would have ever believed that coldness and stodginess are therapeutic or necessary to the therapeutic enterprise. It seems equally remarkable that an entire new theory and new ideology were required to justify a turn away from such coldness and stodginess and toward ordinary empathy, understanding and appropriate human interest and warmth. In short, I am suggesting that one important reason for the widespread appeal of self psychology is its emphasis on empathic understanding which served as a corrective to a therapeutic stance of aloofness and coldness that had come to be associated with traditional psychoanalytic theory and practice.

An emphasis, however, on the importance of empathic understanding in treatment hardly constitutes a distinctive psychoanalytic contribution. If this was the gist of what self psychology had to say, it would represent a rather flimsy and insubstantial basis for a new movement, let alone a psychoanalytic movement. However, Kohut's empathic understanding is no ordinary empathic understanding. Placed in the context of his etiological theory, the therapeutic provision of empathic understanding as well as its special curative efficacy could now be understood as the *fulfillment in the treatment of a traumatically unmet universal need for empathic mirroring* (just as the idealization of the therapist could be understood as the fulfillment in the treatment of a traumatically unmet universal need to idealize a parental figure). What I am trying to depict here is how a commonsense appreciation of the likely therapeutic importance and desirability of an attitude of empathic understanding became transformed—rather arbitrarily—into a combined etiological and treatment theory in which the traumatic lack of empathic understanding is the primary cause of pathology and the provision of such understanding is the primary curative agent. I would also suggest that self psychology theory as a whole trades on the intuitive and commonsensical reasonableness and appeal of stressing the important role of empathic understanding in treatment. That is, I suspect that the emphasis in self psychology on empathic understanding appeals to many who, mainly on the basis of that understandable appeal, go on to accept uncritically the remaining baggage of self psychology's etiological theory and theory of treatment.

Finally, I come now to the question of whether selfpsychology, presumably as an expression of contemporary psychoanalysis, escapes the criticisms Grünbaum has directed to Freudian theory and is thereby on more secure epistemological ground. With regard to etiological theory, there appears to be no basis for concluding that Kohut's environmental trauma-defect hypothesis is on any firmer evidential ground than Freud's repression-conflict hypothesis. As for the status of the clinical data, it does not appear that they are any less subject to the influence of suggestion and the vagaries of memory in Kohut's self psychology than in Freud's theory. One would have good reason to be as wary of a patient's reports that purportedly pertain to early parental failures as of material that purportedly points to infantile repressed wishes and conflicts. Further, even if one could take this patient material at face value, this would be a far cry from demonstrating that parental failures in one case and repression in the other are *causally* implicated in the patient's difficulties.

To sum up, an examination of those theoretical developments—ego psychology, object relations theory, and self psychology—that constitute contemporary psychoanalytic theory indicates that, however interesting they may be in their own right, they do not escape the criticisms that Grünbaum has directed against Freudian theory. Hence, the claim that Freudian theory is no longer representative of contemporary psychoanalytic theory, even if it were an unproblematic claim, does not constitute an adequate response to Grünbaum's most cogent and central criticisms of Freudian theory.

Strenger's Version of Contemporary Psychoanalysis

Recently, Strenger (1991), who is both a philosopher and an analyst, has written a manuscript which is, in effect, a sustained attempt to demonstrate that contemporary psychoanalysis *does* render at least some of Grünbaum's criticisms irrelevant and to argue that other criticisms that contemporary psychoanalysis does not escape are not especially pertinent. What Strenger views as contemporary psychoanalysis is not specifically identified with any of the discrete developments I have just discussed, but is instead a sort of stripped down, "generic" version of what, according to him, many analysts think of as contemporary psychoanalysis. To deal adequately with all of Strenger's arguments in his lengthy book, would require and merit a separate essay devoted only to his work. However, I will attempt to address at least some of his central arguments.

Strenger (1991) argues that Grünbaum's negative verdict regarding the clinical method of investigation is based on the paradigm of the explanatory structure exemplified in Breuer and Freud's *Studies on Hysteria* ([1893–1895] 1955) and that the current view of psychoanalytic interpretative work will substantiate the idea that "psychoanalytic propositions have changed since 1893" (p. 66).[6] According to Strenger, in contrast to the Freud of 1895, contemporary psychoanalysis is no longer concerned with making "causal claims between infantile events and present repressed memories or interpreting "the causal nexus between symptoms" (p. 73). (We are told that today's typical psychoanalytic patient tends not to present discrete symptoms anyway). If the contemporary psychoanalyst is not especially interested in uncovering repressed memories (because he or she no longer believes that that is how analysis works), then, Strenger reasons, the *veridicality* of reported memories is not especially important. But, Strenger continues, if the veridicality of memories is no longer an essential element of psychoanalytic theory, then the "tally argument" (briefly, the claim that only interpretations that "tally with what is real in [the patient]" [Freud, 1916–1917, 452] will be therapeutically effective), which plays such a central role in Grünbaum's critique, loses its force.

A number of replies to Strenger's argument are in order: (1) On the basis of what kind of evidence or other considerations did contemporary psychoanalytic theorists conclude that the uncovering of repressed memories is not essential to the work of analysis? Is this theoretical shift empirically warranted or is it mainly a sociological phenomenon, a matter of shifting fashions? Strenger does not address any of these or related questions. (2) In discussing repression, Strenger refers only to the *treatment* context. Does so-called contemporary psychoanalysis retain or reject Freud's claim that repression plays an *etiological* role in the origin and maintenance of psychopathology? Strenger does not address this issue directly either. But it is this claim that is the primary focus of Grünbaum's critique. Grünbaum brings in the therapeutic issue because, he argues, therapeutic success (resulting from lifting repression) is Freud's only legitimate justification for making that etiological-causal claim. (3) Strenger discusses only the repression and lifting of repression of *memories*. However, even in 1895 Freud did not limit his concern to repressed memories, but also, at least implicitly, referred to patients' ideas, desires, fantasies, and wishes. Surely any adequate description of Freudian theory would recognize the importance (even the primacy) of repressed and conflicted *wishes*, not simply repressed memories. Furthermore—and this seems to me a critical point—Grünbaum's discussion of the "tally argument" is in no way limited to repressed memories, but applies equally to repressed wishes and desires. That is, when one claims that only "interpretations that tally with what is real in the patient" are therapeutically effective, "real" refers not only to veridical memories, but to the patient's desires and wishes. Thus, the "tally" claim applied to the latter is that only interpretations that correspond to the patient's actual or "real" unconscious desires and wishes (and thereby serve to lift repression of these desires and wishes) will be effective. Now, does Strenger really want to say that contemporary psychoanalysis is not concerned with whether or not interpretations tally with what is real in the patient when "real" includes the patient's current wishes, desires, fantasies, and conflicts? After all, Strenger himself tells us that contemporary psychoanalysis is concerned with inferring an adequate and accurate picture of the patient's "present versions of his autobiography" (1991, 88), what Sandler & Rosenblatt (1962) refer to as the patient's "representational world." "Representational world" surely includes wishes, fantasies, and conflicts. Most important, does not contemporary psychoanalysis, as much as traditional Freudian theory, maintain that interpretations that correspond accurately to that representational world (that is, "that tally with what is real in the patient") are more likely to be effective? If this is so, we are back to the question of veridicality, even if it is not veridicality of early repressed memories. We are therefore also back to the questions of how one knows that (1) one's interpretations are veridical (e.g., correspond to the patient's "representational world"), and (2) that it is veridicality rather than other factors, such as suggestion, that account for any effectiveness that these interpretations may have.

Strenger does attempt to respond to Grünbaum's suggestion charge, in particular, the criticism that the patient's productions, pattern of free associations, acknowledgements, responses to interpretations—in short, the clinical data—may be, in large part at least, markedly influenced by the therapist's suggestive pressures, which in turn, reflect his or her theoretical biases and expectations. Hence, as Marmor (1962) notes, and as Grünbaum cites, Freudian analysts get Freudian data from their patients, Jungian analysts get Jungian data, and so on.

One of Strenger's responses to the suggestion charge is to acknowledge that the therapist *does* influence his or her patient's productions. However, Strenger argues, "the patient cannot bring any associations he doesn't have" (1991, 168) and "even the patient's yielding to his analyst's suggestive pressure will still be the reflection of some aspects of the patient's personality" (ibid., 169). It goes without saying that in some sense the patient cannot bring any associations that he does not have. This is simply a truism. But to what degree the therapist is selectively shaping and influencing the patient's associations remains the issue. Furthermore, the degree to which the pattern of the patient's associations reflects the therapist's expectations and "suggestive pressure" as opposed to "what is real in the patient" remains equally as an issue.

Consider Truax's (1966) demonstration that even the avowedly nondirective Rogers, through subtle variations in expression and interest, could, after a number of sessions, "shape" his patient's behavior so that the patient was talking more and more about self (a topic close to Rogers's theoretical heart) and less and less about sex (a topic distant from Rogers's theoretical heart). Clearly, Rogers could not get the patient to talk about material that "he doesn't have." He had concerns with issues of self and sex. But that misses the point. The point is that of the many contents, concerns, and thoughts the patient "had," Rogers encouraged one set and discouraged another. An additional point is the epistemological status and probative value of the data that Rogers obtained.

As for the argument that "even the patient's yielding to his analyst's suggestive pressure will still be the reflection of some aspects of the patient's personality" (Strenger 1991, 169), this too is a truism that misses the point. The analyst does not know that or when he or she is exerting suggestive pressure and, hence, will not know which particular aspect of the patient's personality is being revealed by a particular response. For example, a patient's assent to a particular interpretation may indicate compliance with the therapist's suggestive pressure or a conviction that the interpretation tallies with what is real in him or her or some combination of the two. Or a rejection of an interpretation may betoken rebellious defiance or a conviction that the interpretation has missed the point or some combination of the two.

Strenger's most general reply to the suggestion criticism consists of an appeal to pluralism and to the theory-laden nature, not only of data, but of practice. According to Strenger, "the suggestion charge in its most general form rests on a naive picture of the relation between theory, practice and the world. The theory *always* guides the practice (whether experimental, clinical or otherwise) and thus determines what kind of data will be obtained" (Strenger 1991, 170). It follows, Strenger tells, that of course, the patient's productions will conform to the therapist's theoretical orientation, "It is undeniable that the analyst's interventions guide and influence the patient's associations" (ibid., 172). But, according to Strenger, "the very idea that this might not be the case is incoherent" (ibid.).

If different theoretical approaches elicit different data, then one will find precisely what Marmor (1962) observed, namely, Freudian analysts obtaining Freudian data, Jungian analysts obtaining Jungian data and other analysts obtaining data compatible with their therapeutic approach and theoretical point of view. To expect otherwise, we are told, is to have a "naive" view of these matters. According to Strenger, "every therapist is forced to choose some aspect under which he will think about his patients and also get them to think and feel about themselves" (ibid., 170). Furthermore, that different therapists with different orientations elicit different data is not anything to worry about if one adopts a pluralistic position. According to that position, different "psychotherapy approaches and the personality theories associated with them are not necessarily competing theories about the same subject matter" (ibid., 171). Rather, one should regard them "as conceptual frames which organize phenomena in different ways. . . . A radically pluralist position must recognize that there is no neutral substratum which all theories deal with. Since the methods of observation of therapeutic schools are different, they work on different data" (ibid.).

Obviously, to reply in detail to Strenger's constructionist-pluralist position would require a separate essay. I do, however, want to make a few comments. Earlier in the manuscript, Strenger objects to Schafer's view which suggests that the analyst *constructs* the analytic data. But, as far as I can tell, Strenger comes to essentially the same position, for, if "there is no neutral substratum which all theories deal with," what does it mean to say that different therapeutic schools "work on different data?" If, as Strenger maintains, different therapeutic schools deal with different aspects of a patient's life this suggests that there is a complex "neutral substratum" (i.e., the patient's life) from which different therapeutic schools *select* these different aspects. This is also suggested by the statement that different therapeutic schools *elicit* different data, unless what Strenger means by eliciting different data is *constructing* different data—a position with which, Strenger already has indicated, he is uncomfortable. It seems to me that throughout the manuscript Strenger cannot make up his mind whether he wants to be merely pluralistic, that is, press the view that different theories and different therapeutic schools deal with *different aspects* of the complex reality of a patient's life, or whether he wants also to be radically constructionist, that is, press the view that different therapists of different therapeutic schools *construct* and create different data. These seem to me very different views.

The argument that different therapists of different therapeutic schools focus on different aspects of the patient's life seems to me to beg the question. Strenger appears to assume that each theory adequately captures *some aspects* of the patient's life. That is, each theory is generously awarded a prize for being right in some important way. But how does one know that the central tenets of a given theory correspond, in a significant way, to *any significant aspect* of the patient's life? And how does one know how adequately the theory deals with the aspect(s) of the patient's life on which it chooses to focus—where "how adequately" refers to questions, not only of internal coherence, and intelligibility, but also of empirical support. For example, how does one know that early traumatic failures of empathic mirroring is a critical factor in the patient's current difficulties? It is not simply a matter of self psychology—and I am merely using self psychology as illustrative—focusing on a different aspect of the patient's life than, say, Freudian theory. It is also a matter of how adequately self psychology deals with the aspect on which it chooses to focus. Thus, if it could be shown through controlled longitudinal and prospective studies that early parental failures of empathic mirroring are *not* etiologically linked (at least, not in any simple linear way) to so-called self-pathology in adulthood, then one would have to conclude that self psychology is not dealing adequately with that aspect of psychological life on which it chooses to focus. As Strenger acknowledges, a given theory must have not only internal coherence, but what he calls "external coherence," that is, it must be compatible with and not contradict what is known outside that particular theory.

Strenger seems to believe that if psychoanalytic theories *A* and *B* are not necessarily *competing* theories that contradict each other, it follows that each automatically has a claim on a different aspect of the truth. But while theories *A* and *B* may not contradict *each other,* one or both of them may be mistaken or represent a dead end in its or their own right. It may or may not be true that Freudian and Jungian theories are not competing theories and do not contradict each other. But even if it were true, that fact would not endow either theory with any truth-value. They could both be seriously flawed in significant ways.

So far, in order to examine his arguments, I have been accepting Strenger's claim that different "psychotherapeutic approaches and the personality theories that accompany them are not necessarily competing theories about the same data" and that they may instead represent different "conceptual frames which organize phenomena in different ways." However, whether, to what degree, and in what ways different personality theories are or are not, in fact, competing theories about the same data are not matters that can be settled by proclamation. One would need to examine carefully the content and claims of the different personality theories in question in order to deal meaningfully with these issues. My own impression—for whatever it is worth—is that while the relation among different psychoanalytic theories (as well as different personality theories, in general) is, in part, and as Strenger suggests, analogous to the blind men arguing about the elephant (each feeling a different part of the elephant), in significant ways they *are* competing theories and *do* contradict each other. In any case, as I have noted, the matter cannot be settled by competing assertions. It is work that remains to be done.

Although I have given much attention to questions of theory, Strenger's manuscript is concerned mainly with the *treatment* context. He defends psychoanalysis against its critics mainly by describing for us what analysts *do* in their everyday work and by trying to convince us that in this work they function *rationally*—much like the lawyer, the judge, the historian, and the person in the street all of whom try to *make sense* of events, primarily of other people's actions. In short, Strenger tells us again and again that the analyst simply employs common sense (even if it is a disciplined common sense), a common sense that relies on the "principle of [common] humanity" (Blackburn 1984, 277–81). In his most explicit statement of this position, Strenger writes that "the expertise of the analyst does not consist in the mastery of a well-established theory with strong predictive and retrodictive power but rather in the richness of the analyst's view of the *condition humaine,* the nuances and varieties of human emotional life and a keen sensitivity for human affairs in general" (1991, 133).

Strenger may well be accurate in his description of therapeutic practice. He presents a remarkably frank account of the qualities likely to be necessary in order to be a good therapist and I have much sympathy for his point of view. But what does his description have to do with *psychoanalytic theory, contemporary or otherwise*? It is, after all, psychoanalytic *theory* to which Grünbaum addressed his critique. And it is psychoanalytic *theory*—a theory of psychological development, of psychopathology, of mental functioning, and of human nature—that, Freud believed, constituted psychoanalysis's major claim on posterity. As we know, Freud ([1926] 1959, 254) feared that the therapy would destroy the science and it appears that the kind of defense of psychoanalysis that Strenger and others have mounted—that focuses entirely on the therapy—suggests that Freud's fear was prophetic.

Implicit in the just quoted passage from Strenger's manuscript is the assumption that any mature person who has a rich view of the *condition humaine* and who is sensitive to the nuances and varieties of human emotional life will make a good analyst and a good therapist. There may be a good deal of truth in this view and at least some supporting evidence for it. For example, the Rioch et al. (1963) study carried out many years ago suggests that mature women with as little as one year of training could be effective therapists. Strupp and Hadley's (1979) report shows a similar and more recent finding that mature and sensitive college professors do as well in helping clients as professional therapists. If, as these studies suggest and Strenger believes, it is the therapist's richness of view and sensitivity to the nuances of emotional life alone that counts and if "the mastery of a well-established theory" is irrelevant to one's effectiveness as an analyst, then why bother with all the didactic training one major purpose of which *is* mastery of psychoanalytic theory. Is there any relationship at all between the theory and the therapy? If Strenger is correct, one would have to conclude that psychoanalytic therapy works in a placebo manner when placebo is defined as effectiveness that is attributable to processes and factors *not* identified by the theory (Grünbaum 1986). But, if, as Strenger suggests, there is little real connection between the therapy and the theory (also, see Berger 1985), why discuss and describe the *therapy* as a way of responding to criticisms of the *theory*? Strenger directs little or no attention to any of these questions and issues. But without these questions and issues being addressed, Strenger's defense of psychoanalysis is seriously wanting.

Even if one limited one's interest in psychoanalysis to a theory of therapy, one would want to understand the specific therapeutic *processes* by which the richness of the therapist's view and his or her sensitivity to the nuances and varieties of emotional life would be translated into effective therapy. One would also want to know whether the theoretical conception of psychotherapy implicit in this view is compatible with a psychoanalytic theory of psychotherapy and, if it is, *which* psychoanalytic theory of psychotherapy.

Does the irrelevance of mastering a well-established theory refer to psychoanalytic etiological theory, to a psychoanalytic theory of development, or to psychoanalytic theory of treatment? Strenger does not seem to be aware that his observations and hunches regarding the irrelevance of mastering a theory and the importance of the therapist's richness of view and sensitivity to emotional nuances for therapeutic effectiveness themselves constitute the beginnings of a theory of psychotherapy. That is, one can develop a theory of therapy, a central component of which is the claim that the factors that make for effective therapy are the operation of common sense, knowledge of the "*condition humaine,*" the richness of one's view, and sensitivity to emotional nuances, rather than mastery of theory. Thus, such a theory would suggest that a therapist could know next to nothing about theories of therapy and nevertheless function effectively as a therapist. (This theory would, of course, have important implications for selection and training of therapists.) However, the theorist—as opposed to the therapist—would have an obligation to develop and spell out the specific therapeutic *processes* which "permit" common sense, richness of view, and so on, to constitute the effective ingredients in psychotherapy. Furthermore, the theorist—as opposed to the therapist—would also have an obligation to test the theory in some relatively systematic, rigorous, and empirical way. These latter steps are, to so speak, "metacommonsensical," that is, they attempt to determine whether and in what way "common sense" factors are critical for effective psychotherapy. In short, the defense of psychoanalytic *theory*—even a theory limited to the treatment context—cannot rest on the argument that analysts merely rely on refined common sense in their work.

The defense of psychoanalysis as a "rational discipline" that relies mainly on common sense and general (actual or purported) knowledge of the "*condition humaine*" is a defense that does away with any psychoanalytic theory, whether Freudian theory, ego psychology, object relations theory, self psychology, or any other variant of psychoanalytic theory. Strenger has essentially given up the ghost by stripping so-called psychoanalytic theory to the bare generic essentials of refined common sense. Is there anything left of psychoanalytic theory?

According to Strenger, the repression hypothesis is not central to contemporary psychoanalytic theory. That is, in his version of contemporary psychoanalysis, repression is not etiologically important and lifting repression is not therapeutically important. What, then, is etiologically and/or therapeutically critical in contemporary psychoanalytic theory? Strenger does not address the etiological component of the question (that would involve him in a discussion of an aspect of psychoanalytic theory that is relatively independent of the treatment context), but replies to the latter by observing that "the aim of interpretation . . . is to provide a narrative which gives intelligibility to the patient's previously inexplicable patterns of action, thought, and emotion" (1991, 86). Let us assume that Strenger is correct in his assertion that the aim of interpretation is to provide an intelligibility-enhancing narrative. What, one wants to know, are the *theoretical sources* of the analyst's narrative? As we have seen, Strenger suggests that these sources are primarily refined common sense. However, in that case, in Strenger's version of contemporary psychoanalysis there is neither an etiological nor a therapeutic theory that goes beyond folk psychology and common sense. There is no systematic theory at all.

As was shown in the earlier part of this essay, contrary to Strenger's description, contemporary psychoanalysis—object relations theory and self psychology, for example—*is* characterized by both etiological and therapeutic theories, however adequate or inadequate they may be. Furthermore, again contrary to Strenger, these contemporary theories, no less than Freudian theory, *do* take as critical in their etiological formulations "the causal nexus" between infantile events and present symptoms (ibid., 73)—even if they emphasize different aspects of infantile experiences. Self psychology, for example, posits a causal nexus between infantile experiences of empathic failure and adult self pathology. Hence, I do not believe that his characterization of contemporary psychoanalytic theory as forgoing any claim of a causal nexus between infantile experience and adult pathology is accurate.

There is no systematic presentation in Strenger's manuscript of precisely what he takes contemporary psychoanalytic theory to be. Instead, there are casual comments scattered throughout the manuscript regarding Strenger's conception of the central features of such a theory. I have found the following features mentioned in the paper:

1. The core of psychoanalysis "is a theory of development and of mental functioning under conditions of conflict . . ." (ibid., 17);
2. "psychoanalytic principles" (ibid., 126) include the tendency of people to avoid thinking about aspects of their life that evoke anxiety and guilt;
3. they also include the tendency to avoid conflict by falsifying one's perception of self and others (ibid.);
4. the tension between the pleasure and reality principles is the primary source of neuroses—that is, the essence of neurosis lies in clinging to the pleasure principle and avoiding or subverting the reality principle (ibid., 174).

If one examines these four features, it is difficult to see what is especially "contemporary" about them. They do not include any of the central tenets, for example, of object relations theory and self psychology. Furthermore, they are all distinctively Freudian. Statements (2) and (3), namely, the tendencies, respectively, to avoid thinking about aspects of one's life that evoke anxiety and guilt and to avoid conflict by falsifying one's perception of oneself can be understood as a "generic" watered down version of the Freudian repression hypothesis.

As already noted, when Strenger refers to contemporary psychoanalysis, he has in mind mainly psychoanalytic *practice* rather than theory. The latter receives only a few casual comments and, as I have shown, the aspects of psychoanalytic theory commented on hardly qualify as contemporary. As for contemporary psychoanalytic practice, as I have also already noted, it can be characterized, according to Strenger, mainly as the operation of refined common sense, "The modernist conception of psychoanalysis allows for the justification of analytic interpretations by ways of reasoning analogous to those to be found in other humanistic disciplines and common sense psychology" (ibid., 65). At another point in the manuscript, Strenger notes that contemporary psychoanalysis is not especially concerned with uncovering repressed material, but with understanding the ego's way of coping with inner and outer threat. At still another point, Strenger observes that "the patient's present mental functioning—mainly in the transference—is the object of most of the analyst's attention" (ibid., 74).

Finally, I come to Strenger's view that the central tenets of psychoanalysis should not be construed "as hypotheses to be tested directly" (ibid., 195). Instead, they should, according to Strenger, "be regarded as the *hard core of a research program in Lakatos' (1978) sense*" (ibid., 195; emphasis in the original), and "[psychoanalysis] should be assessed according to its overall fruitfulness in stimulating research and as a form of therapy" (ibid., 186). Following Lakatos, Strenger notes that the "hard core of a research program" refers to the "protected" status of the central tenets of a complex theory. Even contrary evidence does not easily lead to the relinquishment of these central tenets. Instead, these central tenets continue to be preserved through the use of auxiliary hypotheses that are intended to "protect" them.

It seems to me that Strenger may be involved in a contradiction or at least a serious inconsistency here. If anything could serve as the "hard core" of the psychoanalytic research program and if anything in psychoanalytic theory has stimulated research, it is Freud's repression hypothesis, which he viewed as the "cornerstone" of psychoanalysis (the latter is precisely the major reason that Grünbaum chose to focus on it). Yet, one of Strenger's main claims is that contemporary psychoanalysis has essentially abandoned Freud's "hard core" repression hypothesis, both in its etiological and therapeutic contexts. However, following Lakatos, Strenger tells us that the hard core of a research program is not even overthrown by contrary data. How, then, did it get overthrown in the so-called contemporary version of psychoanalytic theory and practice? On the basis of what kind of considerations? Strenger does not adequately address these questions.

I believe that after examining the different variants of so-called contemporary psychoanalytic theory, including Strenger's "stripped-down" version, one must conclude that the theoretical formulations and claims that constitute contemporary psychoanalytic theory are on no firmer epistemological ground than the central formulations and claims of Freudian theory. Thus, complaints such as those voiced by Flax (1981) that Grünbaum's critique of Freudian theory "is like throwing out physics because there are unresolved problems in Newton's theory" (p. 564) simply have no force. There is no evidence that contemporary psychoanalytic theories have remedied the epistemological and methodological difficulties that are associated with Freudian theory.

NOTES

1. Note how similar these ideas are to the Freudian assumption, "corrected" by ego psychology, that were it not for the fact that thinking and other cognitive functions are necessary for drive gratification, that is, were it not for the facts that wishing does not make it so and that planning and action in the real world are necessary for drive gratification, thinking and other cognitive functions would never develop, "[A]ll this activity of thought merely constitutes a roundabout path to wish fulfillment which has been made necessary by experience. Thought is after all nothing but a substitute for a hallucinating wish . . ." (Freud [1900] 1953, 605–06).

2. This issue is complex and much more can be said about it than space here allows. I do, however, want to make a few additional observations. Freud ([1916–1917] 1963, 445) did comment that patients get better, not through intellectual understanding, but out of love for and desire to please their analyst. However, far more frequently, Freud stressed the role of insight and lifting of repression. Most important, however *given the logic of the Freudian theory of mental functioning, psychopathology and treatment* (including such concepts as analytic neutrality and the "blank screen" role of the analyst), it is clear that it is insight and the lifting of repression, not a benevolent therapeutic relationship per

se, that are the critical vehicles for psychoanalytic change. If the therapeutic relationship is important in the Freudian theory of treatment, it is so as a context and vehicle for the achievement of insight and the lifting of repression.

One question that arises with regard to the object relations concept of treatment (and, as we will see, with regard to the self psychology conception of treatment) is what is distinctively psychoanalytic about a point of view in which a benevolent relationship (a "good object situation,"—Fairbairn 1952) is the *primary* therapeutic vehicle of treatment. How would such a point of view differ, for example, from Gendlin and Roger's (1967) claim that "unconditional positive regard" (pp. 10–11) is the primary vehicle in effective psychotherapy or the general commonsense "chicken soup" (psychological "chicken soup" that is) conception of psychotherapy?

Freud ([1914] 1957a, 16) remarked that if transference and resistance play a central role in one's approach to treatment, then one is doing psychoanalysis. Perhaps this is so in a *technical* sense, that is, if one limits one's concerns to treatment techniques and is not concerned with theory. If one is concerned with theory, then clearly the concepts of transference and resistance are embedded in a theoretical structure in which assumptions are made, for example, about the role of infantile instinctual wishes and repression in psychic functioning. If transference and resistance do play a central role in the object relations theory (or self psychology theory) of treatment, at best they bear a "family resemblance" to Freud's concepts of transference and resistance. Certainly, one needs to unpack the relationship among the various definitions of these concepts, including the nature and degree of overlap as well as discrepancy.

3. Fairbairn (1952) also refers to the repression of "intolerably bad internalized objects" (p. 62), which means the banishing into unconscious status of "bad" experiences and memories of deprivation and rejection at the hands of the object. Of course, here repression no longer refers to the banishment of wishes and impulses, but it adheres to a general conception of repression in which its hallmark is the removal into the unconscious of mental contents that are too painful to bear consciously. Fairbairn presents a complex scheme in which internalized objects and the subsidiary egos attached to these objects are repressed.

4. There is one way in which the therapeutic importance of lifting repression (and the accompanying awareness and insight) would appear to be minimized in object relations theory. Fairbairn (1952) remarks that "the actual overcoming of repression as such would . . . appear to constitute if anything a less formidable part of the analyst's difficult task than the overcoming of the patient's devotion to his repressed objects . . ." (p. 75), meaning that awareness and insight (i.e., lifting of repression) are less difficult than the patient's affective connection to her or his early repressed objects. Something parallel to this may also hold in Freudian theory. Thus one may become aware of and acquire insight into the infantile nature of one's repressed wishes and strivings and yet continue to find it difficult to give up the fantasy (and hope) that these wishes will be gratified—either in the transference or elsewhere.

5. The relatively recent clinical theory versus metapsychology debate (e.g., Klein 1976, Gill 1976) as well as the construal of psychoanalysis as a herme-

neutic discipline can both be understood in the context of the given developments. With regard to the first issue, one response to the barrage of cogent criticisms of Freudian theory was to, in effect, concede that while much of Freudian theory was admittedly untenable, this was largely true only of the more abstract and speculative metapsychology. By contrast, the purportedly more concrete and observationally grounded *clinical* theory enjoyed a more secure evidential and epistemological status. This is a complex issue and I do not intend to discuss it at any great length here (see Rubinstein 1976 for an excellent discussion of this issue).

However, it seems to me that this argument represents, in part at least, a "salvaging" operation—that is, an attempt to concede the legitimacy of many key criticisms of Freudian theory while, at the same time, maintaining that the core of Freudian (the so-called clinical theory) enjoyed a more secure and defensible status. However, it was precisely the clinical theory—at the core of which is the repression hypotheses—to which Grünbaum (1984) directed his cogent criticisms. Hence, the defense of psychoanalysis had to take a somewhat different direction, namely, the argument that neither Freudian clinical theory nor the metapsychology represented *contemporary* psychoanalysis.

6. Strenger's argument can clearly be seen to be precisely an example of a typical reaction to Grünbaum's criticisms that was characterized at the beginning of this essay.

REFERENCES

Balint, A., and M. Balint. 1965. *Primary Love and Psychoanalytic Technique.* New York: Liverwright.

Berger, L. S. 1985. *Psychoanalytic Theory and Clinical Relevance: What Makes a Theory Consequential for Practice?* Hillsdale, NJ: The Analytic Press.

Blackburn, S. 1984. *Spreading the Word.* Oxford: Oxford University Press.

Bowlby, J. 1969. *Attachment and Loss.* Vol. 1, *Attachment.* London: Hogarth Press.

———. 1973. *Attachment and Loss.* Vol. 2, *Separation.* London: Hogarth Press.

———. 1980. *Attachment and Loss.* Vol. 3, *Loss.* London: Hogarth Press.

Breuer, J., and S. Freud. [1893–1895] 1955. "Studies on Hysteria." *Standard Edition.* Vol. 2. London: Hogarth Press, pp. 19–305.

Cushman, P. 1990. "Why the Self is Empty: Toward a Historically Situated Psychology." *American Psychologist 45:* 599–611.

Eagle, M. 1986a. "Critical Notice: A. Grünbaum's 'The Foundations of Psychoanalysis: A Philosophical Critique.' " *Philosophy of Science 53:* 65–88.

———. 1986b. "Theoretical and Clinical Shifts in Psychoanalysis." *American Journal of Orthopsychiatry 57:* 175–85.

———. 1987. *Recent Developments in Psychoanalysis: A Critical Evaluation.* Cambridge, Mass.: Harvard University Press.

Fairbairn, W. R. D. 1952. *Psychoanalytic Studies of the Personality.* London: Tavistock Publications and Routledge & Kegan Paul.

Firestein, S. 1978. *Termination in Psychoanalysis*. New York: International Universities Press.
Flax, J. 1981. "Psychoanalysis and the Philosophy of Science: Critique or Resistance?" *Journal of Philosophy 78:* 561–69.
Freud, S. [1900] 1953. *Standard Edition*. Vols. 4 and 5, *The Interpretation of Dreams*. London: Hogarth Press.
———. [1914] 1957a. "On the History of the Psychoanalytic Movement." *Standard Edition,* vol. 14. London: Hogarth Press, pp. 3–66.
———. [1914] 1957b. "On Narcissism." *Standard Edition,* vol. 14. London: Hogarth Press, pp. 67–102.
———. [1915] 1957. "Instincts and Their Vicissitudes." *Standard Edition,* vol. 14. London: Hogarth Press, pp. 110–40.
———. [1926] 1959. "The Question of Lay Analysis." *Standard Edition,* vol. 20. London: Hogarth Press, pp. 179–258.
———. [1916–1917] 1963. "Introductory Lectures on Psychoanalysis." *Standard Edition,* vol. 16. London: Hogarth Press, pp. 243–496.
Gendlin, E. T., and C. R. Rogers. 1967. "The Conceptual Context." In C. R. Rogers, ed., *The Therapeutic Relationship and Its Impact: A Study of Psychotherapy with Schizophrenics*. Madison: University of Wisconsin Press, pp. 3–21.
Gill, M. M. 1976. "Metapsychology is Not Psychology." In M. M. Gill and P. S. Holzman, eds., *Psychology versus Metapsychology: Psychoanalytic Essays in Memory of George S. Klein*. New York: International Universities Press, pp. 71–105.
Goldberg, A. 1978. *The Psychology of the Self: A Casebook*. New York: International Universities Press.
Greenberg, J. R., and S. A. Mitchell. 1983. *Object Relations in Psychoanalytic Theory.* Cambridge, Mass.: Harvard University Press.
Grünbaum, A. 1984. *The Foundations of Psychoanalysis: A Philosophical Critique*. Berkeley and Los Angeles: University of California Press.
———. 1986. "The Placebo Concept in Medicine and Psychiatry." *Psychological Medicine 16:* 19–38.
Guntrip, H. 1969. *Schizoid Phenomena, Object Relations, and the Self.* New York: International Universities Press.
Harlow, H. F. 1958. "The Nature of Love." *American Psychologist 13:* 673–85.
Hartmann, H. 1958. *Ego Psychology and the Problem of Adaptation*. New York: International Universities Press.
———. 1964. *Essays on Ego Psychology: Selected Problems in Psychoanalytic Theory.* New York: International Universities Press.
Holt, R. R. 1962. "A Critical Examination of Freud's Concept of Bound vs. Mobile Cathexis." *Journal of the American Psychoanalytic Association 10:* 475–525.
Holzman, P. 1976. "The Future of Psychoanalysis and Its Institutes." *Psychoanalytic Quarterly 65:* 250–73.
Horney, K. 1945. *Our Inner Conflicts*. New York: Norton.
Klein, G. S. 1976. "Freud's Two Theories of Sexuality." In M. M. Gill and P. S. Holzman, eds., *Psychology versus Metapsychology: Psychoanalytic Essays in*

Memory of George S. Klein. New York: International Universities Press, pp. 14–70.
Klein, M. 1948. *Contributions to Psychoanalysis, 1921–1945*. London: Hogarth Press.
———. 1975. *Envy and Gratitude and Other Works, 1946–1963*. London: Hogarth Press.
Kohut, H. 1971. *The Analysis of the Self*. New York: International Universities Press.
———. 1977. *The Restoration of the Self*. New York: International Universities Press.
———. 1979. "The Two Analyses of Mr. Z." *International Journal of Psychoanalysis 60:* 3–27.
———. 1984. *How Does Analysis Cure?* Chicago: University of Chicago Press.
Lakatos, I. 1978. *The Methodology of Scientific Research Programs*. Cambridge, England: Cambridge University Press.
Marmor, J. 1962. "Psychoanalytic Therapy as an Educational Process." In J. Masserman, ed., *Science and Psychoanalytic Series,* Vol. 5, *Psychoanalytic Education*. New York: Greene & Stratton, pp. 286–99.
Mitchell, S. A. 1979. "Twilight of the Idols: Change and Preservation in the Writings of Heinz Kohut." *Contemporary Psychoanalysis 15:* 170–89.
Rioch, M. J.; C. Elkes; A. A. Flint; B. S. Usdansky; R. G. Newman; and E. Silber. 1963. "National Institute of Mental Health Pilot Study in Training Mental Health Counsellors." *American Journal of Orthopsychiatry 33:* 678–89.
Rubinstein, B. B. 1976. "On the Possibility of a Strictly Clinical Psychoanalytic Theory: An Essay in the Philosophy of Psychoanalysis." In M. M. Gill and P. S. Holzman, eds., *Psychology versus Metapsychology: Psychoanalytic Essays in Memory of George S. Klein*. New York: International Universities Press, pp. 000–000.
Sandler, J., and B. Rosenblatt. 1962. "The Concept of the Representational World." *Psychoanalytic Study of the Child 17:* 128–45.
Shanberg, S. M., and C. M. Kuhn. 1980. "Maternal Deprivation: An Animal Model of Psychosocial Dwarfism." In E. Usdin, T. L. Sourkes, and M. B. Youndin, eds., *Enzymes and Neurotransmitters*. New York: John Wiley, pp. 374–93.
Storr, A. 1988. *Solitude: A Return to the Self*. New York: Ballantine Books.
Strenger, C. 1991. *Between Hermeneutics and Science: An Essay in the Epistemology of Psychoanalysis*. Madison: International Universities Press.
Strupp, H., and S. W. Hadley. 1979. "Specific vs. Nonspecific Factors in Psychotherapy." *Archives of General Psychiatry 36:* 1125–36.
Sullivan, H. S. 1953. *The Interpersonal Theory of Psychiatry*. New York: Norton.
Truax, C. 1966. "Reinforcement and Nonreinforcement in Rogerian Psychotherapy." *Journal of Abnormal Psychology 71:* 1–9.
Winnicott, D. W. 1958. *Collected Papers: Through Pediatrics to Psychoanalysis*. New York: Basic Books.

16

Philosophers on Freudianism

An Examination of Replies to Grünbaum's *Foundations*

Edward Erwin
Department of Philosophy, University of Miami

In the first round of replies to Grünbaum's *Foundations of Psychoanalysis: A Philosophical Critique* (1984), most of the critics were nonphilosophers. Philosophers now have had their turn. In what follows, I discuss some of the issues they have raised.

The Tally Argument

Although *Foundations* (Grünbaum 1984) is mainly concerned with the clinical support for the psychoanalytic theory of repression, Grünbaum also makes a number of significant scholarly points about Freud's work. One of these concerns the interpretation of the following claim, "After all, (a patient's) conflicts will only be successfully solved and his resistances overcome if the anticipatory ideas he is given (by his analyst) tally with what is real in him" (Freud 1953–1974, vol. 16, 452). Grünbaum interprets this remark as implying the causal indispensability of (veridical) psychoanalytic insight into the etiology of a psychoneurosis if it is to be permanently cured: no correct psychoanalytic insight, no cure. On this interpretation, Freud need only add the assumption that some of his patients have been cured of their psychoneuroses, and he will have an argument for the truth of some psychoanalytic interpretations. Grünbaum dubs the resulting argument "The Tally Argument," but appeals to spontaneous remission and certain successes of behavior therapists to refute the first premise. David Sachs (1989) replies, however, that Grünbaum has misinterpreted Freud's "tally" remark; correctly interpreted, it implies nothing about the failure or success of nonanalytic modes of therapy or the possibility of spontaneous remission. He gives two reasons for this contention. First, in the very same lectures in which the "tally" remark appears, Freud says that sometimes his earlier use of hypnotic suggestion resulted in success that was "complete and permanent." Taken in isolation, what Freud says here seems clear enough, but other evidence needs to be considered. To begin with, later in the same 1917 lectures, Freud contrasts hypnotic treatment with psychoanalysis. The former, he notes, acts like a cosmetic, leaves the patient inert and unchanged, and unable to resist any fresh occasion for falling ill. That sounds like a description of a "cure" that is neither complete nor permanent. More tellingly, in his earlier papers, Freud explicitly denies hypnosis's capacity to produce permanent cures, "I gave up the suggestive technique, and with it hypnosis, so early in my practice because I despaired of making suggestion powerful and enduring enough to effect permanent cures. In every severe case I saw the suggestions which had been applied crumble away again; after which the disease or some substitute for it was back once more" (Freud 1953–1974, vol. 7, 261) and "It is true that suggestion can bring about a cessation of the *symptoms* of an illness—but only for a short time. At the end of this time they return and have to be repelled once again by renewed hypnosis and suggestion" (ibid., 301; emphasis added).

Given Freud's repeated insistence that hypnosis is incapable of producing permanent, satisfactory cures of neuroses, it is unreasonable to take one isolated remark, as Sachs does, as definitive of Freud's view of the effectiveness of hypnosis. Still, some explanation should be given of the remark that Sachs relies on. Grünbaum's (1984, 156–57) explanation is that Freud expressed himself misleadingly. In the cases to which Freud was referring, the "cure" was only symptomatic; the underlying cause of the neurotic symptoms was not eliminated. For example, in one such case, a woman's hysterical symptoms prevented her from breast-feeding her first child. Her symptoms were eliminated by hypnosis, but the cure was "permanent" only in the sense that no relapse occurred until she had her second child. On Grünbaum's interpretation, then, Freud meant something like the following: His earlier use of hypnosis sometimes brought about a complete removal of the patient's symptoms—but not their underlying cause—and the symptom removal was "permanent" in that relapse did not occur unless and until a fresh occasion developed for a reactivation of the neurosis. If one rejects this interpretation and takes Freud, as Sachs does, to be saying that hypnosis sometimes brought about a permanent elimination of the neurosis, that is, the symptoms *and* the underlying cause, then one will find difficulty in making sense of so much else that Freud said about the shallowness of hypnosis.

Sachs's second reason for his interpretation pertains to spontaneous remission. In a paper published in 1913, Freud agrees that certain disorders accessible to psychoanalytic treatment, including hysterical convulsions, paralyses and symptoms of obsessional neurosis, are occasionally subject to spontaneous recovery. However, this passage, too, is not decisive. Freud appears to be talking about remission of *symptoms,* leaving open the possibility that their underlying causes remain intact.

I conclude that Sachs fails to show that Freud believed, at the time he made the "tally" remark, that a permanent cure of psychoneuroses could be had without psychoanalytic treatment.

Consider, next, Sachs's paraphrase of the Tally remark: Unless the suggestions analysts make to their patients correspond to facts about them, an understanding of their conflicts will not be attained, and their resistances will not be defeated. Thus understood, Sachs contends, the Tally remark has no implications concerning the failure or success of nonpsychoanalytic modes of therapy or the possibility of spontaneous remission. If it has no such implications, however, then how does it serve the epistemic purpose it was designed to serve: to fend off the charge, as Sachs himself notes, that psychoanalysis works only by suggestion? Suppose, for example, that nonanalysts can effect permanent cures without giving correct interpretations, or any interpretation at all. If Freud concedes that this can happen, how does he know that the ingredients that make up psychoanalytic therapy—for example, the use of free association, the analysis of dreams and the transference—make any difference as to whether his patients are cured? It could very well be that *both* psychoanalytic and nonpsychoanalytic therapies work primarily by instilling in the patient the belief that he or she will be cured. Indeed, on Sachs's interpretation,

Freud is not even eliminating the possibility that all of his so-called "cures" were really spontaneous remissions. In that case, he is not warranted in believing that the "roundabout road by way of memories of the past" (Freud 1953–1974, vol. 16, 446–47) and so on—that is, the use of psychoanalysis—is therapeutic. In short, on Sachs's interpretation, the Tally remark *even if true* fails altogether to answer the skeptical charge that Freud was trying to answer; on this interpretation, Freud would be guilty of a serious and rather obvious blunder. The absence of supporting evidence is reason to reject Sachs's interpretation. I should note, however, that what I reject is not his paraphrase of the Tally remark, but his claim that it has no implications concerning nonanalytic modes of therapy or spontaneous remission. Even this is not exactly right: What I take Sachs to be saying is that the Tally remark has no such implications even when combined with other Freudian assumptions. If he is not saying this, if his point is merely that the Tally remark *by itself* does not have the relevant implications, then his interpretation is consistent with Grünbaum's.

Other philosophers' interpretations of Freud's Tally remark encounter difficulties similar to the one generated by Sachs's interpretation. For example, Robert Richardson (1990, 675) argues that in the Tally passage, Freud is concerned to argue only that psychoanalysis is not a "suggestive therapy," by showing that the means it uses to gain symptomatic relief differ from hypnotic treatments. Freud does hold, according to Richardson, that insight is necessary to the success of *analytic* techniques, but not that insight is necessary for therapeutic success *per se*. So, Freud's remarks have no implications for either alternative treatment modalities or spontaneous remission.

Richardson's reconstruction suffers from several problems. How, on Richardson's view, did Freud show that psychoanalysis differs from hypnotism in not being a (purely) suggestive therapy? The argument is this: Freud held that because hypnotic treatments worked through suggestion, their symptomatic relief was transient and instable. Correspondingly, if psychoanalysis also worked solely by suggestion, its "cures" would have the same pattern of transience and instability. They do not; so, psychoanalysis does not work (merely) by suggestion (ibid., 674). This is a nice argument, but where in the context of the Tally remark does it appear? Nowhere. Richardson refers to belief in the given argument as Freud's "motivation" for the Tally passage, but adds that Freud "did not make it explicit" (ibid., 672).

The first problem, then, with Richardson's interpretation is that even if Freud were trying to show only that psychoanalysis differs from hypnotism in not being (merely) a suggestive therapy, he would have clearly failed in his purpose. He would have omitted the very argument that Richardson claims he is relying on. It might also be noted that the argument attributed to Freud is unconvincing unless he had some evidence for the assumption that the causal explanation of why hypnosis produced only transient therapeutic effects is that it is a purely suggestive therapy. The key point, however, is that in the Tally passage Freud nowhere gives the argument that Richardson attributes to him. A second problem is that Freud is not merely trying to show that his therapy is not a suggestive therapy; he is also trying to answer the skeptical charge (that he himself raises—Freud 1953–1974, vol. 16, 446–47) that the status of his "important psychological discoveries" are uncertain. Even if Freud were to use the argument that Richardson attributes to him, the argument does not address the status of his psychological discoveries. Finally, even if Freud were concerned only with his therapy in the Tally passage—which he clearly is not: "we should have to attach little weight to what it (i.e., psychoanalytic theory) tells us about what influences our lives, the dynamics of the mind or the unconscious" (ibid., vol. 16, 452)—he wants to answer the charge that the effective ingredients in psychoanalysis are not those specified by his theory (see his earlier 1917 remarks quoted in Grünbaum 1984, 133). He is obviously not answering this charge if, as Richardson claims, he is implying nothing about other treatments or spontaneous remission. I have already argued for this conclusion in discussing Sachs's identical claim, but I will repeat one point. If Freud were conceding the possibility of spontaneous cures, then he would be undercutting his grounds for saying that his patients were cured because of what he calls "the roundabout road by way of memories of the past, discovering the unconscious, etc." For all he would know, his patients might all have been cured because of life events taking place outside of the context of the therapy. The psychoanalyses might well have been mere epiphenomenal trappings.

Richardson, in a footnote (1990, 671), also gives a reason for rejecting Grünbaum's interpretation, but he relies entirely on Sachs's point that Freud admitted the occurrence of spontaneous remission. I have already argued that Sachs fails to support this point.

Another interpretation is offered by two psychiatrists (Grant and Harari 1990). Although theirs is not the same as Richardson's, they, too, deny that Freud is claiming that correct psychoanalytic insights are required to cure a neurosis (ibid., 312). Their account, then, has the same weakness as that of Sachs and Richardson. On their interpretation, even if Freud's Tally remark were true, it would provide no satisfactory answer to the skepticism that Freud is trying to rebut.

Still another interpretation of the Tally passage is offered by Donald Levy (1988). Levy (ibid., 195) claims that Freud seems to be thinking mainly of sexual experiences and their importance, especially childhood experiences. What Freud is saying, then, is this: The fact that the patient's resistance to their recall is overcome and his conflict over their importance is solved is evidence that they were not suggested to the patient by the analyst. In other words, what must "tally with what is real" in the patient is not the therapist's interpretations about the unconscious or the dynamics of the mind, but rather the memories of the patient about the existence of his or her childhood sexual experiences and their importance. On this interpretation, as with those of Sachs and Richardson, Freud is utterly failing to answer the skepticism about his theory and therapy that he himself has just raised and is obviously trying to refute. He is saying merely, according to Levy (ibid.), that experience will show that the analyst is not able to suggest to the patient pseudomemories of his or her childhood sexual experiences. That leaves entirely open, however, whether Freud's theorizing about the causal role of these experiences and about the dynamics of the mind is correct, or whether the "roundabout road" of psychoanalytic therapy was required to cure the patient's neurosis.

Levy also gives reasons to doubt the accuracy of Grünbaum's interpretation. One is that Grünbaum's paraphrase employs the words "therapeutic conquest of his neurosis" and Freud does not use this exact phrase in the Tally passage. This difference, however, is too superficial to make any real difference. When Freud talks of what "affects his intelligence, not his illness" (1953–1974, vol. 16, 446–47), he is clearly talking about a *neurotic* illness, that is, a neurosis, and, as suggested by the next sentence, about what is necessary to eliminate it. Another point raised by Levy concerns a second passage quoted by Grünbaum (1984, 139) in support of his Tally Argument interpretation. In part of the passage (not the part relevant to Grünbaum's point), Freud says "To get the better of such an element another person must be brought in, and in so far as that other person can be of assistance the neurosis will be curable" (1953–1974, vol. 10, 104). Levy (1988, 196) counters that Freud knew very well of at least one person whose neurosis was cured without undergoing psychoanalysis: namely, himself. However, Freud's belief that he cured himself is not evidence that he did not endorse the Necessary Condition Thesis (NCT); the alleged cure followed his *psychoanalysis* of himself. What *is* troublesome, if he believed he cured himself of a neurosis, is his saying that such a cure always requires bringing in another person. That remark is puzzling, however, whether or not we agree with Grünbaum's attribution of NCT to Freud. Rejecting Grünbaum's interpretation does not eliminate the trouble.

Levy has other objections that are too numerous to answer here, so I will conclude hypothetically: If these other objections are no more cogent than the two just discussed, then he has not made an adequate case against Grünbaum's interpretation; in any event, his own interpretation encounters the difficulties I mentioned whether or not Grünbaum is right.

My overall conclusion is this, subject to the caveat about some of Levy's unanswered objections: Grünbaum's interpretation of Freud's Tally passage is plausible given all of the relevant evidence that he and his critics have cited, and it is more plausible than their interpretations. I doubt that Grünbaum would want to claim much more than this concerning the solidity of his evidence; I doubt, at least, that he would want to claim, given the complexities of interpreting Freud, that he has met the "very high order of proof" (Levy 1988, 197) that Levy demands he meet. Levy argues that such a standard should be met given that Grünbaum is resting so much on his attribution of the Tally Argument to Freud. He reasons as follows: According to Grünbaum, a component of the Tally Argument, namely, the Necessary Condition Thesis (NCT), constitutes the foundations of psychoanalysis. His arguments are invalid (Levy 1988, 193), however, if he is wrong in attributing NCT to Freud. There are two mistakes here. First, in speaking of the foundations of psychoanalysis, Grünbaum is not referring to the Tally Argument; he is talking about Freud's clinical evidence and his cornerstone theory—the theory of repression. The second, and more important, mistake that Levy makes is to infer that Grünbaum's arguments would be invalid if Freud did not use the

Tally Argument. This point requires a brief characterization of Grünbaum's arguments.

In Part 1 of *Foundations,* Grünbaum argues that the clinical data that Freud, and so many other analysts, have relied on are unreliable because of the intermingling of suggestibility factors. Freud's solution to this difficulty was his appeal to the Tally Argument. That argument, however, is unsound, as evidenced by spontaneous remission and the success of certain behavior therapies. In Part 2, Grünbaum argues that other epistemic liabilities exist besides suggestibility *and* that even if the data were assumed to be epistemically uncontaminated, Freud's arguments for his interpretation of his clinical data would still not succeed (see Grünbaum 1984, 172). Suppose that Grünbaum were now to concede that Freud never endorsed the Tally Argument. Would that concession affect his criticisms of the Freudian interpretation of the clinical data? It would not. Subtracting the Tally Argument from the Freudian corpus would do absolutely nothing to answer the criticism that the clinical data are contaminated or that Freud's arguments are flawed even assuming no contamination; nor would keeping the argument in Freud's hands help his case, given that the argument is now known to be unsound. In short, Grünbaum's case against the alleged probative value of the Freudian clinical data does not rest, even in part, on his attributing the Tally Argument to Freud. Consequently, Paul Robinson's (1993) main reply to Grünbaum's critique fails; for it rests on the false assumption that the critique "hinges on" (p. 259) Grünbaum's construction and criticism of the Tally Argument. Robinson also cites some arguments of David Sachs (1989), but I criticize these later (see the section on Extraclinical Evidence).

Other philosophers also seem to have misunderstood the role of the Tally Argument in Grünbaum's work. For example, referring to psychoanalytic therapy, Richardson (1990) writes, "Nonetheless, we should be reluctant to accept Grünbaum's more negative assessment of psychoanalysis based on his examination of the 'Tally Argument'; or, at least, if we embrace it, we should not do so on the basis of his argument" (p. 671). Grünbaum's (1984, 161) negative assessment of the therapy is based on the results of outcome studies, *not* on his criticism of the Tally Argument. The negative assessment, moreover, is not so bold as Richardson (1990, 670) says: that psychoanalysis must be deemed a "placebo" therapy. In the passage Richardson refers to,

Grünbaum (1984, 165) is making a hypothetical claim: that if the "ghost" symptom hypothesis of Rhoads and Feather (1974) is true, then psychoanalysis must be deemed a placebo hypothesis. He does not endorse, however, the antecedent of this proposition. Grünbaum's weaker negative judgement is that outcome studies have failed to show that psychoanalysis is superior to other treatments. The bolder assessment is still somewhat hedged, "But, if analytic treatment is thus not superior to its rivals in the pertinent diagnostic categories, it becomes quite reasonable—though *not* compelling—to interpret its therapeutic achievements as placebo effects" (Grünbaum 1984, 161).

Another misunderstanding of the Tally Argument's role is evidenced in the following: "But nor can we go whole-heartedly along with Grünbaum, who argues that much of the scientific respectability stands or falls, and in fact falls, with the Tally Argument; as Hopkins points out [reference omitted], there is more to Freud's claims—specifically 'interpretative considerations'—than the Tally argument *alone* can justify" (Wilkes 1990, 245). This comment is likely to mislead the reader to infer the following: so, if Wilkes has him right, Grünbaum thinks he can demonstrate merely by refuting the Tally Argument that much of the scientific respectability of Freudian theory falls. On the contrary, if Grünbaum really believed this, he need not have written part 2 of his book; the Tally Argument is rejected in part 1. He does not, of course, hold the view that Wilkes attributes to him. He agrees (how could he not?) that Freud and his followers have based their theoretical claims partly (in fact, mainly) on clinical evidence, including interpretative considerations such as thematic affinities. Rejecting the Tally Argument is obviously insufficient to dispose of such evidence. As I noted earlier, Grünbaum tries to show via a variety of specific arguments that the clinical data have little probative value. He nowhere claims that if the Tally Argument falls, then we can skip the hard job of demonstrating the weaknesses of the clinical data.

Freud's Tally Argument does not strengthen the case for his theory, given that the NCT has been disconfirmed, but does his use of it at least enhance his reputation as a methodologist? A. A. Derksen (1992) argues that it does not. The issue here is a relatively minor one. Even if Grünbaum's partially favorable verdict on Freud as a methodologist were wrong, this would have no direct bearing on

Grünbaum's appraisal of the Freudian evidence, nor does Derksen claim that it would. Consequently, I will be brief.

Derksen gives three reasons for his verdict:

a. Freud was not justified in believing NCT;

b. if the argument and the considerations for it were evidence of methodological sophistication, they would equally well demonstrate the methodological sophistication of modern-day evangelists; and

c. NCT is trivial given Freud's theory about the unconscious and given the medical dynamics model that he used.

I agree with (a), but I have doubts about (b) and (c). In support of (b), Derksen constructs a case of modern-day evangelists who preach that only knowledge of God provides real happiness. When confronted with challenges analogous to those faced by Freud, they employ an argument containing their NCT: Without knowledge of God, one cannot be really happy (ibid., 94). Contrary to what Derksen suggests, however, their argument, unlike Freud's Tally Argument, fails to dispose of the epistemological challenges to their view *even if* its premises are warranted. Derksen (ibid., 77) labels one challenge to Freud "the placebo charge": How could Freud exclude the possibility that his therapeutic success was just a placebo effect? The analogous question for the evangelists is: How can they rule out the possibility that their clients' happiness was merely a placebo effect, and did not result from the "therapy"? (Derksen ibid., 95, mentions preaching and communal praying as part of the evangelists' therapy.) Establishing the evangelists' NCT *and* that some clients are really happy, however, does not dispose of the "placebo charge," for it does not follow from these two assumptions that the evangelists' therapy made any difference to the happiness of the clients. All that follows is that some clients have knowledge of God, something that they might have had before they entered the therapy. It also fails to follow that knowledge of God causes their real happiness. The evangelists' NCT, after all, states only a necessary condition for being happy. It is consistent with their NCT that many people who have knowledge of God are unhappy and need some secular therapy, or even money, fame and freedom from disease, to become happy. Clear and important disanalogies, then, lie between Freud's Tally Argument and the argument of the evangelists.

One could try to repair the argument that Derksen attributes to the evangelists to make it adequate for meeting the epistemological challenges they face. However, it might then turn out that the use of such a reconstructed argument might be evidence of methodological sophistication. We cannot assume that merely because an evangelist employs an argument that its use is no mark of sophistication.

I am also puzzled about (c). Even if one accepts Freud's medical model and his theory about the unconscious, one could reasonably question whether a cure requires *true* interpretations on the part of the analyst, as NCT requires. Indeed, some modern-day analysts argue that false interpretations can be curative. So, I question whether NCT follows trivially from the conjunction of Freud's medical model and his theory of the unconscious. However, I will not argue the point further, but instead will try to show that even if Derksen's (a), (b) and (c) were all correct, that would do nothing to counter Grünbaum's point that "the epistemological considerations that prompted Freud to enunciate his Tally Argument make him a sophisticated scientific methodologist, far superior than is allowed by the appraisals of friendly critics like Fisher and Greenberg (1977) or Glymour (1980), let alone by very severe critics like Eysenck" (1984, 128).

What Grünbaum contends—and Derksen (1992, 75) quotes this passage—is that the evidence for Freud's methodological sophistication is *not* his mere use of the Tally Argument, but the epistemological considerations that prompted him to use it. Consider an analogy. Suppose a modern Freudian claims without further explanation that his therapy cures some of his patients. Without additional information, we might accuse the analyst of naivete, for he ignores the considerable evidence that spontaneous remission or placebo factors might well have accounted for the therapeutic outcomes. Suppose, however, that we discover a publication in which the analyst had addressed both of these issues. First, he argues, as some sophisticated non-Freudian methodologists have, that a placebo control makes no sense when studying the effects of psychological therapies and that the role of placebo factors can be safely ignored when gauging the effects of such therapies. Second, he agrees with Eysenck's claim of a two-thirds spontaneous remission rate for neuroses within two years of onset, but argues that this is an average for all neuroses and that the kinds of neuroses he treats (say, phobias and sexual disorders) have less than a 30 percent remission rate—far less than his own 50 percent cure rate. Let us assume that at least one premise of both arguments lacks support, that similar arguments could be used by an evangelist faith healer, and that at least one premise in each argument is trivial given some theory the analyst holds. Still, should we not withdraw, or at least modify, our original complaint? The analyst's arguments are not cogent, but the published paper shows that he was aware of the issues we claimed he neglected. Analogously, some critics accuse Freud of wrongly inferring that his interpretations are true without considering the possibility that his patient's behavior resulted largely from suggestion. As the 1917 lectures clearly show, (see Grünbaum 1984, 138), however, Freud was well aware of the epistemological problems posed by the possible role of suggestion and tried to resolve them. Specifically for this reason, and not merely because he used the Tally Argument, Grünbaum credits Freud with more methodological sophistication than do some of his critics.

Psychoanalysis as Natural Science

Grünbaum (1984, 2–9) argues that on Freud's view the psychoanalytic enterprise has the status of a natural science. This point, even if correct, does not show that psychoanalysis should be judged by natural science evidential standards, but it is relevant to the issue of what standards Freud himself believed to be appropriate for his theories. If Grünbaum's interpretation is correct, it is highly unlikely, unless Freud held a blatantly inconsistent view, that Freud would have asked that his theory be judged by nonnatural science's standards. Grünbaum's reading of Freud, however, has been challenged by John Forrester (1986), who accuses Grünbaum of "shoddy [scholarship]" (p. 671), "the hard-sell" (p. 670) and "tendentiously" (p. 671) quoting out of context. Are these rather extreme accusations warranted? One of Forrester's prime examples of "tendentiously" quoting out of context is Grünbaum's use of a quotation from Freud concerning the natural science issue, "Psychoanalysis is a part of the mental science of psychology. . . . Psychology, too, is a natural science. What else can it be?" (1953–1974, vol. 23, 282). Forrester contends that Freud is concerned here to indicate not just the similarities but also the differences between physics and psychoanalysis, as suggested by Freud's next sentence which Grünbaum does not quote, "But its case is different" (ibid.). Contrary to what Forrester suggests, however, this last quoted sentence is of no relevance to the issue Grünbaum discusses. In saying that its case is different, Freud is not affirming any epistemological difference between physics and psychology; he is noting a difference between people who think they are qualified to make judgements about hypotheses in the respective disciplines. The entire quotation from Freud reads, "Psychology, too, is a natural science. What else can it be? But its case is different. Not everyone is bold enough to make judgements about physical matters; but everyone—the philosopher and the man in the street alike—has his opinion on psychological questions and behaves as if he were at least an *amateur* psychologist" (ibid., 282–83; emphasis in the original). Clearly in saying that its case is different, Freud is not modifying or in any way backsliding on his insistence that psychology, of which psychoanalysis is a part, is one of the natural sciences.

Forrester also notes that Grünbaum neglects to mention another point made by Freud: That because the science of psychoanalysis has as its subject matter the psychical apparatus, the analogy between psychoanalysis and the other sciences "ends here." The analogy being referred to by Freud is this: Every science, including psychoanalysis, is based on observations arrived at through the medium of our psychical apparatus. The disanalogy is that only psychoanalysis takes the psychical apparatus as its subject matter. Is Freud's point relevant to Grünbaum's thesis that Freud regarded the psychoanalytic enterprise as part of the natural sciences? It is not. Grünbaum does not say, nor does his thesis imply, that on Freud's view, psychoanalysis is in no way different from physics. Of course, the subject matter of psychoanalysis and physics is different; Freud could hardly have failed to notice *that* fact. However, saying that *this* difference exists hardly commits Freud to the view that the theories in the respective disciplines should be judged by different standards.

Forrester's third example of Grünbaum's alleged scholarly mistake concerns the latter's interpretation of Habermas. The issue concerns the patient's application of interpretations to himself or herself. Forrester's point is that such an application is a *necessary* condition of their validity on Habermas's view, not, as Grünbaum contends, either a sufficient condition or the unique necessary condition. Forrester is correct about the interpretation of Habermas, but wrong in his interpretation of Grünbaum. The latter's position is that Habermas holds that according to Freudian theory, the patient's assent to an interpretation is a *necessary* condition of its being correct (Grünbaum 1984, 27). If this were truly a necessary condition, then the patient would be (in Grünbaum's words) "epistemically preeminent" in the sense that he or she would have a veto power over any of the analyst's interpretations. By refusing to accept an interpretation, the patient would guarantee its falsity. Grünbaum does not say, and need not say in order for his criticism to work, that the patient's acceptance of an interpretation on the Habermas view is also sufficient for the truth of the interpretation. He is correct, moreover, in challenging Habermas's attribution to Freud of the view that the patient's agreement is necessary if the analyst's interpretation is to be correct (as opposed to being necessary for making therapeutic progress). On this issue, Freud writes, "We do not pretend that an individual construction is anything more than a conjecture which awaits examination, confirmation or rejection. We claim no authority for it, we require no direct agreement from the patient, nor do we argue with him if at first he denies it" (1953–1974, vol. 23, 367).

I conclude, then, that on each of the aforementioned scholarly points, Forrester is mistaken; in each case, he misinterprets Freud or Grünbaum or both.

Suggestion and Placebos

The issues discussed so far have been primarily exegetical; I turn now to issues about the quality of the Freudian clinical evidence. One concerns the possibility of contamination by suggestibility factors. For example, Fine and Forbes (1986) argue that Grünbaum makes too much of this possibility, "What we find problematic is that this generic possibility, like Descartes' demon, relies on a general, always-available doubt, independent of any specific mode of instantiation or mechanism of operation. Used in this generic way, suggestibility, like its Cartesian counterpart, is just a place-holder for skeptical doubt or fallibility" (p. 238). Other philosophers have either seconded their criticism (Wollheim 1991) or made a similar complaint (Sachs 1989, 355). One response is to note that criticisms of the Freudian clinical data in Part 2 of *Foundations* make no appeal to contamination by suggestibility. Thus, Grünbaum (1986), in reply to Fine and Forbes, writes, "If need be, I could and would rest my case on Part 2 alone"

(p. 277). This response is important, given the tendency of some critics to overlook its availability, but, of course, it does not render Fine and Forbes's criticism altogether irrelevant or unimportant. Grünbaum does make additional comments about their points (1986, 277; see as well 275–76), but I believe there is more to say about what is really a large and complex issue.

To begin with, what is meant by "suggestion"? In a narrow sense, a therapist suggests an idea or course of action to a client by explicitly mentioning it in the course of the therapy. For example, an analyst may tell a client that her problem is due to penis envy. In hypnosis, the hypnotized patient may be instructed to stop smoking or to eat less. Freud, however, did not always use the term in this narrow sense; as Richardson notes (1990, 674), he seems to have meant a variety of things by "suggestion." Clinical psychologists and psychiatrists have followed Freud's practice of using "suggestion" in both a narrow and a wider sense. For example, Fisher and Greenberg (1977, 363), in speaking of suggestion, refer to such items as the reinforcement of certain types of patient communications, the presence of a prestigious practitioner, and the offering of an interpretation (whether or not it is correct). Grünbaum, too, uses "suggestion" sometimes to refer to the analyst's presentation of "anticipatory ideas," but also to conditioning factors. Finally, Fine and Forbes (1986, 238), although they mention no specific factors, note that "placebo effect" is sometimes used as an alternative to "suggestibility."

When "suggestion" is used in the wider sense, it means something like the presentation of a stimulus to produce an uncritical response, such as an action or the acquisition of a belief. So long as it is understood, however, that "suggestion" is not being used in what I called the "narrow" sense, it is not critical that we have an exact definition. If someone asks what we mean by speaking of suggestion in a clinical context, it is sufficient that we can specify the factors we are referring to; giving a general, precise definition is not necessary. Even if we can specify the factors, however, is the suggestibility hypothesis like Descartes's postulation of an evil demon, that is, in Fine and Forbes's phrase, "just a place-holder for skeptical doubt, or fallibility"? In at least one very crucial respect, the two hypotheses are different. Descartes's hypothesis has at least a surface credibility no matter what the data are like. The theorist who appeals to suggestibility, in contrast, runs a real risk of invoking a hypothesis that is patently unwarranted, or worse. If I say that the elimination of the patient's brain tumor was possibly due to suggestion rather than his brain surgery, then what I say may be too incredible to take seriously. Appeals to suggestion, then, are not seemingly plausible in some circumstances.

So far, I have not said anything incompatible with Fine and Forbes's comments. When they speak of the strategy evolved by "real science" (1986, 238), they seem to agree that the suggestibility hypothesis might have to be taken seriously in cases where it can be made specific, concrete and testable. I have argued only that the hypothesis need not be irremediably vague and that unlike Descartes's demon hypothesis, it cannot *reasonably* be invoked, come what may (of course, any hypothesis can be brought in if plausibility is of no concern).

Still remaining is the crucial question of whether the appeal to suggestibility in a psychoanalytic context is warranted. I will begin with Freud's therapy. Fine and Forbes may wish to restrict their skepticism about suggestibility to discussions of his theory, but some of their points are sufficiently general that they should apply to both theory and therapy *if* applicable to either. The reason that suggestion needs to be taken seriously in evaluating Freudian therapy is that a solid body of evidence exists that placebo factors often do account for the sorts of therapeutic changes observed in psychoanalysis. For a wide variety of (nonpsychotic) psychological problems and clients, it has been demonstrated that clients receiving a credible "placebo," in some cases a sugar pill, will improve more than those in no-treatment or wait-list controls and as much as those receiving some standard treatment. In fact, some of the most effective treatments yet devised have sometimes performed no better than a placebo. For example, Beck's cognitive therapy for depression was studied in a series of well-designed experiments and initially produced better results than a placebo and, in some cases, a standard drug treatment. Advocates of the therapy thus anticipated favorable results in the recently completed National Institute of Health study (NIH), probably the most sophisticated and most expensive experimental study of psychotherapy done to date. The preliminary findings were announced at the 1990 meeting of the Association for the Advancement of Behavior Therapy and shocked some supporters of Beck's treatment. Clients receiving a sugar pill did approximately as well as those receiving the cognitive therapy. Furthermore, the sugar pill did as well as the remaining two treatments, a standard drug treatment and interpersonal therapy. Psychoanalysts, unlike cognitive therapists, can point to no well-controlled study in which analysis has outperformed a credible placebo. There have been no experimental studies (that I know of) of long-term, classical psychoanalysis. There have been a few of short-term psychoanalytically oriented psychotherapy and other insight therapies, but the results have not been encouraging. In the Brill et al. (1964) study, for example, the placebo subjects received a sugar pill and brief interviews which were limited to 15 minutes so as to minimize any inadvertent psychotherapeutic effect. The placebo patients improved to approximately the same extent as those receiving the psychoanalytically oriented psychotherapy. Several other studies of insight therapies reviewed by Prioleau et al. (1983) have also shown a failure to outperform a placebo. What accounts for the comparative effectiveness of placebos? Plausible candidates include the following: the client's expectation of being helped, the presentation of plausible sounding (even if false) interpretations, talking about one's problems to a credible therapist, the therapeutic relationship, and the therapist's demand for improvement. It is possible that some of these factors are generally more important than others and that different factors make a causal difference with certain clients and problems and not others. Still, it would hardly be "real science" (Fine and Forbes's phrase) to bracket it off as a *mere* possibility that such factors make a clinical difference, given the repeated findings that sugar pills and various pseudotherapies can do as well with real patients as virtually any kind of psychotherapy studied to date (Prioleau et al. 1983), with the possible exception of behavior therapy and cognitive therapy. In the absence of a placebo control, the appeal to placebo factors as a rival explanation of psychotherapy outcomes can no more be dismissed as a Descartes-demonlike hypothesis than can such an appeal be similarly rejected in the evaluation of drug treatments.

Binns (1990, 534) claims that the concept of a placebo seems curiously misplaced in the context of setting up a control group to test Freudian theory. He offers this observation as a reply to both Grünbaum and me, who are alleged to use as one of our main grounds for "refuting Freudian theory" (ibid., 533) the absence of evidence that Freudian therapy is superior to "therapies based on placebos" (ibid.). However, the main point of the complaint to which Binns is referring is to support skepticism about a *therapeutic* rather than a *theoretical* hypothesis, namely, the hypothesis that psychoanalysis is more than a placebogenic therapy (even if the point also has theoretical implications). Finally, if it were true that the needed placebo-control studies could not be done, as Binns appears to be suggesting (see ibid., 532), then this would strengthen rather than undermine our skepticism about Freudian therapy.

Why, according to Binns, is doing a proper placebo-control study so difficult (and perhaps impossible)? His reason is this: "To exclude what Freudian therapy consists of—including such things as beliefs, attitudes, etc.—from our control group, the latter clearly must not believe that the placebo they are being given is curative of anything" (ibid., 534). This claim is odd. Research has shown that if the placebo subjects do not believe in the efficacy of their treatment, the placebo control is decidedly inadequate (Erwin 1978). This does not mean that subjects never improve when they do not believe in their placebo treatment's efficacy. In a paper on the placebo concept, Grünbaum (1986) cites a report by Park and Covi (1965) that three of their patients suffering from anxiety improved after receiving capsules that the patients correctly believed to be pharmacologically inert. These patients might have improved as a consequence of receiving a placebo treatment despite their disbelief in its efficacy *or,* as Grünbaum (1986, 25) notes, they might have experienced spontaneous remission. Even if we knew that the former were true, that would not undermine the following epistemological point. Given the evidence from other studies (see Erwin 1978, 8–9), the failure to make the placebo treatment credible to the patients in a control group generally ruins an experiment; for, if those in the treatment group show greater improvement than those receiving the placebo, that might plausibly be attributed to the greater confidence in the (nonplacebo) treatment. So, to design a proper experiment, we should do just the opposite of what Binns recommends; we should insure that the placebo subjects *do* believe that the placebos are effective treatments. That would defeat our purpose only if we were trying to show that beliefs and attitudes make no difference in psychoanalytic therapy. Why, however, would we want to show *that*? What we are trying to test is the hypothesis that belief in the therapy and other placebo factors make most or all of the therapeutic difference, that is, that the working through of presumed conflicts by means of free associations, dream interpretation and resolving the transference neurosis matter little or not at all. To test that hypothesis, we can compare the psychoanalysis to a "believed in" placebo lacking the psychoanalytic ingredients, such as a sugar pill. As noted earlier, that has already been done in the Brill et al. (1964) study of psychoanalytically oriented psychotherapy.

I suspect that Binns is confused about a proper placebo design because he has noticed an important difference between medical and psychological experiments (1990, 533), but has drawn the wrong moral. In drug testing, we are trying to rule out the possibility that all improvement is due to psychological factors such as belief rather than to chemical or physiological causes; in studying psychological therapies, however, we are obviously not trying to rule out the operation of psychological factors. Because of this difference, some commentators have claimed that the concept of a placebo, as used in medicine, cannot be used properly in studies of psychotherapy. It is precisely for this reason that Grünbaum (1981, 1985, 1989) has developed a precise explication of the concept (or a very similar concept) that can be used in both disciplines. In particular, this explication serves to distinguish placebos from nonplacebos *within* the class of *psychological* therapies, no less than it makes that distinction within somatic treatments. Suppose that a certain therapeutic theory specifies that its characteristic treatment factors F_1, F_2, and F_3 (e.g., lifting repressions, working through conflicts by means of free association, and resolution of the transference neurosis) are remedial for a disorder *D* (say, depression). On Grünbaum's analysis, *T* is a placebo for *D* if and only if none of the characteristic factors *F* is remedial for *D*. So, if confidence in the therapy typically explains all of the therapeutic gains of psychoanalysis in treating *D*, then relative to psychoanalytic theory, these gains are placebogenic in treating *D*. If we use the concept of a placebo in this way, then we are not forced to say that all psychological therapies are "placebos" merely because their effective ingredients, if any, are psychological in nature. Difficult problems remain in designing and executing a proper placebo experiment which are addressed in Grünbaum's definition of a placebo *control* (1985, 19), but using the concept in Grünbaum's explicated sense would dispose of the difficulty raised by Binn and others.

I return now to Fine and Forbes's comments. Whether or not they had originally wished to include Freudian *therapy* within the scope of their skepticism about suggestion, they might now exclude it and talk only about suggestibility in the validation of the theory. What are the suggestibility factors, they might ask, that supposedly vitiate the probative value of Freud's clinical data for his *theoretical* claims? As with his therapy, there are more than one. For example, investigators have found that individuals will enthusiastically accept bogus interpretations as accurate descriptions of their personalities, especially if the interpreter is perceived as prestigious (see Grünbaum 1984, 240–41). Second, inaccurate interpretations have also been found to correlate with positive changes in the patient's behavior (see Grünbaum 1980). Third, and very important, operant studies of verbal behavior show how the data of free association are influenced by verbal cues, such as the therapist's pausing, making certain reinforcing sounds (such as "un-hunhs") and giving evidence of paying or not paying attention. Judd Marmor also reports that the theoretical allegiance of the therapist influences the type of phenomenological data yielded in free association. Freudians elicit material about oedipal complexes; Horneyites about idealized images; Alderians about feelings of inferiority, and so on (for reviews see Grünbaum 1984, and Fisher and Greenberg 1977). Before assessing the epistemic relevance of these factors, I want to address an objection pressed by Sachs (1989, 353) and Levy (1988, 198). Both point out that Freud tried to meet the charge of suggestibility by relying on the coincidence of the free associations of neurotics with the avowal of psychotics. Why do such data provide "guarantees" (Sachs 1989, 354) that the free associations of neurotics are not due to suggestion? Freud's answer is that the avowals of psychotics cannot be influenced by suggestion, "Nor must we fail to point out that a large number of the individual findings of analysis, which might otherwise be suspected of being products of suggestion, are confirmed from another and irreproachable source. Our guarantors in this case are the sufferers from dementia praecox and paranoia, who are of course far above any suspicion of being influenced by suggestion" (1953–1974, vol. 16, 453).

At least two problems arise with Freud's "guarantee." First, we have no independent evidence for Freud's claim that the associations of psychotics and his neurotic patients did coincide. Given the great variety of ways in which the verbal responses of both groups could have been both similar and different, how do we know that Freud did not mistakenly emphasize the wrong similarities? The second and more important problem concerns his claim that psychotics cannot be influenced by suggestion. The only reason he gives for this claim is that psychotic patients are "of course far above any suspicion of being influenced by suggestion." Sachs and Levy give no reason at all. They apparently take the claim to be obviously true. The lack of empirical support for Freud's claim, however, is not the only problem. Solid evidence has been available for at least 25 years that the verbal behavior of psychotic patients, including schizophrenics, can be greatly influenced by suggestion. In fact, their verbal behavior has been shown to be influenced by the very same sorts of stimuli that affect neurotics, such as uttering "un-hunhs," moving about, and so on. In one early study, Krasner (1958) used a storytelling procedure to study the verbal responses of two males at a veteran's hospital who were diagnosed as having "schizophrenic reactions in remission" (p. 364). In 25 sessions, each subject was asked to make up a story about at least four characters: a mother, a father, a child and an animal. Krasner found that the experimenter could systematically increase the targeted verbal responses by nodding his head, smiling and emitting an "mmm-hmm" sound. One interesting feature of the experiment, from a psychoanalytic point of view, was the targeted verbal responses: all references to the mother figure in the story.

Krasner's study might be of limited relevance to the present issue because his subjects were in remission and perhaps Freud's psychotic subjects were not. That problem does not arise, however, in certain other studies of psychotics. For example, Ayllon and Haughton (1964) studied three hospitalized female mental patients: Two were chronic schizophrenics and one suffered from involutional depression. None were in remission. In one case, a schizophrenic woman's delusional and neutral verbal responses either increased or decreased as a function of the reinforcement or lack of reinforcement of the staff. The reinforcement consisted of listening to or taking interest in the patient's verbalizations, a point of obvious relevance to the practice of psychoanalysis. In a second experiment, the somatic verbal responses of two women (one a schizophrenic and the other a depressed patient) increased after reinforcement and were virtually eliminated when ignored. Salzinger and Pisoni (1961) were also able to condition affect responses in ten female and four male hospitalized schizophrenics through the experimenter's use of such words as "mm-hmm," "uhha," "I see," "yeah," and so on. In the light of the results of these and other studies (e.g., Salzinger and Pisoni 1958), it can no longer be credibly maintained that psychotics cannot be influenced by suggestion. (In fact, the thesis was questionable even in Freud's time; see the reply by Grünbaum [n.d.]).

I conclude that for both Freud's theory and therapy, suggestibility factors are an important source of rival hypotheses. The hypotheses cannot be reasonably dismissed by saying that they are vague or that they are like Descartes's demon hypothesis. Questions remain, however, about the nature of the rival hypotheses and their epistemic role. For example, as to the hypothesis concerning Freudian theory and clinical data, what does it say? On Sachs's (1989, 357) interpretation, Grünbaum is saying that *all* of the Freudian clinical data *may* be due to suggestion. If that were being said, then the suggestibility hypothesis would be compatible with Freudian theory having strong clinical support. If true, the hypothesis would show only what is not at issue: that Freud's hypotheses lack certitude. If, however, "may" were replaced by "probably" in regard to all clinical hypotheses, then the hypothesis is implausible. It is unlikely that everything a patient says and does in an analysis is due to suggestion. Something in between these extremes is needed. Perhaps the following will do (see Grünbaum 1984, 245, 246): Because of the ubiquity in the clinical setting of those sorts of suggestibility factors found to influence clients' verbal behavior—the prestige of the analyst, the analyst's paying or not paying attention, or uttering reinforcing noises, or moving about, agreeing or disagreeing with the client's statement, and so on—there is no general, reliable way to tell which client responses are due to suggestion (in a broad sense) and which are not, unless experimental controls or some adequate substitute are introduced. Consider the following case. A schizophrenic woman drags a broom around a hospital ward all day and a psychiatrist gives a Freudian-like interpretation of her apparently ritualistic behavior. He says that she drags the broom because it is a phallic symbol. Suppose we know, however, that the behavior was conditioned deliberately by attendants who gave her cigarettes as a reward for dragging the broom. This, in fact, is what did happen (see Ayllon et al. 1965): The psychologists were trying to expose the baselessness of many psychiatric interpretations. If we know what caused the behavior, then we can know in advance that if the psychiatrist offers an argument for a rival hypothesis, the argument will lack cogency. Now, suppose that many psychotic patients are dragging a broom around and we know that the psychologists had access to all of them. Even if it were unlikely that conditioning caused *all* of the responses, we could not appeal to observations of the broom dragging to confirm *nonconditioning* hypotheses if we had no reliable way of telling which responses were conditioned and which were not. Analogously, in the absence of experimental controls, if a client improves in respect R_1 and placebo factors were present and are of the sort that often bring about that sort of change, then observation of the change is unlikely to confirm that the therapy caused the improvement. In the case of Freudian theory, matters are more complex. Despite the ubiquity of conditioning factors, clinical data might disconfirm a Freudian hypothesis. To take a standard example, the observation of an overtly homosexual paranoid would disconfirm the proposition that all paranoia is due to *repressed* homosexuality. In addition, there might be some cases where a Freudian can argue cogently for some Freudian hypothesis or other by appealing to clinical data. Grünbaum, too, agrees that the clinical data are not altogether probatively irrelevant, but argues that they cannot bear the weight put on them by those who claim that such data typically can confirm or disconfirm Freudian hypotheses (1984, 245, 265).

It may be useful to summarize several points. First, a suggestibility hypothesis could be analogous to an evil demon hypothesis if it were invoked to show merely that a rival hypothesis lacks certitude. That may be the only comparison intended by Fine and Forbes (1986). There is also, however, an important disanalogy, which is relevant to Grünbaum's argument. Insofar as the evil demon hypothesis has epistemic force beyond demonstrating a lack of certitude, it has this force no matter what our data are. *That is not true of a suggestion hypothesis.* As I noted in the case of brain surgery, an appeal to suggestion may lack all credibility given certain background evidence. Second, although unspecified talk of suggestion may be unclear, we can eliminate or at least diminish the unclarity by specifying to which sorts of factors are being referred. We can then address the empirical question of whether such factors are likely to be causally relevant in a specific context. Third, the suggestibility factors found to be causally relevant in the verbal conditioning studies—for example, taking interest in or ignoring a patient's verbalizations, moving about, saying "mm-hmm," "yeah," and so on—are commonly found in the psychoanalytic setting. For that reason, they are an important source of epistemic contamination in the absence of experimental controls. Fourth, contrary to Freud, Sachs (1989) and Levy (1988), the verbal behavior of psychotics can be conditioned. The final point is the most difficult to state precisely. How much weight should we accord suggestion as an epistemic contaminant in a clinical setting? It is easier to give a precise answer if we are asking only about therapeutic outcomes. Although certain qualifications are needed, in particular, relativization to type of clinical problem is required, a rule can be stated. In the absence of a satisfactory placebo control, we are generally unwarranted in inferring that overcoming resistances, working through the transference, and other such ingredients of psychoanalysis produced the *so-called* "cure." It is much more difficult to state an analogous plausible general rule concerning nontherapeutic Freudian hypotheses. One problem is that the presence of suggestibility factors may not prevent the confirmation of any and all claims that might be called a "Freudian hypothesis." For example, suppose we define "repression" in the following way: Its essence lies simply in the function of rejecting and keeping something out of consciousness (see Kline 1981, 195). If that is all Freudian repression is, if there is no necessary link to the id or ego, or even to a dynamic unconscious, then we may be able to confirm its mere existence on the basis of clinical data. In contrast, showing that infantile repressions typically cause and maintain what Freud called "psycho-neuroses" is obviously much more difficult. In trying to confirm the latter hypothesis in a clinical setting, the problem posed by suggestibility factors is much more formidable. It would be nice if we could sum matters up via a general rule, but I at least do not know how to do that. Even so, when evaluating alleged clinical confirmation of Freud's repression etiology (which is the subject of *Foundations*), the role of suggestibility factors needs to be taken seriously. The absence of an algorithm telling us how seriously to take it is insufficient reason for discounting it.

In discussing the role of suggestion, I have so far begged an important question: Is confirmation of Freud's hypotheses threatened *if* a suggestibility hypothesis constitutes a credible rival? I am assuming that it is, but Fine and Forbes (1986) question that assumption. This issue is discussed in the next section.

Eliminative Inductivism and Causal Hypotheses

Although *Foundations* does not presuppose any general theory of confirmation, it is not clear that Grünbaum can avoid altogether philosophical issues about confirmation. For example, he does need some rationale for appealing to suggestibility factors as a rival explanation. In fact, he has endorsed a view for the validation of causal hypotheses which requires that the purported causal factor be indeed *causally relevant,* referred to sometimes as "Eliminative Inductivism" and also as "Neo-Baconian Inductivism": What does this view consist of and is it justified? Jim Hopkins (1988) challenges (what he takes to be) Grünbaum's requirement that confirmation of a causal claim always requires satisfaction of neo-Baconian standards that are appropriate for causal hypotheses in any field. What are these neo-Baconian standards? They require at least this much, according to Hopkins (1988, 35, 51): There must be a collation of positive instances from both experimental and control groups if there are to be inductively supportive instances. In commonsense psychology, however, we do not employ this standard in judging that a certain motive was a causal factor. Hopkins (ibid., 38) gives as an example our inferring that someone's moving a glass toward a tap derived from a motive to drink. Another example is Freud's inference that in certain cases his wish for a drink had caused him to dream that he was drinking delicious cool water. In these examples, commonsense psychology regards the inferences as warranted despite the absence of any controlled study bearing on the conclusion. In these and other cases, including Freud's interpretation of the Irma dream, we infer a causal connection, Hopkins holds, from a *thematic affinity* between a motive and its effect (ibid., 50). (On thematic affinity, see the next section). Generally, Hopkins (ibid., 39) contends that connection in psychological content is a mark of causal connection. Although Freudian interpretations go beyond common sense, according to Hopkins (ibid., 37, 51), they rely on the same sort of thematic affinity argument used in commonsense psychology. They do not require, at least not always, a collation of instances from both experimental and control groups. Grünbaum's methodology and Freud's actual method partly "pass one another by" (ibid., 51).

One question prompted by Hopkins's account is this: Does Grünbaum (1984) really require that all causal hypotheses be confirmed by controlled experimentation? That does appear to be too high a standard. In a book on behavior therapy (Erwin 1978, 11–12, 14–15), I argued that in certain cases impressive evidence was presented of a causal connection between shock treatment and the modification of the behavior of autistic children; yet, in these cases, no control group was employed. Grünbaum (1984, 259–60) quotes my description of one such case and agrees about the confirmation. Thus, he actually does not hold that confirmation of a causal claim always requires a control group. He rightly makes the point that in the case of the autistic child, background evidence about the natural history of the behavior of such children obviates the need for experimental controls. Grünbaum gives other examples, including causal hypotheses in astronomy, where background evidence supplies, as he puts it, "the probative equivalent of experimental controls" (1984, 259).

As Hopkins (1988, 35) suggests, Grünbaum holds that confirmation of Freudian theory generally requires the use of controlled experiments (or epidemological studies) since the need to demonstrate causal relevance cannot be plausibly met otherwise. Certain qualifications are needed here (see, e.g., Grünbaum 1984, 265), but they are irrelevant to the following point. Insofar as Grünbaum holds that experimentation is necessary to confirm the Freudian hypotheses he discusses, this is not an *assumption* of his argument; it is rather one of his central *conclusions* (see ibid., 278). In demonstrating causal relevance, I do think it reasonable to require the following: To confirm some causal hypothesis *H*, it is necessary to discount any incompatible rival that on our current evidence is at least as plausible as *H*.

The above standard, which may or may not be acceptable to Grünbaum, is challenged by some remarks of Fine and Forbes (1986). They point out that in reasonably complex scientific investigations, one is not able to eliminate all the plausible rivals to a particular hypothesis, "It follows from the fact that science goes on in this situation that for data to count as supporting a hypothesis it is not necessary to rule out all competitors" (ibid., 238). Clearly, without some additional premise, nothing of the sort follows. Freudian investigations may continue for decades without plausible rivals to psychoanalytic hypotheses being ruled out and yet no support may accrue to the latter hypotheses. Perhaps Fine and Forbes meant that it follows if one also assumes that in at least some of the cases where science goes on in the described situation, there is confirmation of a particular hypothesis. To add that assumption, however, would be to assume what they are trying to show that it is a methodological myth that support accrues only in the absence of rivals, that is, plausible rivals. Furthermore, if their view is to be relevant to Grünbaum's assessment, they must add that the confirmation they are talking about is not merely of a very weak sort. Otherwise, Grünbaum (1984, 278) could accept their point as he concludes that the warrant provided by the Freudian clinical data is remarkably weak, not that it is nonexistent.

Although Grünbaum can allow very weak confirmation even when a Freudian hypothesis has an equally plausible and incompatible rival, I am inclined to say that there is no confirmation at all. The argument for this view is developed in Erwin and Siegel (1989). Here, I will be content to say that Fine and Forbes (1986) do not give a satisfactory reason for denying the need to discount plausible rivals.

Kathleen Wilkes (1990) also raises the issue of the need to discount plausible rivals. She writes (referring to papers in Clark and Wright 1988), "Many of the contributors, particularly perhaps Erwin and Cioffi, complain that experimental evidence in support of Freudian hypotheses is weakened because one could always find an alternative explanation. This is not a good argument" (p. 249). I agree; in fact, it is a silly argument. As Wilkes notes, the ingenious mind can "dream up" a rival explanation. However, I have nowhere even tacitly relied on such an argument. The argument on which I do rely runs like this: Suppose that a Freudian appeals to data *D* to support the claim that *A* caused *B*, but there is an incompatible hypothesis (say, that *C* caused *B*, and *A* was not implicated) and our total evidence makes the rival just as credible or even more credible. For example, suppose that an analyst cites client improvement as evidence of therapeutic effectiveness, but spontaneous remission is at least as likely to have caused the favorable results. In *that* case, the Freudian hypothesis is not adequately supported. As I noted earlier, I would also add that *D* does not confirm the Freudian hypothesis at all, but if that proves wrong, I would be content to settle for the verdict of inadequate support.

Wilkes (1990, 249) also adds that the existence of alternative and incompatible theories is a wholly healthy feature of all sciences. Although I would state the thesis more guardedly, I do not want to dispute it altogether. Rather, I ask: How does that bear on the issue of confirmation? Having rival hypotheses at hand may generally encourage experimentation, as Wilkes says—although many Freudian supporters have not been so encouraged—but confirmation may still be lacking.

I conclude that neither Wilkes (1990) nor Fine and Forbes (1986) give us any good reason to give up the standard scientific practice of looking for evidence that discounts not just any possible rival to a Freudian hypothesis, but one that, all things considered, is at least as likely to be true as the Freudian one.

Extraclinical Evidence

Foundations is primarily concerned with data obtained in clinical situations. Some philosophers, however, have raised issues about other sorts of evidence. For example, Binns (1990) refers to reviews of Freudian experimental studies and notes without comment that the reviewers (Fisher and Greenberg 1977, and Kline 1981) take some of the evidence to corroborate central parts of Freudian theory. In contrast, I believe it can be demonstrated that that same evidence is virtually nonconfirmatory (see Erwin 1986 and, for more details, 1992; but see as well Kline's (1986) reply).

Sachs (1989) also refers to extraclinical evidence, but of a nonexperimental kind; unlike Binns, he makes a lot of it. In section 4 of his paper, Sachs points out that Freud tried to support his doctrines by appeal to diverse phenomena such as psychotic manifestations, the vagaries of sexual orientation, jokes, taboos, myths, and so on. Some of these sources of support, he claims, are impressive and some negligible. At this point, he does not identify the "impressive" sources, nor give any argument that they exist. However, in subsequent sections he does give illustrations of the type he has in mind.

Sachs's first example concerns what he terms "accumulated parapraxes" (ibid., 365). A woman acquaintance of his was anxious about an impending appointment with a gynecologist; the appointment concerned a question, Sachs says, that was "charged with anxiety" (ibid.). To some degree, the anxiety was not consciously experienced. While the person anticipated the appointment with a barely felt anxiety, she, untypically, forgot first her purse and later her car keys. She then made an engagement that conflicted with her doctor's appointment. When she remembered the latter, she tried to postpone it; or she recalled it when the time for keeping it had passed. This is only one case, but, Sachs suggests, our acquaintances may well tell us of similar ones.

The case discussed, and others like it, provide no support for Freudian theory. Given the description of the case, there is no reason to believe that a repressed motive was present. Even if there were, evidence would still be needed that the repressed motive caused the forgetting. How is the forgetting to be explained if we do not postulate a repressed motive? One plausible answer is that the woman's anxiety interfered with her ability to concentrate, and the lack of concentration caused the forgetting. Some or most of her anxiety may have been below the threshold of consciousness, but that alone would not support any distinctively Freudian thesis. Freudians and non-Freudians alike can agree, for example, that a student's nervousness can affect test performance even if the student is not aware of being nervous, or that feelings of anxiety that exist in the so-called "preconscious" can affect mental functioning.

Sachs's next example, or rather class of examples, concerns the Freudian interpretation of symbols. As Freud notes (in a passage quoted by Sachs), his views on the interpretation of dream symbols was based not only on his studies of dreams, but also on his purview of fairy tales, myths, jokes, folklore, and poetic and colloquial linguistic usage. After Freud developed his views on symbolism, many investigators tried to confirm them through experimental tests. The results have been equivocal (Erwin 1991). However, what is at issue here is not what these experiments show, but rather, apart from the experimental studies, what grounds we have for thinking that Freud's theory of symbols is correct. The only grounds offered by Sachs is a statement by Freud that he partly italicizes, "If we go into these sources in detail, we shall find so many parallels to dream-symbolism *we cannot fail to be convinced of our interpretations*" (quoted in Sachs 1989, 364). Non-Freudians have not generally been convinced of Freud's interpretations, but even if they had, that would not have provided firm support. Given that the Freudian theory of symbolism is not self-evident, agreement about its correctness hardly suffices without independent corroboration. Perhaps Sachs's use of italics was not intended to signal an endorsement of Freud's reason for accepting the Freudian view; however, he gives no other reason.

Sachs's next two examples involve what he calls "tendentious" (ibid., 367) forgetting. The first example is Freud's "Signorelli" case and the second is his "aliquis" case. Both of these cases, Sachs contends, have the same phenomenological features (except for the first, which is absent in the "aliquis" case): (1) the context in which the name was forgotten was directly preceded by talk or thought on another topic, which the forgetter broke off or suppressed; (2) the substitute names were displacements of the forgotten one; (3) besides those external associations, an internal, disagreeable content associatively linked the later topic, especially the bearer of the forgotten name and the vivid detail, to the early, broken-off topic; and (4) the disagreeable content motivated forgetting the name (ibid., 365–66). Sachs contends that Grünbaum fails to notice this distinctive ensemble of phenomenological features and, consequently, fails to see that it may require explanation as a whole.

I have several questions about Sachs's contention. First, if the goal is merely to explain the forgetting of the name "Signorelli" and the Latin word "aliquis," why insist that the above ensemble of features be explained as a whole? As Sachs notes, Freud claims that this ensemble appears with uncommon frequency in cases where a name is forgotten and other names come to mind. However, as Sachs concedes, Freud offers no evidence for this claim (Sachs suggests a way of obtaining such evidence). In the absence of such evidence, I ask once more: Why insist that this particular ensemble of features be explained *as a whole*? Why believe that what caused the forgetting is also causally relevant to the other features that Sachs isolates? Second, and of more importance, are all four of the features that Sachs mentions present in the Signorelli case? We can take Freud's word concerning the first feature in the Signorelli case, the breaking off of discussion of a previous topic, but what about the second feature? How does Freud know that the substitute names that came to mind were *displacements* of the original name in Freud's *technical* sense of being a defense product? He says the following about this displacement, "The process that should lead to the reproduction of the missing name has been so to speak *displaced* and has therefore led to an incorrect substitute" (1953–1974, vol. 6, 2; emphasis in the original). Freud is making a causal claim here. How does introspection suffice for him to know that it is true? Contrary to what Sachs says, features (3) and (4) also fail to qualify as "phenomenological" features if this means that their presence can be confirmed by introspection or casual observation. In fact, to assume that the last feature was present, that the disagreeable content motivated the forgetting of the name, is to assume question-beggingly without argument that part of Freud's *explanation* of the forgetting is correct. In short, even if we concede feature (1), neither Freud nor Sachs establish the presence of *any* of the remaining three features alleged to be present in the Signorelli case.

The same comment applies to the aliquis case, except here matters are even worse. As Sachs notes, the first feature was absent; so, as far as we know, *none* of the four features cited by Sachs were present. In short, his whole argument for Freud's explanation in either case collapses. Finally, even if Freud or Sachs knew that the forgetting in the Signorelli and Aliquis cases was motivated, how does either know that repression was involved? Sachs does not claim to know; he interprets Freud as talking about material that was suppressed rather than repressed. Sachs may be right; as he notes, Freud sometimes uses the term "suppression" in the *Psychopathology of Everyday Life* (1953–1974, vol. 6). However, when Freud tries to state his view with care, he uses the term "repression", "We shall, I think, have stated the facts of the case with sufficient caution if we affirm: *By the side of simple cases where proper names are forgotten there is a type of forgetting which is motivated by repression*" (Ibid., 7; emphasis in the original). Suppose, however, that we go along with Sachs and interpret Freud as requiring only that the motive for forgetting be outside of consciousness, not that it be repressed into the unconscious. In that case, however, we would be left without a distinctively Freudian claim. Non-Freudians can agree that *sometimes*—how often or even when may be unknown—a motive we are unaware of may contribute to our forgetfulness. Let us take one of Sachs's examples: I may forget to return a borrowed object because I want to keep it, and yet I may be unaware of that desire (at least until someone brings it to my attention). To take another example from Charles Darwin (cited by Sachs, 1989, 368): I may be more motivated to forget a fact that disagrees with my theory than one that supports it. Darwin apparently had this motive, but it was not repressed; on the contrary, he was sometimes aware of it. In sum, Sachs's examples of tendentious forgetting provide no support for any distinctively Freudian theoretical hypothesis unless evidence is provided that a repressed wish preceded *and* caused the forgetting.

Sachs's next two examples are parapraxes borrowed from Grünbaum (1984):

a. "in the course of giving a lecture on human sexuality, a person misspeaks himself by saying 'orgasm' instead of 'organism' " (p. 199);

b. "the man who turns from the exciting view of a lady's exposed bosom muttering, 'Excuse me, I have got to get a *breast* of *flesh* air!' " (p. 200).

Grünbaum points out that neither of these mistakes is a genuine Freudian slip. The first can be plausibly explained in terms of what the lecturer is conceptually preoccupied with, combined with an appeal to phonetic similarity. There is no reason to say that the slip was motivated. The second may well be motivated, and in that respect resembles a Freudian slip, but the motive, as far as is known, is not repressed.

Sachs replies as follows. In the cases of (a) and (b), Grünbaum relies heavily (Sachs contends) on the thematic affinity of the parapraxis to a wish in order to infer that the wish prompted the parapraxis. (Grünbaum concedes, according to Sachs 1989, 374, that each of these parapraxes *shows* that a wish or affect that was prior to it had occasioned it.) In that event, Grünbaum's criticism of Freud's procedure for interpreting what Sachs calls "opaque" parapraxes is baseless. In his use of free association to discover the causal antecedents of the opaque mistakes, Freud, too, is relying on the thematic affinity between the wish and the parapraxes: Each one, on its own, conveys an indication of having been motivated. Grünbaum, then, on Sachs's interpretation, is stuck with a dilemma: Either he rejects his own reasoning in cases (a) and (b) or he admits that Freud's similar reasoning in so-called "opaque" cases is not fallacious. The way out of this dilemma, however, is not difficult to see: In *Foundations*, Grünbaum does not and need not make any appeal to thematic affinity in either case (a) or (b).

In case (a), the slip hardly *shows* that a preceding wish caused it. As already noted, the slip can be plausibly explained without assuming that it was motivated at all. In case (b), Sachs says that he will assume that the slip was prompted by a wish to caress, although he also expresses doubt that the motive is obvious. Indeed, it is not obvious. Might not heterosexual women or gay men sometimes make the same sort of mistake (under similar circumstances) as that made in (b) even when they lack the desire to caress a woman's breast? Even if we knew that the man in question had such a desire, how would we know, in the absence of further information, that it caused the slip? It is just as plausible to believe that the man's embarrassment caused him to be momentarily confused and his confusion engendered the slip. So, clearly thematic affinity alone is not evidence that the error was motivated by the wish to caress. Could we not, however, in principle have information that would make it plausible to accept the "wish to caress" explanation? We might, for example, discover, contrary to my earlier suggestion, that only men with a wish to caress a woman's breast make the type of error in question under similar circumstances. If we had other empirical evidence, we might also be able to rule out embarrassment as a precipitating factor, and so on. However, to do these things, we would need to appeal to background evidence of the requisite sort and not merely to thematic affinity. In principle, a Freudian could also appeal to relevant background evidence in motivationally opaque cases, but Grünbaum's complaint is that Freud fails to supply such evidence. The same complaint can be made about Sachs's reply. He relies on thematic affinity alone when he says that each of the parapraxes, *on its own,* conveys an indication of having been motivated (1989, 377). He does add that circumstances of their occurrence tend to support these judgements, but he gives no argument for this questionable claim.

Much of Sachs's paper is concerned with subtracting from the Freudian corpus arguments that he believes are mistakenly attributed to Freud and adding others that Grünbaum does not discuss. As I noted in connection with the Tally Argument, however, subtracting bad arguments does nothing by itself to improve Freud's case, although in some instances, it might enhance his reputation. As to the undiscussed arguments, I do not question the reasonableness of Sachs's bringing them to our attention. However, given the vast amount of arguments that Freud employed in his long career, the additional clinical arguments used by his many followers, and the hundreds of Freudian experiments, it is always possible to confront a critic of Freudianism with arguments that he or she has not discussed. What needs to be ascertained is this: Are the newly cited arguments cogent? Those that Sachs appeals to are not: Either singly or in combination, they do virtually nothing to strengthen Freud's case.

Sachs could reply that he did not intend to endorse any of the arguments that I have criticized; his point might be merely that *Foundations* fails to take these arguments into account. In that case, however, he should withdraw his main conclusion, "Freud's clinical data, so far as Grünbaum's criticisms of them go, continue to constitute evidence that can be marshalled on behalf of psychoanalysis" (1989, 378). This conclusion is based on the arguments I have criticized; if they are without merit, his conclusion is left unsupported (see Grünbaum [N.d.]).

Thematic Affinity

Several philosophers have appealed to thematic affinities (or meaning connections) in defending Freudian causal hypotheses. For example, Donald Levy (1988, 212) claims that to speak of an affinity of any kind (including a thematic affinity) between things signifies generally a causal relation between them. Stated in this bold fashion, Levy's thesis has obvious counterexamples. If I speak of an affinity of meaning between two words, I need not signify a causal connection between them; an affinity of content between two wishes in two different people is not evidence, I think even Levy would grant, that one wish caused the other. Last night's dream may have a manifest content very similar to that of a dream I had last year, but that is no reason to think that one dream caused the other. What Levy means, I think, is that generally a thematic affinity between certain *kinds* of items, such as one person's motive and action, or wish and verbal slip, or desire and dream image, is some evidence of a causal connection between the two. However, even this less general thesis seems wrong.

Consider some examples Grünbaum (1988, 1990) discusses. In the first case, I have a dream that includes the image of a house. There is a thematic affinity between my seeing a house the previous day and my now dreaming about a house, but there may be no warrant for thinking that the first event caused the second. I see at least one house almost everyday (e.g., my own) regardless of whether I then dream about a house. That is true of many other people as well. So, in the absence of additional evidence, I am not warranted in believing that seeing a house the previous day made any *difference* to what I dreamt—despite the presence of a thematic affinity.

In contrast, consider a woman who sees for the first time Frank Lloyd Wright's house Falling Water and then dreams about a house containing the same details. Here we have a strong thematic affinity and a warrant for believing in a causal connection, but what supplies that warrant is specific background information rather than thematic affinity. For example, Grünbaum stipulates that the woman had, until her visit, never heard of Falling Water or seen a picture or description of it. Another crucial piece of information is that the woman's dream occurred the night of the day she visited Wright's house. Without such background information, the causal inference is unwarranted. Finally, consider a case where a student writes a term paper by copying from an old encyclopedia article. Here we have an extremely high degree of meaning affinity—the two texts are exactly the same—and yet our background evidence about, for example, the very low probability of two papers written by different people being identical if there is no copying supports the charge of plagiarism. These and other cases (Grünbaum 1990) support the following contentions.

First, thematic affinity *by itself* is not generally evidence of a causal connection, and, second, it is not generally evidence even when the degree of meaning kinship is very high. Third, in some cases of thematic affinities, there is evidence of a causal connection, but it is supplied by background evidence. Fourth, this background evidence need not be either experimental or epidemiological. Consequently, there is a way around the dilemma Levy (1988, 213) presents to Grünbaum. Either insist on experimental (or epidemiological) evidence to warrant any causal claim or allow that thematic affinities themselves warrant causal claims. There is, however, a third choice that Grünbaum can and would accept. Allow that nonexperimental (and nonepidemiological) background evidence sometimes warrants causal hypotheses.

Even though thematic affinity by itself is not generally evidence of a causal connection, a Freudian could argue that it is evidence under certain conditions and try to specify these conditions. James Hopkins (1988, esp. 1991) has tried to do this in an interesting series of papers. What Hopkins argues is pertinent not only to Freudian psychology, but also to the philosophy of mind and the philosophy of psychology generally (e.g., it is relevant to the traditional problem of other minds and to the defense of folk psychology against criticism of eliminative materialists.)

As Hopkins (1991, 123, f.n. 5) uses the term "motive," it applies to a variety of types of psychological causes including beliefs, wishes, or desires and affects such as love, hatred, greed and lust. A close connection holds between language and motives, Hopkins contends, in the sense that motives characteristically have, or can be given, what he calls a "linguistic articulation" (ibid., 88). For example, if we say that John believes (hopes, fears or whatever) that Freud worked in Vienna, we can articulate John's motive by using the sentence "Freud worked in Vienna." The motive, then, can be said to have a *content,* which is given by the sentence used to articulate it (ibid., 89). Hopkins claims (incorrectly, I believe) that such a sentence states what he alternatively calls a "truth-condition" or a "satisfaction condition"; "As I shall be using these terms, the truth condition of 'Snow is white' is that snow is white, of 'Grass is green' that grass is green and so on, ad infinitum. The notion is used for motives by the way of the sentences that articulate them. Thus the sentence that articulates the motive of belief in 'John believes that snow is white' is 'Snow is white.' The truth-condition of this sentence, and hence of the belief itself, is that snow is white" (ibid., 124, f.n. 9). As further examples, Hopkins claims that the satisfaction condition of the *hope* that snow is white is that snow is white, and of the *desire* that snow be white is that snow is white (or be white). Finally, Hopkins claims that a logical or conceptual connection lies between a motive and its satisfaction condition; "The condition of satisfaction, realization, or whatever, of a given motive stands in a relation to that motive that is logical or conceptual. It is a norm or rule, given in language, that having a drink of water satisfies a desire to have a drink of water, or that a belief that grass is green is true if grass is green" (ibid.).

Given this truth-condition relationship, Hopkins claims that the linguistic articulation of a desire serves to describe it as a cause, and in grasping that articulation, we already know a central feature of its causal role, that is, what it is supposed to do (ibid., 92). Furthermore, the way we interpret one another, according to Hopkins (ibid., 95), is by assigning sentences to motives (sentences specifying their content). In doing this we also understand one another in terms of causes; "The finding of sense or meaning, the articulation of object- and satisfaction-directedness, and the establishing of commonsense causal order, are one and the same" (ibid.). And so, our natural criteria for sound interpretation, which are based on content, are at the same time, Hopkins contends, criteria for good causal explanation. Thus, the better a desire and action match in content, the better we take the former to explain the latter. For example, singing the national anthem and a desire to sing the national anthem overlap in content (there is what Grünbaum calls "a thematic affinity") and this makes the desire particularly well suited to explain the action (ibid., 95–96).

I will not explore here the prospects for using Hopkins's analysis, *if* it were correct, to support the more contentious of Freud's dream interpretations or in warranting his hypotheses about the origin of neurotic symptoms (see ibid., 97, 116). Instead, I want to challenge the analysis itself.

To begin with, I deny Hopkins's (ibid., 124, f.n. 9) claim that there is the stated logical or conceptual connection of being a truth-condition between the assertion of the existence of a motive and its condition of satisfaction. That snow is white is, of course, a truth-condition for "snow is white," but it is not a truth-condition for "John believes that snow is white." Clearly, John may *not* believe that snow is white, although it actually is. Similarly, it is not a truth-condition for "John desires to have a drink of water" that John actually has a drink of water. Hopkins does realize that John may desire to drink and yet not drink, but the point is that John may drink water without having desired to do so. What might have misled him is the logical or conceptual connection between the following:

1. "John has a desire to have a drink of water," and
2. "Having a drink of water satisfies that desire."

Given what Hopkins means by "satisfies," (1) implies (2), but that lends no support to his claim that the motive and its satisfaction condition (desiring the drink and having a drink) are logically or conceptually connected by way of the action being a "truth-condition" for the assertion of the existence of the motive. They are not so connected.

Hopkins claims that a connection between a motive and its satisfaction condition enables the articulation of the content of the motive to establish its causal role. Thus, in his view, when we specify the satisfaction condition of a desire (i.e., articulate its content), we thereby "describe it as a cause" (ibid., 92). But suppose I articulate John's desire to act always in a purely selfless manner. The satisfaction condition is that John always acts in that manner. I state this condition, but it does not follow that I have described John's desire as a cause. I would not contradict myself if I said, for example, that John has such a desire, and that the aforementioned condition would satisfy it, and yet the desire has no effect on his actual behavior: Whenever he thinks he acts selflessly, he deceives himself. Nor would understanding the content of John's desire guarantee that I know whether the desire plays any causal role in explaining his behavior. Hopkins speaks of knowing what a desire is "supposed to do" (ibid.). This means what a desire *should* do if acted on intentionally. For example, he claims that a desire to get a drink, if someone acts on it intentionally, *should* produce an action of getting a drink. The basis for this claim, or even what it means is unclear. However, even if Hopkins is right about what desires should do, knowing what a desire is "supposed to do" is not sufficient for knowing its actual causal role. Do people ever act on the desire to get a drink? Does such a desire ever make any difference at all to the way people behave? These are empirical questions, even if they have obvious answers; they cannot be answered merely by articulating the content of the desire. The overlap in content between the desire to get a drink and the action of getting a drink may be grounds, then, for believing that the desire is, in Hopkins's words "supposed to" bring about that action, but it is not grounds for believing that it actually does so.

Even if Hopkins's claim were only about what a desire is supposed to do, it does not follow that "hermeneutic understanding, and a grasp of the causes of behavior, form a unity" (ibid.). At most, what follows is a unity between the understanding of the content of a desire and knowing what it is *supposed to do*. Knowing the actual causes of behavior requires empirical information, not merely hermeneutic understanding.

It also does not follow from Hopkins's argument that the finding of motivational "sense" or "meaning" and the establishment of commonsense causal order are one and the same (ibid., 95). It is one thing to make sense of what people do by postulating motives; it is another thing *to establish* that a motive featuring content overlap with an action really did cause it. The desire to sing the national anthem may be "particularly well suited" to explain someone's singing the national anthem, if this means that if the desire is actually present, then singing the anthem would satisfy it, but noticing this overlap in content between the desire and the action is no evidence that the desire did cause the action. Perhaps the action was performed to curry favor with someone who is very patriotic or because the failure to sing would incur penalties. Or perhaps the singer did act voluntarily, but had no desire to sing the anthem sincerely, and sang it only to mock the audience. In brief, unless I have seriously misunderstood him, Hopkins's claims about the articulation of the content of motives fail to support his thesis that an overlap in content between a motive and action or motive and alleged manifest dream symbol is conceptual evidence of a causal link between them (ibid., 97).

Besides appealing to linguistic articulations, Hopkins presents examples in which overlap in content purportedly provides grounds for a causal hypothesis. However, in all of his cases, either there is no evidence of causality or if there is, the warrant is provided by other background evidence and not thematic affinity.

Hopkins's first example is that of Joseph Breuer's patient, Anna O, who suffered from an aversion to drinking. Under hypnosis, she traced this symptom to an episode in which she watched a dog drink water from a glass. Hopkins claims that the causal link between episode and symptom seems marked in the content of the symptom itself since both were concerned with such topics as drinking water, disgust, anger, and refusal (ibid., 87). Indeed we have an overlap in content, but why is that grounds for thinking that the episode caused the symptoms to appear? As Grünbaum (1990, 572–73; see also 562–663) notes, Breuer's treatment of Anna O's aversion to drinking water was a failure; that is reason to doubt that he had identified and removed the cause of her phobic reaction.

Another example Hopkins uses concerns Freud's Irma dream. In his preamble to his dream analysis, Freud tells us that his treatment of his patient, Irma, was only partly successful; some of her somatic symptoms remained. On the day preceding the dream, a junior colleague, Otto, made a remark about her condition, one that Freud interpreted as a reproof. The next morning Freud had a dream about Irma in which the cause of her problems is attributed to an injection given by Otto, who probably, it is suggested in the dream, used an unclean syringe. Later, when reflecting on his dream, he writes, "It occurred to me, in fact, that I was actually *wishing* that there had been a wrong diagnosis; for if so, the blame for my lack of success would have been got rid of" (1953–1974, vol. 4; 109). This hypothesis may have occurred to Freud, but he nowhere gives any evidence that it is true. He was not aware prior to the onset of the dream that he ever had such a wish. Even if he had, he gives no reason to think it had any effect on the content of his dream. Hopkins says that the hypothesis "fits with" (1991, 101) material in the rest of the dream, in particular with Freud's dreaming that the illness was caused by Otto's injection. If "fits with" means "coheres with," that is true of indefinitely many rival hypotheses and is not evidence for Freud's particular hypothesis. Hopkins, however, presumably means that Freud's conjecture *explains* the dream segment involving Otto. But how satisfactory is the explanation? Freud would not have been responsible for Irma's pain if it were due to Otto's injection, but he would then have been responsible for a misdiagnosis and the ensuing consequences; for his treatment plan was based on his contention that her pains had a psychological origin. So one has to assume not only that Freud did not want to be blamed for Irma's continued problems, but also, and implausibly, that he did not mind being blamed for a gross misdiagnosis. A more plausible explanation of the Otto dream segment is that Freud had no wish for a misdiagnosis, but merely wished to shift the blame for Irma's problems to the thoughtless Otto. (He does report remembering that Otto was thoughtless, or something like that, the previous afternoon [1953–1974, vol. 4; 117].) In short, without some supporting evidence or argument, the hypothesis that Freud puts forth (rather tentatively) concerning the causal relevance of his alleged wish for a misdiagnosis is unsupported. Other things that Freud says about his Irma dream are more plausible, but as Grünbaum (1984, 229) notes, they are supported by grounds from commonsense psychology, as opposed to thematic affinities.

In contrast to the two cases discussed, it is plausible that Freud's dreaming of drinking cool water (another case Hopkins cites) was caused by his being thirsty. At the very least, there is reason to believe that he was thirsty while dreaming. He reports having eaten anchovies before going to bed and waking up thirsty. A clear alternative, however, to mere thematic affinity can be given to what makes the hypothesis plausible. We have background evidence of a Millian kind (see Grünbaum 1992, 154) that events or conditions such as being thirsty or having a desire to urinate often make a difference as to what people dream.

To take another of Hopkins's examples, when we see someone moving toward a tap with glass in hand, we are often in a position to infer that he or she wants a drink. However, that is not always true. I often perform the same action in order to wash a glass and there is no reason to think that I desire a drink. Where the inference is warranted, it is so because we have background evidence that the act is not intended to serve some other purpose, that it is voluntary (I am, e.g., not hypnotized), that the agent is thirsty, that people often drink water to satisfy their thirst, and so on. This background information, not mere thematic affinity, provides the warrant. That is also true of someone singing the national anthem. If we did not have grounds for thinking that a particular rendition of the anthem was voluntary, or that people generally sing the anthem because they want to, we might very well have no reason to believe that the desire to sing caused the action.

Hopkins does have some additional examples, but they, too, fail to help his case. In each example, either it is not clear that the inference to a causal hypothesis is warranted *or* it is very questionable that thematic affinity rather than background evidence supplies the warrant. This questionability harks back to Grünbaum's counterexamples to causal inferences based on thematic affinity.

A proponent of thematic affinities' evidential value could try another tack. Instead of trying to find convincing examples, we can formulate a cogent general rule that allows us to separate cases where thematic affinities are signs of causation from those where they are not. It is clearly insufficient, however, to say merely: Thematic affinities are evidence of causation *given* the proper background evidence. We can also say: If *B* follows *A*, that is evidence that *A* caused *B given the proper background evidence*. Such "rules" are too trivial to be of any use to the Freudian cause. Hopkins does try to state more informative general conditions than these (for thematic affinities being evidence of causation), "These conditions include, at the outset, the accurate ascription of base motives, and also a degree of connection between motive and dream that is significant enough effectively to rule out coincidence" (1991, 106). By "degree of connection," Hopkins means degree of *meaning* as opposed to *causal* connection. At least two problems arise with his conditions. The first, and minor, one is that, on at least one natural reading of "coincidence," we may rule out a coincidental relationship between events *A* and *B*, but be unable to discount the operation of some third factor causing *B*, or causing both *A* and *B*. We can avoid this problem by stipulating that ruling out a "coincidental" relationship between a motive and dream (or action) requires guaranteeing that the relationship is causal. The major problem concerns the applicability of Hopkins's conditions. In the light of Grünbaum's arguments that even a strong meaning connection is not by itself evidence of causation, it is question begging to assume without argument that cases exist in which the degree of (meaning) connection between motive and dreams is significant enough to rule out coincidence. In addition, to apply the rule to Freudian cases of dream interpretation or purported explanation of neuroses, we would have to obtain evidence that

a. the postulated motive did exist, and
b. that the degree of meaning connection is significant enough to rule out coincidence.

There is no reason to think that we can obtain such evidence without using Millian methods. (Hopkins, ibid., 128, claims that Mill's methods are inapplicable to a psychology of motives because motives are unobserved. However, in employing Mill's methods, one need not claim to use them to demonstrate that a putative cause *exists*, but rather to argue that a hypothesized event makes a difference. For a discussion of this point, see Grünbaum 1992, 154).

One final word about thematic affinities. In downgrading their evidential value in establishing causal hypotheses, one need not assert that certification of a meaning connection can never be part of an overall argument for causation. Consider an analogy concerning temporal sequence. A researcher who wants to show that a certain chemical in a patient's urine produced some effect may first have to argue that the alleged effect followed rather than preceded the appearance of the chemical. That the researcher might have to do this, however, is no basis for saying: Establishing temporal sequence (e.g., that *B* followed *A*) is often evidence of causation—but, only, of course, if we have proper background evidence. Analogously, those who want to appeal to thematic affinities to solidify the case for Freud's ideas have to argue more than the triviality that they are like temporal sequence in that sometimes their certification can play a role in an overall argument for a causal connection. Such a triviality is not likely to be of any use in repairing faulty Freudian arguments.

Miscellaneous Issues

Several replies to *Foundations* have been written that for various reasons, including lack of time and space, I cannot discuss in detail.

Without trying to do justice to all of the issues, I will explain briefly why the arguments of these papers fail to undermine any of the major arguments or conclusions of *Foundations.* (Grünbaum [1988, 1992] replies to some of the issues raised in what is now a very large critical literature; he will reply fully to Sachs [1989] and Hopkins [1991] in a forthcoming book.)

First, in his reply to *Foundations,* Richard Miller (1988) raises several interesting issues in the philosophy of clinical psychology. Concerning Freudian theory, however, he defends only what he takes to be a core component consisting of four hypotheses. The confirmation of all four is compatible with the lack of any solid clinical support for Freud's repression theory discussed in *Foundations.* For that reason, I have not discussed Miller's arguments here. I have doubts, moreover, whether Miller's core theory is distinctively Freudian at all, although in a broad sense, it might be considered "psychoanalytic." Of his four hypotheses, the second two make no causal claim. The first is causal but is so vague that it could be accepted by many non-Freudians, including some cognitive behavior therapists. The second hypothesis might be distinctively Freudian, but that too is unclear. A lot depends on precisely what types of problems, threats and defenses are specified.

Second, Matthias Kettner (1991) provides a partial defense of appeals to thematic affinities. He suggests (ibid., 174) that an analyst need not base causal inferences on the strength of meaning affinities alone so long as an extra premise is available: that meanings and their affinities are causally embedded in interpretative processes belonging to a single causally connected mental life, namely, that of the person analyzed. I have no doubt that this extra premise can be confirmed, but what is left unclear is how this premise will help warrant Freud's repression theory. Anna O, for example, may well have seen an affinity between her watching a dog drink water from a glass and her hydrophobia. Assume that her drawing of this connection was "causally embedded in interpretative processes" belonging to a single causally connected mental life (ibid., 174). Yet, we are still left without rational grounds for inferring that the remembered episode caused her symptom to either appear or persist (see my discussion in the previous section).

Kettner (ibid., 175–76) does give an example of an analyst's causal inference allegedly involving meaning affinities that he claims is valid, although it does not bear directly on the repression theory. A patient is said to understand by what means she can assert her authority over her pupils, but is unable to do it. After speaking of the necessity of self-discipline, she remembers being a rebellious child. From these facts alone, the analyst infers, "Her difficulty now is particularly related to her unconscious identification with her pupils" (quoted in Kettner, ibid., 177). Kettner claims that the analyst's argument can be represented by a pattern of abductive inference based on thematic affinity. He presents two versions of that pattern and claims that both are valid. His second version, however, does not contain a premise and conclusion (it consists of a single statement). For that reason, I will examine only the first (ibid., 176):

Abductive premise: It is surprising that *S*'s (= the woman's) behavior manifests features *F*1, *F*2 . . . *Fn*.
Abductive premise: It is unlikely (= not to be expected) that *S*'s behavior should show so many features *F*1, *F*2 . . . *fn* that could derive from *M* without being so derived.
Abductive conclusion: That *M* is causally relevant for *S*'s behavior is a reasonable hypothesis.

This pattern of reasoning makes no mention of thematic affinity, so even if it were valid, it is unclear how its introduction would serve Kettner's stated purpose of rebutting Grünbaum's charge of a "thematic affinity fallacy." Whether the pattern is valid depends on the interpretation of "could derive" in the second premise. If Kettner means "could possibly derive," then the conclusion fails to follow. The features of the woman's behavior could possibly have derived from (i.e., resulted from) her unconscious identification with her pupils, but the features may also possibly have resulted from her *conscious* sympathy for the students. Indeed, many causes are logically possible. Kettner's premises fail to guarantee that it is reasonable to believe that any particular one was causally relevant. What Kettner may mean, however, is that the features could *plausibly* be said to derive from *M* (see Kettner's ibid., 165, discussion on the role of background conditions in sifting out unreasonably "adduced" explanations). On this interpretation, the schema is trivial. Suppose that we first establish that the teacher's unassertiveness is unlikely if it is not caused by her unconscious identification with her pupils *and* that it is reasonable to think that it was so caused; then, of course, the hypothesis that this factor was causally relevant is reasonable. It is hard to see how such trivial arguments are to be of use in defending the repression theory, or in showing how thematic affinities warrant causal claims.

Third, Charles Nussbaum (1991) offers a partial defense of Habermas's views on the logic of psychoanalytic explanations and a rebuttal to Grünbaum's view of the subject. The confirmation and disconfirmation of the repression hypothesis, he holds, require an explanatory model which makes use of reasons (or intentions) rather than causes, although causal explanations cannot be dispensed with altogether (ibid., 193).

According to Nussbaum, the connection between intention and action cannot be causal because in order to explain an action, the connection *must* be described by an analytic statement. It must be so described because to explain an event, we must first identify it, and that requires that we construct a practical syllogism of the following form:

P intends to bring about *b;*
P believes (considers, thinks) he cannot bring about *b* unless he does *a;*
P does *a*.

The above "syllogism" is best understood as a hypothetical statement which, Nussbaum claims, is analytically true (ibid., 195–96).

To illustrate, Nussbaum considers the behavior of a man looking in a field; "When we explain (by reasons) that the man in the field is looking for his lost watch, we construct a practical syllogism which states that he intends to find his watch, that he believes that he will not find it unless he looks about the field, and that he therefore looks about the field" (ibid., 195). However, the corresponding hypothetical—if *P* intends to find his watch, and believes he cannot do so unless he looks about the field, then he looks about the field—is analytic, according to Nussbaum (ibid., 195–96). Consequently, the connection between intention and action is not causal.

Nussbaum's argument has at least two serious problems. The trouble begins with one of the first steps: Why believe that to identify an action we need to construct a practical syllogism? If I see a man apparently looking about a field, to identify his action, it is enough to know that he intends to be looking about a field. To know what he is doing, I need not construct a syllogism specifying a belief and further intention (say, to find his watch). Nussbaum concedes, in fact, for certain types of behavior, such as that of neurotics, it is not even possible to identify them by constructing a practical syllogism (ibid., 198). Yet, in his example of Freud's analysis of an obsessional woman, the patient performed various actions (such as ringing a bell for her housemaid) and Freud was able to ascertain what they were. He had to do that in order to accomplish what he next attempted: to explain the woman's behavior in terms of her preconscious desire to mitigate her husband's embarrassment at being impotent on their wedding night. Nussbaum says of this case that Freud's explanation consists of offering a redescription: What the patient was "really doing," according to Freud, was restoring her husband's honor, or protecting his pride (ibid., 200–01). I will return to Nussbaum's contention shortly, but even if it is accepted without objection, my point remains: Freud was trying to explain the woman's obsessional actions; but to do that, he first had to identify them, and he did that without the help of a practical syllogism.

Without support for this crucial premise, Nussbaum's entire argument collapses, for his conclusion that intentions are not causes depends on the assumption that the connection between intention and action *must* be described by an analytic statement, which in turn is dependent on his initial assumption about the need to invoke a practical syllogism to identify an action.

Even if the initial assumption were granted, however, Nussbaum would need support for a second controversial claim: that the hypothetical proposition that expresses the practical syllogism is analytic. To see the problematic nature of this claim, consider an example given by G. H. von Wright, "We have the premises of a practical argument: an agent intends to bring about something and considers the doing of something else necessary for this end. It is time for him to act. He thinks so himself. Perhaps he has resolved to shoot the tyrant. He stands in front of the beast, aiming at him with his loaded revolver. But nothing happens" (1971, 116). Nussbaum (1991, 195) allows that an intended action may not occur because of interfering conditions; so he might respond that such conditions must be present in von Wright's case that prevent the man from acting. Why, however, *must* there be such conditions? It might turn out that determinism is not true at the level of human action, and that the causation of action is irreducibly stochastic. Even if determinism is true, at least at the macrolevel, it is not analytically true: We have no logical or conceptual guarantees that when an action fails to follow when someone has an appropriate intention and belief, that there will always be interfering conditions or a competing intention.

Nussbaum (ibid.) cites von Wright as a supporter of the logical connection argument, but von Wright does not argue that it is logically or conceptually impossible for the premises of a practical syllogism to be true and the conclusion false. Thus, in discussing the man intending to shoot the tyrant, he denies that there is a logical compulsion to say that either he gave up his intention or there were interfering conditions (such as his being paralyzed). On the contrary, von Wright says that we can just as well say that if the case can be imagined, it shows that the conclusion of a practical inference does not follow with logical necessity from the premises. To insist otherwise would constitute "dogmatism" (von Wright 1971, 117). He adds, "Thus, despite the truth of the Logical Connection Argument, the premises of a practical inference do *not* with logical necessity entail behavior" (ibid.; emphasis in the original).

Nussbaum concedes that the issue of the logic of the practical syllogism is difficult, so he is somewhat hesitant in expressing his view. However, for his argument about intentions not being causes to work, he needs to defend his hardly self-evident claim that the hypothetical proposition corresponding to the practical syllogism is analytic. He does not do that, nor does he answer opposing arguments in *Foundations* (Grünbaum 1984, 69–83).

I conclude that Nussbaum fails to show that intentions (and reasons) are not causes. The failure of his argument prevents him from reaching his stated goal of showing, contrary to Grünbaum's (1984) view, that some Freudian explanations make use of reasons that are not causes.

Nussbaum (1991, 200, 204) also contends that psychoanalytic explanation involves a *redescription* of a patient's behavior. In support of this general claim, he cites but the one case in which Freud attempts to explain a woman's obsessional behavior. It would be risky to rest much on this single case given that, as Nussbaum realizes, Freud did not invoke his repression theory to explain the woman's behavior; rather, he appealed to a noninfantile wish in the woman's preconscious. Furthermore, the woman herself discovered the wish; it was not uncovered through the use of psychoanalysis. Putting aside doubts about how representative the case is, does Freud attempt to explain by offering a redescription? Does he say, as Nussbaum puts it, that what the woman was "really doing" was restoring her husband's honor or pride? He does not; the "really doing" locution is Nussbaum's. Instead, Freud explains the obsessional behavior by referring to the woman's preconscious wish to mitigate her husband's embarrassment (1953–1974, vol. 17, 263). In doing so, Freud is giving a causal explanation. He is not saying merely that the wish was present when the action was performed; he is also implying that it made a difference to what the woman did. Furthermore, if it had made no difference, then what would be the point of saying that what she was "really doing" was trying to restore her husband's pride? Many descriptions of her actions may be given. Why deem Nussbaum's description privileged unless her desire to restore her husband's pride made a difference to what she did?

I am not objecting to the use of the phrase "what the woman was really doing." My point is that redescribing the woman's action by means of Nussbaum's locution is compatible with also giving a causal explanation of her behavior, which is what Freud was attempting. I conclude that Nussbaum's attempt to show that some psychoanalytic explanations are noncausal does not succeed.

Nussbaum appeals to his claim about redescription not only to show that psychoanalytic explanations are sometimes noncausal, but also to demonstrate that they do not meet the requirements of the deductive-nomological model of explanation. He makes this additional point in alleged opposition to Grünbaum, who is said to hold that nothing in Freudian explanations is not captured by the D-N model. Grünbaum, however, does not hold this view. On the contrary, he states more than once in *Foundations* (1984, 11, 79, 179–80) that in Freudian theory, repression is alleged to be only a necessary condition for the development of psychoneuroses, which automatically disqualifies repression explanations from meeting the conditions of the D-N model. Grünbaum, however, nowhere criticizes Freudian explanations on the grounds that they fail to meet the D-N model.

Nussbaum also defends Habermas against some criticisms made in *Foundations*. On the issue of "dissolving causal connections" (Grünbaum 1984, 9–15), he makes the point that unlike the removal of gall stones, the lifting of repression, according to Habermas, is not a process of removal. The repressed wish is not removed, according to

Nussbaum, but rather is "unrepressed" (1991, 205). This point fails to answer Grünbaum's objection. Psychoanalysis is a theory of *unconscious* causation. What is alleged to be pathogenic, according to Freudian theory, is not the mere possession of an infantile wish; it is the *repression* of the wish that causes the neurosis. The lifting of the repression is not necessarily a removal of the wish, but it is a removal of the purported cause of the neurosis and not, as Grünbaum demonstrates contra Habermas, a "dissolving" of the causal connection between repression and neurosis.

Nussbaum also disputes Grünbaum's contention that Habermas makes the patient the "ultimate epistemic arbiter" of a psychoanalytic interpretation. What Habermas is really saying is that the patient is "cognitively involved" (ibid., 208) in a way he or she would not be when treated for a somatic illness. That is, the analysis can be therapeutically successful only if the patient accepts the analytic interpretations as correct (ibid., 202). However, this is not the claim that Grünbaum criticizes à propos of Habermas. What he objects to concerning Habermas is a claim not about the therapy, but about the purported validation of the interpretation. As Habermas says, "Only the patient's recollection decides the accuracy of the construction" (quoted in Grünbaum 1984, 25) and "analytic insights poses validity for the analyst only after they have been accepted as knowledge by the analysand himself" (ibid.). If Habermas were making a point only about the patient's involvement in the therapy, he could not use it, as he clearly intends, to support his contention that psychoanalytic interpretations are not intersubjectively verifiable. Even if the patient's agreement with the analyst's interpretation were necessary for a cure, that alone would not prevent two or more investigators from confirming the accuracy of an interpretation. Finally, as I noted earlier, Freud himself denies the view that Habermas attributes to him, "We do not pretend that an individual construction is anything more than a conjecture which awaits examination, confirmation or rejection. We claim no authority for it, we require no direct agreement from the patient, nor do we argue with him if at first he denies it" (1953–1974, vol. 23, 265).

In sum, although philosophers have raised a number of interesting issues about *Foundations,* Grünbaum's powerful arguments about the weaknesses of the Freudian clinical data and validation have not been challenged successfully.

REFERENCES

Ayllon, T., and E. Haughton 1964. "Modification of Symptomatic Verbal Behavior of Mental Patients." *Behaviour Research and Therapy 2:* 87–97.

Ayllon, T.; E. Haughton; and H. Hughes; 1965. "Interpretation of Symptoms: Fact or Fiction?" *Behaviour Research and Therapy, 3:* 1–7.

Binns, P. 1990. "Experimental Evidence and Psychotherapy." *British Journal for the Philosophy of Science, 41:* 531–51.

Brill, N.; R. Koegler; L. Epstein; and E. Forgy 1964. "Controlled Study of Psychiatric Outpatient Treatment." *Archives of General Psychiatry 10:* 581–95.

Derksen, A. A. 1992. "Does the Tally Argument Make Freud a Sophisticated Methodologist?" *Philosophy of Science 59:* 75–101.

Erwin, E. 1978. *Behavior Therapy: Scientific, Philosophical and Moral Foundations.* New York: Cambridge University Press.

———. 1986. "Psychotherapy and Freudian Psychology." In S. Modgil, and C. Modgil, eds., *Hans Eysenck: Consensus and Controversy.* London: Falmer Press, pp. 179–203.

———. 1992. "Was Freud Right? Philosophical and Experimental Issues." Unpublished manuscript.

Erwin, E., and H. Siegel 1989. "Is Confirmation Differential?" *British Journal for the Philosophy of Science 40:* 105–19.

Fine, A., and M. Forbes 1986. "Grünbaum on Freud: Three Grounds for Dissent." *The Behavioral and Brain Sciences 9:* 237–38.

Fisher, S., and R. Greenberg 1977. *The Scientific Credibility of Freud's Theories and Therapy.* New York: Basic Books.

Forrester, J. 1986. Review of *The Foundations of Psychoanalysis: A Philosophical Critique,* by Adolf Grünbaum. *Isis 77:* 670–74.

Freud, S. 1953–1974. *The Standard Edition of the Complete Psychological Works of Sigmund Freud.* Edited by J. Strachey. London: Hogarth Press.

Grant, D., and E. Harari 1990. "Misreading Freud—Some Thoughts on Adolf Grünbaum's 'Tally Argument.' " *Psychoanalysis and Contemporary Thought 13:* 311–19.

Grünbaum, A. 1980. "Epistemological Liabilities of the Clinical Appraisal of Psychoanalytic Theory." *Nous 14:* 307–85.

———. 1981. "The Placebo Concept." *Behaviour Research and Therapy 19:* 157–67.

———. 1984. *The Foundations of Psychoanalysis: A Philosophical Critique.* Berkeley and Los Angeles: University of California Press.

———. 1985. "Explication and Implications of the Placebo Concept." In L. White, B. Tursky, and G. Schwartz, eds., *Placebo: Theory, Research and Mechanisms.* New York: Guilford Press, pp. 9–36.

———. 1986. Author's Response. *Behavioral and Brain Sciences 9:* 266–81.

———. 1988. "The Role of the Case Study Method in the Foundations of Psychoanalysis." *Canadian Journal of Philosophy 18:* 623–58.

———. 1989. "The Placebo Concept in Medicine and Psychiatry." In M. Shepherd and N. Sartorius, eds., *Non-Specific Aspects of Treatment.* Toronto: Hans Huber Verlag, pp. 7–38.

———. 1990. " 'Meaning' Connections and Causal Connections in the Human Sciences: The Poverty of Hermeneutic Philosophy." *Journal of the American Psychoanalytic Association 38:* 559–77.

———. 1992. *Validation in the Clinical Theory of Psychoanalysis.* Madison: International Universities Press. In Press.

———. N.d. "Reply to Sachs's 'In Fairness to Freud: A Critical Notice of the *Foundations of Psychoanalysis,* by Adolf Grünbaum.' " *The Philosophical Review.* Forthcoming.

Hopkins, J. 1988. "Epistemology and Depth Psychology: Critical Notes on *The Foundations of Psychoanalysis.*" In P. Clark, and C. Wright, eds., *Mind, Psychoanalysis and Science.* New York: Blackwell, pp. 33–60.

———. 1991. "The Interpretation of Dreams." In J. Neu, ed., *The Cambridge Companion to Freud.* New York: Cambridge University Press, pp. 86–135.

Kettner, M. 1991. "Peirce's Notion of Abduction and Psychoanalytic Interpretation." In B. Litowitz, and P. Epstein, eds., *Semiotic Perspectives on Clinical Theory and Practice.* New York: Mouton de Gruyter, pp. 163–79.

Kline, P. 1981. *Fact and Fantasy in Freudian Theory.* New York: Methuen.

———. 1986. "Kline Replies to Erwin." In S. Modgil, and C. Modgil, eds., *Hans Eysenck: Consensus and Controversy.* 2d ed. London: Falmer Press, pp. 230–32.

Krasner, L. 1958. "A Technique for Investigating the Relationship Between the Behavior Cues of the Examiner and the Verbal Behavior of the Patient." *Journal of Consulting Psychology 22:* 364–66.

Levy, D. 1988. "Grünbaum's Freud." *Inquiry 31:* 193–215.

Miller, R. 1988. "A Clinical Science." *Canadian Journal of Philosophy 18:* 659–80.

Nussbaum, C. 1991. "Habermas and Grünbaum on the Logic of Psychoanalytic Explanations." *Philosophy and Social Criticism 17:* 193–216.

Park, L., and L. Covi, 1965. "Nonblind Placebo Trial: An Exploration of Neurotic Patients' Responses to Placebo When Its Inert Content is Disclosed." *Archives of General Psychiatry 12:* 336–45.

Prioleau, L.; M. Murdock; and N. Brody 1983. "An Analysis of Psychotherapy Versus Placebo Studies." *The Behavioral and Brain Sciences 6:* 275–310.

Rhoads, J., and B. Feather, 1974. "Application of Psychodynamics to Behavior Therapy." *American Journal of Psychiatry 131:* 17–20.

Richardson, R. 1990. "The 'Tally Argument' and the Validation of Psychoanalysis." *Philosophy of Science 57:* 668–76.

Robinson, P. 1993. *Freud and His Critics.* Berkeley and Los Angeles: University of California Press.

Sachs, D. 1989. "In Fairness to Freud: A Critical Notice of the *Foundations of Psychoanalysis,* by Adolf Grünbaum." *The Philosophical Review 98:* 349–78.

Salzinger, K., and S. Pisoni, 1958. "Reinforcement of Affect Responses of Schizophrenics During the Clinical Interview." *Journal of Abnormal and Social Psychology 57:* 84–90.

———. 1961. "Some Parameters of the Conditioning of Verbal Affect Responses in Schizophrenic Subjects." *Journal of Abnormal and Social Psychology 63:* 511–16.

von Wright, G. H. 1971. *Explanation and Understanding*. Ithaca: Cornell University Press.

Wilkes, K. 1990. "Analyzing Freud." *Philosophical Quarterly 40:* 241–54.

Wollheim, R. 1991. "Desire, Belief and Professor Grünbaum's Freud." Paper read at the University of Miami, Coral Gables, Florida.

17

How Freud Left Science

Clark Glymour
Department of Philosophy, Carnegie Mellon University, and Department of History and Philosophy of Science, University of Pittsburgh

Sigmund Freud is remembered as the founder of the psychoanalytic movement. He is not usually remembered as an early computational psychologist or as a cognitivist philosopher of mind, but that is how he began. The historical claim is meant seriously. Freud was a cognitive scientist, with a clearly computational theory of mind and its functioning, and that understanding of his early work is amply justified by his texts. But the picture leaves a kind of historical puzzle: How did Freud the computational mental physiologist become something so apparently different, the founder and leader of the psychoanalytic movement? How did Freud the positivist become the leader of a movement widely regarded today as a marginal science, and regarded by many as a full-fledged pseudoscience?

Answering these questions must lead inevitably to a kind of debunking. Freud has been debunked often enough, and there is little need for another performance. There is some amusement to be had in considering recent silly accounts of Freud's intellectual transition, but that is not itself enough reason for the effort. There is a better reason for considering these questions. If we view Freud as a contemporary, as a cognitive scientist just like many among us, the character of his intellectual career may serve as a warning as to how easy it is to give up the search for truth, and as to how fragile is the pursuit of knowledge. Freud is a moral lesson.

The brief story is this: Freud began with a conception of science he had learned from the physiologists, and he attempted to apply it in a clinical setting. Together with a strong prior disposition for sexual etiologies, and a massive insensitivity to experimenter effects in clinical work, those scientific methods led Freud to publish two complementary accounts of the etiology of the neuroses. When, not long after their publication, Freud came to doubt one of those theories, he could not help but also doubt the reliability of his clinical procedures, and he did so. His professional reputation and his income depended upon the reliability of the clinical methods of which he had repeatedly boasted in lectures and publications. The result was a crisis of confidence, half-hearted attempts to detect experimenter effects in the analytic hour, and a prolonged period of equivocating and dissembling about his own claims regarding the neuroses. The resolution of the intellectual part of this crisis was the real genesis of what is popularly understood as psychoanalysis, and it was marked by the publication of *The Interpretation of Dreams* (1953–1974, vol. 4, 1–338; vol. 5, 339–627). Freud submerged all doubts about experimenter effects, claimed extended powers of interpretation for himself and those who practiced his procedures, and replaced the scientific methods he had learned from the neurophysiologists by a kind of formal caricature of them. This is not simply to say that Freud was a scientist before 1900 and a mountebank thereafter, but it is a history that shows a decline in rigor and candor. Freud never fully abandoned the conception of science he had learned in the 1880s, but it became increasingly hollow.

The Discovery of Sex

In 1887, after visiting Paris and working with Charcot, and a stint of hospital work in Vienna, Freud set up shop as a neurologist in private practice. From Charcot and from Breuer he had become interested in neurotic phenomena, especially hysteria. He had from the very beginning a strong inclination to see all neurotic phenomena as disorders of *sexual functioning*. There was nothing unusual in that attitude, save possibly Freud's tendency to look for sexual etiologies exclusively. This was, after all, the time of Krafft-Ebing and Havelock Ellis, and sex was very much in medical fashion. It is worth observing in passing some false historical accounts, including Freud's own.

Freud himself claimed that he came to the hypothesis of sexual etiologies very reluctantly under the force of overwhelming clinical evidence. That claim is reiterated by various biographers and disciples, including Ernest Jones (1953). The interest of the claim is apparent: Freud's evidence is not available to us, and if we suppose him rational and suppose him originally disinclined to sexual etiologies, then it is a reasonable surmise that if he was driven to sexual explanations the evidence for them must have been very convincing. Richard Wollheim offers the most remarkable argument for this account of things. Wollheim writes that in Freud's paper on the neurophychoses of defense, published in 1894:

> the importance of sexuality is newfound. Originally, under the influence of Charcot, Freud too had depreciated sexuality as an influential factor. In an encyclopedia article on hysteria of 1888, he had written "As regards what is often asserted to be the preponderant influence of abnormalities in the sexual sphere upon the development of hysteria, it must be said that its importance is as a rule over-estimated." Freud's views changed slowly, very much against the grain, under two powerful influences. First, there was the study of the "actual neuroses"—that is, psychic disorders entirely attributable to current or ongoing factors. . . . The other powerful influence in bringing Freud around to a new estimate of sexuality was his clinical practice in the various psychoneuroses, for he increasingly found that the trail of free association, in which he now put his trust, ended in sexual ideas. (1971, 19–20)

Wollheim's text-cropping is world-class. What Freud[1] actually published early in 1888, only a few months after beginning private practice is this:

> As regards what is often asserted to be the preponderant influence of abnormalities in the sexual sphere upon the development of hysteria, it must be said that its importance is as a rule over-estimated. In the first place, hysteria is found in sexually immature girls and boys, just as, too, the neurosis with all its characteristics also occurs in the male sex, although a great deal more rarely (1:20). Furthermore, hysteria has been observed in women with a complete lack of genitalia, and every physician will have seen a number of cases of hysteria in women whose genitals exhibited no anatomical changes at all, just as, on the contrary the majority of women with diseases of the sexual organs do not suffer from hysteria. *It must, however be admitted that conditions related functionally to sexual life play a great part in the aetiology of hysteria (as of all neuroses),* and they do so on account of the high psychical significance of this function especially in the female sex. (1953–1974, vol. 1, 50–51; emphasis added)

I think no one honest can doubt that Freud's disposition towards sexual etiologies preceded the development of psychoanalytic theory and the free association technique.

Freud was ambitious. His correspondence with Wilhelm Fliess in the 1890s fairly smells of scientific ambition. His book on aphasia had not been any great success, and *Studies on Hysteria* (1953–1974, vol. 3), while well received, was regarded as essentially a restatement of Pierre Janet's theories of neurosis. Both Janet and Freud described it that way. To make his mark, Freud needed something that was distinctly his own. He had developed original clinical techniques by the middle of the decade, but they were not the sort of psychological or medical hypothesis that would be noticed unless special results could be claimed from them. In 1895 and 1896 Freud announced such results in a series of papers on the neuroses. These papers, on the "actual neuroses" and on the "seduction theory" of the psychoneuroses, mark his first original psychological hypotheses.

Freud's paper on the "actual neuroses" distinguished two syndromes. One, neurasthenia, was a familiar diagnostic category. Neurasthenics were characterized by fatigue and irritability. The other disorder, anxiety neurosis, was a Freudian novelty. Anxiety neurotics suffer diffuse anxiety with no specific fear. Freud claimed that neurasthenia is caused by masturbation; when masturbation ceases, so will neurasthenia. Anxiety neurosis is caused by unsatisfactory sexual fulfillment, which for Freud included celibacy, withdrawal, the use of condoms, and other "sexual noxae." When the afflicted took up a "normal" sexual life, anxiety neurosis would disappear. Freud's hypotheses about the actual neuroses were contested by a number of physicians, and in print by Lowenfeld, who claimed to have a number of patients who suffered the symptoms of one or another of these disorders, but who very probably had normal sexual lives. Freud's defense was that apparent counterexamples to his theory were in all probability misdiagnosed cases of *psychoneuroses,* hysterics in particular. Psychoneuroses had an entirely different etiology and mechanism. Psychoneurotics, Freud claimed, could be reliably diagnosed by the method of psychoanalysis.

Freud published his account of the causes of psychoneuroses in 1896. He dealt principally with two disorders, hysteria and obsessional neurosis. Each, he claimed, was caused by a sexual assault in early childhood. Hysteria and obsessions were the consequences of sexual child abuse. The difference between the disorders is that hysterics, after their first sexual contact, became sexually passive; they did not seek out further sexual contacts. Obsessives, in contrast, became sexually aggressive and sought further sexual encounters even in early childhood.

Surrendering the Seduction Theory

So by 1896 Freud had announced two very simple but startling hypotheses: neurasthenics and anxiety neurotics always suffer from concurrent sexual dysfunctions; hysterics and obsessives have always been seduced as children. The first hypothesis depended on the second, which had been established by psychoanalytic procedures, because psychoanalytic procedures, Freud claimed, showed that apparent counterexamples to his theory of the actual neuroses were misclassified. In the fall of 1897, in his correspondence with his friend Wilhelm Fliess, Freud announced that he no longer believed his theory of the psychoneuroses, and that he had concluded that the tales of seduction that he received from his patients were not generally true. This period of Freud's life is perhaps the most controversial. There are two principal accounts of what happened, neither of which is true.

One account holds that Freud correctly realized that the reports of seductions he heard from his patients were not correct, and then correctly inferred that they were "fantasies" produced by the patient's neurotic condition, and that these fantasies themselves were part of the cause of neurosis. This is the account of matters that Freud himself published much later in his life, but the second part of the account does not square with the historical facts. An alternative account, proposed by Jeffrey Masson (1984), is that Freud knew that the reports of seduction were in fact veridical, but renounced them for two reasons. First, because his theory was very unpopular with his medical colleagues and he wished to regain respectability; second, because he wished to make one of his patients and students into a kind of liar. The first of Masson's reasons is unconvincing; the second is incoherent and at points hilarious.

I wish to tell a different story. I think Freud came genuinely to suspect that the seduction tales were not correct; at the time he was not under great professional pressure, although his friend Fliess was. I think that Freud suspected that his psychoanalytic methods were unreliable and that the seduction tales were the result of his own suggestions to his patients. He tried, rather equivocally, to test this hypothesis, and came to reject it. He had every interest in rejecting the idea that his methods were unreliable, for he had repeatedly boasted of them in print and in public lectures, and he knew full well that he, like Fliess, would become a laughingstock if he now publicly admitted that his methods were suspect. Freud had every reason to believe that his methods were responsible for producing unreliable testimony from his patients, for the methods were extraordinarily directive. Freud's rejection of the idea that his methods were unreliable was a kind of self-deception, and it was accompanied by a public deception. In the late 1890s Freud did not know whether his seduction theory was true or not, but while he published arguments about the causes of hysteria, he did not disclaim his theory, or state any of his doubts. I think he did not because he feared, rightly, that such an admission would cast doubt not just on his seduction theory, but also on his methods. When, *nine years* after the letter to Fliess admitting his doubts, he published a renunciation of the seduction theory, he did so in the context of misrepresenting in an essential way the content of that theory and the history of his claims. In the meanwhile, he sought for some alternative etiology for psychoneurosis. Contrary to the first conventional history—Freud's own—he did *not* directly replace the seduction theory by the hypothesis that childhood fantasies of sexual intercourse could later cause neurosis in adulthood. Instead, he first supposed that the tales of seduction were displacements of memories of real events that took place in adolescence. Subsequently he supposed that sexual trauma in childhood was in fact the real cause of neurosis in adulthood, but the trauma could be self-imposed, generally in the form of early masturbation. Only in 1909, in the Rat Man case, did he introduce the hypothesis that childhood fantasies of sexual encounters could be etiological factors. He did so in order to save his account of psychosexual development from the counterexample that the case study seemed to pose. By that time the psychoanalytic movement was well established, his early theory long since forgotten, and Freud could safely give up a hitherto central pillar without bringing down the theoretical house.

The story is entangled with Freud's friendship with Wilhelm Fliess. They became acquainted in 1887, and formed a friendship that lasted until after the turn of the century. Freud sent to Fliess his personal news, drafts of manuscripts, long letters discussing his hypotheses and doubts, professional news, and responses to Fliess's own work. Fliess was a Berlin nose and throat specialist. He was also a crank, and Freud became devoted to most of his crankery. Fliess held that the nose was the organ responsible for neurosis, or for at least a widespread kind of neurosis, the "nasal reflex neurosis" as he called it. Parts of the nose affected parts of the body, and difficulties with parts of the nose caused neurotic symptoms that affected the corresponding parts of the body. Recommended procedure was to treat the offending nasal part with cocaine, or in extreme cases to perform surgery on the nose. Fliess liked to remove the turbinal bone.

Freud not only endorsed the theory in word, he endorsed it in practice, and Fliess operated on Freud's nose. Fliess is one source of a contemporary crankery, "reflexology," which differs only in holding the feet rather than the nose responsible, and in generalizing the scope of afflictions accounted for. Fliess was also the inventor of biorhythms. He held that all people were subject to two cycles, a female cycle of 28 days and a male cycle of 23 days, and these cycles determined their mental and physical health. Freud calculated lots of cycles.

Frank Sulloway (1983) has argued that Fliess was not a crank but a man of his time. No doubt he was something of both. Certainly he was widely regarded as a crank, and the views he developed in several books were the subject of considerable ridicule and criticism. A review of his book on the nasal reflex neurosis, for example, concluded that it was a disgrace from which the otherwise estimable publishing house would no doubt recover. Freud reported angrily to Fliess in 1897 that friends of Breuer's were busy applying probability theory to Fliess's published data, and claiming that the coincidences Fliess found were due to chance. Any criticism of Fliess, no matter how well posed, offended and angered Freud.

Freud's personal life was troubled in these years. His last child was born in 1896, after which he ceased (according to his letters to Fliess) having sexual relations with his wife. His father died in the same year, and the death came increasingly to distress Freud. He detected symptoms of hysteria in himself and began to treat himself through "self-analysis." He was addicted to cocaine in this period and he suffered various symptoms of heart disorder which he may have counted among his hysterical symptoms. His mood in his letters alternates between depression and buoyancy. His professional reputation was not, however, in any difficulty. Certainly his views were controversial and not widely accepted, but there is no evidence that he was, like Fliess, widely regarded as a crank. Freud had formed a hatred for his former collaborator and patron, Josef Breuer, to whom he owed a considerable sum of money in repayment for loans Breuer had made to him while he was a medical student. Breuer in turn did what he could to continue to promote Freud's career. In November of 1895, Breuer lectured to the College of Physicians defending Freud's general approach and arguing against the claim that Freud's results were due to suggestion. Breuer continued to send patients to Freud, as did other eminences, such as Carl Wernicke. In May of 1896 Freud wrote to Fliess that he was isolated, "Word was given out to abandon me, for a void is forming all around me . . . this year for the first time my consulting room is empty" (Masson 1985, 185). But in November Wernicke sent him a patient. In February of 1897 Freud wrote to Fliess that Freud was the only one in Vienna who believes Fliess's series, but in the same letter Freud revealed that Nothnagel and Krafft-Ebing had nominated Freud for a professorship at the University of Vienna. He wrote in May of 1897 that he had several new students, including a physician who had come from Berlin to study with him. At the end of May of 1897, he remarked that Breuer had sent him a new patient. Freud was not universally celebrated, and no doubt he had his professional difficulties at the time, but he was respected by some of the most distinguished medical people in Vienna. If anyone had real troubles with his reputation, it was Fliess.

Reputation is one thing, clinical success another. There is reason to think that Freud was not having much of the latter. It is difficult to count cases from publications and correspondence, and Freud did not do any systematic follow up of his patients. In 1892 Freud published an account of a successful treatment of an hysterical patient by hypnosis. There are four of Freud's cases discussed in *Studies on Hysteria,* but one of them consisted of a single informal session. Of the three remaining cases, Freud believed two resulted in cures. In *Obsessions and Phobias* (1953–1974, vol. 3, 69–84), published in 1895, Freud mentioned eleven cases, perhaps eight of them distinct, and explicitly remarked that in two of the cases cures resulted. Presumably, the others did not although one cannot be sure. His correspondence with Fliess in the middle of the 1890s seems to indicate that many of Freud's cases did not go well.

Freud's two theories were announced in several papers in 1895 and 1896 and the theory of the actual neuroses was discussed again in 1898. The development of his views can be traced through his correspondence with Fliess in those years and, beginning in 1897, we can also find Freud indicating that he had doubts about the theory. The doubts came to a head in a letter written on 21 September 1897:

And now I want to confide in you immediately the great secret that has been slowly dawning on me in the last few months. I no longer believe in my *neurotica*. This is probably not intelligible without an explanation; after all, you yourself found credible what I was able to tell you. So I will begin historically where the reasons for disbelief came from. The continual disappointment in my efforts to bring a single analysis to a real conclusion; the running away of people who for a period of time had been most gripped; the absence of the complete successes on which I had counted; the possibility of explaining to myself the partial successes in other ways, in the usual fashion—this was the first group. Then the surprise that in all cases, the *father,* not excluding my own, had to be accused of being perverse—the realization of the unexpected frequency of hysteria, with precisely the same conditions prevailing in each, whereas surely such widespread perversions against children are not very probable. The [incidence] of perversion would have to be immeasurably more frequent than the hysteria because the illness, after all, occurs only where there has been an accumulation of events, and there is a contributory factor that weakens the defense. Then, third, the certain insight that there are no indications of reality in the unconscious, so that one can cannot distinguish between truth and fiction that has been cathected with affect. (Accordingly there would remain the solution that the sexual fantasy invariably seizes upon the theme of the parents.) Fourth, the consideration that in the most deep-reading psychosis the unconscious memory does not break through, so that the secret of childhood experiences is not disclosed even in the most confused delirium. If one thus sees that the unconscious never overcomes the resistance of the conscious, the expectation that in treatment the opposite is bound to happen, to the point where the unconscious is completely tamed by the conscious, also diminishes.

I was so far influenced that I was ready to give up two things; the complete resolution of a neurosis and the certain knowledge of its etiology in childhood. Now I have no idea of where I stand because I haven not succeeded in gaining a theoretical understanding of repression and its interplay of forces. It seems once again arguable that only later experiences give the impetus to fantasies which [then] hark back to childhood, and with this the factor of a hereditary disposition regains a sphere of influence from which I had made it my task to dislodge it—in the interest of illuminating neurosis.

> If I were depressed, confused exhausted, such doubts would surely have to be interpreted as signs of weakness. Since I am in an opposite state, I must recognize them as the result of honest and vigorous intellectual work and must be proud that after going so deep I am still capable of such criticism. . . .
>
> It is strange, too, that no feeling of shame appeared—for which, after all, there could well be occasion. Of course I shall not tell it in Dan, nor speak of it in Askelson, the land of the Philistines, but in your eyes and my own. I have more the feeling of a victory than defeat, (which is surely not right). (Masson 1985, 264–65)

The letter contains five arguments. Freud's therapy is unsuccessful, his patient's flee him and his partial successes can be explained "in the usual fashion"—which means, I take it, as the result of the self-limiting character of the disorders, or placebo effects. Second, it is improbable that the seduction of children is as widespread as his hypothesis would require. Third, Freud's own father would have had to have been a seducer, since Freud and his sisters have hysterical symptoms. Fourth, the belief that the recollections of the unconscious are not generally veridical, whether about seduction or about anything else. And finally, that lunatic ravings do not include stories of seduction.

The first argument may seem a non sequitur, but Freud was still gripped by the idea of catharsis and by his physiological account of hysteria and repression; he firmly believed that once the hidden causes had been brought to the attention of the patient, and the patient had come to accept them, the formerly unconscious thoughts would no longer threaten to release quantities of excitation that would generate hysterical symptoms. He says later in the letter that while he doubts the seduction theory he does not doubt his psychology. The second argument is odd, because Freud had raised the same objection in the paper in which he first announced the theory and he had there rejected the objection without bothering to argue against it. But we must also bear in mind that Freud was coming to believe that hysterical symptoms were extremely common, if only in minor forms, and therefore that according to his theory a large fraction of the population must be the victims of sexual abuse in childhood.

And finally, note that the first alternative hypothesis Freud considers is not that childhood fantasies of seduction are etiological factors, but that they are displacements from real adolescent experiences that together with heredity cause adult neurosis. Freud offered the same hypothesis in a letter to Lowenfeld, and Lowenfeld reported it in 1900 in the introduction to a psychiatric textbook.

Masson's History

Masson takes the letter of 1897 to be dishonest. The renunciation of the seduction hypothesis was done, on his account, in order to regain Freud's damaged professional reputation and for personal psychological reasons. I have argued that Freud's professional reputation was not imperiled by his published theories, and if regaining a lost reputation were the aim, one might have expected Freud to renounce the theory publicly, which he did not. In fact, he had drawn at least as much criticism for his theory of the actual neuroses which he endorsed again in a paper in 1898. The psychological reasons Masson offers in explanation of Freud's change of mind are so bizarre and incoherent that it is hard to believe they were meant seriously or attracted any serious notice. No one ever went broke underestimating the American lite intellectual.

Freud had a patient, Emma Eckstein, who also became a disciple. Fliess visited Freud at Christmas of 1894, met Eckstein, and suggested to Freud that she suffered a nasal reflex neurosis and should be operated on. In February of 1895 Fliess returned to Vienna and performed the operation, which he botched. In the days that followed Eckstein suffered severe hemorrhaging, her nose became infected, and she very nearly died. Freud called in two Viennese surgeons who discovered that Fliess had left in the cavity more than half a meter of gauze, which had promoted infection. She continued to be ill for several months with intermittent hemorrhaging. Freud sent Fliess accounts of her health in March and April. By the end of May she had recovered.

What has this to do with Freud's renunciation of the seduction theory? According to Masson, what happened is this. Freud needed to blame Eckstein for her bleeding, so that his friend Fliess would be free of responsibility for the consequences of the botched surgery. Since she was an hysteric, the bleeding must be hysterical, and Freud suggested as much in his letters. Since Freud believed she was an hysteric, he believed she had been seduced in childhood. But it was better that Eckstein be viewed as a liar than a victim. Hence Freud abandoned the seduction theory.

This is pretty loony. It makes no sense as a psychological story. If Freud wanted to regard Eckstein's bleeding as hysterical and believed she had been seduced, then he had only to keep to his theory. Neither does it make sense of history. Freud's first suggestion of the seduction theory is in correspondence with Fliess in October of 1895, well after Eckstein's recovery, and the seduction theory was not published until May of 1896. Freud had no need to renounce the seduction theory to make Eckstein out a liar; he need only have not published the theory.

Doubt

I take Freud's criticism of his own theory in his letters to Fliess more or less at face value. It is hard to imagine what evidence as to his state of mind could defeat their testimony. What then did the recognition of the falsity of the seduction theory mean for Freud's conception of his own work? I think Freud's immediate and foremost concern was that his psychoanalytic method was untrustworthy. That, of course, would have been a real disaster, for Freud's career and whatever research he could claim to do were premised on the assumptions that the responses he got from his patients were not the effects of suggestion by the analyst, and that Freud could discriminate between veridical and false reports. His conclusions about the seduction theory would undermine any reasonable person's faith in these assumptions and that, I believe, is the principal reason Freud did not publicly acknowledge the difficulties he found with the theory.

Freud had good reason to be worried. His clinical methods were almost dictatorial, and he had repeatedly announced in print that they could not possible lead to false reports from patients. The method Freud used in the cases described in *Studies on Hysteria* included Janet's suggestion therapy. Janet's procedure was to hypnotize the patient and induce the patient to recall some traumatic event associated with a symptom. Then Janet would suggest to the hypnotized patient that the traumatic event had not occurred, or that it was in fact harmless and not frightening. Freud used hypnotic suggestion therapy on Frau Emmy Von N. One might well think that the power of suggestion under hypnosis might have given Freud reason to ponder the power of suggestion when the patient is not under hypnosis. Breuer, in his contribution to *Studies on Hysteria*, worried about the matter; Freud was more forceful:

We learn with astonishment from this that *we are not in a position to force anything on the patient about the things of which he is ostensibly ignorant or to influence the products of the analysis by arousing an expectation.* I have never once succeeded, by foretelling something, in altering or falsifying the reproduction of memories or the connection of events; for if I had, it would inevitably have been betrayed by some contradiction in the material. If something turned out as I had foretold, it was invariably proved by a great number of unimpeachable reminiscences that I had done no more than guess right. We need not be afraid therefore, of telling the patient what we think his next connection of thought is going to be. It will do no harm. (1953–1974, vol. 3, 295)

The smugness this passage reveals continued until 1897, and so did Freud's practice of leading his patients and pressing them until they produced the answers Freud sought. His clinical method did change. By about 1894 Freud had ceased hypnosis and begun instead to press his hand against the forehead of the patient and ask the patient to recall relevant events or to associate freely, saying whatever might pass through one's mind. Just as with hypnosis, the procedure left the analyst with great scope for leading the patient and for pressing until the patient produced the expected response. Freud did both, and had no qualms:

If the memory which we have uncovered does not answer our expectations, it may be that we ought to pursue the same path a little further; perhaps behind the first traumatic science there may be concealed a second which satisfies our requirements better and whose reproduction has a greater therapeutic effect. . . . Well, Gentlemen, this supposition is correct. If the first discovered scene is unsatisfactory, we will tell our patient that this experience explains nothing, but that behind it must be hidden a more significant earlier experience. (Ibid., 195)

And from the same period, 1896, in a paper on the seduction theory, he writes:

Doubts about the genuineness of the infantile sexual scenes can, however, be deprived of their force here and now by more than one argument. In the first place, the behavior of patients while they are reproducing these infantile experiences in every respect incompatible with the assumption that these scenes are anything else than a reality which is being felt with distress and reproduced with the greatest reluctance. Before they come for analysis the patients know nothing about these scenes. They are indignant as a rule if we warn them that such scenes are going to emerge. Only the strongest compulsion of the treatment can induce them to embark on a reproduction of them. (Ibid., 204)

These passages show how heavy Freud's hand must have been and they also show why, when he came to doubt the veracity of the seduction reports his patients gave him, he must have begun to doubt his clinical procedures. The passages also show why Freud was reluctant to publish his doubts. Any of his contemporaries who were aware of these statements—and there were many—would likely have taken a retraction of the seduction theory as evidence that Freud's enterprise was premised on an unreliable procedure, and that Freud himself was insufficiently sensitive to the dangers of his own methods.

Almost immediately after his letter of 21 September 1897, Freud determined to conduct an experiment to test the possibility that the seduction reports were the effects of suggestion. It was just the sort of question he was not intellectually suited to address. Freud had never had any luck with experiments or much understanding of experimental design or any sensitivity to problems of experimenter effects. The same limitations of scientific temper bounded his attempt to investigate the role of suggestion in psychoanalysis. Freud understood how to get examples and counterexamples to universal generalizations but he had only the most general idea about how to test the hypothesis that some or many of the seduction reports he received were due, not to the patients history, but to his methods.

The experiment Freud conducted was to have the now recovered Emma Eckstein, who had moved from patient to pupil, treat a new patient without mentioning the possibility of seduction. On 12 December 1897, Freud wrote to Fliess that

> My confidence in paternal etiology has risen greatly. Eckstein deliberately treated her patient in such a manner as not to give her the slightest hint of what would emerge from the unconscious and in the process obtained from her, among other things, the identical scenes with the father. Incidentally, the young girl is doing very well. (Masson 1985, 286)

Eckstein, of course, knew what outcome was wanted, and Freud did not consider whether she may have influenced the patient inadvertently. Nor did he consider that *this* patient may actually have been seduced while many others who give similar reports were not.

We are left to wonder whether Freud was genuinely convinced that his procedures were reliable; perhaps there is no truth to the matter. What emerged in the following years was a pattern of discussion of the neuroses that is inconsistent with such a conviction, and a kind of dissembling about his own views. It can be explained, as Masson prefers, as a deliberate dishonesty about the fact of seduction and a gradual withdrawal from the seduction theory in order to gain respectability. It can also be explained as a kind of deliberate dishonesty about the reliability of his clinical methods. I prefer to understand it as a kind of self-deception, carried out under the most powerful of motives and aided by the limitations of Freud's scientific capacities and temper.

Dissembling

After 1896, Freud never again endorsed the seduction theory in print, although he did not publicly withdraw it for another decade. In his correspondence through 1897 and 1898, he described one infantile sexual attack, which he took to be real. By March of 1898 he had begun to reformulate the sexual etiology of psychoneurosis: What is experienced sexually in childhood produces the psychoneuroses. But "experienced sexually" is broader than seduction by an adult and may include masturbation, which he increasingly emphasized in his letters. Toward the end of March of 1898 he wrote to Fliess that he was "completely disoriented" (Masson 1985, 304) about hysteria, and in April he wrote that he did not care to lecture on hysteria "because I lack the decisive verdict on two important points" (ibid., 307).

It is difficult to understand how after 1897 Freud could have continued in private to doubt his seduction theory without also privately doubting his methods, but in print and in correspondence he was if anything more confident and as insensitive as before to the role of suggestion or to the thought that the analyst might sometimes be deceiving himself. After the Eckstein experiment the Fliess correspondence contains no more doubts about clinical method on Freud's part. In 1898 in an article on sexual etiology of the neuroses devoted entirely to the actual neuroses, Freud wrote:

> Having diagnosed a case of neurasthenic neurosis with certainty and having classified its symptoms correctly, we are in a position to translate the symptomatology into aetiology; and we may then boldly demand confirmation of our suspicions from the patient. We must not be led astray by initial denials. If we keep firmly to what we have inferred, we shall in the end conquer every resistance by emphasizing the unshakeable nature of our convictions. . . .

Moreover, the idea that one might, by one's insistence cause a patient who is physically normal to accuse himself falsely of sexual misdemeanors—such an idea may safely be disregarded as an imaginary danger. (1953–1974, vol. 3, 269)

Not until 1905 do we find a renewed discussion of the causes of psychoneurosis, and even then Freud's account is equivocal. In *Three Essays on Sexuality* (1953–1974, vol. 3) he addresses the role of infantile sexuality on neurosis. It is striking that he still failed to mention any doubts as to the veracity of the seduction accounts his patients provided, still held that seduction is a principal cause of neurosis, but now allowed that seduction is no longer a specific cause. The role of seduction is to generate premature sexual activity, but if that happens by some other means, neurosis can still result:

The reappearance of sexual activity [in children] is determined by internal causes and external contingencies both of which can be guessed in cases of neurotic illness from the form taken by their symptoms and can be discovered with certainty by psychoanalytic investigation. I shall have to speak presently of the internal causes; great and lasting importance attaches at this period to the accidental *external* contingencies. In the foreground we find the effects of seduction, which treats a child as a sexual object prematurely and teaches him, in highly emotional circumstances, how to obtain satisfaction from his genital zones, a satisfaction which he is then usually obliged to repeat again and again by masturbation. I cannot admit that in my paper on "The Aetiology of Hysteria" I exaggerated the frequency or importance of that influence, though I did not then know that persons who remain normal may have had the same experiences in their childhood, and though I consequently overrated the importance of seduction in comparison with the factors of sexual constitution and development. Obviously, seduction is not required in order to arouse a child's sexual life; that can also come about spontaneously from internal causes. (Ibid., vol. 7, 190)

If Freud's aim in giving up the seduction theory was to reingratiate himself with Viennese morality, this new modification of the theory would not have succeeded much. Note also that the account does not in the least suggest that Freud misjudged the truth value of the reports his patients gave to him. Further, Freud's claim that in 1896 he did not realize that people who had been seduced need not become hysteric is false; the contrary claim is emphasized in his 1896 paper. The account in this passage seems disingenuous.

In 1906 Freud gave the following equivocal account of the development of his views. Speaking of 1896, he wrote:

At that time my material was still scanty, and it happened by chance to include a disproportionately large number of cases in which sexual seduction by an adult or by older children played the chief part in the history of the patient's childhood.

I thus over-estimated the frequency of such events (though in other respects they were not open to doubt). (Ibid., 274)

The parenthetical remark is just false, and represents in all probability a deliberate falsehood on Freud's part. Freud continued:

Moreover, I was at that period unable to distinguish with certainty between falsifications made by hysterics in their memories of childhood and traces of real events. Since then I have learned to explain a number of phantasies of seduction as attempts at fending off memories of the subject's own sexual activity (infantile masturbation). (Ibid.)

This passage contradicts the preceding one. The recollections of seduction cannot both have been "not subject to doubt" and also "phantasies" that were "attempts at fending off the subject's own sexual activity." Nor does he tell the reader how it is that he is now able to distinguish with certainty between truth and falsity in the recollections of his patients.

The whole business is distasteful, and it appears Freud could not bring himself to tell an honest story. These passages help persuade me that Freud was principally concerned with concealing his doubts about his clinical methods, and much less concerned with saving the seduction theory.

The Interpretation of Dreams

In the letter of 21 September 1897 in which Freud told Fliess that he no longer believed in his own seduction theory, there occurs the following paragraph; "I have to add one more thing. In this collapse of everything valuable, the psychological alone has remained untouched. The dream stands entirely secure and my beginnings of the metapsychological work have only grown in my estimation. It is a pity that one cannot make a living, for instance, on dream interpretation!" (Masson 1985, 266). Freud found a way to make a living on dream interpretation.

Freud's theory of dreams began with his *Project for a Scientific Psychology,* which is what is referred to in this passage as the "metapsychology." Freud included a section on dreams in the *Project*, and he claimed there that normal dreams, at least, are always wish fulfillments. He seems to have reached this conclusion entirely on theoretical grounds. The reasoning in the *Project* is roughly the following.

The psyche is run by the distribution and fluctuations of *Q* in the nervous system. We can distinguish two kinds of processes in the system, the "primary process" consisting of endogenously determined impulses that seek for the immediate decrease of *Q*, and the "secondary processes" that serve as a control structure permitting short run increases of *Q* in order to obtain a lower long run or average quantity of *Q*. The secondary processes are occasioned by the distributions of *Q* that Freud calls the Ego; they are brought about by experience and they are associated with conscious thought, deliberation, and belief. Objects of thoughts or images are collections of cathected neurones. They can have two sorts of cathexes, either wishful or painful. The cathexes of the primary process are always wishful; without the Ego, we are systems that desire. In sleep the cathexes of the Ego are nearly impotent and the primary processes rage mostly unhindered, save that they are somehow disconnected from motor nerves. Thus in sleep the primary process is free to activate wishful cathexes, and the wishful cathexis, sufficiently energized, is just a dream. In dreaming we do not consciously *think* of what is wished for; instead the very object of the wish appears to us as an hallucination. Freud was uncertain how or why in dreams the object of the wish becomes conscious—or at least present to the Ego—and he was also uncertain as to why the thought of the wish often remains unconscious, although he suggested protomechanisms in terms of neurones and their cathexes.

Throughout 1896 and 1897 Freud occasionally reported and interpreted to Fliess his own dreams or those of others. Sometime in this period he determined to write a book on the interpretation of dreams and began searching the literature and collecting material. In this period Freud also began trying to treat himself and analyzing his dreams was evidently part of the regimen. In March of 1898 he sent Fliess a draft of a chapter of the book on dreams and a list of titles for later chapters. Work on the book became for Freud a way of not thinking about the problems he had had with the psychoneuroses and with his clinical methods; "The habit of working on the dream [book] is—after the hysteria torture—extremely good for me" (Masson 1985, 314).

If Freud took the conclusion of the seduction episode to be that his interpretative methods were unreliable, then *The Interpretation of Dreams* would have been impossible. Equally, if he had remained wedded to the standards of argument he had learned from Brucke and Brentano, the book would have been impossible. The book was published in November of 1899, and its publication signaled that part of Freud had given up the scientific spirit.

The Interpretation of Dreams is a long, complicated and perhaps bewildering book. Despite its title, it is a grand psychology with large aims. The book has a structure, an argument, and a lot of speculation. I think the book represents a kind of revision of Freud's conception of science, a revision that resulted from his own self-analysis to be sure, but also a revision that represents the resolution of his doubts about his clinical methods. Freud had decided not to air his doubts about his clinical procedures and about his ability to discriminate between true and false reports from his patients. He retained the general Millian conception of science with which he had begun his scientific life, but his *Project* and his own self-analysis led him to an interest in dreams and to substantiating his conviction that dreams are wish fulfillments. His hypotheses about dreams as wish fulfillments could only be substantiated by cases, by examples, and since dreams do not generally *state* a wish but only, according to his hypothesis, *show* or hallucinate its fulfillment, the hypothesis could only be established by somehow detecting the wish, the desire and intent, hidden in the hallucination. Freud became convinced that by a combination of common sense, psychoanalytic hypotheses, and sheer intuition he could find the wish behind each dream. The collection of instances and examples needed to substantiate that dreams are wish fulfillments and to support his many other hypotheses about their structure had to be obtained by extending the capacities of the psychoanalytic researcher. Freud would no longer merely extract confessions and memories; he would infer intent. The inferences were not to be based on generalizations that determine the specific meanings Freud found. Instead these were to be established by a kind of extension of everyday psychology. He came to view the devices of conversation and of the theater as public versions of the private devices that direct the theater of the mind. We are accustomed to puns, to irony, to innuendo, to sarcasm, to parable and parody, to deception, to simile, to metaphor, and we routinely infer a meaning in others' words besides the most literal. Freud would do the same for the reports of dreams, and the meanings he found would then provide the Millian instances for generalizations about dreams and their formation.

The Interpretation of Dreams divides into four parts. It begins with a survey and criticism of the literature on dreams and articulates standards for a theory of the subject. Freud then describes and attempts to illustrate and argue for his procedures for dream interpretation. The largest part of the book is the development and exemplification of generalizations about the content of dreams and the mechanisms of representation that occur in them. The final fragment is a description of the general psychological framework of Freud's *Project,* applied to the phenomena of dreams. The book is in some ways deliberately equivocal. Early claims made about terrifying dreams, for example, are taken back or qualified in the final chapters. The neurological basis of the psychology of the *Project* is suppressed and Freud warns against taking the psychological structures he describes as having specific locations in brain autonomy, but he almost immediately after introduces the conjecture that they may be spatially localized, twice lapses into talk of neurones, and writes throughout of "excitation" as a physiological process.

Freud's own announced criteria for a satisfactory theory of dreams were these: (1) the theory ought to remove the appearance that the content of dreams has no determining causes, but it ought not to do so in such a way as to make it causal claims untested or untestable; and (2) a theory that proposes that dreams have a meaning should provide a general procedure for interpreting them, and a warrant that demonstrates the trustworthiness of the procedure.

This may be good sense, but it is good sense by which Freud found it difficult to abide. Freud's account of his procedure for interpreting dreams is this: The analyst starts with a phrase in the dream report, and asks the patient to associate freely, and speak the thoughts that occur to him. When associations stop, or seem to become unproductive, the analyst takes up another piece of the original dream report and asks the patient to associate to that. If themes in a later train of associations seem related to a previously considered element of the dream report, the analyst is free to direct the dreamer's attention back to the previous phrase or association.

The process continues until the meaning of the dream emerges. There are no rules for stopping, or for when to return to a previous topic. Freud does offer some rules of thumb about choosing the initial piece of the dream report for association.

Freud illustrated his method with a dream of his own—the dream of Irma:

A large hall—numerous guests, whom we were receiving.—Among them was Irma. I at once took her on one side, as thought to answer her letter and to reproach her for not having accepted my 'solution' yet. I said to her: "If you still get pains, it's really only your fault." She replied: "If you only knew what pains I've got now in my throat and stomach and abdomen—it's choking me"—I was alarmed and looked a her. She looked pale and puffy. I though to myself that after all I must be missing some organic trouble. I took her to the window and looked down her throat, and she showed signs of recalcitrance, like women with artificial dentures. I thought to myself that there was really no need for her to do that.—She then opened her mouth properly and on the right I found a big white patch; at another place I saw extensive whitish grey scabs upon some remarkably curly structure which were evidently modelled on the turbinal bones of the nose.—I at once called in Dr. M., and he repeated the examination and confirmed it. . . . Dr. M. looked quite different from usual; he was very pale, he walked with a limp and his chin was clean-shaven . . . My friend Otto was now standing beside her as well, and my friend Leopold was percussing her through her bodice and saying: "She has a dull area low down on the left." He also indicated that a portion of the skin on the left shoulder was infiltrated. (I noticed this, just as he did, in spite of her dress). . . . M. said: "There's no doubt its an infection, but no matter; dysentery will supervene and the toxin will be eliminated." . . . We were directly aware, too, of the origin of the infection. Not long before, when she was feeling unwell, my friend Otto had given her an injection of a preparation of propyl, propyls . . . propionic acid . . . trimetheylamin (and I saw before me the formula for this printed in heavy type). . . . Injections of that sort ought not to be made so thoughtlessly. . . . And probably the syringe had not been clean. (1953–1974, vol. 4, 107)

Irma was a patient of Freud's who had relapsed; during the evening of the dream, Freud had written out an account of the case to give to Dr. M., as Freud says, "in order to justify myself" (ibid., 106). M., Otto, and Leopold were all physician acquaintances of Freud's, and Otto had mentioned the case to Freud the previous day with, Freud thought, implied reproof. Freud's interpretation of the dream is this:

The dream fulfilled certain wishes which were started in me by the events of the previous evening (the news given me by Otto and my writing out of the case history.) The conclusion of the dream, that is to say, was that I was not responsible for the persistence of Irma's pains, but that Otto was. Otto had

in fact annoyed me by his remarks about Irma's incomplete cure, and the dream gave me my revenge by throwing the reproach back on to him. The dream acquitted me of the responsibility for Irma's condition by showing that it was due to other factors—it produced a whole series of reasons. The dream represented a particular state of affairs as I should have wished it to be. *Thus its content was the fulfillment of a wish and its motive was a wish.*
This much leapt to the eyes. (Ibid., 118–19)

The procedure of dream interpretation is rather vaguely characterized and the Irma dream does not even accord very well with it—Freud, acting as interpreter and dreamer, mixes association with commentary and there is no clear demarcation between the two. Given the background Freud's interpretation of his dream is perfectly plausible, and that is the trouble. The burden of the discussion is that the *procedure* of dream interpretation is reliable; the example only gives us a particular dream interpretation that is plausible, largely because given the background the dream wears its meaning on its face, and the interpretation does not follow the procedure in any case. We are left to wonder why, in less obvious cases, we should believe Freud's assignments of intent. We are left to wonder whether, if a dream had no meaning, or did not express a wish, one could nonetheless be manufactured by Freud's procedure. We are left to wonder if we could not as easily make every dream express a fear or disgust or any number of other attitudes. Some of Freud's contemporary readers had the same wonders.

Freud gives a one sentence reply to these qualms, "No doubt I shall be met by doubts of the trustworthiness of 'self-analyses' of this kind; and I shall be told that they leave the door open to arbitrary conclusions. In my judgement the situation is in fact more favourable in the case of *self*-observation than in that of other people; at all events we may make the experiment and see how far self-analysis takes us with the interpretation of dreams" (ibid., 105). Lacking some independent access to the meaning of dreams, there is no experiment. Freud proceeds to consider a variety of apparent counterexamples to the thesis that dreams are always wish fulfillments. He distinguishes the manifest content of a dream from the latent content; when the manifest content does not seem to express a wish, the latent content, revealed by interpretation, does so. The details of infantile material in dreams, recent material, distortion, representation in dreams, displacement, condensation, symbolism, typical dreams, absurd dreams, and so on, are established in much the same way. Many of the interpretations are plausible, many of them interesting, but was the method reliable?

One might charitably compare Freud's technology for dream interpretation with, say, Galileo's telescopic technology. Galileo could not, independently of what was seen through his telescope, establish that Jupiter has moons revolving about it, just as Freud could not independently establish the latent content of dreams. Galileo could only argue that when his telescope was pointed at earthly objects it revealed them enlarged but more or less faithfully. Freud could only argue that his method, when applied to homely obvious dreams, gave the obvious interpretation. The comparison is not apt, for an observer through Galileo's telescope could identify earthly objects without knowledge beforehand of what would appear in the telescopic field, but Freud's method, when applied to homely and obvious cases, was always done with the knowledge, independently of any technique, of the homely and obvious interpretation.

Freud's circumstance in *The Interpretation of Dreams* can usefully be compared with John Dalton's circumstance in his *New System of Chemical Philosophy* (1964). Freud claimed a method for locating the sense of dreams but he lacked any independent access to that sense and thus any means for testing his method. Dalton claimed a method, his rule of simplicity, for determining the atomic composition of a compound substance, but he lacked any independent method of assessing the composition, and so any means of testing his rule. Dalton had the advantage that his rule was at least clear, even if sometimes indeterminate in its application. He had the further advantage that within fifty years his followers had replaced the faulty pieces of his theory with better pieces and produced a theory that led to a wealth of new predictions and robust explanations. In the ninety years since the publication of *The Interpretation of Dreams* Freud has had no such luck.

Dora

Freud did not confine his broadened conception of science to dreams. He applied it to his neurotic patients as well. His case studies near the turn of the century show him at work inferring intent and desire not only from dreams but from all manner of other behavior. The plausibility of Freud's dream interpretations depend heavily on context, and *The Interpretation of Dreams* gives only the tiniest glimpses of that. The case studies are heftier and offer a larger view. Freud's *Fragment of an Analysis of a Case of Hysteria* (1953–1974, vol. 7, 1–122), the case of Dora, was offered as an illustration of the application of the theory of dreams. It offers more, an illustration of Freud's broadened interpretative methods in and out of dreams. Freud began the case in 1900 and published his account of it in 1905.

Dora was brought to Freud by her father when she was eighteen, suffering from depression, irritability and suicidal impulses. She had previously suffered from incidents in which she had difficulty breathing, chest pains, and occasional loss of voice. Her parents were friends with the family of Herr and Frau K. Her father explained to Freud that two years previously, during a joint vacation, Herr K. had taken Dora for a walk by a lake and had made a sexual proposal to her. He was convinced that her more recent disorders, including a suicide note, were due to that encounter. Herr K. had denied making advances, and had claimed that his wife, Frau K., had told him that the girl was interested only in reading and talking of sexual matters, and these activities had overexcited her.

Dora told Freud a story, which he fully accepted, of a romance between two families. She had been very close to Frau K. and they had spent a great deal of time together. Frau K. and her father were lovers, although her father was impotent. Dora believed Frau K. practiced fellatio with him. Herr K. had in fact approached her twice and had on previous occasion embraced her; she had fled. After the scene by the lake she had tried to avoid him, but found him in her room when she awoke. When she discovered that the key to her room was missing, she demanded to be returned home.

Freud's interpretation of the case is partly an inference about Dora's sexual practice, and more significantly an ascription of desire. Her shortness of breath, gastric distress and other symptoms that had begun around the age of eight are attributed to masturbation as a child. Dora neither admitted nor recollected the practice, but for various reasons Freud was nonetheless confident of the conclusion. The desire is more important. Freud regarded her flight from Herr K.'s embrace as an hysterical reaction, and he ascribed her sensations of pressure against her chest as an hysterical symptom: the manifestation of the memory of Herr K.'s erect penis against her abdomen. According to Freud, despite her claims to the contrary, she was in love with Herr K. The principle evidence for that conclusion was that she recollected that a friend had made the same remark, and that the first few occurrences of her aphonia corresponded to Herr K.'s absences. (Later attacks apparently had no such connection.) The aphonia was, according to Freud, hysterical speech, a way of saying that with Herr K. gone, no one worth talking to remained. Not only was Dora in love with Herr K., she was also in love, homosexual love, with Frau K. For she had praised Frau K.'s beauty, her "adorable white body," and had spent much of her time with Frau K. The decisive evidence for Freud, however, was that when Herr K. had defended himself with the information about Dora's reading, information obtained from Frau K., Dora had not been angry with Frau K. If further evidence were needed, there was Dora's virtual obsession with her father's relations with Frau K., a "supervalent" train of thought which covered her love for Frau K. and her resentment at Frau K.'s betrayal. But there is more. Not only was Dora in love with Herr K. and with Frau K., she was also in love with her own father. The supervalent train of thought concerning her father's relations with Frau K., and her conscious bitterness towards her father covered an unconscious Oedipal desire:

> Her behaviour obviously went far beyond what would have been appropriate to filial concern. She felt and acted more like a jealous wife—in a way which would have been comprehensible in her mother. By her ultimatum to her father ('either her or me'), by the scenes she used to make, by the suicidal intentions she allowed to transpire,—by all this she was clearly putting herself in her mother's place. If we have rightly guessed the nature of the imaginary sexual situation which underlay her cough, in that phantasy she must have been putting herself in Frau K.'s place. She was therefore identifying herself both with the woman her father had once loved and with the woman he loved now. The inference is obvious that her affection for her father was a much stronger one than she knew or than she would have cared to admit: in fact, that she was in love with him. (1953–1974, vol. 7, 56)

There is of course much more to the case, and for all I *know,* Freud's interpretation of Dora's desire may be correct. But there is a much simpler explanation of her depression, of her suicidal mood, of her anger toward her father, of the absence of comparable anger towards Frau K. or even towards Herr K., and of her insistence that her father break off her relationship with Frau K. Moreover, the essential fact upon which that explanation turns is not in dispute, at least it is not disputed by Freud. Dora believed, and Freud believed she believed correctly, that her father, the man she most trusted and who was most obliged to protect her interests, had traded her to Herr K. in return for Frau K. Herr K. and her father had struck a deal explicitly or tacitly: Herr K. would not make an issue of her father's affair with his wife, and her father would not make an issue of Herr K.'s attempts to seduce Dora. Dora had been made an object of barter, and she was trapped in that status, cause enough, perhaps, for irritability, depression and a suicidal mood; her bitterness was directed at the responsible party: her father.

Conclusion

There are better case studies than Dora (The Rat Man), and worse (Little Hans), and in any case I was not there to see her and to see Freud with her. But Dora makes me think that Freud had long since forgotten Brentano's warning: Do not settle upon an explanation unless you have excluded all equally good alternatives. I think he had also forgotten, or pushed out of mind, the warning in the failure of his own theory of seduction and neurosis. Caught by his career, limited by his own scientific conception, fascinated by self-revelation, Freud put aside the doubts he did not know how to investigate, and could not endure to have made public. In their place he put a new conviction as to his powers, and a new conception of his science—a conception that made it something else. It is one thing to be faced with scientific problems that one's methods of inquiry are impotent to solve; all but the most fortunate investigators confront that circumstance at some time, whether they recognize it or not. It is another matter to sense such limitations only to hide them from others and, eventually, from oneself. Science is a bit of sociology, a collection of people, traditions, pecking orders, and so on. Freud left the broader scientific community sometime after 1910, in the sense that he no longer published in scientific journals not run by his disciples, and no longer attended scientific meetings other than those organized by analysts. For all the cynicism of sociologists, science is also a system of ideals. In every science, everyone not stupid knows the ideals are different from the practice, but there is an inner community who acknowledges the standards and seeks, so well as it can, to live scientific lives by their command. When this century began, Freud was no longer a scientist in his soul.

NOTES

I thank Patricia Kitcher for her comments on a draft of this essay. She just hated it.

1. This is according to scholarly consensus. The encyclopaedia article is unsigned.

REFERENCES

Dalton, J. 1964. *A New System of Chemical Philosophy.* New York: Philosophical Library.

Freud, S. 1953–1974. *The Standard Edition of the Complete Psychological Works of Sigmund Freud.* Vols. 1–18. Edited by J. Strachey. London: Hogarth Press.

Jones, E. 1953. *The Life and Work of Sigmund Freud.* Vol. 1. New York: Basic Books.

Masson, J. 1984. *The Assault on Truth: Freud's Suppression of the Seduction Theory.* New York: Farrar, Straus, & Giroux.

———. 1985. *The Complete Letters of Sigmund Freud to Wilhelm Fliess, 1887–1904.* Cambridge, MA: Belknap Press of Harvard University Press.

Sulloway, F. 1979. *Freud, Biologist of the Mind: Beyond the Psychoanalytic Legend.* New York: Basic Books.

Wollheim, R. 1971. *Sigmund Freud.* New York: Viking Press.

18

Isomorphism and the Modeling of Brain-Mind State

J. Allan Hobson
Department of Psychiatry, Massachusetts Mental Health Center

In an effort to expose, clarify and advance my own thinking about the brain basis of dreaming I want to discuss two issues which have interested Adolf Grünbaum: (1) the nature of the brain-mind relationship (see Grünbaum 1971), and (2) Freud's theory of dreams, which is the cornerstone of psychoanalysis (see Grünbaum 1984).

Since my own philosophical position is a naive, intuitive, and professionally uniformed version of isomorphism, this contribution should be viewed more as a groping response toward Grünbaum-as-teacher than as a polished tribute to Grünbaum-as-scholar. Like many other scientists working at the exciting interface of neurobiology and psychology I am as much calling philosophy to the rescue as I am pointing out what I take to be some underappreciated philosophical opportunities and problems.

Overview: Adolf Grünbaum's Position

Adolf Grünbaum and Brain-Mind Isomorphism

For the purposes of my particular concern with isomorphism, the most relevant aspects of Grünbaum's work are to be found in parts 2B and 4 of his essay on *Free Will and Human Behavior* (1971) where the paradox of causal determinism and free choice is resolved by what Grünbaum calls "the irreducibly statistical nature of the laws of human behavior" (p. 623). Grünbaum first distinguishes between physical, psychological and statutory laws in ways that I find intriguing and challenging. I understand these distinctions best through concrete examples, "An involuntary wink is not a wink at all but rather a blink" (ibid., 611).

To me, this means that we are conscious of winks before the fact and can command or inhibit them by conscious effort; but we are conscious of blinks, if at all, only after the fact; and we can, by acts of will, only temporarily suppress their occurrence. One reason that I like this example is that sleep and dreaming are more like blinks than winks. Another reason that I like this example is because blinks emanate from the same brain stem oculomotor control system that command the rapid eye movement of our sleep and may even constitute an internal stimulus source for our dreams!

However, my main concern is not with the issue of volition, rather it is with Grünbaum's assertion that "any given substate of the organism at a time t_0 can coexist with any of many different mental states, Sψ rather than uniquely giving rise to only one particular Sψ" (1971, 622).

For reasons I will later try to make clear, this statement of the "one-to-many principle" gives me pause. Indeed it makes me wonder if the statisticality that Grünbaum defines as essential to quantum mechanics is really relevant to the argument. It seems likely that each particular dream state $S\psi$ is uniquely state determined by a particular brain state $S\phi$. This one-to-one assumption is critical to my concept of isomorphism. It allows the unpredictability of brain states on the basis of deterministic noise (i.e., chaos, randomness, statisticality) but it does *not* allow the loose coupling between brain and mind that flows from the one-to-many principle. I therefore reject that principle.

Perhaps what Grünbaum calls "*P*-cum-ψ" determinism is enough for my important assumptions to hold. In this case it is unnecessary (although I still hold it theoretically possible) to predict future $S\phi$ states from the history of all past $S\phi$ states. This seems less difficult to accept than Grünbaum's conclusion (which I do, nonetheless, accept) that "our putative *P*-cum-ψ determinism constitutes a denial of epiphenomenalism: the mental states $S\psi$ have causal efficacy precisely because their copresence with the physical states $S\phi$ transforms the mainly probabilistic linkages among the latter alone into deterministic ones" (1971, 622).

As much as I would like to indulge in further speculative discussion of mind-as-causal, I now turn to a brief comment on Grünbaum's critique of Freud.

Adolf Grünbaum and Psychoanalytic Dream Theory

Grünbaum and I have each drawn conclusions about the inadequacy of the Freudian dream theory and have since been allies-in-opposition to the Freudian dogmata that have for so long dominated psychology. My reasons for rejecting the psychoanalytic dream theory were based upon two points: The first was that Freud had based his psychological hypotheses upon erroneous neurobiological assumptions (McCarley and Hobson 1977); the second was that alternative and contrasting hypotheses could now be derived from modern sleep research (Hobson and McCarley 1977; Hobson 1988a, b). Grünbaum based his critique on inconsistencies within the theory itself. Thus our two views are different but strongly complimentary.

At a deeper philosophical level is my rejection of the strictly deterministic aspects of Freud's reflex model which became a part of his dream theory (Hobson and McCarley, 1977). This may be where Grünbaum's use of term "statisticality" would also impugn Freudian assumptions. For me, the neurophysiological evidence suggests that the REM sleep generator in the brain stem has an intrinsically random (or chaotic) nature. Thus, by the principle of isomorphism, dream image formation is necessarily random (or chaotic) and, hence, unpredictable.

Looking at the dreams themselves, it now seems likely that Freud's disguise censorship idea is not only outmoded but downright mistaken, and that the dream is an unsuccessful—if psychologically informative—attempt to integrate inchoate data into a coherent story. For Grünbaum, the unpredictability of the dream would have to be seen as another example of the unpredictability of the "irreducible statisticality of human behavior". I would very much welcome some discussion of this point. How much of this irreducible statisticality derives from quantum mechanics and how much derives from chaos theory?

One important caveat: Grünbaum and I are not merely Freud-bashers. We are, rather, scientists who are committed to the creation of new theories from newly collected data. Since the field of sleep and dream research is such a rich source of new information bearing on the mind-brain question, our more positive alliance is as comembers of the modern neurophilosophy community.

Brain-Mind Isomorphism

Freud himself was convinced that any scientific explanation of mental phenomena must ultimately involve physiology and biochemistry. In abandoning the "bottom-up" approach of his *Project for a Scientific Psychology* (see Freud 1954) and adopting the "top-down" approach of his 1900 dream theory and clinical psychoanalysis, he never gave up this ideal. In this essay I discuss some new concepts that might serve to support a more supple and versatile mode of inquiry—one that could be *both* "bottom-up" and "top-down" at the same time. This mode of inquiry, which is called mind-brain isomorphism, assumes the same similarity of form across the domains of mind and brain that Freud anticipated in his work. Strategic emphasis is placed on the parallel investigation of states of mind and brain, particularly the states of dreaming and REM sleep.

What is an Isomorphism?

Isomorphism has a strict mathematical definition that represents the functions in two related domains with one set of equations. An isomorphism is a bijective mapping—a mapping which is one-to-one. Put more simply, a set of operations on one domain has both a one-to-one and a projective relation to operations in a second domain. In a personal conversation, Quine offers the following formal definition:

> Two *sets* are isomorphic if there is a one-to-one correlation of the members of the one set to the members of the other. Example: the set of all married men, in our culture, and the set of all married women. The correlation is the husband relation.
>
> Two *relations* are isomorphic if all the things that enter into the one relation can be correlated with all the things that enter into the other relation, in such a way that always x will bear the one relation to y if and only if the correlated object x' bears the other relation to y'. Thus the correlation might be the correlation of people to their respective heads. On the strength of this correlation, the uncle relation is isomorphic to the relation that each uncle's head bears to the head of each of his nephews and nieces.
>
> The uncle relation is dyadic, and best viewed as the set of all ordered pairs (x, y) such that x is uncle of y. There are also triadic relations, e.g., between-

ness, and so on up. The notion of isomorphism carries over to any of these in obvious fashion. The correlation involved is still always dyadic, like the husband relation and the head relation in the above examples.

One speaks of isomorphism not only between a set and a set, as above, or between a relation and a relation, but also between a whole complex or system of sets and/or relations on the one hand and a matching system on the other. Thus, using again the correlation of people to their heads, we can say that the system (*M, L, U*) is isomorphic to the system (*M′, L′, U′*) where *M* is the set of all married men, *L* is the set of all lawyers, *U* is the uncle relation, *M′* is the relation mentioned at the end of my second paragraph.

In practical terms, consider the visual system which quantitatively preserves spatial coordinates in its mapping of the neural transformation of retinal signals. This depends upon a retinotopic organization of the thalamic relay system and the multiple maps of visual space at the level of the cerebral cortex. As far as one can trace the signal, there is isomorphism. This isomorphism is preserved in the perceptual product since one's estimates of length, distance, color, and so on, are reliable. We assume that the same general rules apply to higher order operations of the mind.

Psychophysical Isomorphism: Precedents and Problems

Even before Freud conceived his Project, this approach was taken by Gustav Theodor Fechner (1860), the pioneer of psychophysics, who asserted—and attempted to prove—that the intensity of sensation was a function of stimulus intensity. An implication pursued and documented by such modern psychophysiologists as Gerhard Werner and Vernon Mountcastle (1965) is that the intensity of firing of sensory neurons is the critical intervening variable in this isomorphism.

Such simplistic assumptions often fail due to state dependent changes in perceptual function. For example, sensation is *not* proportional to stimulus intensity during sleep. The threshold for perceived sensory stimuli shifts radically even though there is still an isomorphic encoding of stimulus strength in the rate of firing of peripheral and even central sensory neurons. In fact, the excitability of central neurons may even be amplified during sleep so that the isomorphism not only fails to describe the process but the usual relationship is paradoxically reversed.

How can the isomorphist approach survive this *bouleversement* by the radical state-dependency of even the simplest psychophysical laws? Only by creating a mathematically explicit set of laws describing central state regulation itself. This is the task of neurophysiologists studying brain states with an eye to understanding their associated mental states (or vice versa, since a "top-down" is as possible as a "bottom-up" approach). This essential task requires a revolutionary shift in scientific attitude for reasons that go far beyond the methodological problems already discussed and strike at the heart of the experimental method itself. Two other hurdles must be passed: one is the problem of subjectivity that confounds the characterization of mental states, the other is the unproved assumption of physiological homology between human brain physiology and that of lower mammals.

Characterizing Mental States. One philosophical problem with strong impact upon the isomorphist agenda is the unconfirmable nature of all subjective experience. This is compounded by the difficulty of access to the mind in sleep which reduces the reliability of self-description. The recent development of sleep laboratory techniques has given the study of dreaming a more instrumental and systematic character, and the emerging picture has encouraged psychophysiologic integration.

It may be worth underlining this *problem of subjectivity* which is ultimately irreducible—even under the instrumental conditions of the sleep laboratory. The scientist interested in mind thus has no choice other than to accept the reports of her experimental subjects as honest, accurate and truly retrospective—all three dubious, fallible, and unproveable assumptions. Can we ever be satisfied with such limitations? In order to resolve this issue, I offer the following guarantees.

The first guarantee is normative and statistical. If most humans spontaneously describe mental experiences in sleep which are formally identical and definitive of dreaming, I can be more confident than if such reports were exceptional. Such confidence would be further increased if the subjects were disinterested, or naive (i.e., had no psychological mindedness, no education, and no pet theory). The fact that dreaming occurs, and has the same form, in all cultures suggests that it is phenomelogically independent of belief systems and formally free of contamination by suggestion.

The second guarantee is the predictability found through the instrumental study of sleep. The fact that one can reliably predict the nature of a subjective report by reading an oscilloscope (or a digital voltmeter) attached to a person's head moves us toward objectivity in mental science. It remains theoretically possible, however, that such reports are either fabricated or that they arise *after* the experimental awakening has been initiated, as Malcolm has seriously suggested. Even if this were the case, it would be necessary to postulate a physiologically mediated state dependency of the factitious data since the reports differ formally from one another in a systematic manner according to the preexisting brain state. I therefore believe that Malcolm's objections are unfounded and can be discounted.

The third guarantee is pragmatic. Here I attempt to sidestep the trap of subjective doubt by having recourse to two related considerations. One is the absolute certainty that is associated with some dream recall in some subjects. Disbelieving such reports threatens, unacceptably, the whole philosophic and scientific enterprise. Such a skeptical operation of subjectivity upon subjectivity constitutes a hall of mirrors from which there is no escape. Believing that dream reports are veridical is supported by the recognition that *no* subjective experience, including instrumental operations such as the reading of an oscilloscope or a digital voltmeter has any better claim on what we call objectivity. The status of all such claims also depends ultimately upon statistical reliability and predictability.

Objective predictability, however, has a converse that makes such an explanation unlikely and that is the close correspondence of the awakened subjects estimate of dream duration to the actually elapsed clock time of the preexisting brain state. This accuracy could be ascribed (by obsessive doubt) to post-awakening retrospection were it not for the fact that the subject timing of external stimulus incorporation into a dream narrative also corresponds well with real clock time.

Investigative Strategy and the Assumption of Physiological Homology. The rationale and strategy taken here obviously involve another piece of fancy philosophical footwork. Without raising the unanswerable question "Do the cats dream?" we cannot have scientific certainty even that the brain states of the two species we wish to study are comparable. Indeed, since we can be reasonably sure they are not identical (if only because cats lack the hardware necessary for speech) we must be at least as cautious in extrapolating from the brain state of one species to that of another as we are in moving from the level of brain to the level of mind within a species.

The scientific evidence in favor of homology is indirect but strong. No crucial differences between the REM sleep of cats and humans have yet been documented. The many shared features include the following:

1. the brain activation (implying a homologous tonic cognitive arousal),
2. the motoric inhibition (implying a homologous tonic suppression of output), and
3. the eye movements or REMs (implying a homologous internal phasic signal generator).

The recent finding that brain waves coding for eye movement directionality are common to both species would seem to clinch the phasic signal argument.

Some speculation is admitted in my contention that nature would probably not devise two entirely different sets of neurophysiological mechanisms for such remarkably similar behavioral syndromes. All animal models, upon which the progress of knowledge heavily depends, share this uncertainty, however, so we have the consolation of being in good scientific company. One way to gain confidence in the homology assumption is to test, in humans, physiological hypotheses developed in the animal model. Positive results would at least satisfy the Jamesian pragmatist's definition of truth: It works. Such a case is provided by the demonstration that cholinergic agonists do, as predicted from animal experiments, enhance REM sleep *and* dreaming in humans.

Reflexes and States: A Paradigm Shift?

Hobson (1978) contrasted the stimulus-response paradigm (which dominates behavioral physiology as completely today as it did psychoanalysis a century ago) and the central state paradigm (which is only now emergent through research on sleep). Summarizing the arguments, the following four features contrast the two paradigms.

First, *the stimulus-response paradigm assumes state constancy.* This *may* be true of certain highly artificial conditions imposed upon the subject by the experimenter—such as anesthesia or decerebration—but it is *never* true of intact behaving organisms for whom dynamism is the rule. It is, however, probably not even true of those conditions on which all of modern neurophysiological knowledge is based since anesthetic and decerebrate states clearly evolve over time, both metabolically and functionally. These factors have been ignored by scientists who blindly and naively follow Claude Bernard's essential criterion of experimentation as holding all but one variable constant and then systematically varying that one. In any case, neither the anesthetic nor the decerebrate states of experimental animals can be regarded as either natural (they clearly are not) or of interest to the psychophysiologist (because sensation and/or its motoric expressions are eliminated in such states).

Second, *in a mathematical-statistical sense, stationarity is a problematic assumption* of most experimenters who work in either the stimulus-response or the state domain. Two factors operate to upset the validity of this assumption in the experimental preparation of the greatest interest (the intact, unanesthetized animal).

The first is habituation and is well known to stimulus-response scientists. Some become so impressed with the fact that successive stimuli elicit responses of decreasing strength (and finally no response at all) that they give up the study of sensation for the study of learning.

The second is the constant fluctuation of background activity that appears especially strongly at certain times of day but is present at all times of day, such intrinsic variability flies in the face of the constancy of input conditions that is an experimental requirement of the Benardian investigator. In other words, the very conditions necessary to psychophysical experimentation (input restriction and stimulus control) actually invoke dynamic changes in the system of interest. Habituation is thus the herald of sleep just as sleep is the herald of dreams.

Third, since it is a description of the rule governing the system's own dynamics that are the goal of the state scientist, one must, for a time at least, abandon the classical experimental paradigm. *The system must be allowed to run free so as to express itself clearly.* The experimenter must *do nothing* but wait and watch, a tedious, difficult, and unpopular observational mode known better to ethologists than to physiologists. For this reason, it is to be recommended that future students of behavioral physiology have experience in ethological research. Interestingly, this is exactly what psychiatry, in its concern for natural behavior, could become if only it would give up its own impulse to control states (as it does in psychopharmacology) or to interpret them by philosophically untenable theories (as it does in psychoanalysis). It is comforting to recognize that a single attitude, *expectant observation,* can unite the behavioral clinician with the behavioral scientist at this seemingly divisive moment in the history of the field.

Fourth, the stimulus-response scientist is aided (and hindered) by the obviousness of the order of investigations appropriate to his paradigm. He begins at the receptor, proceeds to the first, then the second synaptic relay describing the changes at each early information processing step (as if they were state-independent, which they are not) and only then becomes lost in the diffusely radiating maze of central interconnections that must, somehow, underlie sensation and perception. Some wise but unambitious colleagues sidestep this issue by following the stimulus out (as motor reflex activity).

The state scientist has no such inherent structural signposts to tell him what to do first. Where (and of course, how) do states begin and end? For states, the order sought is temporal, not spatial (since states are global) and the mechanism is an oscillator (or a clock), not a transformer (or an encoder). But (at least) 100 billion neurons exist in the brain, a daunting figure for anyone who would begin a cell-by-cell inventory! Localizing oscillators depends upon lesion studies, an unattractive method with its own problematic logical pitfalls. Destruction of the nervous system, followed by functional deficit is only suggestive, not conclusive, evidence of cause. These and related concerns have recently been elaborated in detail in a discussion of the scientific tribulations of our own reciprocal interaction hypothesis (Hobson et al. 1986).

Isomorphism and the State Concept

What is a State?

Summarizing extensive discussion of this question a state is any condition of a system (Hobson 1978). More precisely, and following Ashby (1960), I define "the state" of a system at a given instant as the set of numerical values which its variables have at that instant. A completely specified system state would include all possible measurements on the system and predict the effect of all inputs to the system in terms of output and change in state. A complete description of state would also take account of any intrinsically unstable properties of the system leading to state change as well as the means by which such changes are compensated or regulated.

Since a complete specification (as in finite automaton theory) can never be achieved for biological systems, most investigators use some arbitrary selected and limited set of variables which can be objectively measured. In the initial stages of investigation, this is acceptable if the degree of resolution matches the question asked of the system, but, as Ashby points out, any set of variables selected by an observer from the infinite set available may create a picture different from that of the "real machine." An even more critical problem is that the variables chosen to characterize state must be appropriate to the level of analysis intended. Thus a *behavioral* state is defined behaviorally, not physiologically, while a *physiological* state is defined physiologically and not behaviorally. If this criterion is not carefully observed, circularity is introduced into the analysis of the data and one risks viewing states as causing themselves.

Some Formal Considerations of the Given Issues

Being unaware of the possibly extensive mathematical and philosophical literature on state-to-state isomorphism, I draw, from Grünbaum (1971), the following conclusions.

First, the physical state of the brain $S\phi$ is specified by the numerical value of all its variables, $a \ldots n$, where any variable may be a function.

Second, the psychological state of the mind $S\psi$ is specified by the numerical value of all its variables, $a \ldots n$, where any variable may be a function.

Third, the isomorphist assumption states that a limited set of variables in $S\phi$ is associated with a limited set of variables $S\psi$ so that knowledge of the one reliably predicts the other. Because of the necessarily limited set of variables chosen for analysis, reliability can only denote "better than chance," (statistically speaking) and apparent exceptions are to be expected.

Grünbaum points that any one state $S\phi$ may be associated with more than one state $S\psi$. He also allows the converse, that any one psychical state $S\psi$ may be associated with more than one physical state, $S\phi$. For reasons already presented at the beginning of this essay, I am uncomfortable with the looseness and flexibility of Grünbaum's view. The reasons for my ambivalence follows.

Why I Doubt Grünbaum in Principle

First, assuming that the physical state $S\phi$ is an isomorphic function of the state of 10^{10} neurons (a modest estimate) where the state of each neuron can vary continuously from 0–1 (although it tends to be at one extreme or the other) the continuum of states $S\phi$ is at least 10^{10} or virtually infinite, and hence practically unknowable. Apparent resemblances between two physical states, $S\phi$ 1 and $S\phi$ 2, are misleading; they must be different, by definition. The upshot is that no two states of any one organism are ever completely *identical,* even without recognizing that, as time passes, the physical state of all neurons in the brain changes owing to aging.

Second, the same reasoning necessarily applies to psychological states.

Third, association of any $S\phi$ with more than one $S\psi$ is impossible since at any instant a complete specification of $S\phi$ would necessarily predict, exactly, a complete specification of $S\psi$.

Why I Am Willing to Live with Grünbaum in Practice

First, the isomorphist assumption is an unattainably ideal expression of brain-mind identity. While very probably true, we may never know it with certainty.

Second, the assumption of infinite variability is unlikely to be true, on evolutionary grounds. Instead organismic states are likely not to vary indefinitely but rather will vary within specifiable limits. That is, the values of $S\phi$ and $S\psi$, however multiple, will be within certain ranges. Only in this way could an animal's behavior be organized. Put another way, since an animal's behavior is *organized,* its states $S\phi$ and $S\psi$ must be *organized.* Thus, as a first approximation, we study the organization of states, $S\phi$s and $S\psi$s, seeking the limits within which the variables may range (the state space) and the rules by which the sets of variables are temporally ordered (the state sequence).

Third, my own observations on the neurons of the sleeping brain suggest that the limited variability assumption regarding $S\phi$ is justified allowing the ultimate truth of the specifiability of all $S\phi$ and hence of the isomorphist principle.

A set of assumptions concerning $S\phi$, $S\psi$, and their relationship then follows. First, given current limitations in the specification of $S\phi$, a reliable association of $S\phi$ with $S\psi$ (however limited) could not be expected.

Second, one would have to grant that a (necessarily) limited assessment of $S\phi$ could suggest one state of $S\phi$ associated with multiple states of $S\psi$ (and vice versa). For example, it is possible that an individual with an activated brain (low voltage-fast EEG), inhibited muscles (flat EMG), and moving eyes (EOG deflections) could be awake (responsive to external stimuli) or asleep and dreaming (responsive to external stimuli) or both responsive to external *and* internal stimuli (i.e., hallucinating in the strict sense of the word). At this level one state of $S\phi$ would appear to be associated with three states of $S\psi$ and psychophysiological identity (or even parallelism) would be lost.

Third, however, a moment's thought reveals this to be an illusion and a mistaken conclusion related to our currently greater capacity to discriminate $S\psi$s than $S\phi$s; this discrepancy is due to our deep access to the former states ($S\psi$s) and our shallow access to the latter states ($S\phi$s). Since, by definition all natural states of $S\psi$ are currently knowable, the fine description of such states and hence their categorical discriminations remain to be worked out. Unfortunately the introspective tools by which consciousness considers itself cannot be expected to improve a great deal in the foreseeable future. In the same foreseeable future we cannot expect, either, to get the information obviously needed to bring our knowledge of $S\phi$ states into register with our knowledge of $S\psi$ states.

Fourth, in the example given we need to know the states of sensory neurons to define the three correlative $S\phi$ states adequately. They could be biased toward exteroception (awake), interoception (dreaming), or stuck midway between those two settings (hallucinating). Since none of the three $S\phi$ variables used to measure states assess (even indirectly) the state of sensory systems, the state-to-state discrepency is only apparent and the assumption of psychophysiological correspondence-identity cannot be rejected.

The upshot of these considerations is to emphasize the need for a more formal and strict definition of isomorphism that would allow state-to-state mapping in domains that *are* in register on a priori grounds. As a more careful description of psychological states proceeds, and more penetrating physiological state assessments are undertaken, the isomorphist effort must remain tentative and cautious *but especially regarding false negative conclusions.*

At What Level May States Be Identified?

A *psychological state* is a particular set of values of psychological parameters. This domain is scientifically problematical for all the reasons we have detailed in discussing the issue of subjectivity. Accepting and respecting those caveats we would define a report of mental experience as emanating from the dream state if it described

- internally generated perceptions predominantly visual (which were)
- vivid and convincingly real (despite)
- internal inconsistencies, especially in the orientation domain and
- uncertainties about causes and effects,
- violation of physical laws,
- other "bizarre" features of plot, action, and character,
- strong emotions, especially fear, anger, and surprise,
- incomplete and/or fleeting memory for the experience (after the fact), contrasting with heightened recall of past events (within the dream experience itself).

This psychological state is correlated with states at other levels. We conceive it as the set of inward and subjective signs of a given brain state.

A *behavioral state* is a particular set of values of behavioral parameters. We would say that a subject was asleep if he was lying down, his eyes were closed, he was motionless, and a given sensory input no longer produced any output. We can assess the time spent and the timing of such a behavioral state which we conceive of as the outward and objective signs of a given brain state.

An *electrographic state* is any set of physiological parameters. We could have been more certain that our subject was asleep if her electromyogram showed decreased tonus, electroencephalogram showed slowing, high amplitude or spindle waves, and electrooculogram showed slow deflections or no change. Our behavioral assessment would thus be established and measured instrumentally. This physiological state is a set of inward signs objectively and outwardly detectable by recording weak electrical signals from the body surface. This is as close as we can currently get to a direct assessment of brain state in humans.

The *neuronal state* can be assessed at a precisely quantifiable level by measuring the activity of the brain's constituent neurons in an animal model.

If we accept the strong likelihood that our subject's neurons share the same state-dependent features as those of the cat (another mammal), then in slow-wave sleep our subject's large pyramidal tract and reticular giant cells would increase rate, small pyramidal tract cells would decrease rate, reticular thalamic cells would fire in bursts, and locus coeruleus and dorsal raphe neurons would decrease their regular rate. Thus, a neuronal state is a set of values of neuronal discharge parameters.

What are the Characteristics of the States of Biological Systems?

The above criteria concerning the state concept could help investigators to deal with one of the most elusive and important facts of physiological life that is, that *the nervous system is constantly changing state*. If behavioral physiologists remain aware of this fact, they would be able to correlate psychological behavioral, brain, and neural states in potentially meaningful, even explanatory, ways because behavioral states and neuronal states are almost certainly causally related.

The state scientist faces problems related to the fact that the nervous system is continuously changing state. For example, when we diagnose closed eyes, motionless, and nonresponsive behavior as sleep, we suggest that there is only one state of sleep and that it is constant and enduring. This, of course, is not true. Sleep, even as defined behaviorally, changes constantly. If we had watched awhile, we would have observed that periodically there were postural shifts, following which the eyes moved rapidly under the still closed lids, muscle tone became completely flaccid, and our subjects became even more unresponsive than at the onset of sleep.

Had we awakened our subject at such times she might actually have said that she dreamed she was at the beach. We assume that the state of her neurons would also have changed: her pyramidal tract (PT) cells and other sensorimotor neurons would have fired in high-frequency bursts and her reticularis thalami units would now give irregularly spaced single spikes (Mukhametov et al., 1970). Some of her locus coeruleus and raphe neurons would have ceased firing altogether. This behavioral state would have a new set of physiological concomitants: (1) the electromyogram (EMG) would be isoelectric; (2) the electroencephalogram (EEG) would show activation; and (3) the electrooculogram (EOG) would show bursts of activity. However, even if we call this set of properties the rapid eye movement (REM) or "dreaming sleep" state and say that there are (at least) two states of sleep, a problem remains. All the measures that have been mentioned so far, be they behavioral (such as posture), physiological (such as electrographic or neuronal), or psychological (such as dreaming), change their values continuously.

Thus, either the concept of state must be modified to mean a succession of instantaneous values (necessitating an infinite multiplicity of states), or the definition of state must be modified to accept values of measures *within a specified range.* This change involves no quantitative compromise but allows accommodation of theory to fact. Thus, the nervous system never conforms to the statistical ideal of stationarity, but *periods* or *phases* of relative stability, which we call *states,* are important (because the operating principles of the machine are so distinctive) and can be investigated (because it can be objectively determined when that state is entered and when it has been left).

How Can Isomorphism Between State Levels be Established?

As indicated elsewhere, I believe that the current state of knowledge limits the isomorphist approach to the most approximate, global, and statistical correlations of variables from the several levels at which states are assessed. If we take dreams as our starting point, we must then focus on the formal level of analysis, leaving to the indefinite future the analysis of content (in all its narrative and syntactical richness). We thus aim at an interpretation of *dreaming* (as a universal mental process) rather than the interpretation of *dreams* (as individual mental experiences).

This about-face from the Freudian approach deserves some emphasis. I share the conclusions of Clark Glymour (1978) and of Grünbaum (1984) concerning the deficits of Freud's method of dream interpretation that it is anecdotal, ad hoc, internally inconsistent, uncontrolled, gratuitous, and it fails even to do what it claims to do (to specify, by means of free association, the unconscious wish that motivates and determines the dream). I thus reject such an approach as both fundamentally unscientific and, in any case, as unsuited to the early stages of the isomorphic project we have adopted.

It is puzzling that even after 30 years of laboratory sleep research, the formal aspects of dreaming, as a psychological state, remain ill-defined and unquantified. Two powerful traditions work against achievement of this goal.

One is psychoanalysis itself, which still claims the allegiance of many empirically dedicated sleep scientists. Some of these colleagues follow other Freudian revisionists into the adjacent domain of linguistics in a sincere but fatally doomed effort to specify symbolic representations. Here again, I agree with Grünbaum (1984) and Morris Eagle in regarding all the self-styled breakthroughs to hermeneutic paradigms as scientific regressions. I prefer the clarity and the biological inspiration of the early Freud.

The other is a radical and molecular empiricism which catalogs informational items in dream reports and neglects the *process* that organizes the items. This approach is epitomized by the Calvin Hall-Robert van de Castle "dream bank" with its index of ten thousand dream reports, each of which is itemized with respect to descriptions of dream characters (are they men, women, or Martians?), plot features (are they running, jumping, standing still?), and so on. While more useful to the isomorphic project than any efforts at interpretation, at the narrative level this accountant approach is at too low a level to be of real use.

A single example may help make the point. It is necessary and sufficient (for the isomorphist) to know that all well-remembered dream reports describe color and that the common supposition that dreaming is colorless (the "black and white" theory) is an incorrect inference related definitively and exclusively to the problems of recall (an after-the-fact memory defect). This means that no state-specific change in higher order visual processing need be invoked—or sought—in developing a physiological state correlate for dreaming; rather a state dependent change in memory is to be postulated—and its neuronal correlate sought in experimental studies (in animals).

By contrast, it is neither necessary (nor even helpful) for the mind-brain isomorphist to know the incidence, in reports of the words "red," "yellow," or "chartreuse" since the higher order physiological correlates of such specific details are unlikely either to be state-dependent (or discovered within our lifetime). Of course, if all color reports were chartreuse or if a primary color were absent we would sit up and take notice so the approach is not a priori useless.

The choice of level that is likely to be fruitful in a state-to-state correlation is thus governed by the scientific maturity of work in one or both of the states under consideration. This limitation is severe in the case of all mental states, including dreaming, and for most physiological states except, perhaps, REM sleep which may now be the most adequately defined state at the behavioral and the neuronal level. In this case, it is thus the level of knowledge of the neuronal state of REM sleep which directs the contemporary state isomorphist to the appropriate psychological level in the study of dreaming. That level is the *formal* and involves a qualitative and quantitative assessment of the distinctive information processing characteristics of the state.

Having observed certain *formal* features of the brain state in REM sleep, the mind-brain isomorphist then moves from the bottom-up to ask if there is an isomorphic set of formal features of the mind-state in dreaming. Conversely, the presence of distinctive formal features of the mind state in dreaming direct a top-down quest for isomorphic features in the brain state of REM sleep. I conclude that such an approach has already proven useful.

NOTE

Research for this essay was supported by grants from the National Institutes of Health, the National Science Foundation, and the Commonwealth Fund. Some of the material was first presented at the Gifford Conference on Freud and the Philosophy of Mind, St. Andrews, Scotland, April 1985. It was updated and revised for the symposium in honor of Adolf Grünbaum sponsored by the Center for the Philosophy of Science, University of Pittsburgh, 5–7 October 1990.

REFERENCES

Ashby, W. R. 1960. *Design for a Brain*. London: Chapman & Hull.

Bernard, C. 1865. *Introduction á l'étude de la médicine éxpérimental*. Paris: Baillere.

Fechner, G. T. 1860. *Elemente der Psychophysik*. Vol. 3. Leipzig: Druck.

Freud, S. 1954. *The Origins of Psychoanalysis: Letters to Wilhelm Fliess, Drafts and Notes: 1887–1902*. New York: Basic Books.

Glymour, C. 1978. "The Theory of Your Dreams." In R. S. Cohen, and L. Laudan, eds., *Physics, Philosophy, and Psychoanalysis*. Dordrecht: Reidel, pp. 57–71.

Grünbaum, A. 1971. "Free Will and Laws of Human Behavior." *American Philosophical Quarterly 8:* 605–27.

———. 1984. *The Foundations of Psychoanalysis: A Philosophical Critique*. Berkeley and Los Angeles: University of California Press.

Hobson, J. A. 1978. "What is a Behavioral State?" In A. Ferrendelli, ed., *Aspects of Behavioral Neurobiology: Society for Neuroscience Symposia*. Vol. 3. Bethesda: Society for Neuroscience, pp. 1–15.

———. 1988a. *The Dreaming Brain*. New York: Basic Books.

———. 1988b. "Psychoanalytic Dream Theory: A Critique Based Upon Modern Neurophysiology." In P. Clark, and C. Wright, eds., *Mind, Psychoanalysis and Science*. Oxford: Blackwell, pp. 277–308.

———. 1990. "Activation, Input Source, and Modulation: A Neurocognitive Model of the State of the Brain-mind." In R. Bootzin, J. Kihlstrom, and D. Schacter, eds., *Sleep and Cognition*. Washington, DC: The American Psychological Association, pp. 25–40.

Hobson, J. A.; R. Lydic; and H. A. Baghdoyan. 1986. "Evolving Concepts of Sleep Cycle Generation: From Brain Centers to Neuronal Populations." *The Behavioral and Brain Sciences 9:* 371–448.

Hobson, J. A., and R. W. McCarley. 1977. "The Brain as a Dream-state Generator: An Activation-synthesis Hypothesis of the Dream Process." *American Journal of Psychiatry 134:* 1335–68.

McCarley, R. A., and J. A. Hobson. 1977. "The Neurobiological Origins of Psychoanalytic Dream Theory." *American Journal of Psychiatry 134:* 1211–21.

Mukhametov, L. M; G. Rizzolatti; and V. Tradardi. 1970. "Spontaneous Activity of Neurons of Nucleus Reticularis Thalami in Freely Moving Cats." *Journal of Physiology 210:* 651–67.

Werner, G., and V. B. Mountcastle. 1965. "Neural Activity in Mechanoreceptive Cutaneous Afferents: Stimulus-response Relations, Weber Functions, and Information Transmission." *Journal of Neurophysiology 28:* 359–97.

19

Psychoanalytic Conceptions of the Borderline Personality

A Sociocultural Alternative

Theodore Millon
Department of Psychology, University of Miami, and Department of Psychiatry, Harvard Medical School

Controversy is natural to a young science such as psychology and its subdisciplines (e.g., psychoanalytic theory), an inevitable and perhaps desirable sign that intellectual energy is being expended in efforts to explore and understand the further reaches and more subtle intricacies of its subject domain. It seems appropriate, therefore, to find that Adolf Grünbaum, a distinguished philosopher of science, has alloted a portion of his broad-ranging intellectual activities to take issue with its dominant trends, to voice doubts, point to gaps, confront and question, and show that our knowledge is not in a final and neatly packaged form, but in the throes of productive turmoil. Not the first to undertake this formidable task, Grünbaum may, however, be the most assiduous, incisive and controversial of his colleagues to come to grips with the substantive and methodological illusions that undergird both classical and modern renditions of psychoanalytic thought (Grünbaum 1984). In this essay, I will undertake a critique of a recent recourse to analytic theory as explicator of psychic pathology.

The Issue

Despite exegetic brilliance, I am concerned that the Talmudic habit of intricate and abstruse argument within the psychoanalytic community has drawn us into recondite intellectual territories, leading us to overlook the impact of logically more plausible and substantively more palpable forces generative of what has come to be labeled the borderline personality disorder (BPD). Unfortunately, the riveting effects of the internecine struggles between competing analytic theories keep us fixated on obscure and largely unfalsifiable etiologic hypotheses, precluding thereby serious consideration of notions that may possess superior logic and validity. It is toward this latter goal that the present essay is addressed.

By no means should the role of early experience, a key tenet of analytic thinking, be dismissed. Nevertheless, the logical and evidential base for currently popular analytic hypotheses is briefly examined and left wanting. Supplementary proposals that favor the deductive and probative primary of sociocultural factors are posited. They point to a wide array of influences that either set in place or further embed those deficits in psychic cohesion that lie at the heart of the disorder. Specifically, the view is advanced that our contemporary "epidemic" of BPD can best be attributed to two broad sociocultural trends that have come to characterize much of Western life this past quarter century, namely, the emergence of social customs that exacerbate rather than remediate early, errant parent-child relationships and, second, the diminished power of formerly reparative institutions to compensate for these ancient and ubiquitous relationship problems.

To raise questions about either the validity or adequacy of one or another analytic model is not to take issue with all aspects of their formulations; much of what has been proposed concerning the nature and origins of the BPD has both substantive merit and heuristic value. The alternative proposed should be seen, therefore, more in the nature of an addendum than a supplantation.

Substantive Agreements

Overdiagnosed and elusive as the BPD syndrome has been—approached from innumerable and diverse analytic perspectives (e.g., Eriksonian, Kernbergian, Mahlerian, Kohutian), as well as clothed in an assortment of novel conceptual terms (identity diffusion, self-object representations, projective identification)—there remain, nevertheless, certain shared observations that demonstrate the continued clinical astuteness and heuristic fertility of the analytic mind-set. Whatever doubts one may have with respect to either the logical or methodological merit of conjectures currently posed by third-generation analytic thinkers (Eagle 1984), they deserve more than passing commendation for their perspicacity in discerning and portraying the key features of a new and major clinical entity.

Overlooking for the present conflicts between analytic schools of thought, let me note the key borderline features that contemporary theorists appear to judge salient and valid.

A pervasive instability and ambivalence intrudes into the stream of the borderline's everyday life, resulting in fluctuating attitudes, erratic or uncontrolled emotions, and a general capriciousness and undependability. Since they are impulsive, unpredictable, and often explosive, others are commonly uncomfortable in their presence, so that borderlines often elicit rejection rather than the support they seek.

Dejection, depression, and self-destructive acts are common. Their anguish and despair is genuine, but it also is a means of expressing hostility, a covert instrumentality to frustrate and retaliate. Angered by the failure of others to be nurturant, borderlines employ moods and threats as vehicles to "get back," to "teach them a lesson." By exaggerating their plight and by moping about, borderlines avoid responsibilities and place added burdens on others, causing their families not only to care for them, but to suffer and feel guilt while doing so. In the same way, cold and stubborn silence may function as an instrument of punitive blackmail, a way of threatening others that further trouble is in the offing. They are impatient and irritable unless things go their way.

Borderlines may exhibit rapidly changing and often antithetical thoughts concerning themselves and others. The obstructiveness, pessimism, and immaturity which others attribute to them are only a reflection, they feel, of their "sensitivity" and the inconsideration that others have shown. But here again, ambivalence intrudes; perhaps, they say, it is their own unworthiness, their own failures, and their own "bad temper" which is the cause of their misery and the pain they bring to others.

Segmented and fragmented, subjected to the flux of their own contradictory attitudes and enigmatic actions, their very sense of being remains precarious. Their erratic and conflicting inclinations continue as both cause and effect, generating new experiences that feed back and reinforce an already diminished sense of wholeness.

Philosophical Quandaries

The premise that early experience plays a central role in shaping personality attributes is one shared by all developmental theorists. To say the preceding, however, is not to agree as to which specific factors during these developing years are critical in generating particular attributes, nor is it to agree that known formative influences are either necessary or sufficient. Psychoanalytic theorists almost invariably direct their etiologic attentions to the realm of early childhood experience. Unfortunately, they differ vigorously among themselves (e.g., Kernberg, Kohut, Mahler-Masterson, Erikson) as to which aspects of nascent life are crucial to development. In the following paragraphs I will examine both the logical and evidential basis of analytic theories in general, with only limited reference to notions related specifically to the origins of the BPD.

First I want to address the concept of "etiology." Meehl (1972, 1977) makes clear that the concept of etiology is a "fuzzy notion," one that not only requires the careful separation of constituent empirical elements, but one that calls for differentiating its diverse conceptual meanings, ranging from "strong" influences that are both causally necessary and/or sufficient, through progressively "weaker" levels of specificity, in which causal factors exert consistent, though quantitatively marginal differences, to those which are merely coincidental or situationally circumstantial.

There is reason to ask whether etiologic analysis is even possible in psychopathology in light of the complex and variable character of developmental influences. Can this most fundamental of scientific activities be achieved given that we are dealing with an interactive and sequential chain of "causes" composed of inherently inexact data of a highly probabilistic nature in which even the very slightest variation in context or antecedent condition—often of a minor or random character—produces highly divergent outcomes? Because this "looseness" in the causal network of variables is unavoidable, are there any grounds for believing that such endeavors could prove more than illusory? Further, will the careful study of individuals reveal repetitive patterns of symptomatic congruence, no less consistency among the origins of such diverse clinical attributes as overt behavior, intrapsychic functioning, and biophysical disposition? And will etiologic commonalities and syndromal coherence prove to be valid phenomena, that is, not merely imposed upon observed data by virtue of clinical expectation or theoretical bias (Millon 1987)?

Among other concerns is that evidence from well-designed and well-executed research are lacking. Consistent findings on causal factors for specific clinical entities would be useful were such knowledge only in hand. Unfortunately, our etiologic data base is both scanty and unreliable. As noted, it is likely to remain so owing to the obscure, complex, and interactive nature of influences that shape psychopathologic phenomena. The yearning among theorists of all viewpoints for a neat package of etiologic attributes simply cannot be reconciled with the complex philosophical issues, methodological quandaries, and difficult-to-disentangle subtle and random influences that shape mental disorders. In the main, almost all etiologic theses today are, at best, perceptive conjectures that ultimately rest on tenuous empirical grounds, reflecting the views of divergent "schools of thought" positing their favorite hypotheses. These speculative notions should be conceived as questions that deserve empirical evaluation, rather than be promulgated as the gospel of confirmed fact.

Central to this section is the question of whether contemporary analytic hypotheses addressing the developmental origins of the BPD have a more sound epistemic or probative foundation than found among most etiologic notions of psychopathology. The following paragraphs and quotes address this question.

Inferences drawn in the consulting room concerning past experiences, especially those of early childhood, are of limited, if not dubious, value by virtue of having only the patient as the primary, if not the sole, source of information. Events and relationships of the first years of life are notably unreliable, owing to the lack of clarity of retrospective memories. The presymbolic world of infants and young toddlers comprises fleeting and inarticulate impressions that remain embedded in perceptually amorphous and inchoate forms that cannot be reproduced as the growing child's cognitions take on a more discriminative and symbolic character (Millon 1969, 1981). What is "recalled," then, draws upon a highly ambiguous palette of diffuse images and affects, a source whose recaptured content is readily subject both to direct and subtle promptings from contemporary sources, for example, a theoretically oriented therapist. In his recent meticulous examination of the logical and empirical underpinnings of psychoanalytic theory, Grünbaum (1984, 1986) concludes that its prime investigatory method, the case history, and, in particular, data generated "on the couch" through so-called free associations, are especially fallible. Commenting on the influences of the therapist in directing the flow and content of the patient's verbal production, Grünbaum writes:

> The clinical use of free association features epistemic biases of selection and manipulative contamination as follows: (1) the analyst selects thematically from the patient's productions, partly by interrupting the associations—either explicitly or in a myriad of more subtle ways—at points of his or her own theoretically inspired choosing; and (2) when the Freudian doctor harbors the suspicion that the associations are faltering because of evasive censorship, he uses verbal and also subtle nonverbal promptings to induce the continuation of the associations until they yield theoretically appropriate results. (1984, 210–11)

A quarter century ago, the analyst Marmor, commented on the ease and inevitability with which therapeutic colleagues of contending analytic orientations would "discover" data congenial to their theoretical predilections:

> Depending upon the point of view of the analyst, the patients of each school seem to bring out precisely the kind of phenomenological data which confirm the theories and interpretations of their analysts. Thus, each theory tends to be self-validating. Freudians elicit material about the Oedipus complex and castration anxiety, Adlerians about masculine strivings and feelings of inferiority, Horneyans about idealized images, Sullivanians about disturbed interpersonal relationships, etc. (1962, 289)

Arguments pointing to thematic or logical continuities between the character of early experience and late behaviors, no matter how intuitively rational or consonant with established principles they may be, do not provide unequivocal evidence for their causal connections; different, and equally convincing, developmental hypotheses can be and are posited. Each contemporary explication of the origins of the BPD is persuasive yet remains but one among several plausible possibilities.

Among other troublesome aspects of contemporary analytic proposals are the diverse syndromal consequences attributed to essentially identical causes. Although it is not unreasonable to trace different outcomes to similar antecedents, there is an unusual inclination among analysts to assign the same "early conflict" or "traumatic relationship" to all varieties of psychological ailment. For example, an almost universal experiential ordeal that ostensibly undergirds such varied syndromes as narcissistic and borderline personalities, as well as a host of schizophrenic and psychosomatic conditions, is the splitting or repressing of introjected aggressive impulses engendered by parental hostility, an intrapsychic mechanism requisite to countering the dangers these impulses pose to dependency security should they achieve consciousness or behavioral expression.

Not only is it unlikely that singular origins would be as ubiquitous as analytic theorists often posit them but, even if they were, their ultimate psychological impact would differ substantially depending on the configuration of other concurrent or later influences to which individuals were exposed. "Identical" causal factors cannot be assumed to possess the same import, nor can their consequences be traced without reference to the larger context of each individual's life experiences. One need not be a Gestaltist to recognize that the substantive impact of an ostensive process or event, however formidable it may seem in theory—be it explicit parental spitefulness or implicit parental abandonment—will differ markedly as a function of its developmental covariants.

To go one step further, there is good reason, as well as evidence, to believe that the significance of early troubled relationships may inhere less in their singularity or the depth of their impact than in the fact that they are precursors of what is likely to become a recurrent pattern of subsequent parental encounters. It may be sheer recapitulation and consequent cumulative learning that ultimately fashions and deeply embeds the engrained pattern of distinctive personality attributes we observe (Millon 1969, 1981). Fisher and Greenberg (1977), though generally supportive of the scientific credibility of "Freudian" theories, conclude with this thesis following their examination of the etiologic origins of the anal character:

Several investigators have identified significant positive relationships between the anality of individuals and the intensity of anal attitudes present in their mothers. This obviously suggests that anal traits derive from associating with a parent who treats you in certain ways or provides you with models of how the world is to be interpreted. One should add that since a mother's anal traits are probably a permanent part of her personality repertoire, it would be reasonable to assume they would continue to affect her offsprings not only during the toilet-training period but also throughout her contacts with him. . . . The only thing we can state with even moderate assurance is that a mother with anal character traits will tend to raise an offspring with analogous traits. (Pp. 164–65)

Despite the foregoing, I share the analytic view that, unit for unit, the earlier the experience, the likely greater its impact and durability (Millon 1981). For example, the presymbolic and random nature of learning in the first few years often precludes subsequent duplication, and hence "protects" what has been learned. But, I believe it is also true that singular etiologic experiences, such as "split introjects" and "separation-individuation" struggles, are often only the earliest manifestation of a recurrent pattern of parent-child relationships, as suggested by Fisher and Greenberg. Early learnings fail to change, therefore, not because they have jelled permanently but because the same slender band of experiences which helped form them initially continue and persist as influences for years.

A Sociocultural Alternative

What follows contends that societal customs which served in the past to repair disturbances in early parent-child relations have declined in their efficacy, and have been "replaced" over the past two to three decades with customs that exacerbate these difficulties, contributing thereby to what I term our contemporary BPD "epidemic."

The central questions guiding this commentary are, first, what are the primary sources of influence that give rise to symptoms that distinguish the BPD, namely, an inability to maintain psychic cohesion in realms of affect, self-image, and interpersonal relationships; and second, which of these sources has had its impact heightened over the past several decades, accounting thereby for the rapid and marked increase in the incidence of the disorder?

The first question will not be elaborated, other than to note the observation that well-reasoned yet contending formulations have been posited by constitutional, analytic, and social learning theorists; and despite important divergences, there is a modest consensus that biogenic, psychogenic, and sociogenic factors each contribute in relevant ways.

The second question calls for explication; it relates to which of these three etiologic factors, each productive of the borderline's diffuse or segmented personality structure—constitutional disposition, problematic early nurturing, or contemporary social changes—has shown a substantial shift in recent decades. Is it some unidentified yet fundamental alteration in the intrinsic biological makeup of current-day youngsters; is it some significant and specifiable change in the character with which contemporary mothers nurture their infants and rear their toddlers; or is it traceable to fundamental and rapid changes in Western culture that have generated divisive and discordant life experiences while reducing the availability of psychically cohering and reparative social customs and institutions?

Despite the fact that tangible evidence favoring one or another of these possibilities is not accessible in the conventional sense of empirical "proof," it is our contention that the third "choice" is probatively more sustainable and inferentially more plausible. Toward these ends, two sociocultural trends generative of the segmented psychic structures that typify the BPD will be elucidated. Although they are interwoven substantively and chronologically, these trends are separated for conceptual and pedagogic purposes. One adds to the severity of psychic dissonance; it appears to have been on the upgrade. The other has taken a distinct downturn, and its loss also contributes to diminished psychic cohesion.

Increase in Divisive Social Customs

We are immersed deeply both in our time and culture, obscuring thereby our ability to discern many profound changes—often generative of unforeseen psychic and social consequences—that may be underway in our society's institutions. Tom Wicker, the distinguished columnist for the *New York Times,* portrays sequential effects such as these:

> When a solar-powered water pump was provided for a well in India, the village headman took it over and sold the water, until stopped. The new liquid abundance attracted hordes of unwanted nomads. Village boys who had drawn water in buckets had nothing to do, and some became criminals. The gap between rich and poor widened, since the poor had no land to benefit from irrigation. Finally, village women broke the pump, so they could gather again around the well that had been the center of their social lives. (1987, 23)

Not all forms of contemporary change can so readily be reversed. "Progress" wrought by modern-day education and technology is too powerful to be turned aside or nullified, no less reversed, despite "conservative" efforts to revoke or undo their inexorable effects.

It is both intuitively and observationally self-evident that sweeping cultural changes can affect innumerable social practices, including those of an immediate and personal nature such as patterns of child nurturing and rearing, marital affiliation, family cohesion, leisure style, entertainment content, and so on.

It would not be too speculative to assert that the organization, coherence, and stability of a culture's institutions are in great measure reflected in the psychic structure and cohesion of its members. In a manner analogous to the DNA double helix, in which each paired strand unwinds and selects environmental nutrients to duplicate its jettisoned partner, so too does each culture fashion its constituent members to fit an extant template. In societies whose customs and institutions are fixed and definitive, the psychic composition of its citizenry will likewise be structured; and where a society's values and practices are fluid and inconsistent, so too will its residents evolve deficits in psychic solidity and stability.

This latter, more amorphous, state, so characteristic of our modern times, is clearly mirrored in the interpersonal vacillations and affective instabilities that typify the BPD. Central to our recent culture have been the increased pace of social change and the growing pervasiveness of ambiguous and discordant customs to which children are expected to subscribe. Under the cumulative impact of rapid industrialization, immigration, urbanization, mobility, technology, and mass communication, there has been a steady erosion of traditional values and standards. Instead of a simple and coherent body of practices and beliefs, children find themselves confronted with constantly shifting styles and increasingly questioned norms whose durability is uncertain and precarious.

No longer do youngsters find the certainties and absolutes which guided earlier generations. The complexity and diversity of everyday experience play havoc with simple "archaic" beliefs, and render them useless as instruments to deal with contemporary realities. Lacking a coherent view of life, maturing youngsters find themselves groping and bewildered, swinging from one set of principles and models to another, unable to find stability either in their relationships or in the flux of events.

Although transformations in family patterns and relationships have evolved fairly continuously over the past century, the speed and nature of transitions since the Second World War have been so radical as to break the smooth line of earlier trends. Hence, today typical American children no longer have a clear sense of either the character of the purpose of their fathers' work activities, much less a detailed image of the concrete actions that comprise that work. Beyond the little there is of fathers' daily routines to model oneself after, mothers of young children have shifted their activities increasingly outside the home, seeking career fulfillments or needing dual incomes to sustain family aspirations. Not only are everyday adult activities no longer available for direct observation and modeling, but traditional gender roles, once distinct and valued, have become blurred and questionable. Today, there is little that is rewarded and esteemed by their larger society that takes place for children to see and emulate. What "real" and "important" people do cannot be learned from parents who return from a day's work too preoccupied or too exhausted to share their esoteric activities. Lost are the crystallizing and focusing effects of identifiable and stable role models which give structure and direction to maturing psychic processes. This loss contributes, then, to the maintenance of the undifferentiated and diffuse personality organization so characteristic of young borderlines.

With the growing dissolution of the traditional family structure there has been a marked increase in parental separation, divorce, and remarriage. Children subject to persistent parental bickering and family restructuring not only are exposed to changing and destructive models for imitative learning but develop the internal schisms that typify borderline behaviors. The stability of life, so necessary for the acquisition of a consistent pattern of feeling and thinking, is shattered when erratic conditions or marked controversy prevail.

Devoid of stabilizing internalizations, such youngsters may come to prefer the attractions of momentary and passing encounters of high salience and affective power. Unable to gauge what to expect from their environment, how can they be sure that things that are true today will be there tomorrow? Have they not experienced capriciousness when things appeared stable? Unlike children who can predict their fate—good, bad, or indifferent—such youngsters are unable to fathom what the future will bring. At any moment, and for no apparent reason, they may receive the kindness and support they crave; equally possible, and for equally unfathomable reasons, they may be the recipient of hostility and rejection. Having no way of determining which course of action will bring security and stability, such youngsters vacillate, feeling hostility, guilt, compliance, assertion, and so on, shifting erratically and impulsively from one tentative action to another. Unable to predict whether one's parents will be critical or affectionate, they must be ready for hostility when most might expect commendation, assume humiliation when most would anticipate reward.

Other "advances" in contemporary society have stamped deep and distinct impressions as well, ones equally affectively loaded, erratic, and contradictory. The rapidly moving, emotionally intense, and interpersonally capricious character of television role models, displayed in swiftly progressing half-hour vignettes that encompass a lifetime, add to the impact of disparate, highly charged, and largely inimical value standards and behavior models. What is incorporated is not only a multiplicity of selves, but an assemblage of unintegrated and discordant roles, displayed indecisively and fitfully, especially among those youngsters bereft of secure moorings and internal gyroscopes. The striking images created by our modern-day flickering parental surrogate have replaced all other sources of cultural guidance for many; hence by age 18, typical American children will have spent more time watching television than in going to school or relating directly to their parents.

To add to this disorienting and cacophonous melange are aggravations consequent to drug and alcohol involvements. Although youth is a natural period for exploratory behaviors, of which many are both socially adaptive and developmentally constructive, much of what is explored entails high risks with severe adverse consequences in both the short and long run. Viewed, however, from the perspective of youngsters who see little in life that has proven secure or desirable, the risks of these all-too-accessible substances are experienced neither as intimidating nor perilous. While they may be considered by many to be casual and recreational, their psychic effects are hazardous, especially among the already vulnerable. Thus, for the borderline-prone, the impact of these substances will only further diminish the clarity and focus of their feeble internalized structures, as well as dissolving whatever purposefulness and aspirations they may have possessed to guide them toward potentially reparative actions. Together, these mind-blurring effects add fresh weight to already established psychic diffusions.

Decrease in Reparative Social Customs

The fabric of traditional and organized societies not only comprises standards designed to indoctrinate and inculcate the young, but it also provides "insurance," that is, backups to compensate and repair system defects and failures. Extended families, church leaders, teachers, and neighbors provide nurturance and role models by which children experiencing troubling parental relationships can find a means of substitute support and affection, enabling them thereby to be receptive to society's established body of norms and values. Youngsters subject to any of the diffusing and divisive forces described previously must find one or another of these culturally sanctioned sources of surrogate modeling and sustenance to give structure and direction to their emerging capacities and impulses. Without such bolstering, maturing potentials are likely to remain diffuse and scattered.

Therefore, the thesis of this paper is that the changes of the past three to four decades have not only fostered an increase in intrapsychic diffusion and splintering, but have also resulted in the discontinuation of psychically restorative institutions and customs, contributing thereby to both the incidence and exacerbation of features that typify borderline pathology. Without the corrective effects of undergirding and focusing social mentors and practices, the diffusing or divisive consequences of unfavorable earlier experience take firm root and unyielding form, displaying their structural weaknesses in clinical signs under the press of even modestly stressful events.

One of the by-products of the rapid expansion of knowledge and education is that many of the traditional institutions of our society which formerly served as a refuge to many, offering "love" for virtuous behavior and caring and thoughtful role models have lost much of their historic power as a source of nurturance and control in our contemporary world.

Similarly, and in a more general way, the frequency with which families in our society relocate has caused a wide range of psychically diffusing problems. We not only leave behind stability, but with each move jettison a network of partially internalized role models and community institutions such as furnished in school and in friendships.

The scattering of the extended family, as well as the rise of single-parent homes and shrinkage in sibling number, adds further to the isolation of families as they migrate in and out of transient communities. Each of these undermines the once powerful reparative effects of kinship support and caring relationships. Contemporary forms of disaffection and alienation between parent and child may differ in their particulars from those of the past. But rejection and estrangement has been and is ubiquitous, and as commonplace as rivalry among siblings. In former times when children were subjected to negligence or abuse, they often found familial or neighborly parental surrogates—grandmothers, older siblings, aunts or uncles, even the kind or childless couple down the street—who would, by virtue of their own needs or identifications, nurture or even rear them. Frequently more suitable to parental roles, typically more affectionate and giving, as well as less disciplinary and punitive, these healing surrogates have historically served not only to repair the psychic damage of destructive parent-child relationships, but have "filled in" the requisite modeling of social customs and personal values that youngsters, so treated, would now be receptive to imitate and internalize.

For some the question is not which of the changing social values they should pursue, but whether there are any social values that are worthy of pursuit. Youngsters exposed to poverty and destitution, provided with inadequate schools, living in poor housing set within decaying communities, raised in chaotic and broken homes, deprived of parental models of "success and attainment," and immersed in a pervasive atmosphere of hopelessness, futility, and apathy, cannot help but question the validity of the "good society." Reared in these settings one quickly learns that there are few worthy standards to which one can aspire successfully. Whatever efforts are made to raise oneself from these bleak surroundings run hard against the painful restrictions of poverty, the sense of a meaningless and empty existence and an indifferent, if not hostile, world. Moreover, and in contrast to earlier generations whose worlds rarely extended beyond the shared confines of ghetto poverty, the disparity between everyday realities and what is seen as so evidently available to others in enticing television commercials and bounteous shopping malls is not only frustrating, but painfully disillusioning and immobilizing. Why make a pretense of accepting patently "false" values or seeking the unattainable goals of the larger society, when reality undermines every hope, and social existence is so pervasively hypocritical and harsh?

Nihilistic resolutions such as these leave youngsters bereft of a core of inner standards and customs to stabilize and guide their future actions, exposing them to the capricious power of momentary impulse and passing temptation. Beyond being merely "anomic" in Durkheim's sense of lacking socially sanctioned means for achieving culturally encouraged goals, these youngsters have incorporated neither the approved customs and practices nor the institutional aspirations and values of our society. In effect, they are both behaviorally normless and existentially purposeless, features seen in manifest clinical form among prototypical borderlines.

Until a generation or two ago children had productive, even necessary, economic roles to fill within the family. More recently, when the hard work of cultivating the soil or caring for the home were no longer requisites of daily life, youngsters were encouraged to advance their family's fortunes and status via higher education and professional vocations. Such needed functions and lofty ambitions were not only internalized, but gave a focus and a direction to one's life, creating a clear priority to one's values and aspirations, and bringing disparate potentials into a coherent schema and life philosophy.

Coherent aspirations are no longer commonplace today, especially among the children of the middle classes. In contrast to disadvantaged, yet anomic, youngsters, they are no longer "needed" to contribute to the family's economic survival; on the other hand, neither can upwardly mobile educational and economic ambitions lead them to readily surpass the achievements of already successful parents. In fact, children are seen in many quarters today as economic burdens, not as vehicles to a more secure future or to a more esteemed social status. Parents absorbed in their own lives and careers frequently view children as impediments to their autonomy and narcissistic indulgences.

But even among children who are not overtly alienated or rejected, the psychically cohering and energizing effects of "being needed" or of "fulfilling a worthy" family aspiration have been lost in our times. Without genuine obligations and real purposes to create intent and urgency in their psychosocial worlds, such youngsters often remain diffused and undirected. At best, they are encouraged to "find their own thing," to follow their own desires and create their own aims. Unfortunately, freedoms such as these translate for many as freedom to remain in flux, to be drawn to each passing fancy, to act out each passing mood, to view every conviction or ethic of equal merit, and, ultimately, to feel evermore adrift, lost, and empty.

Concluding Comment

The preceding pages describe a number of the contributory elements that comprise a broad mosaic of BPD-disposing influences of our times. Although bypassed in their specifics, there are salient ingredients of a biogenic nature in this multifactorial mix of determinants. Similarly, and prior critiques notwithstanding, this mosaic should be seen as also encompassing the psychogenic role of adverse early nurturing and rearing.

What is troubling to those who seek an "ecumenical" synthesis among rival etiologic models is not the observation that some analytic authors hold the view that this or that ordeal of early life is crucial to the development of a particular personality disorder. Rather, it is when claimants couple empirically unproven or philosophically untenable assumptions with the assertion that it is they alone who possess the sole means by which such etiologic origins can be revealed. Perhaps it is too harsh to draw parallels, but presumptions such as these are similar to biblical inerrantists who claim their construals of the Bible to be "divine" interpretations, or conservative legalists who assert their unenlightened views to correspond to the "original intent" of the Constitution's framers. So, too, do many of our more self-righteous interpreters practice blindly what some have judged our "hopelessly flawed craft" (Grünbaum 1984, 1986), one in which we demonstrably can neither agree among ourselves, nor discover either the data or methods by which a coherent synthesis may be fashioned among myriad conjectures.

REFERENCES

Eagle, M. 1984. *Recent Developments in Psychoanalysis: A Critical Evaluation.* New York: McGraw-Hill.

Fisher, S., and R. P. Greenberg. 1977. *The Scientific Credibility of Freud's Theories and Therapy.* New York: Basic Books.

Grünbaum, A. 1984. *The Foundations of Psychoanalysis: A Philosophical Critique.* Berkeley and Los Angeles: University of California Press.

———. 1986. "What are the Clinical Credentials of the Psychoanalytic Compromise Model of Neurotic Symptoms." In T. Millon, and G. L. Klerman, eds., *Contemporary Directions in Psychopathology.* New York: Guilford, pp. 193–214.

Marmor, J. 1962. "Psychoanalytic Therapy as an Educational Process." In J. Masserman, ed., *Science and Psychoanalysis*. Vol. 5. New York: Grune & Stratton, pp. 286–99.

Meehl, P. E. 1972. "Specific Genetic Etiology, Psychodynamics, and Therapeutic Nihilism." *International Journal on Mental Health 1:* 10–27.

———. 1977. "Specific Etiology and Other Forms of Strong Influence: Some Quantitative Meanings." *Journal of Medicine and Philosophy 2:* 33–53.

Millon, T. 1969. *Modern Psychopathology*. Philadelphia: Saunders.

———. 1981. *Disorders of Personality*. New York: Wiley-Interscience.

———. 1987. "On the Nature of Taxonomy in Psychopathology." In C. Last, and M. Hersen, eds., *Issues in Diagnostic Research*. New York: Plenum, pp. 3–85.

Wicker, T. 1987. "The Pump on the Well." *New York Times 136:* 23.

20

On a Contribution to a Future Scientific Study of Dream Interpretation

Rosemarie Sand
Institute for Psychoanalytic Training and Research

Psychoanalysts typically rely not on one, but on two different theories when they interpret dreams. One of these theories is Freud's. The other is an ancient hypothesis which has increasingly made its way into psychoanalytic practice over the decades.

Freud's theory has been repudiated by scientific dream researchers because they regard its method as yielding pseudoevidence. Aspects of the old hypothesis, on the other hand, have been well accepted by some experimenters. I suggest, therefore, that if psychoanalysts will eschew the Freudian method and rely entirely upon the ancient technique, they will thereby remove a major roadblock which has prevented cooperation between dream researchers and clinicians.

In this essay, I hope to contribute to future understanding by first reviewing once again the fallacy in Freud's method which has been pointed out by many critics and then by describing the merits of the old, or classical, technique.

When Freud presented his theory, in 1900, he specifically rejected what he called the "popular, historic and legendary" ([1900] 1957, 99, 104) method of interpretation. He made it plain that he was breaking with the past by distinguishing between what he called the "manifest" and the "latent" dream. Every previous attempt to solve the problem of dreams, he averred, dealt with the manifest dream, meaning the dream as recalled and reported by the dreamer. "We are alone," he asserted, "in taking something else into account" (ibid.,

277). This "something else" was the latent, or hidden, content of the dream which he believed could be uncovered by the process of free association.

The manifest dream, according to Freud, was not to be interpreted directly, as it had been in the past, because it was merely a facade, a deliberate disguise which hid the real meaning of the dream. The manifest dream was to be split into segments and then free associations were to be obtained for each of these separate segments. In this way, the latent thoughts, the real meaning of the dream, would emerge.

It has often been remarked that, in spite of his own careful instructions, Freud did not always segment the manifest dream before interpreting it. A perusal of his work shows that not only did Freud often rely upon the manifest dream in practice, but it was also incorporated into his theory in which it appears in many guises. Thus, Freud's new theory and the old hypothesis, as well as the methods associated with them, coexisted from the start and have continued to coexist. Now, when psychoanalysts use the old method, that is, when they base interpretations on the whole, unsegmented dream and do not elicit free associations, they refer to it as "manifest dream interpretation." According to one recent estimate, psychoanalytic practice has shifted from Freud's free association method to manifest dream interpretation to a point where in over 95 percent of dreams interpreted, "the manifest content was from 50% to 100% the main constituent" (Warner 1990, 216) of the interpretation. The report noted that many contemporary psychoanalysts are not aware of this shift.

However, the "official" theory remains Freud's and the essential feature of this theory is the distinction between the manifest dream and its latent content, the latter presumably uncovered by free association.

Freud's basic hypothesis was as follows: A dream is produced when an unacceptable, repressed, infantile wish becomes active during sleep. Repression being weakened during sleep, the unacceptable wish threatens to break through into consciousness and to disturb sleep. However, a process which safeguards sleep intervenes. Instead of being wakened by the wish, the sleeper dreams of it, but not in its disturbing form. A transforming process, the "dream work," distorts it so that the wish appears in the dream in an unrecognizable guise. The dream which the dreamer recalls, the "manifest dream," therefore does not picture the real wish but masks it. The essential cause of the dream, the trigger which sets the process in motion, is the hidden, unacceptable wish.

Modern science has cast grave doubts upon Freud's hypothesis that wishes are the causes of dreams. Dream research received a powerful impetus when, in 1953, Aserinsky and Kleitman at the University of Chicago discovered that people rapidly move their eyes at intervals during the night, that this rapid eye movement, or REM period, coincided with specific brain activation patterns and that, during these periods, dreams were in progress. This discovery strongly suggested that the neural mechanisms which caused the REM state also caused the dream. It has since been learned that dreams also occur outside of the REM state but it still remains likely that neural mechanisms, and not repressed infantile wishes, trigger the dream.

Meanwhile, Grünbaum has cogently argued on other grounds that Freud's theory is false. Freud stated that the dream was related to neurotic symptoms because both dreams and symptoms were caused by unacceptable wishes which were repressed and unconscious. He also asserted that symptoms disappear when repression is removed during psychoanalytic treatment because undoing the repression enables unconscious wishes to become conscious, after which they can no longer produce symptoms. It seems necessary to conclude, therefore, that once unconscious wishes become conscious they will no longer be able to cause dreams either. If Freud's theory were true, Grünbaum suggested, then psychoanalytic treatment ought to at least reduce the frequency of dreaming as treatment progressed and as repressions were removed. He has concluded that in the case of long-term psychoanalytic patients, "either their free associations are chronically unsuccessful in retrieving their buried infantile wishes, or, if there is such retrieval, then Freud's account of dream generation is false" (forthcoming; p. 19 of typescript).

Although wishes do not cause dreams, they may still be pictured in dreams, just as anything else may be pictured. In that case, a wish may be regarded as a *source* of the dream. It is useful to distinguish the *cause* of a dream from its *sources*. The cause is the trigger of the dream; the cause makes the dream happen. Scientists suggest this is a neural mechanism. The dream's sources determine the contents of the dream. So, for instance, if an image of my Aunt Mildred appears in a dream, then a previous perception, a memory, of this aunt is the source of the image. My wish to see my aunt did not cause me to dream but my wish to see her may be pictured in the dream. The source of a dream is a kind of cause also, in fact, a necessary cause, because if I did not have a memory of my aunt, her image would not appear in my dream. However, the use of the word "cause" in two senses, one as trigger of the dream and the other as a determinant of dream content, has proven confusing. Therefore, when I refer to the latter, I will use the word "source." The source is pictured, or displayed, in the dream.

Thus, a wish, as well as any other mental state, can be a source of a dream. That we often see our wishes come true in dreams is one of humankind's oldest convictions.

Freud, however, insisted that the wish which produced the dream was *not* displayed in the dream. He averred that his method could show that the real wish was hidden and that only its disguised derivatives appeared in the manifest dream.

This claim was rejected by critics long before the REM state was discovered, based on a fallacy located in Freud's method. Freud thought he could prove that a hidden wish produced the dream because free association to segments of the dream would bring to light the "background thoughts" of the dream and these would be related to the dream-engendering unacceptable wish. He put it this way: If "I put the dream before him [the dreamer] cut up into pieces, he will give me a series of associations to each piece, which might be described as the 'background thoughts' of that particular part of the dream" ([1900], 1957, 103).

How so? Freud's critics asked as soon as *The Interpretation of Dreams* (ibid.) appeared and they have continued to ask this question ever since. Why should a series of associations to pieces of the dream give you the "background thoughts" of those pieces? How can you assume that a series of thoughts which *follow* a dream will lead you to the thoughts which *preceded* the dream? No doubt, a series of thoughts will lead you *somewhere* but you cannot justifiably assume that you have been led to the background thoughts which produced the dream.

When persons are asked to free associate "to" an element of a dream, as psychoanalytic parlance has it, they are instructed to relax their critical faculty and to express "everything that comes to mind," regardless of whether it seems to them inapplicable, nonsensical or embarrassing. This entails *not* focusing on the element which is the take-off point for the associations but rather letting the mind wander from one thought to the next, perhaps in a long series. The assumption is that if conscious control is abandoned, the liberated, or "free" associations will be guided by unconscious motivational forces which, supposedly, are the same forces which produced the dream. This is the much-criticized assumption: Granted that free associations are influenced by some force or other, what licenses the supposition that this is the force responsible for the dream?

I will refer to this erroneous supposition as the "free association fallacy," not because all free association provides misleading information but because a particular error occurs when free association is used to discover specific determinants of symptoms, dreams and parapraxes. The fallacy occurs, for instance, when it is assumed that something—a thought, a feeling, a motive—which turns up during free association to a dream, merely because it turns up, must be a background thought of the dream. In other words, to put it most succinctly, the free association fallacy is that, given a dream *A*, if I associate from *A* to *B*, then *B* must have been a determinant of *A*. This is the reasoning which Freud's critics condemned. It should be noted, however, that a particular circumstance exists in which the assumption that something which turns up in free association is the source of a dream is warranted. This exception to the fallacy will be examined.

A complete account of the fallacy which can occur during free association would have to take into consideration other, more complex, aspects of the error, which sometimes make it hard to detect. Moreover, a complete account would have to deal with Freud's responses to his critics on this issue. It was apparently repeatedly called to his attention. In any case, he addressed the problem on several occasions without, however, successfully dealing with the main point.

An example of the free association fallacy as committed by Freud ([1900] 1957) is the dream known as "Otto was looking ill." Freud had dreamed that "my friend Otto was looking ill. His face was brown and he had protruding eyes" (ibid., 269–71). This was the manifest dream. Freud then free associated to this dream and after a long series of thoughts arrived at what he considered to be the latent dream, two wishes, one that he might become a professor and the other that he might live a long life. The subject matters of the manifest and the supposed latent dreams are entirely different and there is no apparent meaningful connection between them. The assumption that the supposed latent dream, the wishes for a professorship and a long life, gave rise to the manifest dream about Otto's appearance, is based solely on the fact that a string of associations connected them. That is, it is assumed that *B* is a determinant of *A* because *B* occurred in the series of associations.

This conclusion seems entirely unjustified unless it is simply *presupposed* that ideas which emerge in association to a dream preceded it and produced it. It must be remarked that the possibility that such a presupposition could be true cannot be absolutely ruled out because one cannot predict what the science of the future may bring. Perhaps in the near or in the distant future it may be possible to show that even long strings of associations which emerge in association to a dream really are "dream thoughts" which went into the production of the dream.

However, with no such development perceptible on the immediate horizon, psychoanalysts would do well to eschew the claim that they have discovered the source of a dream by means of association of this sort. Psychoanalysts are not being asked to depreciate free association; they are only being requested to recognize one of its limits. It is important to note that the general capacity of free association to provide information about the dreamer is not in question here. When people free associate, to dreams or at other times, expressing their thoughts as they let their minds wander, there is no reason to doubt that they will provide data about *themselves*. The analyst is likely to learn something about the patient and on that account may justly value free association. But the analyst is mistaken if she believes that whenever a patient free associates to a dream that the patient is necessarily on a trail which leads to the source of the dream.

For many psychoanalysts, the usefulness of free association in the clinical situation has likely obscured the fallacy which may occur in it. The patient associates to a dream and during the course of the association important thoughts and feelings may emerge, some perhaps previously unrecognized. The analyst is then likely to attribute any discovery to "analysis of the dream." The conviction that all of the thoughts uttered by the patient following the dream report can be taken to be "associations to the dream" is not uncommon. Indeed, the analyst may even suppose that she is justified in accepting that everything the patient said and did *before* reporting the dream during the session is an "association to the dream."

In order to understand that this assumption is fallacious, the analyst would have to realize that free association can be used in two different ways which are ordinarily not distinguished from each other. First, free association can be used for a general exploration of the patient's condition. This use is not being criticized. Second, it can be employed in the search for specific determinants of symptoms, dreams and parapraxes and it is typically during such a search that the fallacy occurs.

As soon as *The Interpretation of Dreams* appeared, commentators in scientific journals focused on the free association fallacy and rejected Freud's dream theory because of it.

Over the years, objections of this kind have abounded, but no past critic has analyzed Freud's logic as thoroughly, or followed it to its conclusions as consistently, as Grünbaum. Grünbaum has made fully clear, in a number of studies, the extent of the devastation which resulted from the uncritical use of the method of free association. Commenting on the free association fallacy as it occurs in the search for causes of neurotic symptoms, he has said that "not even the tortures of the thumbscrew or of the rack should persuade a rational being that free associations can *certify* . . . causes" (1984, 186).

Psychoanalytic dream interpreters are faced with serious challenges from both the philosophy of science and from scientific dream research. They are being asked to show whether their interpretations are rationally justifiable. I suggest that they will be in a better position to defend interpretations if they give up the search for the supposed latent contents and concentrate instead on the manifest dream. That is, I urge that they acknowledge and accelerate a trend which is apparently already underway.

The manifest dream should be recognized as possessing an independent theory with its own structure, its own concept of the dream-creating process and its own interpretative rules. According to this theory, the dream can reveal circumstances, attitudes and motives of the dreamer by displaying them in the dream content.

This is an ancient theory which for many centuries coexisted with, but was often overshadowed by, the supposition that dreams revealed the future. For instance, in the second century after Christ, the celebrated oneiricritic, Artemidorus of Daldis, whose profession was to predict, assumed that many dreams did not foretell the future but simply revealed the hopes and fears of the dreamer. He stated that "it is natural for a lover to seem to be with his beloved in a dream and for a frightened man to see what he fears" (1975, 14). Centuries earlier, Zeno, founder of the Stoic philosophy, recommended the use of the dream to study progress attained by the dreamer in the pursuit of the Stoic ideal.

The assumption that dreams can picture the dreamer's states of mind should not be thought of as modern. It stands to reason that as soon as a society becomes sophisticated enough to dispense with the theory of divine or demonic creation of the dream, it will come to entertain the supposition that the dream is a creation of the dreamer and that therefore the dream may reveal something about the dreamer.

In the nineteenth century it was widely believed that the dream could picture the life, including the inner life, of the dreamer. In the chapter which introduces *The Interpretation of Dreams* Freud stated that "the preponderant majority of writers" ([1900] 1957, 7) on the subject hold that dreams are related to the dreamers' interests and he quoted several of these. One stated that "the content of a dream is invariably more or less determined by the individual personality of the dreamer, by his age, sex, class, standard of education and habitual way of living, and by the events and experiences of his whole previous life" (ibid.). Another noted that "we dream of what we have seen, said, desired or done" (ibid., 8). It stands to reason, therefore, that a perusal of a dream may teach us something about the dreamer. In this connection, Freud quoted what he called a "familiar saying," "Tell me some of your dreams and I will tell you about your inner self" (ibid., 67).

Modern dream science strongly supports these opinions. Researcher Milton Kramer, surveying experimentation, concludes that "the dream is indeed signal rather than noise. The dream is a structured product and reflects meaningful psychological differences among subjects. The dream responds to or reflects emotionally charged influences, and the domains of waking and dreaming are significantly and straightforwardly related" (1982, 98). Fisher and Greenberg, after summarizing results of a number of dream studies, arrived at a similar estimate and noted that "the manifest dream content carries a great deal of meaning. . . . [T]he weight of the evidence argues against viewing the manifest content as a largely meaningless conglomeration of camouflage devices, such as Freud spelled out" (1985, 46).

That the theory which relies solely upon the manifest dream for purposes of interpretation, the classical, or traditional theory, is an independent theory which can be sharply distinguished from Freud's is demonstrated by a comparison of the basic principles of these two hypotheses. These principles are not merely different for three of them are directly contradictory:

1. *Freudian theory:* The manifest dream is not to be interpreted directly.

Classical theory: The mainifect dream is to be interpreted directly.

2. *Freudian theory:* The manifest dream *disguises* the origin of the dream.

Classical theory: The manifest dream does *not disguise* the origin of the dream. It pictures, or displays, it.

3. *Freudian theory:* The dream-producing process is structured to produce a disguise.

Classical theory: The dream-producing process is not structured to produce a disguise; it is structured to create a picture, or display.

Moreover, there are other important differences between the two theories. According to Freud's theory, an unconscious infantile wish *causes* the dream. The "background" thoughts of the dream may include any other mental state.

According to the classical theory, psychological states of all kinds are *displayed* in the dream. Historically, there is a diversity of opinion about dream causation with unanimity, however, that psychological states are *at least* displayed. There is no particular emphasis on wishes. Other states, perceptions, intellectual puzzles, memories, intentions, fears, worries and so on are equally important dream sources.

The Freudian interpretative technique is *free association* to segments of the dream. Classical interpretation depends upon the *discovery of a pictorial relationship* between the dream and a state of the dreamer. If no such relationship can be found, the dream cannot be interpreted.

In order to further specify features of the traditional mode of interpretation and compare them with the characteristics of free association I will use artificially simple examples. It should be understood that these are designed to illustrate specific principles and should not be regarded as describing actual dreams and interpretations as they are found in clinical practice.

If we want to interpret a dream according to the classical method, we must attempt to find a match between the dream and some circumstances, external or internal, of the dreamer. For instance, suppose that I have gone on a shopping trip with my aunt and then dream about this selfsame trip with the aunt. Then the memory of this trip is a fairly indisputable source of the dream. If I lost my handbag on this trip, worried about it and the worry then appeared in the dream, then this worry is another source. The point being made here is that when I know the source of the dream, I also have its interpretation. The dream *means* that I am worried about the loss of my handbag. Source, meaning and interpretation are all equated. An interpretation is a hypothesis about the source, or meaning, of a dream.

I will use the same method when I interpret the dreams of others. If, for instance, a woman dreams that her husband is in the hospital and adds to this that, in reality, her husband is scheduled to undergo a test at a hospital and that she is anxious about this, I will assume that I have discovered a match and can connect the dreamer's mental state with the content of the dream.

If this were all there was to dream interpretation, it would not be very interesting. All that we could do, according to this exposition, would be to compare the dream with mental states of the dreamer with which we were already familiar. The dream could tell us nothing *new*. In fact, however, whenever the dream has been regarded as picturing mental states, as constituting a "mirror of the soul," it has generally also been regarded as sometimes providing information otherwise not obtainable, or at least not easily obtainable. Why, for instance, would Zeno have urged Stoics to study their dreams if what they could learn from these was also easily available by way of simple introspection? The assumption typically has been that dreams may reveal truths about the dreamer which otherwise remain obscure or altogether hidden. In the eighteenth and nineteenth centuries, this conviction was expressed by philosophers, physicians and writers.

In England, David Hume claimed that "moralists" were urging the study of dreams for purposes of self-discovery. Hume wrote that

> moralists have recommended it as an excellent method of becoming acquainted with our own hearts, and knowing our progress in virtue, to rec-

ollect our dreams in a morning, and examine them with the same rigour, that we wou'd our most serious and most deliberate actions. Our character is the same throughout, say they, and appears best where artifice, fear, and policy have no place, and men can neither be hypocrites with themselves nor others. The generosity, or baseness of our temper, our meekness or cruelty, our courage or pusillanimity, influence the fictions of the imagination with the most unbounded liberty, and discover themselves in the most glaring colours. ([1739] 1969, 268)

In Germany, at the beginning of the nineteenth century, physician Friedrich Carus (1808) asserted that "dreams can be described as each person's secret court" (p. 190) He stated that "frightening dreams are nothing but a continuation of our feelings. They ought not to alarm us but rather rouse us and bring us to full consciousness of ourselves, even if this should be something ugly. . . . Such dreams can warn us, if not against an impending misfortune, then certainly about ourselves and about the direction we have chosen to pursue. . . . The dreamer is frequently the only betrayer of the waking man (ibid.).

In the United States, Nathaniel Hawthorne expressed a similar opinion:

The mind is in a sad state when Sleep, the all-involving, cannot confine her spectres within the dim region of her sway, but suffers them to break forth, affrighting this actual life with secrets that perchance belong to a deeper one. . . . Truth often finds its way to the mind close muffled in robes of sleep, and then speaks with uncompromising directness of matters in regard to which we practice an unconscious self-deception during our waking moments. ([1854] 1970, 51–2)

According to classical dream theory, we may penetrate "unconscious self-deception" by using the dream to make inferences about unconscious sources, that is, given the dream, we frame hypotheses about states of mind of which the dreamer is not aware. Consider, for instance, two hypothetical situations; in the first, the dreamer is conscious of anxiety, in the second, she is not.

A woman dreams, "I was driving alone in the car, speeding to get to the hospital to see my husband. I was very scared." Inquiry reveals that upon the recommendation of a cardiologist, her husband has been scheduled for a special stress test at the hospital. The dreamer tells us that this news terrified her and that she had already pictured her husband as hospitalized and had even gone further to imagine herself widowed.

We regard the dreamer's thoughts about her husband and the hospital test as the source of the dream because a trip to the hospital to see the husband is depicted in the dream. The dreamer's fear is an additional source of the dream for the dream contains the element "I was very scared" which can be matched with the real-life anxiety. The interpretation, that is, the inference about the dream's source, is unproblematic. The dream means that the woman has been thinking about her husband's health and has been worried about it.

The inference is unproblematic because the dreamer is conscious of the dream's sources which can therefore be elicited by inquiry. Suppose, however, that the woman whose husband is facing a stress test denies that she is frightened. The interpreter can match the dreamer's conscious thoughts about the hospital with the hospital content of the dream. However, no conscious source is available to match the dream element "I was very scared." If the interpreter knows the woman well she may suspect that she is simply unaware of her anxiety. Perhaps on a previous occasion the woman apparently remained calm for a while during an emergency and then suddenly panicked. On some such basis, the interpreter may infer that the dream element "I was very scared" is matched by a current, but unconscious, fear.

The assumption on which such inferences are based is that a psychological state of the dreamer of which the dreamer is not aware may nevertheless be displayed in a dream. The unconscious source cannot be elicited by inquiry and may be specifically denied by the dreamer. Nevertheless, on the basis of the dream content and information about the dreamer's character and circumstances, the interpreter may frame suppositions about the unconscious source of the dream. Such inferences, of course, are always only hypotheses.

The traditional method can go no further than this. It is constrained by the content of the dream. Free association, on the other hand, can purportedly supply a great deal of additional information about unconscious dream determinants. For instance, in the case of the woman's hospital dream, it could easily discover an unconscious motive for the dream which would support Freudian theory. Free association to the dream element "I was driving alone in the car" can issue in the series: "automobile," "autoerotic," "masturbation." Or the series could proceed from "driving alone" to "solitary activity" to "masturbation." However, the assumption that such associations can justifiably be regarded as determinants of the dream fragments is mistaken.

However, there is a circumstance in which the use of free association to discover a source of a dream does not result in the fallacy. This occurs when, either deliberately or inadvertently, the interpreter, listening to the flow of associations, picks out as sources just those which are identifiable according to the traditional method. For instance, in the preceding dream example, if the woman had free associated she might well have come up with thoughts about her husband's heart condition. In this case, the association would have been from *A* to *B*, the interpreter would have supposed that *B* was a source of *A*—and would in this instance have been right. However, the supposition that *B* is a source of *A* is warranted here only because the criterion of the traditional theory has been met, that is, *B* depicts *A*. It is not warranted because *B* emerged during free association.

It should be realized, therefore, that, subject to this severe restriction, some free association does reveal the sources of the dream and that a criterion exists which makes it possible to distinguish justifiable from nonjustifiable assumptions about dream sources arrived at through free association. However, to the best of my knowledge, since psychoanalysts have not concerned themselves with this problem, they have been unaware of this criterion.

No discussion of dreams is complete without a consideration of the subject of symbolism and here we arrive at a great parting of the ways. Here the line has long been drawn between those to whom it has seemed clear that the dream contains symbols and others who reject this notion, often with distaste, as "mystical" or "unscientific." And yet, the question as to whether dreams do or do not contain symbols, whether they contain them frequently or only rarely, does not seem to lie beyond the purview of research and could be studied as scientifically as any other. The notion that dreams may be symbolic is one of humanity's oldest beliefs and perhaps deserves consideration for that reason alone.

To begin with, the symbolism associated with the traditional theory must be distinguished from that known as "Freudian." According to the well-known Freudian theory, symbolism is generally, although not exclusively, sexual: elongated objects such as sticks and spears are assumed to stand for the male sexual organ while boxes, chests, and other hollow objects are regarded as representing the female. According to Freud, these symbols were not created by the process which produced the dream; they already existed in the unconscious and were to be found not only in dreams but also in myths, fairy stories and the like. The dream-creating process merely made use of these symbols.

The traditional symbolism, on the other hand, is the symbolism which we use every day, encountering it, for instance, in advertisements and cartoons; it is limited to no particular subject matter; it may represent anything at all. A familiar example of this symbolism is to be found in the biblical dream of Joseph and his brothers. According to Genesis, Joseph was the favorite of his father who rather blatantly preferred him to his eleven brothers, honoring him with a coat of many colors, and so on. Joseph dreamed and told the dream to his brothers, "Behold, we were binding sheaves in the field and lo! my sheaf arose, and stood upright and behold, your sheaves stood round about and made obeisance to my sheaf." The brothers interpreted this dream immediately and were enraged, protesting "shalt thou indeed reign over us?"

In biblical times the dream was viewed as a prediction; the modern dream interpreter would suppose that it reflected some state of mind. However, in either case, the symbol remains the same: the brothers' sheaves bowing down to Joseph's sheaf signified an elevation of Joseph over his brothers. We may picture Joseph's dream in the form of a politically motivated modern cartoon, the bowing down of the sheaves representing the brothers' subservience to Joseph.

The traditional symbolism is metaphorical, or allegorical. A product of the imagination, it is endlessly varied; new symbols are constantly being created.

An interesting source of examples of this symbolism is the ancient dream book of Artemidorus of Daldis. In addition to much that is incomprehensible and seems nonsensical, the work contains a multitude of interpretations based upon symbolism which is immediately intelligible today, thus attesting to the antiquity and universality of these metaphors. For instance, Artemidorus asserts that anything that goes badly in the dream, any accident, malfunctioning or clumsiness, may allegorically represent aspects of real-life situations. This would include such mishaps as falling off a horse, plunging from a cliff, wrecking a boat, being stuck in a swamp or drowning. Thus the cartoonist who wanted to signify that a politician was losing control of his party could show him struggling to manage a horse, or tumbling off a horse. Presumably, if the politician was a Democrat he would be having trouble with a donkey, if Republican, then with an elephant. If the politician's troubles were terminal, then he might be shown falling off, not a horse, but a cliff. Perhaps his opponent might be pictured as pushing him off the cliff. Cartoons which would feature the wrecking of a boat, being stuck in a swamp or drowning can easily be imagined. Since the most ancient times it has been assumed that this kind of symbolism appears in dreams.

Let us stress that interpretations based upon this symbolism are not "Freudian." The fact that Freud, after turning against the traditional theory and its symbolism, nevertheless still often relied upon it thereafter, should not obscure the differences between either the two symbolisms or the two theories. In practice, however, these are often confused. So far as the theories are concerned, this confusion is undoubtedly due in great part to a confusion of terminology. For generations, scholars used simply the expressions "dream" and "meaning." In the case of Joseph's dream, the "dream" consists of the content showing the brothers' sheaves bowing down to Joseph's sheaf and the "meaning" of the dream is that "Joseph reigns over his brothers." When the expression "dream" is replaced with "manifest dream" and "meaning" is rebaptized "latent content," the difference between these theories is obscured. Then the interpretation of Joseph's dream becomes a "Freudian" interpretation. Freud deliberately invented the new expressions in order to distinguish his hypothesis from the classical theory. Therefore, these expressions should be understood to be what in fact they are, technical terms of his theory. As regards symbolism, Freud's restriction of symbolism to primarily sexual themes is often disregarded and symbols of all kinds are thought of as "Freudian."

The hypothesis that the traditional symbolism may appear in the dream, together with the consequence of this hypothesis, that the mind, during sleep, can create such symbols has been welcomed by some and opposed by others for the same reason: It has seemed to grant the sleeping mind mysterious powers.

In the early nineteenth century, a coterie of Romantic philosophers and writers in Germany enthusiastically accepted the concept that nocturnal mental activity could produce something which in some respects resembled a work of art. Chief among these was the physician Gotthilf Heinrich Schubert who referred to the dream-creating process simply as "the dream poet" ([1814] 1968, 3). He, his followers and their intellectual heirs assumed that the "dream poet" had all of the devices of poetry at its disposal and could create allegories at will. Moreover, he believed that the traditional symbolism, because it was recognized both by ancient peoples and by contemporary primitive tribes, constituted a universal language which transcended temporal and geographic bounds. Schubert, a mystic, speculated that dream symbolism was the remnant of an ancient language with which God once had communicated with humankind. Many of Schubert's contemporaries were firmly convinced that dreams could reveal the future and that future-telling dreams were expressed in symbols. Dream prognostication flourished throughout the nineteenth century and the traditional symbolism came to be closely associated with it.

As a result, sober researchers who were struggling to defend the scientific point of view against encroachment by superstition, took a cautious attitude toward symbolism. However, some data seemed to constitute evidence for the validity of the traditional symbols. Innumerable anecdotes described how stimuli which did not waken the sleeper were nevertheless registered and appeared in dreams, strangely transformed. The mewing of a kitten turned into the roar of a lion; the creak of a door produced the booming of a cannon, complete with combat imagery and action; the tapping of a twig on a window was converted into a dream hurricane. During the last half of the nineteenth century, a number of respected researchers produced this phenomenon experimentally.

The phenomenon, incorporation of a stimulus into a dream, has also been recognized by modern researchers and can be further illustrated with a modern example. Experimenters today have the advantage over their nineteenth-century forerunners in that they can determine when a subject is dreaming by objective criteria. In this typical instance, when it was ascertained that a dream was going on, the subject was stimulated with a fine spray of cold water. A few minutes later, when the subject was awakened, he reported a complex dream in which he had been acting in a play when, suddenly, as he was walking behind the leading lady, she collapsed and water was dripping down on her. Water dripped onto him also. "The roof was leaking," he reported, "... I looked up and there was a hole in the roof" (Dement and Wolpert 1958, 550).

In the nineteenth century, the occurrence of this phenomenon was generally accepted; its meaning, however, was disputed. Wilhelm

Wundt, the "father" of experimental psychology, insisted that the representation of the stimulus which appeared in the dream was not a "symbol" of the stimulus but an "association" to it. Champions of symbolism responded that these recognizable representations of the stimulus were not mere associations but were fully developed metaphors. They argued that if one accepts representations of this type, then one must necessarily also grant that the sleeping mind has the capacity to create symbols.

Undoubtedly, for the scientists, much of the dispute was fueled by the underlying fear, very reasonable at the time, that, while explanations based upon the "laws of association" are safely rational, those which invoke a symbol-producing faculty of the sleeping mind seemed to put the theoretician onto a slippery slope at the bottom of which lies the quagmire of supersition. Related considerations are not entirely moribund, even today. The question as to the nature of these representations remains to be settled; there is a need for careful definition and responsible research.

Beyond this there lies a larger question: If it is granted that in the laboratory the sleeping mind evinces a symbol-producing capacity, can it be further assumed that this capacity is employed at other times, for instance, in the creation of personally meaningful dreams? Does the symbol-producing capacity respond to the dreamer's hopes and fears as it does to the laboratory stimulus by constructing allegories depicting those hopes and fears?

A tentative positive answer comes, not from sleep research, but from medicine. Antedating all experiments is a medical tradition described in the Hippocratic writings and transmitted by Galen according to which symbolic dreams can be symptoms of physical and mental ailments. In the case of a few maladies, the symbols recommended show a surprising consistency through the ages. For instance, an Arab physician in the ninth century, an English doctor in the seventeenth and a French medical man in the early nineteenth century would have been in remarkable agreement on the dream imagery which pointed to "melancholia." Unhappy people, afflicted with a superfluity of "black bile," tended to dream of misery and disaster, it was believed. Their dreams were full of frightening events, terrible circumstances, groans and lamentations. This assumption tallies interestingly with the modern finding of Aaron T. Beck and Marvin S. Hurvich (Beck 1967) that there is a statistically significant difference between the dreams of the depressed and those of the undepressed. In comparison with an undepressed control group, the depressed tend to dream of misery and disaster, just those themes recognized by physicians in the past. The dreams of the depressed thus presumably reflect their inner state.

However, this granted, we still face the problem of definition. Are the dreams of the depressed mere "associations" to the inner state, as Wundt suggested, or are they "symbols" of it? A sampling of depressed dreams may facilitate a decision: "I was a bum" (or "mentally defective," "a cripple," "blind," or "had pus oozing out of all my pores"); "leeches were crawling all over me"; "I was strapped down to a table"; "I was buried alive"; "I took careful aim and fired at the deer, but my gun didn't go off"; "I was in a restaurant but the waitress would not serve me"; "I lost all my money"; "I was the only one not invited to the party"; and "my analyst said he didn't want to see me anymore" (ibid., 339–44).

If it is decided that these dreams are symbolic of the depressed mental state rather than mere associations to it, then this decision should be regarded as support—however preliminary—of the old medical hypothesis that dreams can symbolically reflect mental ailments.

Whether or not he was aware of it, Freud was following in the footsteps of generations of physicians when he declared that the dream was a symptom and could be used to probe the psyche of the patient. This hypothesis had not been forgotten in Freud's day. Commenting on it in 1885, the philosoper Carl Du Prel noted that, because emotions are symbolized in dreams, these "often correspond to our states of mind" (p. 161). The dream, he averred, "is a symbolic representation of inner states of the dreamer, often it is a symptom of health or disease" (ibid, 169) and therefore physicians should pay attention to it when thinking about diagnoses.

The dream to which Du Prel referred is obviously the dream as recalled by the dreamer, Freud's "manifest dream." The manner in which the traditional symbolism presumably may represent inner states in the manifest dream is illustrated by the following two vignettes, taken from my practice.

A woman in psychotherapy, torn between rage and regret, had finally decided to divorce her husband, Larry. She dreamed, "There seemed to be a civil war going on, or a revolution. I was on a battlefield with a lot of bomb craters and fire kept falling from the sky. I was wandering around crying. I found Larry. He was wounded and was going to die. He pleaded with me to put him out of his misery. I had a gun. I cried and cried and I shot him."

A patient, "Jordan," had led a confused, unhappy life, successful neither in his career nor in his relationships. The first few psychotherapy sessions visibly made a favorable impression on Jordan. In a hopeful mood, he began to make plans for the future. He dreamed, "I got lost on the subway. I kept getting on and off trains and always ended up going in the wrong direction. Then I found a staircase that led up and out. I got out and looked around to find out where I was and there was a sign saying, 'New Jordan Street.' "

Clinicians assume that if the therapist has intimate knowledge of the dreamer's character and circumstances and is cognizant as well of the dreamer's current state of mind, he or she may hypothesize that dreams such as these symbolically represent that state or an aspect of that state. Moreover, clinicians also believe that such dreams may provide glimpses of *unconscious* psychological states of the dreamer. The requirements for a hypothesis that an unconscious state is symbolically reflected in a dream resemble those described in the case of the nonsymbolic dream. The interpreter who is familiar with the dreamer's character and intimate circumstances may infer that a given dream symbolically pictures a psychological state, or element of that state, of which the dreamer is not aware. Suppose, for instance, that in the preceding vignettes the woman who dreamed of shooting her husband had not been aware of her rage, or of its extent, or that the man who dreamed of "New Jordan Street" had not realized that he was beginning to feel optimistic. In those cases, a clinician might suppose that the dream symbolically depicted an unconscious state. The hypothesis would be confirmed if and when the dreamer became aware of such a state.

That the dream may suggest specific hypotheses about unconscious psychological states is a matter which, like the general postulate of unconscious states, requires demonstration. If in the future experimenters and clinicians were to get together, perhaps on the basis of the traditional dream theory, then we might hope that these hypotheses would rank high on the agenda.

In spite of its great age, the traditional theory has been insufficiently tested. Dream researchers, when they have been occupied with psychoanalytic theory, have focused on Freud's hypotheses. Psychoanalysts, although relying on it, have in the past treated the classical theory as something of a stepchild, employing it ambivalently because of Freud's strictures against interpretation of the manifest dream. Yet, in spite of the paucity of the evidence garnered so far, the traditional dream theory is clearly superior to Freud's for the simple reason that the former does not depend upon free association and consequently, not being involved in the free association fallacy, escapes the charge that it systematically leads to erroneous conclusions. Relinquishing the hypothesis that free association provides a sure passage from a manifest dream to a latent dream concealed beneath it removes a major roadblock to mutual endeavors on the part of experimenters and clinicians. The traditional theory promises to open doors. Researchers are already focused upon it; clinicians are coming to understand its full significance. Perhaps these two groups, so often at odds in the past, may yet be tempted to work together. If they did so, then this hypothesis, whose roots are ancient, could be examined in an unprecedented manner. We might expect that, whatever the outcome of such joint research, the result would be the healing of the breach which, with a few notable exceptions, has separated these disciplines. The study of the dream could only benefit from their collaboration.

REFERENCES

Artemidorus. 1975. *The Interpretation of Dreams*. Translation and commentary by R. J. White. Park Ridge, N.J.: Noyes Press.

Beck, A. T. 1967. *Depression: Clinical, Experimental, and Theoretical Aspects*. New York: Harper & Row.

Carus, F. 1808. *Psychologie*. Leipzig: Barth.

Dement, W., and E. Wolpert. 1958. "The Relation of Eye Movements, Body Motility, and External Stimuli to Dream Content." *Journal of Experimental Psychology 55:* 543–53.

Du Prel, C. 1885. *Die Philosophie der Mystik*. Leipzig: Ernst Günthers Verlag.

Fisher, S., and R. Greenberg. 1985. *The Scientific Credibility of Freud's Theories and Therapy*. New York: Columbia University Press.

Freud, S. [1900] 1957. *Standard Edition of the Complete Psychological Works of Sigmund Freud*. Vols. 4–5, *The Interpretation of Dreams*. Translated by J. Strachey. London: Hogarth Press.

Grünbaum, A. 1984. *The Foundations of Psychoanalysis*. Berkeley and Los Angeles: University of California Press.

———. 1992. "Two Major Difficulties for Freud's Theory of Dreams." In T. Gelfand, and J. Kerr, eds., *Freud and the History of Psychoanalysis*. Hillsdale, N.J.: The Analytic Press, pp. 193–213.

Hawthorne, N. [1854] 1970. "The Birthmark." In *Mosses from an Old Manse*. Freeport, NY: Books for Libraries Press, pp. 47–69.

Hume, D. [1739] 1969. *A Treatise of Human Nature*. Baltimore: Penguin Books.

Kramer, M. 1982. "The Psychology of the Dream: Art or Science?" *Psychiatric Journal of the University of Ottawa* 7. 87–100.

Schubert, G. H. [1814] 1968. *Die Symbolik des Traumes*. Reprint. Darmstadt: Lambert Schneider.

Warner, S. 1990. Review in *Psychoanalytic Books*. In A. Rothstein, ed., *The Interpretation of Dreams in Clinical Work*. Monograph 3, Workshop Series of the American Psychoanalytic Association, Madison, CT.

Wundt, W. 1880. *Grundzüge der physiologischen Psychologie*. Leipzig: Engelmann.

Freedom and Determinism; Science and Religion

21

Indeterminism and the Freedom of the Will

Arthur Fine
Department of Philosophy, Northwestern University

I formed them free, . . . they themselves ordain'd thir fall.
—**Milton,** *Paradise Lost,* Book III

Adolf Grünbaum's writings on the free will problem, although not extensive, have been widely reproduced and influential (see Grünbaum 1953, 1967, and 1972). Characteristically, at the center of Grünbaum's work are clear and forceful arguments—in this instance, for the compatibility of determinism and free will. This compatibilism, I suspect, derives in good measure from Grünbaum's passionate concern to protect the possibility of an adequate human science, which is to say (as he sees it) a causally based science of individual and social behavior. Insofar as libertarian incompatibilism seems to draw boundaries around causal analysis, exempting human action from its reach,[1] incompatibilism may seem to stand in the way of true human science. Grünbaum has been concerned with opening the path. In the course of his writings, at least to a limited extent, Grünbaum considers the bearing of indeterminism and (inevitably) of the quantum theory on the free will issue. Those limited considerations are my starting point here.

I will lay out the first half of a two-part argument intended to deconstruct the metaphysical concept of human freedom. In the first half, which constitutes this essay, I show how the libertarian conception of human freedom self-destructs. In the second half I would take up the antilibertarian (or compatibilist) conception with the same end in view (see also Earman's 1986 discussion of the difficulties with free will). I will not pursue the second half here, however, because in fact I am not yet sure just how the argument goes. It may be useful, nevertheless, to sketch the picture that motivates the whole project. In the usual picture there is some antecedent concept of human freedom and agency, which our moral, legal or social concepts of responsibility track and which they are bound to respect. In my view this has the tail wagging the dog, for it seems to me that things are just the other way around. It is we who manufacture conceptions of responsibility in order to meet the changing conditions of our social lives. We may then try to construct metaphysical concepts of freedom and agency in order to ground our attributions of responsibility. This foundational enterprise, however, (like others) is an unproductive fancy. From the fact that our metaphysical constructs are incoherent, we can see that our conceptions of responsibility actually require no such grounding. This, at any rate, is the important moral I would draw from the argument—if only I had both parts in place!

Freedom and Indeterministic Laws

I begin with an argument that turns on considerations concerning indeterministic laws; that is, laws whose statistical character, as Grünbaum puts it, "is not removeable by the possession of more complete information" (1972, 306).[2] He suggests the statistical laws of the quantum theory as a paradigm. Grünbaum argues that such laws are of no comfort to the libertarian. He argues, that is, that if determinism were truly incompatible with free will, as the libertarian conceives it, then indeterminism would be incompatible as well. Grünbaum draws the libertarian conception of human freedom from C. A. Campbell, who holds that one's act is free only if one could have acted differently under the very same circumstances. In line with a long compatibilist tradition, Grünbaum rejects this conception of the freedom of the will as inadequate. However, he argues for something much stronger than its inadequacy; namely, Grünbaum argues that we would not have this libertarian kind of freedom even if human behavior were governed by indeterministic laws.

Why not? The argument proceeds as follows (Grünbaum 1972). Suppose a community is subject to an indeterministic law according to which, in the long run, 80 percent of the population will commit a certain kind of crime. Can we hold those members of the community who commit the crime morally responsible for their behavior?

Only if, as the libertarian standard would have it, they could have done otherwise—that is, refrained from committing the crime. But, argues Grünbaum, the statistical law does not entitle us to say of an individual *A* who commits the crime that *A* could have done otherwise. To be sure, on the basis of the law we cannot tell which particular individuals in the community will commit the crime. The law does not specify that. But this limitation (epistemic or otherwise) does not entail that, in the very circumstances in which an individual *A* does commit the crime, *A* could have refrained from so doing.[3] This is the consequence that the libertarian needs, and thus Grünbaum concludes that irreducibly statistical laws would not ground human freedom in the special libertarian sense.

If we deal with a statistical law, then—as in the quantum case—we can restate the statistical content in purely probabilistic terms. So, for the example at hand, we can say that each member of the community has a probability of 0.80 to commit the crime. This probabilistic statement can be given what it seems appropriate to call a libertarian interpretation or an antilibertarian one. According to the *libertarian interpretation,* probability 0.80 to commit the crime for an individual *A* entails that in the very same circumstances in which *A* does in fact commit the crime *A* might not have done so. In this sense, *A*'s act is undetermined (or uncaused). According to the *antilibertarian interpretation,* probability 0.80 to commit the crime for an individual *A* entails that if the circumstance were exactly the same, then *A* would do the same again. In the antilibertarian sense *A*'s act, although governed by probabilities, is not fully undetermined. Insofar as the circumstances would entail a repetition of the act, causes seem to operate, at least partially.

These distinctions, in terms of what would or might happen in *exactly* the same circumstances, may seem to be idle and perhaps there is a hint of this attitude in Grünbaum's evident irritation with the libertarian conception of freedom. But, as we have come to learn in connection with foundational studies in the quantum theory, just such counterfactual distinctions may turn out to have unexpected and testable consequences. I think this is precisely the case here. For if we subject the statistical laws of the quantum theory to the antilibertarian interpretation then, given other reasonable assumptions, I think we have all the machinery in place to derive the Bell inequalities, which is to say to contradict the quantum statistics. It follows that, given certain reasonable assumptions, the indeterministic laws of the quantum theory cannot be subjected to an antilibertarian interpretation. They call for a libertarian one. If indeed quantum theory is the paradigm for indeterministic laws more generally, then the centerpiece in the argument against the libertarian fails, for an indeterministic law would entail the libertarian conclusion that *A* could have done otherwise. Thus it seems that one could indeed appeal to indeterministic laws to ground the libertarian conception of human freedom. My agenda will be to carry the discussion up to this last point, and then to see whether an appeal to indeterministic laws really does help the libertarian cause. (To anticipate, the answer is no.)

Libertarianism and the Quantum Theory

I turn now to the conflict between antilibertarianism and the quantum theory (for more of the quantum details see Cushing and McMullin 1989 and Redhead 1987). Consider a typical EPR-type experiment that involves the measurement of four variables in two spatially separate locations: two variables in the "*A*-wing" of the experimental apparatus (say, A and $\overline{A}$) and two in a distant "*B*-wing" (say, B and $\overline{B}$). The reasonable assumption that we need is that the experiment involves no action-at-a-distance. Specifically, we assume a locality principle according to which the circumstances affecting a measurement outcome in one wing of the experiment do not depend on which variable is being measured in the other wing. To confront antilibertarianism consider two experimental runs each consisting of n pairs of measurements. In the first ("AB") run we measure A in one wing simultaneously with (or, at spacelike separation from) B in the other wing. For concreteness, let α be the sequence of outcomes of the n A-measurements and β the sequence of outcomes of the n measurements of B. In the second ("$\overline{AB}$") run we measure $\overline{A}$ with $\overline{B}$ and obtain outcome sequences $\bar{\alpha}$ and $\overline{\beta}$. Suppose, hypothetically, that in the first run we had measured A with $\overline{B}$ instead of with B. According to the locality assumption, in this hypothetical $A\overline{B}$ run, the circumstances affecting the measurement outcome in the A-wing would have been exactly as they were in the actual AB run. Hence in this hypothetical case the antilibertarian interpretation of the quantum statistical laws entails that the A-outcomes would be just as before, namely, the sequence α. Since the circumstances affecting the measurement outcome in the B-wing do not depend on which variable is measured in the A-wing, the results of the $\bar{B}$ measurement in the hypothetical $A\bar{B}$ run might have been the same as they were in the actual $\bar{A}\bar{B}$ run; that is, $\bar{\beta}$. Assuming that no special bias attaches to an outcome sequence due to the fact that one variable rather than another is measured in the opposite wing, the statistics of the experiment for the AB and $A\bar{B}$ runs (that is, the probability distributions for the variables A, B and $\bar{B}$ and the correlations or joint distributions for the pairs $\langle A, B\rangle$ and $\langle A, \bar{B}\rangle$) can be computed from the three outcomes sequences α, β and $\bar{\beta}$. Suppose, again hypothetically, that in the second ($\bar{A}\bar{B}$) run we had measured $\bar{A}$ with B instead of with $\bar{B}$. Then, just as before, locality combined with antilibertarianism entails that in this hypothetical $\bar{A}B$ run the $\bar{A}$-outcomes would be just as in the actual $\bar{A}\bar{B}$ run, namely, $\bar{\alpha}$. Again, as in the first hypothetical case, the results of the B measurement in the $\bar{A}B$ run might have been the same as they were in the actual AB run, that is, β. Once again, assuming our sample of outcomes is fair, we conclude that statistics of the experiment, this time for the $\bar{A}B$ and $\bar{A}\bar{B}$ runs, can be computed from three outcome sequences, namely, from $\bar{\alpha}$, β and $\bar{\beta}$. Thus, taken together, the four outcome sequences that occur in our two actual runs carry statistics for all of the four possible runs of the experiment, that is, the four single distributions and the four A-wing, B-wing pairs. However, it is well known (e.g., Fine 1982) that the single and joint distributions that can be computed from four fixed outcome sequences satisfy the Bell inequalities, which the quantum statistics violate for certain experimental configurations. Hence the antilibertarian reading of the statistical laws of the quantum theory, together with the principle of locality, conflicts with the quantum statistics. Assuming that the statistical laws of the quantum theory are correct, we can conclude that the antilibertarian reading of those laws entails action-at-a-distance; that is, it entails that, at least in certain experiments, measuring one variable rather than another would immediately alter the circumstances affecting some measurement outcome in a distant region of space. If we assume that the quantum theory is correct in its statistical predictions and we hold to the reasonable no action-at-a-distance condition involved in the stated locality principle, then it follows that the statistical laws of the quantum theory cannot be given an antilibertarian interpretation.

The quantum theory, we have argued, requires a libertarian reading of its probabilistic assertions, on pain of action-at-a-distance. In view of that argument, it seems reasonable to conclude that when the quantum theory says that an individual *A* has an 80 percent chance of doing something (e.g., a radioactive atom of decaying within the hour) and then *A* does it, other things being equal, *A* could have done otherwise. Thus the antilibertarian position finds little room to breathe in a statistical world if we take laws of the quantum theory as the exemplars of the statistical laws in such a world. So, it appears that, contrary to what Grünbaum claims, the libertarian's "could have done otherwise" does indeed find support from indeterminism if we take the indeterministic laws to be of the sort found in the quantum theory. It remains to be seen, however, whether such an indeterminism provides refuge for the libertarian position on the freedom of the will from a more general perspective, for the conclusion that Grünbaum draws may turn out to be more robust than the particular argument he gives for it.

Freedom and Indeterminism

The conclusion that Grünbaum set us after is a conditional one: If free will is incompatible with determinism, then it is also incompatible with indeterminism. I will call this *Grünbaum's conditional.* It is a strong conditional assertion and a very important one since, if it holds, then there is no free will in the libertarian sense. For, assuming that determinism and indeterminism are jointly exhaustive, if the will were free, both would fail.[4] Hence, the assumption that the will is free leads to a contradiction if Grünbaum's conditional holds.

The suggestion of an incompatibility between free will and indeterminism *tout court,* however, which is the consequent of that conditional, has a respectable history of its own. In the *Treatise* ([1739] 1902, book 2, part 3, chaps. 1 and 2), Hume questioned whether indeterminism (or "chance") provided a sound basis for the idea of human agency. In Hume ([1777] 1902), he summarized it this way, "Actions are by their very nature, temporary and perishing; and where they proceed not from some *cause* in the character and disposition of the person who performed them, they can neither redound to his honour, if good; nor infamy, if evil. . . . According to the principle, therefore, which denies necessity, and consequently causes, a man is as pure and untainted after having committed the most horrid crime, as at the first moment of his birth" (sec. 8, part 2, sec. 75). As I understand it, the line of thought goes something like this.

If after due deliberation and under circumstances free of any coercion (in the ordinary sense) my choice to do a certain deed, and my doing of it, are at best under the rule of a probabilistic law, then in what sense can I be said to be the author of my deed, and responsible for it? To say that I determined my action would seem to require that under the same circumstances (via my choice, perhaps) *I* made the difference between doing or not doing. But this would only be the case if, under the *very* same circumstances, things were bound to come out the same way. If the determining laws, however, are statistical and if we give them a libertarian reading, then even were circumstances exactly the same (including everything that pertains to me) the outcome might have been different; that is; I might not have done the deed. The libertarian reading leaves room for chance and change. According to that kind of indeterminism, the kind of indeterminism the quantum theory requires, literally nothing determines the outcome. But if nothing determines the outcome, literally, then *I* do not determine the outcome. A freedom of the will that can support attributions of responsibility (that can "redound to [one's] honour, if good; . . . infamy, if evil"), therefore, seems incompatible with indeterminism, under the libertarian conception.[5]

Like the arguments for the incompatibility of free will and determinism, this incompatibilist line too contains a number of moves that can be questioned. Moreover, one might think that the stochastic conception of causality can rescue the idea of agency in an indeterminist setting. One might think, that is, that although I do not determine the outcome of my behavior in the sense of strict causality (no probability or chance), still in the stochastic sense of "cause" I am a causal factor (maybe even the most significant causal factor) in my behaving as I do, and that this is enough for agency. The relevant sense of causal factor here is, roughly, this (see Cartwright 1989 and Humphreys 1989 for the problems with this rough account of stochastic causality, and for different proposals for how to polish and trim it). To say that I cause *B* to happen (or that my willing it does) is to say that my willing *B* makes *B* more likely to happen than would be the case if I did not will it. *I* make a difference. To be sure I do, but concerning what do I make the difference? Unfortunately for the application to agency, it is not with regard to *B*'s happening or not happening. On the libertarian reading, *nothing* makes a difference concerning that. What I (or my will) influences are the odds that *B* happens. My will raises the odds. Is this enough for agency with regard to the act itself? The basic rule of thumb for causal efficacy in the context of human agency (for example in criminal law and the law of torts)[6] is the "but for" test: But for my willing it *B* would not have happened. Of course this is the test whose failure is built into stochastic cases of the kind we are supposing here, since in *no* stochastic case is it true that *B* would not have happened without my willing it. In every case, *B* might have happened anyway. Thus the customary underpinning for agency has no grounding when chance enters in. We can see the difficulty clearly by considering stochastic cases where I will the act (thus upping the odds) but nevertheless the act does not happen. In these cases I may get credit for trying, but I cannot be blamed when the act does not transpire. That result, we might say, is simply the luck of the draw. But then, we need to ask, why are we supposed to be accountable for the act if, in identical circumstances, it does transpire? That is, in the usual case, where is agency supposed to get its grip? Since I play the same part whether the act occurs or not, I do not see how I may be counted out when the act does not occur, but held to account for the act when it does.

The conditions for agency do not fit the stochastic conception of causality. That conception is an attempt to extend the idea of a cause in circumstances where some of the usual concomitants of causality are absent. It shifts the effects of a cause from outcomes to the odds (or probability) for outcomes. This may be a perfectly good extension, which is to say one that is useful in some circumstances for certain purposes. We should not presume its universal applicability, however. Extending concepts is a context-sensitive business. Like making puns, whether it succeeds depends on whether the particular circumstances are just right. When it comes to the circumstances of human agency in an indeterminist setting, the stochastic move from outcomes to odds does not work well enough to rescue the freedom of the will.

The line that traces an incompatibility between indeterminism and free will has been attacked directly. Daniel Dennett, for example, advertises himself as having refuted it, "It has often been claimed that responsibility and indeterminism are incompatible. The argument typically offered is fallacious, as I show in. . . ." (1984, 142, n. 8) Dennett (1978, chap. 15) attempts to show that random elements in the decision process down the line from the act itself can still leave room for the efficacy of judgement and choice. Following this argument Dennett pleads. "The libertarian could not have wanted to place the indeterminism at the end of the agent's assessment and deliberation. . . . It would be insane to hope that after all rational deliberation had terminated with an assessment of the best available course of action, indeterminism would then intervene to flip the coin before action" (ibid., 298). Dennett goes on to urge the standard compatibilist gloss on "could have done otherwise," a gloss he could find in Grünbaum, namely, that the circumstances in which the agent could have done otherwise are *not* exactly the same—in particular, that they do not involve the same beliefs and desires with which rational deliberation ended. If the incompatibility of agency (or responsibility) and indeterminism (in the libertarian sense) rests on a fallacy, Dennett has certainly not displayed it. Instead he concedes the incompatibility and calls it "insane," hoping, no doubt, to draw our attention away from his failure to come to terms with it.

What Dennett might claim to show (at least this is where he engages in argument rather than invective) is that if the indeterminism is limited to a certain place in the deliberative process that leads to action, then the agent might still intelligibly be thought of as the author of the act. (He uses the term "authentic.") Dennett thinks that indeterminism might come into play, somehow, in generating considerations which rational choice then converts deterministically into action. This separatist scheme is similar to ideas that Grünbaum (and others) have entertained with respect to the impact of quantum indeterminacy on human freedom. The common idea is to keep the indeterminacy suitably confined and to hope that deterministic principles, sufficient to support a decent sense of agency, function appropriately somewhere at the level of molar behavior and choice. It seems to me, however, that this form of separatism is not stable unless the indeterminacy is taken out of the intentional stream entirely. That move, however, involves rejecting the attractive Leibnitzian idea that motives (or whatever) incline but do not necessitate. It would take us too far afield to explore the cogency of that rejection. Let me just show that Dennett's idea of placing the indeterminism somewhere downline but still within the intentional stream does not work.

Suppose we grant, with Dennett, that if the coin flips between the end of deliberation and the consequent act, then indeed agency is called into question. So we grant that the action is not authentically mine if, having determined a rational course of action, it is a then not up to me but rather a matter of chance whether the act transpires. To be *my* act, the relation between the act's happening and what we may call its intentional determinant has to be nonprobabilistic, or so we will suppose. Well, what if the coin flips a little sooner; that is, what if chance enters between the considerations that form the background of my choice and the rational decision process itself? Then, surely, application of the same principle of ownership requires that the determination of the course of action is, likewise, not authentically mine. For how can background consideration be said to form a basis for my decision if it is not up to me but rather a matter of chance how I proceed to deliberate, given those considerations? The fact that Dennett does not consider this possibility at all, but simply assumes that the background considerations go hand in glove with a decision process, which in turn fixes the action, seems to corroborate the analysis. To be my deliberations, their determinants have to be nonprobabilistic as well, or so Dennett seems to assume.

Dennett would flip the coin just before the considerations themselves, where he thinks it is harmless. But suppose I am a lobbyist for *Birds First!* (a radical environmental organization) and have been offered a job by *Pollution, Inc.* (a large and notoriously irresponsible oil company). In Dennett's sense, considerations on which I base my decision about accepting the job might include the difference in salaries, the quality of the support staff, the workload, the location of the jobs, the amount of travel involved, and so on. Some or all of these might occur to me and let us say, with Dennett, that whether they (or other considerations) do occur involves an element of chance. In the end I do not determine which considerations occur to me and which do not. Nothing fixes that. However, I am a sincere person, committed to the welfare of birds and the environment, and generally scrupulous in keeping that commitment. I would never wittingly put my talent to work for *Pollution, Inc.,* not, that is, if it occurred to me to consider these commitments of mine to birds and the environment, or if I considered the self-image problem that would be engendered by earning my livelihood out of the ill-gained profits of *Pollution, Inc.*—and so on. But suppose that despite my most sincere efforts in making a good decision, these considerations just never occur to me. Chance intervenes to flip the coin only on the first list above. Pursuing a rational decision policy, I wind up accepting the offer. Now, given the role of chance in that result, I do not see why I am any more the captain of my fate in these circumstances than I would be had chance intervened later down the road, say, to mediate the execution of a more balanced decision. (Indeed, would it be different had the environmental considerations occurred to me by chance and, because the coin landed the other way around, I made a different decision?) For an act to be authentically mine, we have supposed that chance must not divide it from its intentional antecedents. On that basis, it would seem that the coin may not flip to separate me and my characteristic concerns from the considerations on which my choice is based.

I hope that one can see in this the outline of a good inductive argument. It starts with Dennett's concession that agency is compromised if chance enters between the end of deliberation and the act. The argument proceeds via the inductive rule that agency excludes chance between any act and its intentional predecessors. We conclude that chance may not enter at all in any chain of intentional antecedents of the act, if the act is to be mine. This line support's Hume's intuition that responsibility is rooted in connections between character and behavior that leave no room for chance. Nothing in the separatist strategy that we have been looking at suggests that Hume was mistaken, for it appears that regardless of where in the intentional stream you put it, one chancy apple may spoil the whole pile. To get around Hume, one would have to question whether necessity is really required for agency at all. I propose to avoid that issue and return us instead to the conditional incompatibilism with which we started.

Conditional Incompatibility

Is indeterminism of any help to the libertarian? A negative answer follows from Grünbaum's conditional, that if free will is incompatible with determinism, then it is also incompatible with indeterminism. Insofar as the results of the preceding section support the consequent here they support the conditional itself. In this section I will try to provide even stronger support by arguing that if there were an adequate indeterminist account of an agent's behavior there would be an equally adequate determinist account. It follows that if indeed determinism is not compatible with an agent's free will then neither is indeterminism, which is the conditional we are after.

The argument derives from the trivial observation that if the probability of an event is 8/10, say, then one can think of that as involving ten possible cases in eight of which the event occurs. The argument simply gives a formal expansion of that idea, uniformly, for a class of probabilistic assertions. Suppose then that we consider an agent *A*. I will piggyback on the state-observable framework of quantum theory to suppose that we can talk meaningfully about "the state" of our agent at a given time, which I will denote by σ (suppressing a temporal index unless required). We suppose that an indeterminist theory of *A*'s behavior treats possible acts *B* from a well-defined class and yields a set of probabilistic laws of the form

$$\mathrm{Prob}_A(B \mid \sigma) = p$$

("The probability that agent *A* does act *B* in state σ is p."). Indeterminism requires that the probability p be different from 0 or 1 for some acts *B*. Were the probabilities for all acts either 0 or 1, then (relative to the state) nothing would be left to chance and the theory would be deterministic. If the state includes enough relevant information about the intentional situation of the agent just prior to the act (beliefs, desires, or whatever) the indeterminist theory is precisely of the "insane" kind that (*pace* Dennett) interests the libertarian, and which was the starting point for the considerations in the previous section.

We want to show that if an agent's states and behavior are governed by an indeterminist theory, they are also governed by a deterministic one. It is important that the deterministic theory we will produce treat the same states and behavior as the indeterminist one. Otherwise, one might suppose that the determinism depends on a trick of redescription, for example, shifting from an intentional idiom (waves goodbye) to a nonintentional one (arm moves 27 centimeters vertically). To be sure, what is undetermined relative to one sort of description might be determined relative to another, but that has nothing to do with the point here at issue. My trick is different. It connects, rather, with the concept of indeterminism already introduced, that is, with whether the statistics are "removeable by the possession of more complete information." When such a reduction is possible one obtains a deterministic theory, as previously defined. We now show that, in principle, it is always possible to reduce the statistics of an indeterminist theory.

Suppose the state of the agent A is σ. To achieve the reduction we introduce a set of indices, which we take just to be the numbers between 0 and 1 (inclusive). These indices correspond to the "ten possible cases" mentioned, and there are several interpretations that one might give to them. Here I will treat the indices as marking different possible ways of being in a given state, and assume that each agent has just one way. Keeping track of the ways provides the additional information needed to get rid of the statistics. For, corresponding to each act B, we can use the indeterminist theory to introduce a function $B_A(.)$ taking only 0 or 1 as values and defined as follows for any index x between 0 and 1:

$$B_A(x) = 1 \text{ if } 0 \leq x \leq \text{Prob}_A(B);$$
$$\text{otherwise, } B_A(x) = 0.$$

We understand $B_A(x) = 1$(or 0) just in case it is true (or false) that if A were in state σ in the way marked by x, A would do B. Thus our new theory supplements the indeterminist one by adding the indices and the "truth" functions defined with respect to them as just given.

The new theory involves a new class of probabilistic ascriptions of the form

$$\text{Prob}_A(B \mid \sigma,x)$$

where x is an index. This is the probability that A does B, given that A is in state σ in the way marked by index x. The assumption that every agent is in a state in just one way implies that these probabilities are either 1 or 0, depending (respectively) on whether it is true or false that the way A is in σ yields that A does B. In fact

$$\text{Prob}_A(B \mid \sigma,x) = B_A(x)$$

and, since $B_A(x) = 0$ or 1, the new theory is deterministic. The probabilities from the indeterministic theory, however, all follow from the deterministic one as averages. That is,

$$\text{Prob}_A(B \mid \sigma) = \text{AVERAGE}\,[\text{Prob}_A(B \mid \sigma,x)] = \text{AVERAGE}\,[B_A(x)],$$

where the average is taken over all x between 0 and 1. This completes the reduction.

The indeterminist theory distinguishes one agent from another only by their states, treating agents generically as randomly selected from all the ways of being in a state. The deterministic theory individuates in a more precise way, one that reduces the posited statistics. That reduction eliminates chance. No coins flip. Hence, if there is an indeterministic theory of an agent's behavior, there is a deterministic one too. Therefore, if determinism is incompatible with free will, so is indeterminism. This result gives strong support to Grünbaum's conclusion that indeterminism is of no help to the libertarian. Before we rest content with that conclusion, however, we need to study the possibility for a determinist reduction in a little more detail.

Reducibility

In claiming that indeterminism implies determinism, we show that the statistics of an indeterminist theory are always reducible to a determinist base. But we have already argued that the statistics of the quantum theory require a libertarian reading that would prevent their deterministic reduction. Attention to two different factors will help us see how to reconcile these claims. First of all, in the quantum case we deal with a more complicated set of probabilistic assertions than we have just considered. Those all concerned the probability for a single agent to do this, or that. In the quantum theory we deal with joint probabilities: the probability that *A* does *B* and *C* does *D* (to put it in agent terms). So the quantum theory gives us a family of joint (actually multiple) probability distributions, whereas before I deal only with single distributions. It turns out that difficulties with reducibility in the quantum theory (technically, with the introduction of deterministic hidden variables) always relate to the special character of the family of quantum joint distributions (see Fine 1982). The Bell inequalities, which we previously used to block reducibility, are a case in point. They are inequalities constraining joint distributions which the twin requirements of antilibertarianism and locality imply cannot be matched by the quantum joints. This points to the second issue, which is that of external conditions. To get the irreducibility of the quantum statistics we need to go beyond the formal requirements of the quantum theory itself and assume locality. (Other no-go results in the foundational literature; e.g., the Kochen-Specker theorem, or the Heywood-Redhead result, also require external assumptions, although in these cases the assumptions are less well motivated and plausible than locality, at least from a physical point of view—see Redhead 1987 and Elby 1990). Nonlocal reductions of the quantum statistics along the lines sketched in the general argument just given are perfectly possible, indeed trivial from a formal point of view. (Somewhat less trivial is the nonlocal de Broglie-Bohm "pilot wave" theory that also reduces the quantum statistics and offers a challenging deterministic alternative to the usual understanding of the quantum theory—Bell 1987 contains the relevant details.) By taking account of the joint distributions and the external constraints, we can reconcile the claims of irreducibility in the quantum case with the claim of reducibility for agent indeterminism.

Taking these factors into account, however, seems to point to serious shortcomings in the argument of the preceding section, for the claim that agent indeterminism leads to agent determinism appears to be misleading. That result is purely formal and, as we learn by comparison with the situation in the quantum theory, it seems to depend on attending only to a restricted class of statistical theories (namely, ones without joint probabilities) and on ignoring plausible external constraints on the proposed scheme of reduction—or so one might object. Despite the lesson of the quantum theory, however, I am not so sure that these objections are well founded.

The problem concerning joint distributions can be put this way. How does one deal with correlations between the behavior of different agents? I think this question has two good answers. The first is simply to deny that we have to deal with them at all; that is, that we have to consider the behavior of more than one agent at a time. After all, we are not speculating about the reducibility of some general science of human behavior. We are only addressing the theory of agency for a single individual, any *one* of us. At any rate, that is the traditional philosophical setting for the discussion of free will, and it is hard to see how possible difficulties in reducibility that might arise in theories of group behavior would affect that discussion. Even if one is inclined to a social theory of the mind I would have thought that, over the freedom of the will, a person is not to be treated as part of a herd. So the first good answer is to assert that, in the absence of a specific limitation on reducibility that is derived from theories of group behavior, our single distribution framework is adequate (at least prima facie). The second good response to the question of how to treat correlations between agents is to offer the conventional wisdom, which is this. If the correlations are not spurious then either they derive from a direct causal connection between the agents (perhaps a chain of causes) or they can be explained in terms of common background causes that affect the behavior of both agents. Let us look at these alternatives in turn.

Direct causal connections can give rise to the full range of possible correlations in the behavior of agents. To see that, just consider a class of students taking an essay exam. If they are in communication with one another (the causal connection) then they can arrange for any correlation at all, from all their answers being exactly the same (word for word) on every question, to none of them being the same on any question. However, whatever they arrange can be explained deterministically, if it can be explained stochastically, for a stochastic explanation must yield a probability distribution for all the possible answers the class might give to the examination questions. Any such (multiple) distribution, however, can be obtained by averaging over a large number of determinate answers, just as in our given simple proof of reduction. The field of determinate answers, arranged as how the students would have answered (if), provides the deterministic reduction that we were looking for. Thus direct causal connections do not produce any challenge to reduction that goes beyond what we have already treated. What then of common causes?

To explain correlations by means of common causes is to derive the correlations by averaging over the effects of background causes that uncouple the correlated variables. To take a quantitative example, the crime rate in Chicago is inversely correlated with the attendance rate at the movies during mid-week for theaters located in suburban shopping centers. The more suburbanites who attend local movies during the week, the lower the crime rate in the city. This is not because all the criminals live in the suburbs. Nor is there a direct chain of causes linking inner city crime and suburban movie attendance. (No one is planning a mass suburban movie-in to stop crime in the city.) But the conditions that make for less crime in the city (for example, high employment and bad weather) also make for moviegoing in the suburbs. High city employment means business is good and that contributes to the size and affluence of the families of suburban professionals and businesspeople (which, in turn, contributes to their mobility and restlessness), and (within limits) bad weather is good for the movie business. However, for fixed levels of employment and weather conditions, there is no significant correlation between urban crime and suburban movie-going. It is only by averaging over various levels that the correlation appears. If we adapt this example to the behavior of agents, then we would explain correlations in behavior by averaging over independent factors, that is, factors that contribute to the behavior of each agent independently of each other. These factors can be stochastic in the sense discussed in an earlier section; that is, they may merely change the odds ("incline") without necessitating the behavior. The question is whether this kind of explanatory framework for treating joint agent behavior is also subject to a deterministic reduction in the sense previously defined. The answer is "yes." If there is a common cause explanation for correlated behavior, even if the causal factors are only stochastic, there is also a deterministic explanation for the behavior. I refer the reader to Fine (1982) for the details of the proof, but the idea is simple enough. We start with the fact exploited in the discussion of direct causal links, namely, that any multiple probability distribution has a deterministic reduction. In the case of common cause explanations, corresponding to each of the causal factors there is a single probability distribution for each agent. Because these distributions are independent (in the stochastic sense) their product is a well-defined multiple distribution that can be averaged over all the causal factors. That average has a deterministic reduction, which is the deterministic theory that we were seeking.

The second answer to the question of how one deals with correlations between the behavior of different agents, then, is this. If the connections between the behavior of several agents arise from causal links or common causes, we can deal with them by effecting exactly the same sort of deterministic reduction that we gave for the single agent theory. To back up the charge that this framework is too limited, one would have to produce correlations in behavior that could not be treated this way. The quantum theory contains just such correlations, that is, ones that defy direct causal or common causal explanations. In Fine (1989), I have urged that the natural way to understand them is to acknowledge the irreducibility of the quantum statistical laws, in which case the correlations emerge as irreducible as well (see also the Appendix to van Fraassen 1989). Whatever is the right setting in the quantum case, however, we have yet to see any correspondingly "anomalous" correlations in the behavior of human agents. There is a good reason why not. Such correlations can only arise if there is no possible way to integrate the behavior of a number of agents into a single probabilistic framework, or, more specifically, if there is no joint probability function for them all. This happens in the quantum theory where noncommuting observables do not have joint distributions.But it is difficult to see how this could occur in the arena of human behavior, and we certainly have no instances at hand. A simple sort of case would be that of three individuals *X*, *Y* and *Z* where the behavior of *X* and *Y* is jointly describable, and that of *X* and *Z* is also jointly describable, but there is no possible description for *Y* and *Z* taken together. This may remind one of intransitivity in preference rankings. But such intransitivity still allows for a composite preference ranking for all the individuals, which we obtain by simply conjoining all the separate rankings. A much deeper incoherence would have to obtain in order to forestall any conjoint probabilistic description at all. Thus, barring the demonstration of such a deep incoherence in group behavior, the reducibility of indeterminist theories of agency to deterministic ones seems well founded.

This way of viewing the correlation problem is also responsive to the second issue raised in contrasting the reducibility of theories of agency with the irreducibility of the quantum statistics. That was the issue of external constraints. We saw that in order to get quantum irreducibility we needed to impose some extratheoretical assumption, like locality. Could there be plausible constraints of this sort that would block the reduction in the case of human agency? We have just rehearsed what the consequences would be if there were such constraints. The theory of agency could not treat all the agents in one probabilistic framework. There would be incompatibilities between agents that prevented any conjoint probabilistic description. Of course one cannot rule this possibility out a priori. However, one can say that there is nothing in our human experience (including theories of human behavior with which we are familiar) that suggests the plausibility of limitations on theorizing that would issue in this sort of incoherence. On the grounds that one need not be frightened away by skeptical possibilities that run counter to ordinary human experience, I feel comfortable in putting the idea of reduction-limiting constraints on the shelf. If some specific limiting principles were offered, then (of course) we would have to take them seriously.

This last response also addresses the final charge, bound to be brought, that this whole discussion of a deterministic reduction is merely formal. All I have shown, after all, is that nothing stands in the way of determinism from a purely formal point of view. But surely that does not mean that *in reality* we could always find an adequate deterministic reduction to a given indeterminist theory of behavior. To be sure, my proof is merely formal. But I think it shows something relevant to free will nevertheless. For "in reality" there is no indeterminist theory to reduce. As described in an earlier section, our standard practice with regard to human agency employs the "but for" test and other rules of thumb that embody deterministic presuppositions. Likewise, theories of behavior look for causes or determinants in a strict sense. The realistic issue is whether anything stands in the way of these deterministic presuppositions and practices. The example of the quantum theory shows that there can be serious obstacles. By examining the conditions required for such obstacles to arise, the preceding discussion shows that in the case of human agency there are none. This is the sense in which we may suppose that any indeterminist theory would be reducible.

Concluding Remarks

In the first part of this essay I argued that the statistical laws of the quantum theory require a libertarian interpretation. This is a reading that incorporates the libertarian idea that when something happens it could have been otherwise, and under the very same circumstances. If human behavior were governed by statistical laws like those of the quantum theory, this would seem to lend support to the libertarian ("could have done otherwise") picture of a free agent. However, the libertarian also thinks that free agency is incompatible with a deterministic account of human behavior. In the second part of this essay I support Grünbaum's conditional and show that if there were an indeterminist account of the behavior of an agent (that is, an account that made essential use of probability) then there would also be a determinist account of the same behavior (that is, a probability-free account). Thus indeterminism is also incompatible with free will (assuming that determinism is). It follows that on the libertarian conception there is no free will since, as I use them in these demonstrations, determinism and indeterminism are logical opposites.

This is not the conclusion for which the libertarian was hoping. The libertarian would rather have it that human behavior is governed by indeterministic laws that cannot be given a deterministic reduction. For that to be the case, however, there would have to be a deep incoherence in group behavior, one that prevented complete statistical descriptions of the behavior of several agents together. One might contrast this circumstance with the usual understanding of the libertarian position, which is nicely captured in this remark by a well-known legal scholar, "No matter how well or fully we learn the antecedent facts, we can never say of a voluntary action that it had to be the case that the person would choose to act in a certain way. In a word, every volitional actor is a wild card" (Kadish 1985, 360). Now in dealing wild cards (or, to use Dennett's metaphor, tossing coins) we give odds, and make book on the outcomes. If the laws could not be given a deterministic reduction, however, then just this sort of probabilistic description would be ruled out, at least in certain cases. Thus the burden on the libertarian is to find a way of treating human behavior that involves neither probabilistic description nor nonprobabilistic description. The only way that recommends itself is to treat people as arational, that is, as behaving outside the bounds of rational description altogether. This seems a self-defeating move for the libertarian, who wishes to uphold the dignity of humankind. Thus the libertarian conception of free will seems truly incoherent.

The upshot of these considerations leaves room for a causal and deterministic description of human behavior, the kind of theorizing that is widely practiced and familiar. This kind has its own account of free will, a strained version that fits the compatibilist tradition and according to which (although "free") we really could *not* have done otherwise. The libertarian conception of free will turns out to be incoherent. I would argue that this twisted compatibilist conception is hardly better.

NOTES

I want to dedicate this essay to Adolf Grünbaum, whose concern for clear reasoned discussion provides an admirable model for us all. Work on this essay was supported by NSF Grant DIR-8905571. My thanks to Micky Forbes, Richard Manning, and Aaron Snyder for helpful discussions, although it must be said that they do not necessarily agree with my lines of argument. I owe a debt to Elizabeth Dipple for direction in the classics. John Earman made useful comments on an earlier draft and that has helped too, although, I fear, he will think not enough.

1. Witness the formulation by Hart and Honoré, "A deliberate human act is . . . something through which we do not trace the cause of a later event" (1959, 41).

2. Grünbaum calls "irreducibly statistical" the laws that I label *indeterministic*.

3. See Feigl et al. (1972, 614) where Grünbaum shifts from an epistemic to a nonepistemic formulation of this line of argument.

4. In this essay I employ a very minimal sense of determinism and indeterminism, depending on whether (indeterminism) the laws of nature involve an essentially probabilistic element or (determinism) not. In this sense, I take the alternatives here to be mutually exclusive and jointly exhaustive. There are other senses. See Earman (1986). In what follows, I sometimes mix in causal talk. Given the subject matter, the mix is inevitable.

5. For the purposes of the deconstructive *reductio,* I follow the tradition in discussions of the freedom of the will in supposing that responsibility tracks agency, although I do not believe any such thing. As explained in the beginning of the essay, I think agency is crafted out of the need to ground the social practices involved in assigning and judging responsibility. Like other foundational "needs," we can do better.

6. In tort law, assignments of liability require proximate cause and may use several variations of the "but for" criterion, variations like the INUS conditions familiar in philosophical discussions. In a stochastic setting, however, these variations do not fare any better than the "but for" condition itself. There is some discussion among legal scholars concerning how to apportion damages where, say, a polluter is responsible for increasing certain health risks by a determinate percentage. Increasing the risk of disease, here, corresponds to inducing certain medical conditions which would be considered harmful in themselves, and for which damages could be assessed. This is different from holding a polluter responsible for (stochastically) "causing" a disease that may or may not occur, which would be closer to the issue discussed in the text. So far, all this is legal speculation, not (I believe) supported by case law. Moreover, the awarding of damages in torts is not the most reliable guide to responsibility (just think of the strict liability).

REFERENCES

Bell, J. S. 1987. *Speakable and Unspeakable in Quantum Mechanics*. Cambridge, England: Cambridge University Press.

Cartwright, N. 1989. *Nature's Capacities and Their Measurement*. Oxford: Clarendon Press.

Cushing, J., and E. McMullin, eds. 1989. *Philosophical Consequences of Quantum Theory.* Notre Dame: University of Notre Dame Press.

Dennett, D. 1978. *Brainstorms.* Cambridge, Mass.: Bradford Books.

———. 1984. *Elbow Room.* Cambridge, Mass.: MIT Press.

Earman, J. 1986. *A Primer of Determinism.* Dordrecht: Reidel.

Elby, A. 1990. "On the Physical Interpretation of Heywood and Redhead's Algebraic Impossibility Theorem." *Foundations of Physics Letters 3:* 239–47.

Feigl, H.; W. Sellars; and K. Lehrer. 1972. *New Readings in Philosophical Analysis.* New York: Appleton-Century-Crofts.

Fine, A. 1982."Joint Distributions, Quantum Correlations and Commuting Observables." *Journal of Mathematical Physics 23:* 1306–10.

———. 1989. "Do Correlations Need to Be Explained?" In J. Cushing and E. McMullen, eds., *Philosophical Consequences of Quantum Theory.* Notre Dame: University of Notre Dame Press, pp. 175–94.

Grünbaum,A. 1953. "Causality and the Science of Human Behavior." In H. Feigl and M. Brodbeck, eds., *Readings in the Philosophy of Science.* New York: Appleton-Century-Crofts, pp. 766–77. (Originally published in *American Scientist 40:* 665–76.)

———. 1967. "Science and Man." In M. Mandelbaum et al., eds., *Philosophic Problems.* New York: Macmillan, pp. 448–62. (Originally published in *Perspectives in Biology and Medicine 5:* 483–502.)

———. 1972. "Free Will and the Laws of Human Behavior." In H. Feigl, W. Sellars, and K. Lehrer, eds., *New Readings in Philosophical Analysis.* New York: Appleton-Century-Crofts, pp. 605–27. (Originally published in *American Philosophical Quarterly 8:* 299–317.)

Hart, H. L. A., and A. M. Honoré. 1959. *Causation in the Law.* Oxford: Clarendon Press.

Hume, D. [1739] 1902. *A Treatise of Human Nature.* 1st ed. Edited by L. A. Selby-Bigge. Oxford: Clarendon Press.

———. [1777] 1902. *An Enquiry Concerning Human Understanding.* 2d ed. Edited by L. A. Selby-Bigge. Oxford: Clarendon Press.

Humphreys, P. 1989. *The Chances of Explanation.* Princeton: Princeton University Press.

Kadish, S. 1985. "Complicity, Cause and Blame: A Study of the Interpretation of Doctrine." *California Law Review 73:* 323–410.

Redhead, M. 1987. *Incompleteness, Nonlocality and Realism.* Oxford: Clarendon Press.

van Fraassen, B. 1989. "The Charybdis of Realism." In J. Cushing and E. McMullin, eds., *Philosophical Consequences of Quantum Theory.* Notre Dame: University of Notre Dame Press, pp. 97–113.

22

Adolf Grünbaum and the Compatibilist Thesis

John Watkins
Department of Philosophy, Logic and Scientific Method, London School of Economics

Compatibilism has been a long-standing thesis of Adolf Grünbaum (he first defended it in his 1952 article). His version involves three subtheses: (1) all human behavior is open to scientific explanation with the help of causal laws; (2) it makes no difference, with respect to human freedom and responsibility, whether the laws are deterministic or statistical; and (3) there is "*no incompatibility* between the existence of either causal or statistical laws of voluntary behavior, on the one hand, and the feelings of freedom which we actually do have, the meaningful assignment of responsibility, the rational infliction of punishment, and the existence of feelings of remorse or guilt on the other" (1972, 608; emphasis added). I will challenge all three subtheses. I will first set the scene by indicating the direction from which the challenges will come.

Grünbaum and I share a staunchly naturalistic outlook. Although my deductivism obliges me to call myself an agnostic rather than an atheist, since it does not allow me to conclude from our human experience that no God exists, it does allow me to conclude that human affairs are not under the (continuous and effective) supervision of a caring God. I added that parenthetical remark in view of a hypothesis of Martin Buber's, which Grünbaum (1987, 178) mentions, to the effect that there is a caring God who, however, goes into eclipse from time to time—as He did, for instance, during the Nazi holocaust. I accept unquestioningly that we are part of nature, in the sense that we have been evolved by the same mechanisms of natural selection as have other creatures, and that consciousness ceases with death.

Some evolutionists adopt an epiphenomenalist stance and belittle the causal significance of consciousness. Like Grünbaum, I think that that is a great mistake, but I agree that consciousness does pose a dilemma to evolutionists. They will (or should) insist that mind is not superadded, whether divinely or in some other way, but develops naturally, having a physical and genetic basis that enables mental characteristics and capacities to be transmitted by the usual mechanisms of heredity; and to that extent they will incline to a "bottom-up" view. On the other hand, they will (or should) insist that for minds to have been evolved by natural selection, minds must *do* something, have biological utility, render animals fitter to survive than the same physical animals would have been if they had lacked consciousness. (Essentially this argument goes back to Darwin himself.) To that extent evolutionists will side with dualist interactionism against epiphenomenalism.

I try to extricate myself from this dilemma by taking a leaf out of Marx's book. The materialist theory of history, as I understand it, is dualist in sharply distinguishing a society's ideological superstructure from its economic basis to which it gives causal priority. But it adds that, having been generated from below, such a superstructure begins to acquire a certain causal efficacy of its own, and can act back, to a limited but significant degree, on its own infrastructure. For "superstructure," read "mind" and for "economic," read "physical" and you have the core of my view of the mind-body relation. It allows that epiphenomenalism holds in human infancy, but adds that at an early stage the small child's mind begins to acquire a certain capacity, weak and clumsy at first but normally growing stronger and more exact as it matures, to control its limbs, especially its fingers, and—most miraculously—those of its organs involved in the production of speech. In other creatures, consciousness may exhibit its survival value by, for instance, helping to steer the animal away from things, such as bad meat, that smell or taste nasty. I go along with Darwin's view that, in our case, developments on the mental side have produced a degree of critical inventiveness that has made our species preeminent. Finally, I hold that this mental efficacy, especially when coupled with inventiveness, introduces a disturbing element of indeterminism.

I have one more preliminary. Some philosophers, for instance, Spinoza, have sought an absolute either-or distinction between self-determination and other-determination, but it seems obvious to me that it has to be a matter of degree. I think that we can, nevertheless, lay down an absolute precondition or necessary condition for a person X to be at all self-determining when X behaves in a certain way in given external circumstances: Something in X's consciousness must be causally relevant to X's bodily behavior. We could render this idea as follows. Let M be a variable that ranges over items in X's consciousness. Then

$$(\exists M)P(B, AM) > P(B, A)$$

says that something in X's consciousness, call it M, is causally relevant to X's bodily behavior B in external circumstances A in the sense that B is more probable given A and M than given A alone. Although this condition is very weak, it is strong enough to exclude epiphenomenalism, or the assertion that consciousness has no causal influence on bodily behavior, which we may express (relativized to X's bodily behavior B in circumstances A) thus:

$$(\forall M)P(B, AM) = P(B, A).$$

In what follows I will take it for granted that (1) is a necessary condition for a positive degree of self-determination.

Voluntarism-cum-Determinism: The Example of Schopenhauer

Grünbaum is not a determinist in rather the same spirit that he is not a steady state theorist; he judges that the evidence, especially developments in quantum mechanics, tells against it. But he has no metaphysical axe to grind, here; had the evidence told in the other direction, he would have been content to accept determinism, for he holds that a scientific view of human doings is fully compatible with our feelings of freedom, responsibility, and such, whether all its laws are deterministic or some are irreducibly statistical. He is very insistent that the microindeterminacies introduced by quantum mechanics have no relevance to the question of human freedom. Grünbaum (1972) claimed that quantum mechanical indeterminism is compatible with a macrodeterminism and, moreover, with one that repudiates epiphenomenalism (p. 621). He outlined what he called a P-cum-

Ψ determinism (with P-variables ranging over neurophysiological states and Ψ-variables over psychological states) which "constitutes a *denial* of *epiphenomenalism*" (ibid., 622; italics in the original). It is deterministic in that there are laws such that an organism's total P-cum-Ψ state at any one time is uniquely specified by its total P-cum-Ψ state at any other time (ibid., 621), and it denies epiphenomenalism in that the Ψ-values of its state at one time may help to determine the P-values of its state at a later time.

I accept, for the sake of the present argument, the compatibility of quantum mechanics with macrodeterminism. My concern here is the compatibility of determinism (whether macro only, or macro-cum-micro) with the denial of epiphenomenalism. Schopenhauer was at once a thoroughgoing determinist and a strong voluntarist who repudiated epiphenomenalism. We might call him a P-cum-Ψ determinist. I will use his views to make my case. My thesis will be that his determinism rendered his voluntarism virtually indistinguishable from epiphenomenalism.

Schopenhauer gave a central place to the will, "[M]an's will is his authentic self, the true core of his being" ([1841]1960, 21). He had no doubts concerning the sovereignty of a person's will over his bodily behavior, "[m]y body will immediately perform [an action] as soon as I will it, quite without fail" (ibid., 16). He also spoke of the volition's "absolute power over the parts of the body" (ibid., 17). My dualist interactionism involves no such extremist voluntarism, which seems obviously untenable. (A beginner on the nursery slope grimly mutters, "This time I *will* get my weight on the outer ski," and then again fails to do so.)[1] Indeed, in my conception of self-determination, *will* does not play a major role. In place of clockable acts of will, each followed by its own separate packet of will-guided bodily behavior, it puts creative-cum-critical thinking guiding a sustained course of action, perhaps one that issues in some ontological enrichment of our world. Where Ryle asked the voluntarist "how many acts of will he executes in, say, reciting 'Little Miss Muffet' backwards" (1949, 65), I would ask him how many acts of will Michelangelo might have executed in the course of, say, a day's work on the ceiling of the Sistine Chapel. But these disagreements with Schopenhauer do not affect the issue I now raise, which is not whether Schopenhauer's voluntarism is exaggerated, but how far it can be kept separate from epiphenomenalism, given his determinism.

According to Schopenhauer, the belief of philosophically untutored people that the will is free is an illusion, but one which arises in a very understandable way; it is the natural verdict of self-consciousness. My self-consciousness is what would remain if that part of my consciousness that is directed outwards could be subtracted from my total consciousness. What my self-consciousness tells me is that I am a being that *wills* and that my will is sovereign. It tells me, "You can do whatever you will." And that, properly understood, is correct: If I will to do this rather than that, then I shall indeed do this, and if I will to do that rather than this, I shall indeed do that. But are both options—of willing to do this or of willing to do that—open to me? Is my will free? As Schopenhauer put it, "We grant that his *acting* depends entirely on his willing, but now we want to know on what his willing itself depends. Does it depend on nothing at all or on something?" ([1841] 1960, 19) This question baffles self-consciousness; having no cognizance of anything external to my willing, it can only assume that my willing depends on nothing beyond itself (ibid., 20). This natural verdict (on which he quoted Descartes as saying "there is nothing we comprehend more clearly and perfectly") really results, Schopenhauer said, from the myopia of "stupid self-consciousness" (ibid., 81).

The matter stands very differently when investigated from an external rather than an introspective viewpoint. As a disciple of Kant, Schopenhauer considered it incontestable that the law of causality holds sway throughout the phenomenal world (ibid., 28). He answered the question "whether man alone constitutes an exception to the whole of nature" (ibid., 21) with an emphatic *no*. I can do whatever I will, but my willing is causally necessitated (ibid., 24).

Schopenhauer somewhat softened his account by adding that causality operates very differently at different ontological levels. At the mechanical level, effects are exactly proportional to their causes, and this is also the case, though less obviously, at the chemical level. But at progressively higher levels, there is a growing disparity, disproportion, and heterogeneity between cause and effect. At the animal level, the activating causes are *motives* (ibid., 32). Here, the "heterogeneity between cause and effect, their separation from one another, their incommensurability" (ibid., 39), has reached a very high degree, for the cause "which originates the motion of an animal" (ibid.) is *immaterial.* Whatever Schopenhauer's ultimate metaphysical position may have been, at the commonsense level he was a dualist interactionist who held that, in the case of animals as well as people, immaterial motives play the governing role in their behavior. There is a difference, however, between the ways in which motives operate in people and in brutes; the latter lack abstract concepts:

> [T]hey have no ideas other than perceptual ones. . . . [T]hey live only in the present. . . . They can choose only among the things which spread themselves out perceptually before their limited field of vision. . . .
>
> On the other hand, by virtue of his capacity for non-perceptual ideas . . . man has an infinitely larger field of vision, which includes the absent, the past, and the future. Because of this he can . . . exercise a choice over a much greater range than the animal. . . . His action . . . proceeds rather from mere *thoughts* which he carries around everywhere in his head and which make him independent of . . . the present. (ibid., 35)

He gave this picturesque summary of the difference. People's 'actions are guided, as it were, by fine invisible threads (motives consisting of mere thoughts), whereas the actions of animals are pulled by rough, visible strings of what is perceptually present" (ibid., 36).

Schopenhauer was here touching on something important. Near the end of his book he spoke of the "consciousness of self-determination and originality which undeniably accompanies all our acts, and by virtue of which they are *our* acts" (ibid., 98). But this line of thought was virtually stifled by his determinism. He insisted that we must not be "misled by the just described immaterial nature of abstract motives which consist of mere thoughts" (ibid., 42), for, in a case where "mere thoughts [are] struggling with other thoughts, until the most powerful of them becomes decisive and sets the man in motion, [the process] takes place with just such a strict causality as when purely mechanical causes work on one another" (ibid., 46). If Schopenhauer seemed briefly to hold out the prospect that a man might, by dint of his distinctively human capacity for a kind of thinking that is not tied to his current perceptions and which may have a touch of imagination and originality, break free from the fetters of his past, and perhaps enrich his little corner of the world by contributing something new to it, it was soon snuffed out. The reign of determinism means that there is nothing new under the sun:

> [A] man's character remains unchanged and . . . the circumstances whose influence he had to experience were necessarily determined throughout and down to the last detail by external causes, which always take place with strict necessity and whose chain, entirely consisting of likewise necessary links,

continues to infinity. Could the completed life course of such a man turn out in any respect, even the smallest, in any happening, any scene, differently from the way it did?—No!

... Our acts are indeed not a first beginning; therefore nothing really new comes into being through them. (Ibid., 62)

Let us now consider the bearing of Schopenhauer's determinism on the idea of self-determination. Suppose (in line with Schopenhauer's view but, I believe, contrary to fact) that the sustained course of action of a person *X* could be chopped up into a series of separate, voluntary acts each of which *X* could individuate, and such that, immediately prior to each of them, *X* is conscious of *willing* so to act. Then a full-fledged determinism allows two possibilities. One is what Schopenhauer envisaged, namely, that such a volition determines *X*'s body to move accordingly, the volition itself being an effect of a causal chain stretching back indefinitely: The chain predetermines the volition which in turn determines the act. The other possibility is that what *X* experiences as a volition is only an epiphenomenon: The chain predetermines a volition-experience which, however, does not determine the act, but is a side effect of physical processes on which it has no effect. Is there any significant difference between these two possibilities?

From that antecedent causal chain, select a segment, call it *C*, that terminates before *X* was born; denote by *V* and *B* the volition-experience and the ensuing bodily movements involved in *X*'s act. In both scenarios *B* is precisely predetermined by *C*, which is absolutely external to *X*. Does it matter whether the causal arrows go from *C* to *B* via *V*, or from *C* to *B* and *V* with no arrow from *V* to *B*? The two cases would be empirically indiscernible. As David Armstrong put it, suppose that "a physical event in the [central nervous] system brings about a spiritual event which brings about a further physical event. . . . There are then laws to be discovered which permit us . . . to predict in principle the second physical event from the first physical event. Now how in this case do we decide that a spiritual event does play a part in the causal chain?" (1968, 360).

The evolutionist argument for the efficacy of the mental that I touched on earlier calls for a *robust* interactionism, one which affirms that an animal's consciousness helps it to perform better, with respect to survival and reproduction, than it would have done if its consciousness made no difference to its behavior. Determinism excludes robust interactionism. Suppose that we *knew* determinism to be true, but did not know whether epiphenomenalism or interactionism is true. Then it would be certain that *V* is determined by *C* and likewise that *B* is determined by *C*; it would be uncertain whether the causal processes that lead to *B* by-pass *V* or make a little detour through *V*. Suppose they make that detour; then there is interaction, but the behavior is what it would have been if *V* had been by-passed and were merely an epiphenomenal volition-experience. I conclude that, although Schopenhauer and Grünbaum were formally correct in holding that determinism can be combined with a denial of epiphenomenalism and the acceptance of dualist interactionism, determinism renders the latter virtually indistinguishable from epiphenomenalism.

The Switch to Indeterministic Laws

Are what Grünbaum calls irreducibly statistical laws, namely, "laws whose statisticality does not arise from human ignorance of relevant hidden parameters" (1972, 613), any more compatible with human freedom and responsibility than deterministic laws? He claimed that they are on a par with deterministic laws in this respect, both kinds of law being fully compatible with feelings of freedom, assignments of responsibility, and so on. Here he diverged from Hume, who had said that only deterministic laws are compatible with responsibility, because there is no middle possibility between causal necessity and sheer chance (1739, 171), and the introduction of chance factors into volitional processes would sever the connection between a man's actions and his character (p. 411). I will now suggest that irreducibly statistical laws may be of two kinds: a very strong kind and a weaker kind. Against Hume I will then argue that laws of the weaker kind open up a middle possibility between chance and necessity (which will be a development of an argument first presented in Watkins 1975). Against Grünbaum I will argue that a statistical law of the weaker kind may be compatible, whereas both a deterministic law and a statistical law of the strong kind would be incompatible, with a degree of freedom or self-determination.

Let a law assign a probability of one half to the outcome *B* at time t_2, given only that initial conditions *A* obtain at the earlier time t_1. We may abbreviate this to $P(B, A) = 0.5$. Then I say, very much in line with what Grünbaum says about irreducible statisticality, that this law is *indeterministic* if the theory in which it is embedded rules out the possibility that further, as yet unspecified, conditions Y obtain at t_1 such that $P(B, AY) = 1$ or 0; in other words, even a Laplacean Demon who took into account not just condition A but *all* conditions obtaining at t_1 would still be unable to predict either B or $\neg B$. (Restrictions have to be placed on the range of Y. As well as obtaining at or before t_1 and being compatible with A, the conditions over which Y ranges must be what I call future-independent; being now pregnant is future-independent whereas being now a mother-to-be is not—see Watkins 1984, sec. 6.22.) A very strong kind of indeterministic law adds that, no matter what other conditions may coexist with A at t_1, the probability of the outcome B *is always the same*—say 0.5. We may represent such a law by

$$(\forall Y)P(B, AY) = 0.5. \tag{22.1}$$

Disintegration laws are perhaps of this strong kind (see Watkins, 1984, 231–32). It would be an indeterministic law of the other, weaker kind if it said only that, no matter what other conditions may coexist at t_1, the probability of the outcome B never sinks to 0 or rises to 1. We may represent this by

$$(\forall Y)\ 0 < P(B, AY) < 1. \tag{22.2}$$

In order to bring deterministic laws into line with these equations, we might represent one that says that the presence of condition A at time t_1 necessitates outcome B at time t_2, by

$$(\forall Y)P(B, AY) = 1, \tag{22.3}$$

with restrictions of the kinds indicated always being placed on the range of Y.

Now if all scientific laws were of type (22.1) or type (22.3), Hume would have been right: There would be no middle possibility between chance and necessity. But with laws of type (22.2) a new possibility opens up. You may be said to have an "iron control" over some outcome B if you can send the probability of its occurring to 1 or 0 (an on-off switch may make such an "iron control" possible). At the opposite extreme, you have no control over the outcome *B* if you cannot vary its probability at all. Now consider an intermediate possibility. Let a knob control the width of a slit in a screen, on one side of which is a source that emits electrons, and on the other a photographic plate. Assume that at one extreme setting of the knob, the probability that an electron that passes through the slit hits a certain area of the plate is raised to 0.99, while at the opposite extreme it is lowered to 0.01. I am going to fire electrons, one at a time, until 1000 have passed through the slit. What sort of control do I have over the number of hits in that area? Not an "iron control," obviously; I cannot determine the aggregate outcome exactly. But I can influence it rather strongly. Popper (1972, chap. 6) introduced the idea of what he called "plastic control," intermediate between "iron control" and no control; this knob gives me a "plastic control."

I now turn to the human scene. Let *A* now denote certain conditions, obtaining at a certain time, that are external to me in the strong sense that I had no hand or say in their presence (thus the presence on my shelves of *Philosophical Problems of Space and Time* and *The Foundations of Psychoanalysis* is not in this sense external to me); let *B* denote possible behavior of mine at a later time; and let *M* be a variable ranging over items in my consciousness. Suppose, first, that *B* is bound to *A* by a deterministic law of type (22.3). Then I am not at all self-determining when I engage in *B*, for the supposition that there is something in my consciousness that is causally relevant to *B* is excluded by the following consequence of (22.3):

$$(\forall M)P(B, AM) = P(B, A) = 1,$$

which tells us that any such *M* is causally irrelevant to *B*, which is fully determined by external factor *A*. Suppose, second, that *B* is related to *A* by an indeterministic law of the strong type (22.1); a shift from a deterministic to an indeterministic coupling of this kind cannot bring about any enhancement of freedom since (22.1) implies

$$(\forall M)P(B, AM) = P(B, A) = 0.5,$$

which again tells us that any such *M* is causally irrelevant to *B*. Now suppose that the switch is to an indeterministic law of the weaker type (22.2). That would be consistent with, say,

$$(\exists M)1 > P(B, AM) > P(B, A) > 0,$$

which says that there is an M that is causally relevant to B although it does not suffice to determine B. Thus M might be a *strong resolve* to do B. So a switch to indeterministic laws may have a positive significance for the question of human self-determination, contrary to Hume's claim that the introduction of chance would only worsen the human condition and to Grünbaum's claim that it would not improve it.

The Amenability of Human Behavior to Scientific Explanation

I turn now to the naturalistic premise of Grünbaum's compatibilism, namely, the thesis that all human behavior is open to scientific explanation with the help of causal laws. I mentioned Darwin's view that *inventiveness* has been decisively important for our species, and I will now suggest that this inventiveness defeats that naturalistic premise. By way of example, consider Michelangelo and the painting of the ceiling of the Sistine Chapel. Gombrich reports how reluctant Michelangelo was to take on "this thankless commission" ([1950] 1984, 232). When the pope insisted, "he started to work out a modest scheme of twelve apostles in niches, and to engage assistants from Florence to help him with the painting. But suddenly he shut himself up in the chapel, let no one come near him, and started to work alone on a plan which has indeed continued to 'amaze the whole world' from the moment it was revealed" (ibid.).

Obviously, neither Michelangelo nor anyone else could have accurately predicted, before he shut himself up in the chapel, the ceiling's future appearance. You cannot predict something which you do not have the conceptual resources to describe; and the marvellous new ideas that would inspire the painting of the ceiling did not yet exist. But as Grünbaum (1972), with Popper (1950) and MacKay (1967) in mind, rightly insisted, "Determinism must be distinguished from predictability, since there are . . . situations in which there may be no predictability for special *epistemic* reasons even though determinism is true" (p. 606).[2] So let us ask whether, Michelangelo's painting of the creation of Adam, say, is open, at least in principle, to retrospective scientific explanation.

Kant had a view about creativity which seems right and which implies (though I wonder how Kant reconciled this with his own determinism) that Michelangelo's achievement was *not* open to scientific explanation. Let x be the occurrence of some unprecedented event. Thus x might be the first explosion of an atom bomb. A scientific explanation of x, if there is one, consists of premises depicting laws of nature together with a (perhaps very large) set of initial conditions. If true, such an explanation shows both how x was in fact produced and, at least in principle, how it could have been produced at an earlier (or later) time; for it implies that x occurs whenever such a set of initial conditions is assembled. In short, a scientific explanation of x would provide a recipe for the production of x (see Briskman 1980, 84–85). It might have been impracticable to assemble such a set of initial conditions at any earlier time; but a recipe for a cake does not stop being a recipe just because some of the ingredients it calls for are presently unobtainable.

So a scientific explanation of Michelangelo's painting of the creation of Adam would be a (perhaps very cumbersome) recipe, not for duplicating the picture subsequently, which would be without philosophical significance, but for its first creation. Now Kant insisted that there can be no such recipe, "Where an author owes a product to his genius, he does not himself know how the *ideas* for it have entered into his head, nor does he have it in his power to invent the like at pleasure, or methodically, and communicate the same to others in such precepts as would put them in a position to produce similar products. . . . [It is something] for which no definite rule can be given" ([1790] 1928, sec. 46; see also D'Agostino 1986, 163f.).

If Kant was right, the products of artistic creativity constitute a (perhaps the) major crux for determinism.[3] A determinist is obliged to hold that such "creativity" is not truly creative; it cannot involve any bringing into existence ex nihilo. As we saw, Schopenhauer asserted that nothing new ever comes into being through human deeds. Before that, Spinoza had repudiated the suggestion that in the creation of paintings, temples, and such, the human body is under the direction of an initiating mind; he likened their creators to sleepwalkers whose feats are achieved "without the direction of Mind," (Curley 1985, 496). Spinoza's combination of determinism and parallelism implied that his fingers absentmindedly wrote his books for him. True, when they, for example, crossed out a passage, this would be paralleled by the thought, say, that this passage is inconsistent with an earlier one, *but this thought had no causal influence on those finger movements*.

More recently, Ted Honderich has insisted that determinism, though consistent with the idea that human actions are voluntary, is incompatible with the idea that they ever *originate* anything (1988, 385–91). According to determinism, the future appearance of the ceiling of the Sistine Chapel was already causally predetermined when Michelangelo was still resisting that "thankless commission", which implies that the marvellous new ideas for it that he began to form soon afterwards were only late links in an indefinitely long causal chain, and have no more claim to a seminal, formative role that antecedent factors much earlier in the chain. The determinist's view of human creativity is like a divine-inspiration theory in subordinating artists to something external to them, though taking the divinity out of this. By contrast, a robust interactionism, coupled with a belief in human inventiveness, involves a macroindeterminism because of its insistence that novel ideas, the occurrence of which is not open to scientific explanation, may trigger and guide bodily performances.

Epicurus's Challenge

In conclusion I turn to Grünbaum's handling of an old and much-discussed argument against determinism,[4] which he presented as follows: "[I]t is first pointed out rightly that determinism implies a causal determination of its own acceptance by its defenders. Then it is further maintained, however, that since the determinist could not, by his own theory, help accepting determinism under the given conditions, he can have no confidence in its truth. Thus it is asserted that the determinist's acceptance of his own doctrine was forced upon him" (1972, 617). His answer was, "The causal generation of a belief does not, of itself, detract in the least from its truth" (ibid.). But on my reading of it, this argument says that determinism impugns, not its own *truth*, but the *rationality* of accepting it. Epicurus declared, "The man who says that all things come to pass by necessity cannot criticize one who denies that all things come to pass by necessity: for he admits that this too happens of necessity" (Bailey 1926, 113); I read that as saying that there could be no rational discrimination between determinism and indeterminism if all intellectual choices were causally predetermined.

Grünbaum's answer to this was, in effect, that causes and reasons may run in tandem, "[T]he decisive cause of the acceptance of determinism by one of its adherents may have been his belief that the available evidence supports this doctrine" (1972, 617). An interactionist can readily agree that reasons may be causes, but this poses a severe problem for a deterministic interactionism. To my earlier claim that a deterministic interactionism is virtually indistinguishable from epiphenomenalism I now add that it calls for a preestablished harmony. Consider the following example (see also Watkins 1985, 17f.). Two airline passengers strike up a conversation. This is not like one of Harold Pinter's dialogues with people talking past each other; each responds appropriately. Thus when she asks, "Do you know what time it is in New York," he does not say, "I wish they would bring the drinks trolley", he glances at his watch and says, "It'll be 6:30 in the morning over there." Later, she laughs when he says something funny. Let e_1 be movements of her lips, tongue, larynx, and so on when asking him what time it is in New York, and let e_2 be the corresponding movements when he replies. According to determinism, e_1 and e_2 are end-results of indefinitely long causal chains. Let C be the physical state of the universe at t_0, say a billion years ago. The following coordination problem arises: Why should C, in conjunction with the laws of nature, as well as predetermining the occurrence of e_1 a billion years later, *also* predetermine the subsequent occurrence of e_2 which expresses a *meaningful and appropriate response to the meaning* expressed by e_1? Unless God was taking a hand, there was no prevision at t_0 of the causal consequences of C a billion years later, nor was there any cognizance of meanings then.

A determinist might reply along the following lines: "In conjunction with the laws of nature, C determined the future existence of macrostructures of various kinds; in particular, it determined that certain atoms would form into giant molecules, including nucleic acid ones that are self-replicating; that these would result in the evolution of biological structures among which there would eventually be language-using animals; and among these there would be ones who, subjected to stimulus e_1, would make response e_2."

Yes, but according to determinism, no indeterminacy entered during this long progression. A Democritean atomist would have said that, while the macrostructures formed by interlocking atoms do all sorts of things that individual atoms do not do, the career of each atom remains perfectly determinate, and a sophisticated present-day determinist should agree with the underlying idea here. According to determinism, the physical states of the universe during the dialogue between our two airline passengers were predetermined by C at t_0 *in every detail*. Among those details were the factors that constituted e_1 and e_2; that these were constituents of larger wholes does not alter this. The question remains: Why should C have predetermined an e_2 that was, in the sense indicated, *appropriate* to e_1?

So far as I can see, there are only two ways of surmounting this difficulty. One is to combine interactionism with an indeterminism that allows for the physically disturbing effects of human decisions, especially ones inspired by novel ideas.[5] The other is to combine determinism with the theological postulate of a divinely preestablished harmony; and that is surely something that the dear old atheist we are celebrating here would never do.

NOTES

1. I borrow this example from Noretta Koertge.

2. John Earman (1986, 9, 65) takes Popper to task for confounding the ontological issue of determinism with the epistemological issue of predictability.

3. I would have included the products of scientific creativity as well, but Kant excluded even Newton's "immortal work on the Principles of Natural Philosophy" ([1790] 1928, sec. 47) on the erroneous ground that what is accomplished in science "all lies in the natural path of investigation and reflection according to rules" (ibid.).

4. There is an impressively long list of thinkers who have discussed it in Honderich (1988, 361). Popper (1982, 81) suggested that there are anticipations of this argument in Descartes and St. Augustine, but I did not find them.

5. For example, the decision reached by President Roosevelt in October 1939 after Alexander Sachs had managed to get him interested in a letter, composed by Szilard and signed by Einstein, about the need for atomic research: "Pa," he exclaimed to his aide General "Pa" Watson, "this requires action!" See Jungk (1960, 106–07).

REFERENCES

Armstrong, D. 1968. *A Materialist Theory of the Mind*. London: Routledge & Kegan Paul.

Bailey, C. 1926. *Epicurus: The Extant Remains*. Oxford: Clarendon Press.

Briskman, L. 1980. "Creative Product and Creative Process in Science and Art." *Inquiry 23:* 83–106.
Curley, E., ed. 1985. *The Collected Works of Spinoza.* Vol. 1. Princeton: Princeton University Press.
D'Agostino, F. 1986. *Chomsky's System of Ideas.* Oxford: Clarendon Press.
Earman, J. 1986. *A Primer on Determinism.* Dordrecht: Reidel.
Gombrich, E. H. [1950] 1984. *The Story of Art.* 14th ed. Oxford: Phaidon Press.
Grünbaum, A. 1952. "Causality and the Science of Human Behavior." *American Scientist 40:* 665–76.
———. 1972. "Free Will and Laws of Human Behavior." In H. Feigl, W. Sellars, and K. Lehrer, eds., *New Readings in Philosophical Analysis.* New York: Appleton-Century-Crofts, pp. 605–27.
———. 1987. "Psychoanalysis and Theism." *The Monist 70:* 152–92.
Honderich, T. 1988. *A Theory of Determinism: The Mind, Neuroscience, and Life-Hopes.* Oxford: Clarendon Press.
Hume, D. 1739. *A Treatise of Human Nature.* Edited by L. A. Selby-Bigge. Oxford: Clarendon Press.
Jungk, R. 1960. *Brighter Than a Thousand Suns.* Translated by J. Cleugh. Harmondsworth: Penguin Books.
Kant, I. [1790] 1928. *Critique of Judgement.* Translated by J. C. Meredith. Oxford: Clarendon Press.
MacKay, D. M. 1967. *Freedom of Action in a Mechanical Universe* (Eddington Memorial Lecture). Cambridge, England: Cambridge University Press.
Popper, K. R. 1950. "Indeterminism in Quantum Physics and in Classical Physics." *The British Journal for the Philosophy of Science 1:* 117–33, 173–95.
———. 1972. *Objective Knowledge.* Oxford: Clarendon Press.
———. 1982. *The Open Universe: An Argument for Indeterminism.* Edited by W. W. Bartley, III. London: Hutchinson.
Ryle, G. 1949. *The Concept of Mind.* London: Hutchinson.
Schopenhauer, A. [1841] 1960. *Essay on the Freedom of the Will.* Translated by K. Kolenda. Indianapolis: Bobbs-Merrill.
Watkins, J. 1975. "Three Views Concerning Human Freedom." In R. S. Peters. ed., *Nature and Conduct.* London: Macmillan, pp. 200–28.
———. 1984. *Science and Scepticism.* Princeton: Princeton University Press, London: Hutchinson.
———. 1985. "Second Thoughts on Landé's Blade." *Journal of Indian Council of Philosophical Research 2:* 13–19.

23

Creation, Conservation, and the Big Bang

Philip L. Quinn
Department of Philosophy, University of Notre Dame

In a recent paper Adolf Grünbaum has argued that those who think recent physical cosmology poses a problem of matter-energy creation to which there is a theological solution are mistaken (see Grünbaum 1989). In physical cosmology, creation is, he claims, a pseudoproblem. My aim in this essay is to refute that claim.

The essay is divided into five sections. First, I provide some background information about the theological doctrine of divine creation and conservation. By citing both historical and contemporary philosophers, I try to make it clear that there is widespread and continuing agreement among philosophers committed to traditional theism that all contingent things depend for their existence on God whenever they exist and not just when they begin to exist if they do have beginnings. I explicate the particular doctrine of creation ex nihilo set forth by Thomas Aquinas to account for the existence of contingent things that do begin to exist. Second, I propose a fairly precise formulation of the doctrine of divine creation and conservation. My way of stating it is meant to be faithful to the leading ideas in the historical and contemporary sources I have cited, and it has the Thomistic doctrine of creation ex nihilo as a special case. Third, I discuss two classical big bang models of cosmogony, one including an initial singularity and the other not. I argue that only the latter model is inconsistent with the Thomistic doctrine of creation ex nihilo. I go on to contend, however, that neither model is inconsistent with the general account of creation and conservation I have formulated. Fourth, I criticize the view that neither big bang model leaves anything that could be explained unexplained and that therefore there is no explanatory work that the doctrine of creation and conservation could do. I argue that in each of these cosmogonic models there is something that either has no explanation at all—if only scientific explanation is allowed—or is explained in terms of causes external to the physical cosmos—if the appeal to theological explanation succeeds. Fifth, and finally, I sketch an extension of the discussion from big bang cosmogonies to steady state and quantum cosmologies. I suggest that they too leave open the question of whether the existence and persistence of matter-energy are inexplicable or are explained by something like the traditional doctrine of divine creation and conservation.

Historical Background

How are we to conceive the manner in which contingent things are supposed to depend for their existence on God? A minimalist account of this dependence is deistic in nature. On this view, God brings all contingent things into existence by creating them. Once having been created, such things continue to exist and operate on their own, without further support or assistance from God. In other words, God is like a watchmaker: Once the watch has been made and wound up, it persists and goes on ticking without further interventions by its maker. Grünbaum assumes this view will shape a creationist reading of big bang theories. It is presupposed in his remark that "God has been thus unemployed, as it were, for about 12 billion years, because the big bang model of the general theory of relativity features the conservation law for matter-energy, which obviously precludes any nonconservative formation of physical entities" (1989, 376).

It is not now and has not in the past been the view typical of philosophically reflective theists. Going well beyond deistic minimalism, they hold that all contingent things are continuously dependent upon God for their existence. On this view, God not only creates the cosmos of contingent things but also conserves it in existence at every instant when it exists. And divine creation and divine conservation both involve the same power and activity on the part of God. In a famous passage from the *Meditations,* which Grünbaum quotes, Descartes says, "It is as a matter of fact perfectly clear and evident to all those who consider with attention the nature of time, that, in order to be conserved in each moment in which it endures, a substance has need of the same power and action as would be necessary to produce and create it anew, supposing it did not yet exist, so that the light of nature shows us clearly that the distinction between creation and conservation is solely a distinction of the reason" (Descartes 1955, 168). Bishop Berkeley writes to Samuel Johnson, "Those who have all along contended for a material world have yet acknowledged that *natura naturans* (to use the language of the Schoolmen) is God; and that the divine conservation of things is equipollent to, and in fact the same thing with, a continued repeated creation: in a word, that conservation and creation differ only in the *terminus a quo*" (1950, 280).

Leibniz repeatedly endorses this idea. In the *Theodicy* he says, "[W]e must bear in mind that conservation by God consists in the perpetual immediate influence which the dependence of creatures demands. This dependence attaches not only to the substance but also to the action, and one can perhaps not explain it better than by saying, with theologians and philosophers in general, that it is a continued creation" (1952, 139). In the *New Essays Concerning Human Understanding* he remarks, "Thus it is true in a certain sense, as I have explained, that not only our ideas, but also our sensations, spring from within our own soul, and that the soul is more independent than is thought, although it is always true that nothing takes place in it which is not determined, and nothing is found in creatures that God does not continuously create" (1949, 15–16). In the correspondence with Clarke he claims, "The soul knows things, because God has put into it a principle representative of things without. But God knows things, because he produces them continually" (1956, 41).

Joining the chorus, Jonathan Edwards insists, "God's upholding created substance, or causing its existence in each successive moment, is altogether equivalent to an *immediate production out of nothing,* at each moment, because its existence at this moment is not merely in part from God, but wholly from him; and not in any part, or degree, from its antecedent existence" (1970, 402). So there was striking agreement among the major theistic philosophers of the seventeenth and eighteenth centuries that God not only creates all contingent things but also conserves them in existence, moment by moment, in a way that is tantamount to continuously creating or recreating them.

Nor is this view of divine creation and conservation of merely historical or antiquarian interest. It also commands the loyalty of leading contemporary theistic philosophers. After asking his reader to consider as an example a physical object, a part of the natural world such as a tree, George Mavrodes observes, "If Christian theology is correct, one of the characteristics which this tree has is that it is an entity whose existence is continuously dependent on the activity of God. The existence of something with this characteristic is a perfectly reliable sign of the existence of God" (1970, 70). David Braine makes the point vivid by comparing God to a painter:

> It is a corollary of the reality of time that the past and present can have no causal role in that continuance of the substance of things—the continuance of the stuff of the world and the regularities or "nature" on which its continuity depends—which is involved in the future's becoming present. It follows from this that the same power is involved in the causing of this continuance of things (if any cause is needed) as would be involved in the causing of the coming to be of things from nothing. As it were, if the continuance of the world is the result of the continuance of a divine brush stroke, then the same power is exercised at each stage in the continuance of the brush stroke as would have been exercised in the beginning of the brush stroke—if the brush stroke had a beginning. (1988, 180)

As one would expect, Braine then proceeds to offer an argument which, he claims, "compels acceptance that continuance into the future must be caused" (ibid., 198), and from this he concludes that "so-called secondary or natural causation has its reality grounded in the primary causality exercised immediately to every time and place by the First Cause" (ibid.). Richard Swinburne considers the doctrine of creation and conservation to be constitutive of the theistic conception of God. The first volume of his impressive philosophy of religion trilogy begins, "By a theist I understand a man who believes that there is a God. By a 'God' he understand [*sic*] something like a 'person without a body (i.e., a spirit) who is eternal, free, able to do anything, knows everything, is perfectly good, is the proper object of human worship and obedience, the creator and sustainer of the universe' " (1977, 1).

Because he thinks God would hardly be less worthy of worship if he on occasion permitted some other being to create matter, Swinburne attributes to the theist the belief that "God's action (or permission) is needed not merely for things to begin to exist but also for them to continue in existence" (ibid., 128–29). He acknowledges that theists like Aquinas have held that revelation teaches us that the universe is not eternal but had a beginning of existence, but he supposes that this doctrine is not as important in their thought as the doctrine that God is responsible for the existence of the universe whenever it exists. So he attributes to theists the view that "God keeps the universe in being, whether he has been doing so for ever or only for a finite time" (ibid., 129).

The views of Aquinas about creation and conservation are worth examining in more detail. He holds that all contingent things are preserved in being by divine action and not merely by divine permission. All creatures stand in need of such divine preservation, he thinks, because "the being of every creature depends on God, so that not for a moment could it subsist, but would fall into nothingness were it not kept in being by the operation of Divine power" (Aquinas 1981, 511). An interesting objection to this view can be based on the fact that even finite agents can produce effects that persist after they have ceased to act. Thus, to use the examples Aquinas gives, houses continue to stand after their builders have ceased building them, and water remains hot for some time after fire has ceased to heat it. Since God is more powerful than any finite agent, it would seem to be within his power to make something that continues to exist on its own after his activity has ceased. In reply to this objection, Aquinas denies the legitimacy of the comparison of finite agents, which merely cause changes in things that already exist and so are causes of becoming rather than of being, and God, who is the cause of the being of all contingent things. He insists that "God cannot grant to a creature to be preserved in being after the cessation of the Divine influence: as neither can He make it not to have received its being from Himself" (1981, 512). However, just as God acted freely and not from natural necessity in creating contingent things in the first place, so also his conserving activity is free, and any contingent thing would forthwith cease to exist if he were to stop conserving it. This leads Aquinas to return a positive answer to the question of whether God can annihilate anything, "Therefore, just as before things existed, God was free not to give them existence, and not to make them; so after they have been made, He is free not to continue their existence;

and thus they would cease to exist; and this would be to annihilate them" (ibid., 514). Annihilation would, of course, not involve a positive destructive act on God's part; it would consist in the mere cessation or withdrawal of his positive conserving activity.

Since Aquinas believes that contingent things have existed for only a finite amount of past time, it might be thought that his view is that the contingent cosmos was created after a prior period of time in which it did not exist. Some of the things he says about creation ex nihilo suggest this doctrine. In an article devoted to the question of whether angels were produced by God from eternity, he maintains that "God so produced creatures that He made them *from nothing;* that is, after they had not been" (ibid., 302). An objection to the view that it is an article of faith that the world began consists of a purported demonstration that the world had a beginning which concludes, "Therefore it must be said that the world was made from nothing; and thus it has being after not being. Therefore it must have begun" (ibid., 242). And, in replying to this objection, Aquinas remarks, "Those who would say that the world was eternal, would say that the world was made by God from nothing, not that it was made after nothing, according to what we understand by the word creation, but that it was not made from anything" (ibid., 243). Again, an objection to the thesis that creation is not a change goes as follows, "Change denotes the succession of one being after another, as stated in *Phys.* v, 1: and this is true of creation, which is the production of being after non-being. Therefore creation is a change" (ibid., 1952, 88). Such passages as these give the impression that, for Aquinas, being created ex nihilo entails existing after a time at which one did not exist. If this impression were accurate, Aquinas would be committed to the view that there was a time when nothing contingent existed before the first contingent things existed.

But Aquinas says things flatly inconsistent with the view that such an entailment exists. Commenting on the first verse of Genesis, according to which in the beginning God created heaven and earth, he contends that time itself was one of the contingent things God created in the beginning. As he interprets scripture, "[F]our things are stated to be created together—viz., the empyrean heaven, corporeal matter, by which is meant the earth, time, and the angelic nature" (ibid., 1981, 244). To the objection that time cannot be created in the beginning of time because time is divisible and the beginning of time is indivisible, Aquinas replies, "Nothing is made except as it exists. But nothing exists of time except *now*. Hence time cannot be made except according to some *now;* not because in the first *now* is time, but because from it time begins" (ibid., 245). This reply commits Aquinas to the view that one of the things God made in the beginning was a unique first *now,* from which time began. If this is right, there were no times prior to that time. And since that *now* was made by God together with the empyrean heaven, earth and angels, those contingent things existed at that time. Hence the four first contingent things are, according to Aquinas, such that God made them, and yet there was no time prior to their being made and so no such prior time at which they did not exist.

It is therefore easy to see why Aquinas holds that the creation of the four first things is not a change. In the course of setting forth his views on this topic, he distinguishes three kinds of change. The first involves a single subject being changed from one contrary to another and covers the cases of alteration, increase and decrease, and local motion. The second involves a subject which is actual at one end of the change but only potential at the other and covers the cases of ordinary generation and corruption. The third involves no common subject, actual or potential, but merely continuous time, in the first part of which there is one of two contraries, and in the second part of which there is the other. This last kind is not change properly speaking because it involves imagining time itself as the subject of things that take place in time; it is the sort of thing we are referring to when we say that afternoon changes into evening. According to Aquinas, the creation of the four first things belongs to none of these kinds. His reason for excluding it from the last of them is that "there is no continuous time, if we refer to the creation of the universe, since there was no time when there was no world" (ibid., 1952, 90). But we can conceive of or imagine possible times before the first actual *now.* Thus Aquinas makes this concession:

> And yet we may find a common but purely imaginary subject, in so far as we imagine one common time when there was no world and afterwards when the world had been brought into being. For even as outside the universe there is no real [spatial] magnitude, we can nevertheless picture one to ourselves: so before the beginning of the world there was no time and yet we can imagine one. Accordingly creation is not in truth a change, but only in imagination, and not properly speaking but metaphorically. (Ibid.)

The creation of the four first things is thus not literally but only metaphorically to be counted as a case of change.

Davidsonian charity dictates that we attribute a consistent view to Aquinas if we can find one. I suggest that the way to accomplish this is to take the slogan that creation ex nihilo involves being after not being figuratively in its application to the four first things. What is literally true of them is that they are such that they have existed for only a finite amount of past time and began to exist at the first time. It is therefore literally false that there were times before they existed and, a fortiori, that their creation ex nihilo involves existence temporally after nonexistence. But because we can in imagination extend time back beyond the first real time, we may speak metaphorically of the creation ex nihilo of even the four first things as involving being after not being if we keep in mind that we really have in view something like being (imagined to be) after (imaginary) nonbeing. On this reading of Aquinas, his view of time is the same in some respects as the one embodied in the big bang model that includes the initial singularity. In both cases, time is metrically finite in the past and is also topologically closed because there is a unique first time.

An Account of Divine Creation and Conservation

In order to lend some precision to the subsequent discussion, I draw on an account of divine creation and conservation I have worked out in Quinn (1983, 1988). I begin by stating my ontological assumptions and notational conventions. I assume time is a linear continuum composed of point instants, with t, t', and t'' as variables of quantification ranging over point instants of time. I also assume that there are concrete individuals, with x as a variable of quantification ranging over them. I leave open the question of which things are genuine individuals. Perhaps items of middle-sized dry goods such as tables and chairs are individuals, but maybe they are mere aggregates composed of genuine individuals. For the sake of convenience, I suppose that in discussions of cosmological models in which matter-energy is conserved it is permissible to speak of the sum total of matter-energy as if it were an individual. Finally, I assume that there are states of affairs. Some states of affairs obtain, and others do not.

My account of creation and conservation rests on the further assumption that there is a special two-place relation of divine bringing about defined on ordered pairs of states of affairs. The primitive schematic locution of the account, which expresses that relation, is this: God willing that *x* exists at *t* brings about *x* existing at *t*. I leave open the question of whether God and his volitions are timelessly eternal by not building into this locution a variable ranging over times of occurrence of diving willings. Since it is a primitive locution, I have no definition of it to offer, but I can provide a partial informal characterization of it. Obviously it must express a relation of metaphysical dependence or causation. Beyond that, I think this relation must have the following marks in order to serve its theological purposes: totality, exclusivity, activity, immediacy, and necessity. By totality, I mean that what does the bringing about is the total cause of what is brought about; nothing else is required by way of causal contribution in order for the effect to obtain. In particular, divine volitions do not work on independently existing matter, and so no Aristotelian material cause is required for them to produce existence. By exclusivity, I mean that what does the bringing about is the sole cause of what is brought about; causal overdetermination is ruled out. By activity, I mean that the state of affairs that does the bringing about does so in virtue of the exercise of some active power on the part of the individual involved in it. By immediacy, I mean that what does the bringing about causes what is brought about immediately rather than remotely through some extended causal chain or by means of instruments. By necessity, I mean that what does the bringing about in some sense necessitates what is brought about.

Using the primitive locution I have adopted, my account of divine creation and conservation is easy to state. It consists of a single, simple postulate:

POSTULATE 23.1. (P) *For all x and t, if x is contingent and x exists at t, then God willing that x exists at t brings about x existing at t.*

According to this account, then, divine volition brings about the existence of every contingent individual at every instant at which it exists and so brings about the existence of every persisting contingent individual at every instant of, and thus throughout, every interval during which it exists. Since this postulate is consistent with there having been contingent individuals for either an infinite past time or only a finite past time, it nicely captures Swinburne's claim that what is important for theists to hold is that God keeps the universe in being, whether he has been doing so forever or only for a finite time.

How are we to distinguish creation from conservation? According to Scotus, "Properly speaking, then, it is only true to say that a creature is created at the first moment (of its existence) and only after that moment is it conserved, for only then does its being have this order to itself as something that was, as it were, there before. Because of these different conceptual relationships implied by 'create' and 'conserve,' it follows that one does not apply to a thing when the other does" (1975, 276). These remarks motivate the following definitions:

DEFINITION 23.1. (D1) *God creates x at t = def. God willing that x exists at t brings about x existing at t, and there is no t′ prior to t such that x exists at t′;*

and

DEFINITION 23.2. (D2) *God conserves x at t = def. God willing that x exists at t brings about x existing at t, and there is some t′ prior to t such that x exists at t′.*

Notice that the first conjunct in the definiens is exactly the same in both definitions. This feature captures the Cartesian claim that creation and conservation require the same power and action on the part of God as well as Berkeley's claim that conservation is in fact the same thing as continued repeated creation. It also captures Braine's claim that, if the continuance of the world is the result of a conserving divine brush stroke, as it were, then the same power is exercised at each stage as would have been exercised at the beginning of the brush stroke if it had a beginning.

Taken together, (P), (D1) and (D2) have these consequences. God creates every contingent individual at the first instant of its existence, if there is one, and only then; God conserves each contingent individual at every other instant at which it exists. So if there are any contingent individuals whose existence lacks a first instant, either because they have existed for infinite past time or because the finite interval throughout which they have existed is topologically open in the past, God never creates those individuals, but he nevertheless conserves them whenever they exist. Thus even such individuals would depend on the divine will for their existence at every instant at which they existed.

It is worth noting that an individual whose existence was brought about by God at the first instant of time, supposing there to be one, would satisfy (D1). Hence being created by God does not entail first existing after a temporally prior period of nonexistence. A definition that does yield such a consequence is this:

DEFINITION 23.3. (D3) *God brings x into existence at t after a prior period of its nonexistence = def. God willing that x exists at t brings about x existing at t, there is no t′ prior to t such that x exists at t′, and there is some t″ prior to t such that x does not exist at t″.*

Being brought by God into existence at a certain time after a prior period of one's nonexistence entails being created by God at that time, but not vice versa. If Aquinas is right that God made the empyrean heaven, the earth and the angels at the first actual *now,* then he created all three of them then but did not then bring any of them into existence after a prior period of its nonexistence, there having been no earlier times to compose such prior periods.

Armed with this account of divine creation and conservation, I next attack the problem of applying it to classical big bang cosmogonic models. Enough has been said by now to make those applications straightforward.

Two Big Bang Models

I follow Grünbaum's presentation in dividing the material to be discussed into two cases. The second but not the first of them is a genuine case of general relativity. In both, matter-energy obeys a conservation law.

As Grünbaum describes it, "*case* (*i*) features a cosmic time interval that is *closed* at the big bang instant $t = 0$, and furthermore, *this instant had no temporal predecessor*" (1989, 389). I agree with Grünbaum that in this case it is not legitimate to ask such questions as these: What happened before $t = 0$? What prior events caused matter to come into existence at $t = 0$? Such questions contradict the assumptions of the case by presupposing that there were times before

$t = 0$. I also agree that the claim that a sudden violation of matter-energy conservation occurred at $t = 0$ presupposes that there were times before $t = 0$ at which there was either no matter-energy at all or, at least, some amount of it different from that which, according to the assumptions of the case, exists at $t = 0$ and thereafter.

Clearly, however, the application of my account of divine creation and conservation to this case need involve no such presuppositions. Treating the constant sum total of matter-energy as a large contingent individual for purposes of discussion, we can deduce from (P) the claim that, for $t = 0$ and every time thereafter, God willing that this matter-energy exists at that time brings about it existing then. On the basis of (D1), we may claim that God creates this matter-energy at $t = 0$ and only then, and, on the basis of (D2), we may also claim that he conserves it at all times thereafter. All these claims seem to be consistent with the assumptions of the case. We would deny an assumption of the case if we claimed that God brings this matter-energy into existence at $t = 0$ after a prior period of its nonexistence, for those assumptions rule out the existence of any such prior period. But we need not make this claim, and, indeed, had better not do so. Theists should not find this constraint disconcerting. Because the past time of this case is isomorphic to the past time of Thomistic cosmology, the example of Aquinas shows us that theists can and sometimes do claim that God creates things ex nihilo at $t = 0$ without then bringing them into existence after a prior period of their nonexistence. So it appears that the Thomistic cosmology, at least to the extent that it applies to physical things such as matter-energy, is wholly consistent with the assumptions of this case.

This is not so in the other case, which Grünbaum describes as follows:

> *Case (ii).* This subclass of big bang models differs from those in Case (i) by excluding the mathematical singularity at $t = 0$ as not being an actual moment of time. Thus, their cosmic time interval is *open* in the past by lacking the instant $t = 0$, although the duration of that past interval in years is finite, say 12 billion years or so. But just as in Case (i), no instants of time exist *before* $t = 0$ in Case (ii). And despite the equality of finite duration of the time intervals in the two models, the crucial difference between Case (ii) and Case (i) is the following: In Case (ii), *there is no first instant of time at all,* just as there is no leftmost point on an infinite Euclidean line that extends in both directions. (1989, 391)

The assumption that there is no first instant of time is, of course, inconsistent with the Thomistic claim that there is a first *now.* However, this inconsistency does not preclude the application of my account of divine creation and conservation to the case. Once again taking the constant sum total of matter-energy to be a large contingent individual, we can deduce from (P) the claim that, for every time after the mathematical singularity at $t = 0$, God willing that this matter-energy exists at that time brings about it existing then. On the basis of (D2), we may go on to claim that God conserves this matter-energy at each of those times. However, because in this case the mathematical singularity is not a time, we must not claim that God creates it at $t = 0$, and a fortiori we must also deny that God brings it into existence at $t = 0$ after a prior period of its nonexistence. So the upshot in this case is that, according to my account, God conserves the sum total of matter-energy whenever it exists, but there is no time at which he creates it or brings it into existence after a prior period of its nonexistence. This claim too seems to be consistent with the assumptions of the case under consideration. In a way, it should not even sound surprising. As Grünbaum has noted elsewhere, Lévy-Leblond has introduced an alternative time-metrization of the case that "confers an infinite duration on the ordinally unbounded past of the Big Bang universe" (Grünbaum 1990, 821). It would, after all, be natural for theists to describe a situation in which the sum total of matter-energy remains constant throughout infinite past time as one in which God always conserves it but never creates it or brings it into existence after a prior period of its nonexistence.

So far I have spoken in a guarded fashion and said only that the claims which result from applying my account of creation and conservation to the two cases being discussed *seem* consistent with their assumptions. But Grünbaum offers an argument which purports to show that this is not so. If it succeeds, the appearances are deceiving. So I must next rebut that argument.

In order to motivate the argument, we may appeal to an historical contrast, which is also cited by Grünbaum, that illustrates the point that physical theories may differ over what they take the unperturbed state or behavior of a system to be. According to Aristotle, Grünbaum tells us:

> When a sublunar body is not acted on by an external force, its *natural,* spontaneous unperturbed behavior is to be at rest at its "proper place", or—if it

is not already there—to move vertically toward it. Yet, as we know, Galileo's analysis of the motions of spheres on inclined planes led him to conclude that the empirical evidence speaks against just this Aristotelian assumption. As Newton's First Law of Motion tells us, uniform motion never requires any external force as its cause; only accelerated motion does. (1989, 386)

Grünbaum invokes this notion of freedom from perturbing influences in the following passage, which refers to the remark by Descartes I quoted earlier:

Similarly, if—as we first learned from the chemist Lavoisier—there is indeed matter-conservation (or matter-energy conservation) in a closed finite system on a macroscopic scale *qua spontaneous, natural, unperturbed behavior of the system,* then Descartes was *empirically* wrong to have assumed that such conservation requires the intervention of an *external cause.* And, if he is thus wrong, then his claim that external divine intervention in particular is needed to keep the table from disappearing into nothingness is based on a false presupposition. (Ibid., 378)

So Grünbaum holds that the presence of a conservation law for matter-energy in our two cases rules out divine conservation of the sum total of matter-energy. If this were so, the claim that God conserves the sum total of matter-energy at all times after $t = 0$ in both cases, which results from applying my account of creation and conservation to them, would be inconsistent with their assumptions.

However, in my opinion, it is not so. We can see why if we reflect a bit on the limits of what we might have learned from Lavoisier. Suppose the empirical evidence supports the hypothesis that the sum total of matter-energy in a closed system remains constant throughout a certain interval of time. The assumption that the system is closed implies that no external physical causes act on it during that interval. But does it also imply that God does not act on the system then? As theists conceive of God, there is no such implication. Divine causality is, as Braine puts it, exercised immediately to every time and place on that conception, and Leibniz too insists that divine causation, when it operates to conserve, is a perpetual immediate influence. That is why, in characterizing the causal relation in my account of creation and conservation, I specified that what does the bringing about causes what is brought about immediately rather than remotely by means of instruments such as secondary physical causes. On this view, it is absurd to suppose that any physical system or region of spacetime is closed in the sense of being isolated from divine influence. Hence what we first learned from Lavoisier does not count as an empirical refutation of Descartes. On the contrary, the view that the sum total of matter-energy in a system remains constant, in the absence of external physical influences on the system, only when God conserves it, and would not continue to do so if he did not, is perfectly consistent.

Of course, one might stipulate that a closed system is one unperturbed and uninfluenced by any external cause, natural or supernatural. But then no empirical evidence can suffice to prove that there are in nature any closed systems in this sense, and so theists can consistently maintain that there are none.

In correspondence, Grünbaum has argued that what I have said so far does not get to the heart of the matter. His key thesis, he says, is that the mere physical closure of a system is causally sufficient for the conservation of its matter-energy because the conservation of matter-energy is a matter of natural law. However, this thesis rests on an understanding of the conservation law that theists of the sort I have have been discussing would reject. They would insist that the sum total of matter-energy in a physically closed system remains constant form moment to moment only if God acts to conserve it from moment to moment. Because they hold that divine conserving activity is causally necessary for the conservation of matter-energy even in physically closed systems, such theists would deny that the mere physical closure of a system is causally sufficient for the conservation of its matter-energy. So they would take the true conservation law to contain an implicit ceteris paribus clause about God's will. When spelled out in full detail, the law is to be understood as implying that if a system is physically closed, then the sum total of matter-energy in it remains constant if and only if God wills to conserve that sum total of matter-energy.

It is important to realize that it does not lie within the competence of empirical science to determine whether Grünbaum's understanding of the conservation law is rationally preferable to the theistic understanding I have sketched. The empirical methods of science could not succeed in showing that divine activity does not conserve the matter-energy in physically closed systems. Of course metaphysical naturalism entails that there is no such divine activity, but metaphysical naturalism is not itself entailed by any of the empirically established results of science. All those results are consistent with both theistic and naturalistic metaphysical worldviews. Hence, since the claims about divine conservation made by Descartes are part of a theistic metaphysics, they have not been shown to be empirically wrong by Lavoisier or by any other scientist.

Thus Grünbaum's argument fails. Knowing of no better argument to the same effect, I believe I am entitled to conclude that my claim that God conserves the sum total of matter-energy at all times after $t = 0$ in both big bang cases not only seems to be, but actually is, consistent with the assumptions of those cases. I further conclude that the application of my account of divine creation and conservation to the two big bang cases yields no results that are inconsistent with their assumptions.

Theism and Explanation

Vindicating the consistency of applications of my account of divine creation and conservation to big bang cosmogonic models with the assumptions of those models will contribute to a positive case for theistic belief only if theism, if true, could explain something that has no scientific explanation. Hence the next question I need to address is whether divine creation and conservation are bound to be explanatorily idle or superfluous in physical cosmology. Is there something about the matter-energy of big bang models that has no explanation internal to them? I think there is.

For the sake of simplicity, I assume that scientific explanation in these classical cosmological models is ideally deductive-nomological in form (see Hempel 1965). Roughly put, the idea is that particular facts are explained scientifically by being deduced from laws together with initial or boundary conditions and laws by being deduced from more general laws. Thus, for example, one might explain why there is a certain total amount of matter-energy at a certain time by deducing a statement of that fact from the conservation law for matter-energy and the statement that the same amount existed at a certain earlier time. Of course, one cannot appeal to the amount of matter-energy that existed at an earlier time in order to explain why there is a certain amount of it at $t = 0$ in Grünbaum's Case (i), there being no times earlier than $t = 0$ in that case. But, for two reasons, I do not think that this is a decisive objection to the claim that science leaves nothing which could be explained unexplained. In the first place, because deductive-nomological explanation can proceed either predictively in terms of conditions that obtain before the explanandum or retrodictively in terms of conditions that obtain after the explanandum, the existence of a certain amount of matter-energy at $t = 0$ in Case (i) can be explained retrodictively in terms of the conservation law and the existence of the same amount at a later time. And, in the second place, the existence of a certain amount of matter-energy at any particular past time can be explained predictively in Grünbaum's Case (ii), which is the interesting case from the point of view of general relativity, in terms of the conservation law and the existence of the same amount at an earlier time, because in this case, for every time, there is an earlier.

So the crucial question is whether there would be anything science could not explain even if, because time is unbounded in the past, it is the case that, for every time, the existence of a certain amount of matter-energy at that time can be deduced from the conservation law and the existence of the same amount at an earlier time. Of course, the question will arise both for the unbounded but metrically finite past time of Grünbaum's Case (ii) and for the case in which matter-energy obeys a conservation law throughout metrically infinite, unbounded past time. In my view, there would be in both these cases something that cannot be explained in deductive-nomological terms. Leibniz (1969) gives a striking example that clarifies this:

For a sufficient reason for existence cannot be found merely in any one individual thing or even in the whole aggregate and series of things. Let us imagine the book on the *Elements of Geometry* to have been eternal, one copy always being made from another; then it is clear that though we can give a reason for the present book based on the preceding book from which it was copied, we can never arrive at a complete reason, no matter how many books we may assume in the past, for one can always wonder why such books should have existed at all times; why there should be books at all, and why they should be written in this way. What is true of books is true also of the different states of the world; every subsequent state is somehow copied from the preceding one (although according to certain laws of change). No matter how far we may have gone back to earlier states, therefore, we will never discover in them a full reason why there should be a world at all, and why it should be such as it is. Even if we should imagine the world to be eternal, therefore, the reason for it would clearly have to be sought elsewhere, since we would still be assuming nothing but a succession of states, in any one of which we can find no sufficient reason, nor can we advance the slightest toward establishing a reason, no matter how many of these states we assume. (P. 486)

Just before he cites this passage, though in a different translation, Swinburne puts the point in terms of explanation and with greater precision in a discussion of the case in which time is metrically infinite in the past. Letting *L* symbolize the laws of nature, he writes:

> Further, the universe will have during its infinite history, certain constant features, F_1 which are such that given that the universe has these features at a certain time and given *L*, the universe will always have them. But they are such that the universe could have had a different set of features F_2 equally compatible with *L*. What kind of features these are will depend on the character of *L*. But suppose for example that *L* includes a law of the conservation of matter, then given that there is a quantity M_1 of matter at some time, there will be M_1 at all times—and not merely that quantity of matter, but those particular bits of matter. Yet compatible with *L* will be the supposition that there was a different quantity M_2 made up of different bits. Then it will be totally inexplicable why the quantity of matter was M_1 rather than M_2. If *L* does not include laws of the conservation of matter, it is hard to see how it could fail to include laws formulable as conservation laws of some kind (e.g., of energy, or momentum, or spin, or even the density of matter). And so a similar point would arise. Why does the world contain just that amount of energy, no more, no less? *L* would explain why whatever energy there is remains the same; but what *L* does not explain is why there is just this amount of energy. (1979, 124–25)

Thus, to return to the Leibnizian example, if there is a law of the conservation of kinds of books, then, given that there are copies of Euclid's *Elements* at some time, there will be copies at all times. But it is compatible with this conservation law to suppose that there were different books than there are. Hence it will be inexplicable by appeal to this law why what is conserved includes Euclid's *Elements* and not Pierre Menard's *Don Quixote* or includes any books at all rather than being a state of booklessness in which no books exist. Similarly, in the big bang cases I have been discussing, given that there is a certain quantity of matter-energy at some time, there will be that quantity at all times. But it is compatible with the conservation law for matter-energy to suppose that there was a different quantity of matter-energy than there actually is. Thus it will be inexplicable by appeal to that law why what is conserved is this particular quantity of matter-energy rather than some other or, for that matter, a state in which there is no matter-energy.

The correct conclusion to draw from these considerations is Swinburne's. Applied to the unbounded but finite past time of Grünbaum's Case (ii), it is that there is no scientific explanation of why there is a certain amount of matter-energy rather than some other amount or none at all, even if, for every time, a statement that this amount exists at that time can be deduced from the conservation law for matter-energy plus a statement that the same amount existed at an earlier time. This is an inexplicable brute fact if only scientific explanation of a deductive-nomological sort is allowed.

There is more. The conservation law for matter-energy is logically contingent. So if it is true, the question of why it holds rather than not doing so arises. If it is a fundamental law and only scientific explanation is allowed, the fact that matter-energy is conserved is an inexplicable brute fact. For all we know, the conservation law for matter-energy may turn out to be a derived law and so deducible from some deeper principle of symmetry or invariance. But if this is the case, the same question can be asked about this deeper principle because it too will be logically contingent. If it is fundamental and only scientific explanation is allowed, then the fact that it holds is scientifically inexplicable. Either the regress of explanation terminates in a most fundamental law or it does not. If there is a deepest law, it will be logically contingent, and so the fact that it holds rather than not doing so will be a brute fact. If the regress does not terminate, then for every law in the infinite hierarchy there is a deeper law from which it can be deduced. In this case, however, the whole hierarchy will be logically contingent, and so the question of why it holds rather than some other hierarchy will arise. So if only scientific explanation is allowed, the fact that this particular infinite hierarchy of contingent laws holds will be a brute inexplicable fact. Therefore, on the assumption that scientific laws are logically contingent and are explained by being deduced from other laws, there are bound to be inexplicable brute facts if only scientific explanation is allowed.

There are, then, genuine explanatory problems too big, so to speak, for science to solve. If the theistic doctrine of creation and conservation is true, these problems have solutions in terms of agent-causation. The reason why there is a certain amount of matter-energy and not some other amount or none at all is that God so wills it, and the explanation of why matter-energy is conserved is that God conserves it. Obviously nothing I have said proves that the theistic solutions to these problems are correct. I have not shown that it is not an inexplicable brute fact that a certain amount of matter-energy exists and is conserved. For all I have said, the explanatory problems I have been discussing are insoluble. But an insoluble problem is not a pseudoproblem; it is a genuine problem that has no solution. So Grünbaum's claim that creation is a pseudoproblem for big bang cosmogonic models misses the mark.

Steady State and Quantum Cosmologies

Grünbaum's treatment of physical cosmology also contains discussions of the steady state theory of Bondi and Gold and of an account of quantum cosmology by Weisskopf. I conclude with brief remarks on these two cases.

The steady state theory on which Grünbaum focuses his attention postulates conservation of matter-density in an expanding universe and so requires a nonconservative accretion of matter over time. According to Grünbaum, the natural, spontaneous and unperturbed behavior of the physical universe presented by this theory conserves matter-density rather than the total quantity of matter and so requires matter-accretion because the universe is expanding. The key point he makes is this, "*Just as a theory postulating matter-conservation does not require God to prevent the conserved matter from being annihilated, so also the steady-state theory has no need at all for a divine agency to cause its new hydrogen to come into being!*" (Grünbaum 1989, 388).

But neither does the steady state theory rule out a divine cause for the coming to be of its new hydrogen. Empirical evidence would warrant the claim that new hydrogen comes into being spontaneously only if spontaneity precluded no more than external physical causes for the postulated matter-accretion. Applied to the new hydrogen of the steady state theory, my account of divine creation and conservation tells us that each new hydrogen atom is either both created and brought into existence after a prior period of its nonexistence by God at the first instant of its existence, if there is one, and thereafter conserved by God, or, if there is a last instant of its nonexistence but no first instant of its existence, conserved by God at all times after the last instant of its nonexistence at which it exists, though neither created nor brought into existence after a prior period of its nonexistence by God. This claim is consistent with the assumptions of the theory. Moreover, the steady state theory, too, poses explanatory problems that science cannot solve. If only scientific explanation is allowed, it will be a brute inexplicable fact that the matter-density has the particular value it does rather than some other value, and the same goes for the fact that a conservation law for matter-density holds or the fact that the deeper laws from which it can be deduced rather than alternatives hold if the conservation principle is a derived law. And, again, if the theistic doctrine of creation and conservation is true, such facts as these have explanations in the divine will. So if the steady state theory were true and such things were facts, theism would not be explanatorily superfluous or idle.

The same is true in the case of quantum cosmology. Following Weisskopf, Grünbaum lays out a cosmogonic scenario. The initial state of the physical cosmos is the true vacuum, which is subject to energy fluctuations but devoid of matter and energy proper. When such fluctuations take place, there is a transition to the false vacuum, which contains energy but not matter. The false vacuum undergoes a rapid inflationary expansion until it reaches a certain size. When it does, the inflationary expansion stops and a true vacuum emerges, but within a microsecond the energy contained in the false vacuum shows up as light and in the form of various particles and antiparticles. Thereafter the universe expands relatively slowly in ways previously familiar, forming in due course atoms, stars, galaxies and so forth. Speaking of the transition from true to false vacuum, Grünbaum says, "Thus, according to quantum theory, this sort of emergence of energy, which is ex nihilo only in a rather Pickwickian sense, proceeds in accord with pertinent physical principles, rather than as a matter of inscrutable external divine causation" (1989, 392).

However, from the fact that, according to quantum cosmology, the emergence of energy proceeds in accord with physical principles it does not follow that there is nothing in this scenario for theism to explain. The litany should by now be familiar. What, if anything, explains the fact that the universe is initially a true vacuum rather than being in some other state? And what, if anything, explains the fact that the particular physical principles in accord with which its evolution from this initial state proceeds rather than others hold? Perhaps these things, if they are facts at all, are inexplicable brute facts. They are if only scientific explanation is allowed. But, then again, perhaps not. If theism is true, they are not.

I confess to a bit of uncertainty about how to apply my account of divine creation and conservation to this scenario, but I will venture a guess. Because the initial true vacuum seems not to be a genuine state of nonbeing, I think it should be described as created by God at the first instant of its existence, if there is one, and conserved by God thereafter for as long as it exists, or, if there is no such first instant, conserved by God as long as it exists. As new individuals come into being later in the scenario, those that have a first instant of existence are then both created by God and brought into existence by God after a prior period of their nonexistence and subsequently conserved by God, and any that have a last instant of nonexistence but no first instant of existence are conserved by God after the last instant of their nonexistence for as long as they exist. And, of course, physical principles describe any empirical regularities there are in the effects of all this divine activity.

I submit that two conclusions are warranted on the basis of the foregoing examination of cosmological theories. First, all the theories surveyed are consistent with my account of divine creation and conservation, and, what is more, none of them gives us any reason for thinking that the theistic doctrine of creation and conservation is false. Second, each of the theories considered gives rise to genuine explanatory problems that science alone cannot solve but that theism, if true, does solve.

In this essay I have not discussed the question of whether physical cosmology provides positive evidential support for the theistic doctrine of creation and conservation. To raise the question is to ask whether there is a successful cosmological argument for the existence of God, and I must reserve for another occasion an exposition of my views on that difficult topic. In closing I will make four brief remarks on the prospects for a successful cosmological argument. First, the version of the argument Grünbaum (1989) criticizes is a particularly simple one, and he makes no effort to show that it was endorsed by any of the historical figures who have made important contributions to philosophical theology. After explicitly noting that there are other versions of the argument, he says "I do *not* claim that my charge of pseudo-problem applies necessarily to all of the questions addressed by these other versions" (ibid., 378). Second, some of these other versions do not suffer from the flaws Grünbaum detects in the version he attacks and are very much worthy of consideration on their own merits. One, to which Grünbaum does refer, is a deductive argument formulated by William Rowe, who finds some of its ideas in an earlier argument constructed by Samuel Clarke. Though he concludes that this argument does not prove the existence of God, Rowe leaves open the possibility that it renders theistic belief reasonable, "I am proposing to the theist that in seeking rational justification for his belief in the conclusion of the Cosmological Argument he would do well to abandon the view that the Cosmological Argument is a proof of theism, and, in its place, pursue the possibility that the Cosmological Argument shows the reasonableness of theistic belief, even though it perhaps fails to show that theism is true" (1975, 269). Another version, to which Grünbaum does not refer, is a Bayesian argument constructed by Swinburne. He argues that the hypothesis of theism is more probable given the existence over time of a complex physical universe than it is on tautological evidence alone, and he further contends that this argument is part of a cumulative case for theism whose ultimate conclusion is that "on our total evidence theism is more probable than not" (1979, 291). Third, because such versions of the argument appeal only to very general features of the cosmos of contingent things, for example, the fact that it consists at least in part of a complex and enduring physical universe, the details of scientific cosmological theories are apt to have little if any bearing on whether such arguments lend theistic belief evidential support, render it reasonable, or even demonstrate its truth. This is, of course, not to say that such arguments are immunized against philosophical criticism or capable of surviving it in the long run. But, fourth, it does suggest that those who are concerned with the fate of cosmological arguments would do better to study philosophical theology and its history than to immerse themselves in contemporary physics.

At the very end of his paper, Grünbaum says that the traditional cosmological argument for divine creation "dies hard" (1989, 393). My reply is that the tradition contains many such arguments and that it is premature to predict the death of them all.

NOTE

I am grateful to Alfred J. Freddoso, Adolf Grünbaum, and Ernan McMullin for helpful suggestions. Grünbaum (1991), which appeared in print after this essay had been completed, contains further discussion of the topic of creation as does the forthcoming Grünbaum (1993); I hope to respond to these articles on another occasion.

REFERENCES

Aquinas, T. 1952. *De Potentia.* Translated by English Dominican Fathers. Westminster, Md.: Newman Press.

———. 1981. *Summa Theologiae.* Translated by English Dominican Fathers. Westminster, Md.: Christian Classics.

Berkeley, G. 1950. *Works.* Vol. 2. Edited by A. A. Luce and T. E. Jessop. London: Nelson.

Braine, D. 1988. *The Reality of Time and the Existence of God.* Oxford: Clarendon Press.

Descartes, R. 1955. *Philosophical Works.* Vol. 1. Edited by E. S. Haldane and G. T. R. Ross. New York: Dover.

Edwards, J. 1970. *Works.* Vol. 3. Edited by C. A. Holbrook. New Haven: Yale University Press.

Grünbaum, A. 1989. "The Pseudo-problem of Creation in Physical Cosmology." *Philosophy of Science 56:* 373–94.

———. 1990. "Pseudo-creation of the Big Bang." *Nature 344:* 821–22.

———. 1991. "Creation as a Pseudo-explanation in Current Physical Cosmology." *Erkenntnis 35:* 233–54.

———. 1993. "Theological Misinterpretations of Current Cosmology." Preprint.

Hempel, C. G. 1965. *Aspects of Scientific Explanation.* New York: Free Press.

Leibniz, G. W. 1949. *New Essays Concerning Human Understanding.* Translated by A. G. Langley. La Salle, Ill.: Open Court.

———. 1952. *Theodicy.* Translated by E. M. Huggard. New Haven: Yale University Press.

———. 1956. *The Leibniz-Clarke Correspondence,* Edited by H. G. Alexander. Manchester: Manchester University Press.

———. 1969. *Philosophical Papers and Letters.* Edited by L. E. Loemker. Dordrecht: Reidel.

Mavrodes, G. 1970. *Belief in God.* New York: Random House.

Quinn, P. L. 1983. "Divine Conservation, Continuous Creation, and Human Action." In A. J. Freddoso, ed., *The Existence and Nature of God.* Notre Dame: University of Notre Dame Press.

———. 1988. "Divine Conservation, Secondary Causes, and Occasionalism." In T. V. Morris, ed., *Divine and Human Action.* Ithaca: Cornell University Press.

Rowe, W. 1975. *The Cosmological Argument.* Princeton: Princeton University Press.

Scotus, J. D. 1975. *God and Creatures: The Quodlibetal Questions.* Translated by F. Alluntis and A. B. Wolter. Princeton: Princeton University Press.

Swinburne, R. 1977. *The Coherence of Theism.* Oxford: Clarendon Press.

———. 1979. *The Existence of God.* Oxford: Clarendon Press.

Moral Problems

24

Moral Obligation and the Refugee

Nicholas Rescher
Department of Philosophy, University of Pittsburgh

Half a century has gone by since the outbreak of war closed the exits from Nazi Germany, so that the particular group of refugees to which Adolf Grünbaum and I belong is now passing from the scene. Jews and non-Jews alike, these people left their homeland under very difficult and painful circumstances. And this group has made impressive contributions to the United States—especially in the areas of science, social thought, scholarship, and medicine (see Cosher, 1984). Still, these refugees represent simply one more link in a long chain extending across the history of this country, past, present, and undoubtedly future as well. It is this general phenomenon that will concern us here.

The issue to be addressed is a straightforward-looking question of applied moral philosophy: Do refugees owe any special debt of moral obligation to the nation and society that provide them with refuge? I observe, with some surprise, that this question is one which I have never seen discussed in the pages of popular moralists or of moral philosophers.

It must be stressed from the start that this discussion concerns itself with duties and obligations that relate to the *moral* rather than *legal* aspects of the issue—not with what the law requires, but with those duties and obligations that inhere in a proper concern for what people owe to one another within the context of a rightly ordered life.

Note to begin with that political or religious refugees comprise a very special sort of immigrant. They are not escapists seeking to escape a pressing creditor or a difficult spouse. They are not comfort-seekers looking for a kinder climate or more congenial surroundings. Nor are they opportunists seeking a better economic environment for employment or investment—or bright opportunities for a flourishing career or an appealing lifestyle. Refugees come to their new homeland not so much because they want to, but because they must. For them, the new homeland is, in the first instance, not an opportunity but a necessity, because it is their refuge against the storm of oppression, discrimination, or persecution.

But what about obligations? When refugees arrive in their new homeland, they enter upon a stage whose action is already well under way, a going concern of facilities and institutions to whose creation—all too obviously—they themselves have made no contribution. Their country provides them with a ready-made environment for living at no further cost or sacrifice to themselves. This, clearly, is a benefit for which new refugee-residents are indebted to others—and they certainly have no moral right or entitlement to expect as a free gift.

Or is this indeed so? It could be said, after all, that this happens to all of us at birth. All of us are, in a way, refugees in this life, emigrating from the state of nonbeing into this ready-made world to whose creation we have nowise contributed.

But there is scope for counterargumentation here, based in the ideal of inheritance. In one's native land, one's forebears have in the past contributed their efforts and energies to making the society a "going concern." Whatever property they accumulated privately they disposed of in their wills as best the laws of the land would let them. And the public fruits of their efforts and energies—their general contribution to the society and its resources—become part of the social legacy of the community at large to be inherited by their fellow citizens. It would clearly be inappropriate to deny the claims of inheritance of those whose families have struggled for generations to bring into being an economy, a society, and a culture of whose resources refugees are beneficiaries as of the day they step across the borders of the land. The "native" can accordingly claim a share of the public goods by right of inheritance. The immigrant can make no such claim. So it is not implausible to argue that refugees owe a significant debt—not merely of gratitude but also of justice—to the nation and society that provide them refuge.

Refugees have found the prior setting of their lives made intolerable for political or social reasons. Casting about for a more viable alternative, they succeed in the end in finding a new "homeland." For refugees, the immediate object of concern is survival. Given the distractions of the moment, it is, no doubt, only with the passage of years that refugees come to realize in distant retrospect what the citizenry of that new homeland are entitled to ask of them—what sorts of expectations the latter are warranted in entertaining and how the entire transaction looks from *their* point of view. But sooner or later conscientious refugees will come around to this recognition. For even the most rudimentary attention to the facts cannot but bring home to refugees that their welfare and well-being—their livelihoods and perhaps their very lives—are due to the willingness of the host country to take them into its midst. The host society extended its hospitality in the hour of need.[1] Owing nothing, it gave much—not only safety, but opportunity as well.

What can be said of the special obligation that results? What can the host society rightfully expect of the refugee?

The debt which refugees owe to the new homeland can be characterized straightforwardly, for it is simply the general obligation of good citizenship—albeit in a particularly substantial degree.

This simple single idea, of course, ramifies out into a score of special obligations: putting one's hands to the work of communal production, obeying the laws of the land, shouldering the burdens of good citizenship (military service, jury duty, voting, and the like). The least that conscientious refugees can do for the society that affords them refuge is to assume cheerfully the obligations of conscientious communal membership. And in particular, conscientious refugees also owe it to the new homeland to support the values and traditions that characterize its social order. It is, after all, these values and traditions that have shaped the society which sheltered and nourished refugees in the hour of need. A combination of gratitude and obligation thus binds them in an implicit social contract of sorts not only to the laws but also to this underlying "spirit of the laws" of the host society.

The obligations at issue also impact in a somewhat diminished way upon that special subcategory of refugees, the exiles, whose "refugeedom" is only temporary—who are, as it were, transit passengers awaiting the possibility of return to a place they continue to regard as home. (In the particular case of American refugees from Nazi Germany, this category was for understandable reasons statistically diminutive.) In point of the demands and obligations incumbent upon refugees at large, their condition differs only in degree and not in kind from that of those not leaving their native land on a more permanent basis.

Do refugees owe the new homeland unquestioning approbation? Surely not. Clearly no moral impediment need prevent refugees from criticizing the host society where it fails to live up to its own ideals. To the contrary, this generic duty of good citizenship should for them be something of a special obligation. However, it should never be allowed to serve as an excuse for a failure of loyalty and attachment to the new-found homeland. So, morally conscientious refugees must in due season shoulder the burden of a social conscientiousness that makes them, insofar as possible, an asset to the community rather than a liability.

There arises, to be sure, the interesting theoretical question of refugees who arrive in a country that goes bad—white Russian refugees, say, who arrive in Germany in time to be caught up in the Nazi disaster. Are we to see the refugees as obliged to become more Nazi than the Führer himself on the preceding principles of gratitude and tacit agreement? Clearly not! The moral weight of such obligations can never override the larger duties of humanity and humaneness that lie at the very core of the moral duties we owe to one another.

These considerations lead to the question: What is the *ground* (basis, rationale) of the obligations of refugees toward the nation and society that provided them with a refuge (asylum) in time of need?

The basic fact is that refugees are a very special sort of immigrant—the kind who, by the nature of their situation, have incurred particularly pressing responsibilities and obligations. These ultimately root in a social contract of sorts, an implicit agreement between refugees and the nation and society that gives them the refuge. A personalized social contract is in operation through a tacit and, as it were, hypothetical bargain: "You take me in; I will bend my efforts to ensure that you will be the better off for my presence."

The analogy of guests is relevant here. They are invited into someone's home and treated as part of the family. To fail to be helpful and cooperative, good natured and patient, and so on, is to run afoul not only of the requirements of good manners, but those of morality as well. Failure in gratitude is a defect that is particularly unseemly in the case of refugees. In their case one can say that patriotism in the highest and most positive sense of the term—as concerned dedication to the country's best interests—is something mandatory.

To be sure, a small-scale tempest of discussions has recently brewed about whether "political obligations" arise from considerations of gratitude (see Plamenatz 1968; Simmons 1979, esp. chap. 7; Card 1988; Walker 1988, 1989; and Klosko 1989, 1991). Several points deserve to be stressed in this connection.

First, the very idea of a "political obligation" has its problems. We all realize what *legal* obligations and what *moral-ethical* obligations are, but the idea of "*political* obligations" confronts us with a puzzle. The only reasonably straightforward way to solve this puzzle is to construe *political* obligation as encompassing the total of our legal and moral obligations to act in a certain way in matters connected with politics. We would, accordingly, class as "political obligations" both a public official's (legal) obligation to provide an accounting for the disbursement of public funds placed at her disposal, and an elected representative's (moral) obligation to respond truthfully to a constitutent's question as to whether or not he has decided to vote for some item of pending legislation.

Given such an understanding of "political obligation" it becomes clear that in many instances gratitude does not actually underwrite a political obligation. For example, one is not obliged—either legally or morally—to vote, in circumstances where a better-qualified rival is on the scene, for a legislator who has, even at one's own urgings, promoted a piece of legislation (a protective tariff, say) from which one has benefited. On the other hand, there will—clearly—be *some* instances where gratitude can engender a "political obligation" in the specified sense. For example, if you and I represent neighboring districts, and I found it helpful to enlist your support in campaigning on my behalf during a difficult election year, it would clearly be morally questionable—in the absence of strong and cogent countervailing reasons—if you were to deny me similar aid on another occasion. In such circumstances you are, clearly, placed under a "political obligation" (in the moral mode) by considerations of gratitude.

Clearly from this standpoint that while various sorts of "political obligation" *cannot* appropriately be grounded in considerations of gratitude, there will indeed be others that can be so grounded. In particular, it would clearly be (morally) inappropriate as an act of ingratitude toward benefits extended for refugees to set at naught those (political) burdens which are part and parcel of good citizenship in the country which extended them the benefit of refuge.[2]

What is at issue here is thus not only primarily a matter of gratitude, but one of justice. That refugees' presence should prove needlessly burdensome for those who sheltered the refugees in the hour of need is clearly unjustifiable. Until they take in refugees, the citizenry of the new homeland owes them nothing; the debts of these hosts—such as they are—are to their predecessors, their neighbors, their fellow citizens. Refugees are an extraneous element in the equation. That refugees should place avoidable burdens upon those who have, alike in generic intent and specific effect, been benefactors would clearly be unconscionable from the moral point of view.

It is illuminating in this regard to ask the question: Just what goes amiss in the moral order when the moral obligations of gratitude that are incurred by refugees are not honored?

1. A violence to virtues: lack of gratitude, of due appreciation for benefits extended;

2. a failure in duties: the violation of tacit agreement, the breech of a social contract—of the tacit bargain struck when the refugee arrived in the new homeland ("You take me in; I will foster your interests and support your cause");

3. a fostering of ill consequences: in particular, a poisoning of the well for others. (How can the nation or society as readily accommodate new refugees when the earlier ones have shown themselves to be facilitators of evil rather than good in perpetrating a kind of social vandalism that diminishes the quality of life for all?)

Refugees who wittingly fail to honor the moral obligations that are incumbent on those who are circumstanced as they are are thereby morally reprehensible in ways that may differ in degree but not in kind from the conditions of those involved in other sorts of moral transgression.

A further issue fraught with moral overtones is the question of refugees' relationship to the former homeland. At first, of course, one can expect little but distaste and dismay. One does not lightly tear up one's roots and abandon the comfortable familiarities of the land of one's ancestors; terrible things must have happened to drive one away. But gradually time works its inevitable changes. A new generation comes to the fore that had no part in the wrongs of the past.

New conditions come about; persons and attitudes, and programs and policies change. In due course (if all goes well) the land of a later generation is a very different land. It becomes inappropriate "to visit the sins of the fathers upon the children." (In point of inheritance the situation of goods and evils is asymmetric from the moral point of view; gratitude and recrimination stand on a different footing.)

In these circumstances, it would clearly be improper for the refugee to let the increasingly irrelevant wrongs of a bygone time intrude upon the changing scenes of a living present. Irrespective of any present feelings, it would be morally improper for the refugee to foster hostility between the new homeland and the abandoned one, letting the bitter experiences of one's own personal past be the occasion for impeding the important and salutary work of a reconciliation among peoples and nations. Among the many obligations that refugees owe their new homeland is surely to avoid making its relations with other countries and peoples less benign and less mutually beneficial than make good sense in the prevailing circumstances. Working actively for a reconciliation is perhaps more than can be asked of them, but they should surely not impede it. Hatred is generally a bad counselor in human affairs—to say nothing of its negativity from the moral point of view that is at issue here.

A particularly complex issue is posed by the question of the refugees' stand with respect to other, later refugees. Just what are the former's moral obligations in this regard?

The issue involves a delicate balance. Having themselves found refuge, it would clearly be inappropriate for refugees to take the negative line on others, encouraging a policy of pulling up the drawbridge and slamming the gates on those who find themselves in a position similar to that which the former refugees occupied a brief time ago. To some extent, it is clearly refugees' duty to help to assure that the opportunities which have proven so critical for themselves are also available to others. On the other hand, it would also be unjustifiable to have a country go overboard with the admission of refugees, to admit more of them than it can absorb without injury to its quality of life and the viability of its cultural, social, and political traditions. In this regard, then, the stance of conscientious refugees must be one of a sensible balance between undesirable extremes.

In summary, the moral dimension of "refugeeship" requires that the attitudes and actions of conscientious refugees should be conditioned by three fundamental principles:

(i) a gratitude that is properly due to those who have extended a helping hand to oneself in one's hour of need—a gratitude of the sort that makes the advancement of their interests part of one's own;

(ii) a due sense of the obligations implicit in the social contract between the refugees and citizenry of the sheltering country that they will not only try to avoid being a burden but will bend their efforts to ensure that the country of refuge would be the better off for their presence;

(iii) a sympathetic and sensitive concern for the condition of others who find themselves in similar distressing circumstances, requiring the safety of a refuge much as the earlier refugees once did.

The fundamental moral factors that are at work here—gratitude, good citizenship, fellow-feeling—are universal virtues. But the particular circumstances of refugees mean that the former exert their bearing upon the refugees in a special and characteristically emphatic way.

Given the nature of the situation, refugees are people to whom the sheltering country has extended benefits "above and beyond the call of ordinary duty." Clearly, then, from the moral point of view at any rate, an acceptance of these special benefits also demands the acceptance of certain special responsibilities, and requires of its beneficiaries an exertion toward productive contributions that go above and beyond this call. The law, of course, makes no distinction between the duties and obligations of native-born citizens and those who came as refugees. In this regard, then, as in various others, the demands of morality will exceed those basic demands that the law makes on everyone alike. It would, of course, be very questionable to contemplate transmuting those moral obligations into legal ones. However, this does not render them any less cogent. (It is one of the regrettable features of the present ethos that many people think—very wrongly indeed—that those obligations that are not subject to the sanctions of the law are for this reason somehow rendered null and void.)

NOTES

1. The case of *illegal* refugees is a different one. Here, that "take them in" and "hospitality" are, at best, a euphemisms. But in *any* of these circumstances there will be moral burdens.

2. Simmons (1979) dismisses the idea of political obligations on grounds that ultimately root in "doubts about benefits provided by groups of persons [because] where a group of persons is concerned, there is very seldom anything like a reason, common to them all, for which the benefit was provided" (pp. 187–88), considerations which lead Simmons to reject the idea of debts of gratitude to institutions. What is mysterious here is (i) why a uniformity of reason should be crucial, and (ii) why a benefit extended by social (rather than individual) decision should not engender correspondingly socially oriented indebtedness. Admittedly, the extent to which one should be grateful to a benefactor depends to some extent on the intention of the benefactor in performing the action that provides the benefit at issue. But there can be little doubt that a nation that adopts and implements an immigration policy which provides asylum and refuge to the citizens of other countries in situations of necessity thereby deserves credit for good intentions.

REFERENCES

Card, C. 1988. "Gratitude and Obligation." *American Philosophical Quarterly 25:* 118–27.

Coser, L. A. 1984. *Refugee Scholars in America.* New Haven: Yale University Press.

Klosko, G. 1989. "Political Obligation and Gratitude." *Philosophy and Public Affairs 18:* 353–58.

———. 1991. "Four Arguments Against Political Obligation from Gratitude." *Public Affairs Quarterly 5:* 33–48.

Plamenatz, J. P. 1968. *Consent Freedom and Political Obligation.* 2d ed. Oxford: Oxford University Press.

Simmons, A. J. 1979. *Moral Principles and Political Obligations.* Princeton: Princeton University Press.

Walker, M. 1988. "Political Obligation and the Argument for Gratitude." *Philosophy and Public Affairs 17:* 191–211.

———. 1989. "Obligations of Gratitude and Political Obligation." *Philosophy and Public Affairs 18:* 359–72.

Name Index